MELZAK—Mathematical Ideas, Modeling and Applications (Volume II of Companion to Concrete Mathematics)

MEYER—Introduction to Mathematical Fluid Dynamics

NAYFEH—Perturbation Methods

NAYFEH and MOOK—Nonlinear Oscillations

ODEN and REDDY—An Introduction to the Mathematical Theory of Finite Elements

PAGE—Topological Uniform Structures

PASSMAN—The Algebraic Structure of Group Rings

PRENTER—Splines and Variational Methods

RIBENBOIM—Algebraic Numbers

RICHTMYER and MORTON—Difference Methods for Initial-Value Problems, 2nd Edition

RIVLIN—The Chebyshev Polynomials

RUDIN—Fourier Analysis on Groups

SAMELSON—An Introduction to Linear Algebra

SCHUMAKER—Spline Functions: Basic Theory

SCHUSS—Theory and Applications of Stochastic Differential Equations

SIEGEL—Topics in Complex Function Theory

Volume 1—Elliptic Functions and Uniformization Theory

Volume 2—Automorphic Functions and Abelian Integrals

Volume 3—Abelian Functions and Modular Functions of Several Variables

STAKGOLD—Green's Functions and Boundary Value Problems

STOKER—Differential Geometry

STOKER—Nonlinear Vibrations in Mechanical and Electrical Systems

STOKER—Water Waves

WHITHAM—Linear and Nonlinear Waves

WOUK—A Course of Applied Functional Analysis

D1582324

SPLINE FUNCTIONS

SPLINE FUNCTIONS: BASIC THEORY

LARRY L. SCHUMAKER

A WILEY-INTERSCIENCE PUBLICATION

JOHN WILEY & SONS New York · Chichester · Brisbane · Toronto

ACCESSION No. 208993

CLASS No. 511·42 SCH

17 NOV 1982

O/S | N | CATEGORY N.L.

Copyright © 1981 by John Wiley & Sons, Inc.

All rights reserved. Published simultaneously in Canada.

Reproduction or translation of any part of this work beyond that permitted by Sections 107 or 108 of the 1976 United States Copyright Act without the permission of the copyright owner is unlawful. Requests for permission or further information should be addressed to the Permissions Department, John Wiley & Sons, Inc.

Library of Congress Cataloging in Publication Data

Schumaker, Larry L 1939-
Spline functions.

(Pure and applied mathematics ISSN 0079-8185)
"A Wiley-Interscience publication."
Bibliography: p.
Includes index.
1. Spline theory. I. Title.

QA224.S33 511′.42 80-14448
ISBN 0-471-76475-2

Printed in the United States of America

10 9 8 7 6 5 4 3 2 1

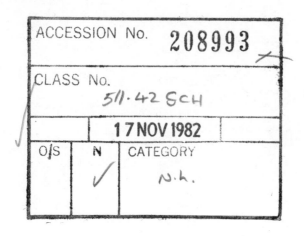

IN MEMORY OF
STEPHEN ANTHONY SCHUMAKER

PREFACE

The theory of spline functions and their applications is a relatively recent development. As late as 1960, there were no more than a handful of papers mentioning spline functions by name. Today, less than 20 years later, there are well over 1000 research papers on the subject, and it remains an active research area.

The rapid development of spline functions is due primarily to their great usefulness in applications. Classes of spline functions possess many nice structural properties as well as excellent approximation powers. Since they are easy to store, evaluate, and manipulate on a digital computer, a myriad of applications in the numerical solution of a variety of problems in applied mathematics have been found. These include, for example, data fitting, function approximation, numerical quadrature, and the numerical solution of operator equations such as those associated with ordinary and partial differential equations, integral equations, optimal control problems, and so on. Programs based on spline functions have found their way into virtually every computing library.

It appears that the most turbulent years in the development of splines are over, and it is now generally agreed that they will become a firmly entrenched part of approximation theory and numerical analysis. Thus my aim here is to present a fairly complete and unified treatment of spline functions, which, I hope, will prove to be a useful source of information for approximation theorists, numerical analysts, scientists, and engineers.

This book developed out of a set of lecture notes which I began preparing in the fall of 1970 for a course on spline functions at the University of Texas at Austin. The material, which I have been reworking ever since, was expanded and revised several times for later courses at the Mathematics Research Center in Madison, the University of Munich, the University of Texas, and the Free University of Berlin. It was my original intent to cover both the theory and applications of spline functions in a single monograph, but the amount of interesting and useful material is so large that I found it impossible to give all of it a complete and comprehensive treatment in one volume.

This book is devoted to the basic theory of splines. In it we study the main algebraic, analytic, and approximation-theoretic properties of various spaces of splines (which in their simplest form are just spaces of piecewise polynomials). The material is organized as follows. In Chapters 1 to 3 background and reference material is presented. The heart of the book consists of Chapters 4 to 8, where polynomial splines are treated. Chapters 9 to 11 deal with the theory of generalized splines. Finally, Chapters 12 to 13 are devoted to multidimensional splines. For the practical-minded reader, I include a number of explicit algorithms written in an easily understood informal language.

It has not been my aim to design a textbook, *per se*. Thus throughout the book there is a mixture of very elementary results with rather more sophisticated ones. Still, much of it can be read with a minimum of mathematical background—for example calculus and linear algebra. With a judicious choice of material, the book can be used for a one-semester introduction to splines. For this purpose I suggest drawing material from Chapters 1 to 6, 8, and 12, with special emphasis on Chapters 4 and 5.

The notation in the book is quite standard. In order to keep the exposition moving as much as possible, I have elected to move most of the remarks and references to the end of the chapters. Thus each chapter contains sections with remarks and with historical notes. In these sections I have attempted, to trace the best of my ability, the sources of the ideas in the chapter, and to guide the reader to the appropriate references in the massive literature.

I would like to take this opportunity to acknowledge some of the institutions and individuals who have been of assistance in the preparation of this book. First, I would like to thank Professor Samuel Karlin for introducing me to spline functions when I was his graduate student at Stanford in the early sixties. The Mathematics Research Center at the University of Wisconsin gratiously supported me at two critical junctures in the evolution of this book. The first was in 1966 to 1968 when the lively research atmosphere and the close contact with such experts as Professors T. N. E. Greville, M. Golomb, J. W. Jerome, and I. J. Schoenberg sharpened my interest in splines and taught me much about the subject. The support of the Mathematics Research Center again in 1973 to 1974 gave me a much needed break to continue work on the book.

In 1974 to 1975 I was at the Ludwig-Maximilians Universität in Munich. My thanks are due to Professor G. Hämmerlin for the invitation to visit Munich, and to the Deutsche Forschungsgemeinschaft for their support. Since January of 1978 I have been at the Free University of Berlin and the Hahn-Meitner Atomic Energy Institute. I am grateful to Professors K.-H.

Hoffmann and H.-J. Töpfer for suggesting and arranging my visit, and to the Humboldt Foundation of the Federal Republic of Germany for their part in my support. Finally, I would like to express my appreciation to the U.S. Air Force Office of Scientific Research and the Center of Numerical Analysis of the University of Texas for support of my research over the past several years.

Among the many colleagues and students who have read portions of the manuscript and made useful suggestions, I would especially like to mention Professors Carl deBoor, Ron DeVore, Tom Lyche, Charles Micchelli, Karl Scherer, and Ulrich Tippenhauer. The task of tracking down and organizing the reference material was formidable, and I was greatly assisted in this task by Jannelle Odem, Maymejo Moody Barrett, Nancy Jo Ethridge, Linda Blackman, and Patricia Stringer. Finally, I would like to thank my wife Gerda for her constant support, and for her considerable help in all stages of the preparation of this book.

LARRY L. SCHUMAKER

Britton, South Dakota
October 1980

CONTENTS

SPLINE FUNCTIONS

1
INTRODUCTION

The first three chapters of this book are devoted to background material, notation, and preliminary results. The well-prepared reader may wish to proceed directly to Chapter 4 where the study of spline functions *per se* begins.

§ 1.1. APPROXIMATION PROBLEMS

Functions are the basic mathematical tools for describing and analyzing many physical processes of interest. While in some cases these functions are known explicitly, very frequently it is necessary to construct approximations to them based on limited information about the underlying processes. Such approximation problems are a central part of *applied mathematics*.

There are two major categories of *approximation problems*. The first category consists of problems where it is required to construct an approximation to an unknown function based on some finite amount of data (often measurements) on the function. We call these *data fitting problems*. In such problems, the data are often subject to error or noise, and moreover, usually do not determine the function uniquely. Data fitting problems arise in virtually every branch of scientific endeavor.

The second main category of approximation problems arises from mathematical models for various physical processes. As these models usually involve operator equations that determine the unknown function, we refer to them as *operator-equation problems*. Examples include boundary-value problems for ordinary and partial differential equations, eigenvalue–eigenfunction problems, integro–differential equations, integral equations, optimal control problems, and so on. While there are many theoretical results on existence, uniqueness, and properties of solutions of such operator equations, usually only the simplest specific problems can be solved explicitly. In practice we will usually have to construct approximate solutions.

The most commonly used approach to finding approximations to unknown functions proceeds as follows:

1. Choose a reasonable *class of functions* in which to look for an approximation.
2. Devise an appropriate *selection scheme* ($\equiv$*approximation process*) for assigning a specific function to a specific problem.

The success of this approach depends heavily on the existence of convenient classes of approximating functions. To be of maximal use, a class $\mathcal{Q}$ of approximating functions should possess at least the following basic properties:

1. The functions in $\mathcal{Q}$ should be relatively smooth;
2. The functions in $\mathcal{Q}$ should be easy to store and manipulate on a digital computer;
3. The functions in $\mathcal{Q}$ should be easy to evaluate on a computer, along with their derivatives and integrals;
4. The class $\mathcal{Q}$ should be large enough so that arbitrary smooth functions can be well approximated by elements of $\mathcal{Q}$.

We have required property 1 because functions arising from physical processes are usually known to be smooth. Properties 2 and 3 are important because most real-world problems cannot be solved without the help of a high-speed digital computer. Finally, property 4 is essential if we are to achieve good approximations.

The study of various classes of approximating functions is precisely the content of *approximation theory*. The design and analysis of effective algorithms utilizing these approximation classes are a major part of *numerical analysis*. Both of these fields have a rich history, and a voluminous literature.

The purpose of this book is to examine in considerable detail some specific approximation classes—the so-called spline functions—which in the past several years have proved to be particularly convenient and effective for approximation purposes. Because of space limitations, we shall deal only with the basic theoretical properties of spline functions. Applications of splines to data fitting problems and to the numerical solution of operator equations will be treated in later monographs.

§ 1.2. POLYNOMIALS

Polynomials have played a central role in approximation theory and numerical analysis for many years. To indicate why this might be the case, we note that the space

$$\mathcal{P}_m = \left\{ p(x) : p(x) = \sum_{i=1}^{m} c_i x^{i-1}, \qquad c_1, \ldots, c_m, x \text{ real} \right\} \tag{1.1}$$

of *polynomials of order m* has the following attractive features:

1. $\mathcal{P}_m$ is a finite dimensional linear space with a convenient basis;
2. Polynomials are smooth functions;
3. Polynomials are easy to store, manipulate, and evaluate on a digital computer;
4. The derivative and antiderivative of a polynomial are again polynomials whose coefficients can be found algebraically (even by a computer);
5. The number of zeros of a polynomial of order m cannot exceed $m-1$;
6. Various matrices (arising in interpolation and approximation by polynomials) are always nonsingular, and they have strong sign-regularity properties;
7. The sign structure and shape of a polynomial are intimately related to the sign structure of its set of coefficients;
8. Given any continuous function on an interval $[a,b]$, there exists a polynomial which is uniformly close to it;
9. Precise rates of convergence can be given for approximation of smooth functions by polynomials.

We shall examine each of these assertions in detail in Chapter 3, along with a number of other properties of polynomials.

While this list tends to indicate that polynomials should be ideal for approximation purposes, in practice, it has been observed that they possess one unfortunate feature which allows for the possibility that still better classes of approximating functions may exist; namely,

10. Many approximation processes involving polynomials tend to produce polynomial approximations that oscillate wildly.

We illustrate this feature of polynomials in Section 3.6. It is a kind of *inflexibility* of the class $\mathcal{P}_m$.

§ 1.3. PIECEWISE POLYNOMIALS

As mentioned in the previous section, the main drawback of the space $\mathcal{P}_m$ of polynomials for approximation purposes is that the class is relatively inflexible. Polynomials seem to do all right on sufficiently small intervals, but when we go to larger intervals, severe oscillations often appear—particularly if m is more than 3 or 4. This observation suggests that in order to achieve a class of approximating functions with greater flexibility, we should work with polynomials of relatively low degree, and should divide up the interval of interest into smaller pieces. We are motivated to make the following definition:

DEFINITION 1.1. Piecewise Polynomials

Let $a = x_0 < x_1 < \cdots < x_k < x_{k+1} = b$, and write $\Delta = \{x_i\}_0^{k+1}$. The set Δ partitions the interval $[a,b]$ into $k+1$ subintervals, $I_i = [x_i, x_{i+1})$, $i = 0, 1, \ldots, k-1$, and $I_k = [x_k, x_{k+1}]$. Given a positive integer m, let

$$\mathcal{P}\mathcal{P}_m(\Delta) = \begin{array}{l}\{f\text{: there exist polynomials} \\ p_0, p_1, \ldots, p_k \text{ in } \mathcal{P}_m \text{ with } f(x) = p_i(x) \\ \text{for } x \in I_i,\ i = 0, 1, \ldots, k\}.\end{array} \tag{1.2}$$

We call $\mathcal{P}\mathcal{P}_m(\Delta)$ the *space of piecewise polynomials* of *order m* with *knots* $x_1, \ldots, x_k$.

The terminology in Definition 1.1 is perfectly descriptive—an element $f \in \mathcal{P}\mathcal{P}_m(\Delta)$ consists of $k+1$ polynomial pieces. Figure 1 shows a typical example of a piecewise polynomial of order 3 with two knots.

While it is clear that we have gained flexibility by going over from polynomials to piecewise polynomials, it is also obvious that at the same time we have lost another important property—piecewise polynomial functions are not necessarily smooth. In fact, as shown in Figure 1, they can even be discontinuous. In most applications, the user would be happier if the approximating functions were at least continuous. Indeed, it is probably precisely this defect of piecewise polynomials which accounts for the fact that prior to 1960 they played a relatively small role in approximation theory and numerical analysis—for an historical account, see Section 1.6.

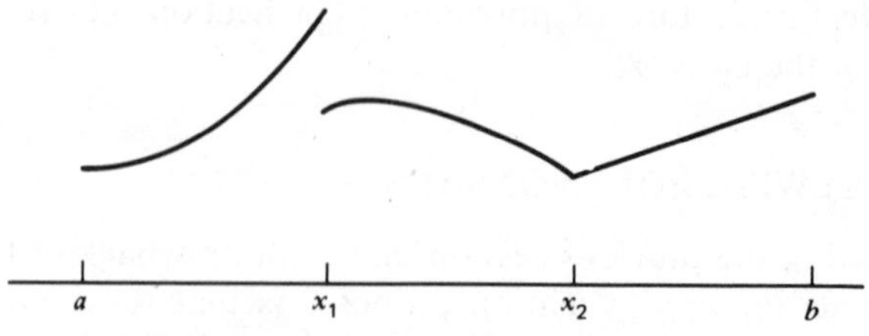

Figure 1. A quadratic piecewise polynomial.

§ 1.4. SPLINE FUNCTIONS

In order to maintain the flexibility of piecewise polynomials while at the same time achieving some degree of global smoothness, we may define the following class of functions:

DEFINITION 1.2. Polynomial Splines With Simple Knots

Let Δ be a partition of the interval $[a,b]$ as in Definition 1.1, and let m be a positive integer. Let

$$\mathcal{S}_m(\Delta) = \mathcal{P}\mathcal{P}_m(\Delta) \cap C^{m-2}[a,b], \tag{1.3}$$

where $\mathcal{P}\mathcal{P}_m(\Delta)$ is the space of piecewise polynomials defined in (1.2). We call $\mathcal{S}_m(\Delta)$ the space of *polynomial splines of order m with simple knots at the points* $x_1,\dots,x_k$.

It is easy to define various related classes of piecewise polynomials with varying degrees of smoothness between the pieces. We refer to such spaces of functions as *polynomial splines*, and we study them in detail in Chapters 4 to 8. We shall see that polynomial splines possess the following attractive features:

1. Polynomial spline spaces are finite dimensional linear spaces with very convenient bases;
2. Polynomial splines are relatively smooth functions;
3. Polynomial splines are easy to store, manipulate, and evaluate on a digital computer;
4. The derivatives and antiderivatives of polynomial splines are again polynomial splines whose expansions can be found on a computer;
5. Polynomial splines possess nice zero properties analogous to those for polynomials;
6. Various matrices arising naturally in the use of splines in approximation theory and numerical analysis have convenient sign and determinantal properties;
7. The sign structure and shape of a polynomial spline can be related to the sign structure of its coefficients;
8. Every continuous function on the interval $[a,b]$ can be approximated arbitrarily well by polynomial splines *with the order m fixed*, provided a sufficient number of knots are allowed;
9. Precise rates of convergence can be given for approximation of smooth functions by splines—not only are the functions themselves approximated to high order, but their derivatives are *simultaneously* approximated well;

10. Low-order splines are very flexible, and do not exhibit the oscillations usually associated with polynomials.

These properties of polynomial splines are shared by a wide variety of other piecewise spaces. We discuss spaces of nonpolynomial splines in Chapters 9 to 11. A class of multidimensional splines is treated in Chapter 12.

It is perhaps of some interest to explain the origin of the terminology "*spline function*." It was introduced by Schoenberg [1946a, b] in connection with the space $\mathcal{S}_m(\Delta)$, which he used for solving certain data fitting problems. Schoenberg states that he was motivated to use this terminology by the connection of piecewise polynomials with a certain mechanical device called a spline.

A *spline* is a thin rod of some elastic material equipped with a groove and a set of weights (called *ducks* or *rats*) with attached arms designed to fit into the groove. (See Figure 2.) The device is used by architects (particularly naval architects) to draw smooth curves passing through prescribed points. To accomplish this, the spline is forced to pass through the prescribed points by adjusting the location of the ducks along the rod. It was discovered in the mid-1700s by Euler and the Bernoulli brothers that the shape of the centerline of such a bent rod is approximately given by a function in $\mathcal{S}_4(\Delta)$. In particular, suppose that the points of contact of the ducks with the spline are located at the points $(x_i, y_i), i = 1, 2, \ldots, k$ in the Cartesian plane. Then the centerline of the spline is approximately given by the function s with the following properties:

1. s is a piecewise cubic polynomial with knots at $x_1, \ldots, x_k$;
2. s is a linear polynomial for $x \leqslant x_1$ and $x \geqslant x_k$;

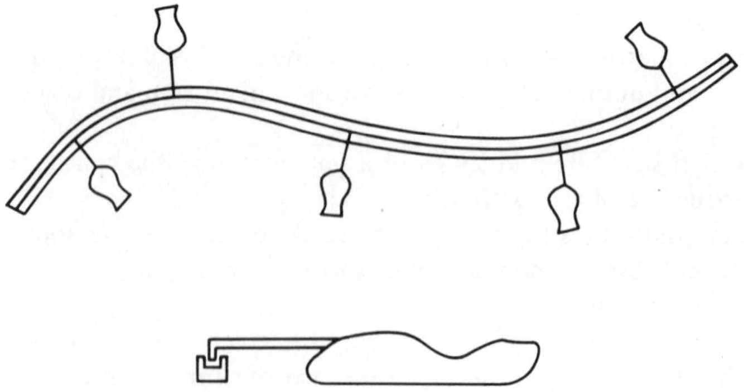

Figure 2. The mechanical spline.

3. s has two continuous derivatives everywhere;
4. $s(x_i)=y_i, i=1,2,\ldots,k$.

The function s is a kind of *best interpolating* function.

§ 1.5. FUNCTION CLASSES AND COMPUTERS

In this section we show that because of certain fundamental limitations of digital computers, only a rather restricted set of approximating spaces are compatible with digital computers. This discussion will provide further motivation for our selection of piecewise polynomials for intensive study.

We begin by recalling the capabilities of a modern digital computer. Such a computer is capable of storing a rather large (but finite) number of "words" consisting of real numbers or integers. It is capable of very rapid access to these words. Finally, it can perform the following *five basic arithmetic operations*: (1) addition, (2) subtraction, (3) multiplication, (4) division, and (5) comparison of the size of two numbers.

Suppose now that $\mathcal{Q}$ is a class of functions which we desire to handle with a computer. For $\mathcal{Q}$ to be *compatible* with a digital computer, we must require the following two properties:

> The class $\mathcal{Q}$ should be defined by a finite number of real parameters; that is, each function s in $\mathcal{Q}$ should be uniquely identifiable with a vector $\mathbf{c}=(c_1,\ldots,c_n)$ of real numbers. (1.4)

> Given the parameter vector $\mathbf{c}$ defining a particular member s_c in $\mathcal{Q}$, it should be possible to compute the value of s_c at any point in its domain using only the five basic arithmetic operations mentioned above. (1.5)

Property (1.4) assures that we have a way of identifying and storing individual elements of $\mathcal{Q}$ on the computer, while property (1.5) guarantees that we can evaluate a given function in $\mathcal{Q}$ at a given point.

To illustrate how requirements (1.4) and (1.5) severely limit the structure of computer-compatible functions classes, consider the case where $\mathcal{Q}$ is to be a class of real-valued functions defined on an interval $I=[a,b]$. Clearly, one such computer-compatible class is the space $\mathcal{P}_m$ of *polynomials* of order m defined in (1.1). Indeed, each polynomial p can be associated with the vector $\mathbf{c}=(c_1,\ldots,c_m)$ of its coefficients (and it is proved in Theorem 3.1 that this association is unique). Thus to store a polynomial on the machine, we need only store its m coefficients. On the other hand, a given polynomial can be evaluated using only addition, subtraction, and multiplication (see Algorithm 3.2).

If we admit the division capability of a digital computer, we see immediately that another class of computer-compatible functions is given by the set of *proper rational functions*:

$$\mathcal{R}_{l,m} = \left\{ r(x) = \frac{p(x)}{q(x)} : p \in \mathcal{P}_l, q \in \mathcal{P}_m, \right.$$
$$\left. \text{and } q(x) \neq 0 \text{ for all } a \leqslant x \leqslant b \right\}. \tag{1.6}$$

There remains one major capability of the digital computer that we have not yet exploited; namely, the ability to compare two numbers. Using it, we see that the space of *piecewise polynomials* $\mathcal{P}\mathcal{P}_m(\Delta)$ defined in (1.2) is also computer compatible. Indeed, to store a particular element $s \in \mathcal{P}\mathcal{P}_m(\Delta)$, we need only store the partition points $x_1, \ldots, x_k$ and the coefficients of each of the polynomial pieces. To evaluate $s(x)$ for a given x, we need only decide in which interval x lies (using the comparison capability of the machine), and then evaluate the appropriate polynomial at x.

If we combine all five capabilities of the digital computer, we see that the class of *piecewise rational functions* is also computer compatible. It follows that this class is the largest class of real-valued functions defined on an interval $[a,b]$ that is computer compatible.

This conclusion is, in fact, a bit of an overstatement. Because of two further limitations of digital computers (which are due to the finite word length), even the classes mentioned above are not truly computer compatible in the strict sense defined above. The limitations are (1) not every real number can be represented exactly in a digital machine, and (2) the result of arithmetic operations may be subject to round-off error. Thus, for example, irrational coefficients cannot be stored exactly, nor can polynomials be evaluated exactly at irrational arguments.

Because of the round-off problem, it can be argued that several other classes of functions should be admitted as computer compatible. For example, there are a number of elementary functions (such as the trigonometric, exponential, and logarithmic functions) for which computer routines are available, permitting their evaluation to machine accuracy for any argument. In a way, such functions are just as compatible as are the polynomials. In view of this observation, in Chapters 9 to 11 we shall consider some classes of rather general piecewise functions. Special attention will be focused on the cases where the pieces are trigonometric, exponential or logarithmic.

§ 1.6. HISTORICAL NOTES

Section 1.1

Approximation theory has a long and rich history which we need not review in detail here. Some early books on the subject include dela Valleé Poussin [1919], Bernstein [1926], Jackson [1930], and Zygmund [1959]. Interest has remained high, and some of the more recent books on general approximation theory include Achieser [1947], Natanson [1949], Davis [1963], Sard [1963], Timan [1963], Todd [1963], Rice [1964, 1969b], Cheney [1966], Lorentz [1966], Meinardus [1967], Butzer and Berens [1967], Rivlin [1969], and Laurent [1972]. There are many more specialized books dealing with approximation, as well as a large number of books containing the proceedings of conferences on approximation theory.

High-speed digital computers were first produced in the early 1950s, and this was followed by the rapid development of numerical analysis. We do not bother to give the long list of books on the subject which have been written in the last 30 years. Typical examples include the texts by Ralston [1965], Isaacson and Keller [1966], and Conte and deBoor [1972].

Section 1.2

Polynomials have been important in approximation theory and numerical analysis from the beginning. Their importance can be judged by examining any of the books mentioned above. For example, see Ralston [1965, p. 25] where a special effort is made to extoll the advantage of polynomials.

Section 1.3

We have not attempted the difficult task of tracing the complete history of piecewise polynomials. Certainly they have been useful in mathematics for a long time. It may be of some interest to mention just a few early appearances. We have already discussed the work of Euler and the Bernoulli brothers on the shape of an elastica in § 1.4. Various classical quadrature formulae (e.g., Newton-Cotes, Euler-MacLaurin, and composite Gauss formulae) are based on piecewise polynomials. Piecewise polynomials also arose in connection with solution of initial-value problems for ordinary differential equations (e.g., in Euler's method). They have also proved to be useful tools in analysis. For example, Lebesgue used piecewise linear polynomials to prove Weierstrass' Approximation Theorem 3.11.

Section 1.4

The terminology "spline function" was introduced by Schoenberg [1946a, b]—see the discussion in § 1.4. As far as we can determine, the only other papers mentioning splines explicitly prior to 1960 were those by Curry and Schoenberg [1947], Schoenberg and Whitney [1949, 1953], Schoenberg [1958], and Maclaren [1958].

Although the papers mentioned above seem to be the only ones published prior to 1960 that actually mention splines by name, there were a number of papers dealing with splines without using the name. Here we can mention Runge [1901], Eagle [1928], Quade and Collatz [1938], Favard [1940], Sard [1949], Meyers and Sard [1950a, b], Holladay [1957], and Golomb and Weinberger [1959]. In the early 1900s there was also extensive development of interpolation formulae based on piecewise polynomials. These methods were called osculatory interpolation methods—for a survey of their development, see Greville [1944]. Our list is surely not complete.

The theory of spline functions (whether called by that name or not) had a rather modest development up until 1960. After that the development was nothing short of explosive. The main impetus for the intense interest in splines in the early 1960s seems to have been provided primarily by the fact that (in addition to I. J. Schoenberg) a number of researchers realized that spline functions were a way to mathematically model the physical process of drawing a smooth curve with a mechanical spline. Papers in which this connection was noted include Maclaren [1958], Birkhoff and Garabedian [1960], Theilheimer and Starkweather [1961], Asker [1962], Fowler and Wilson [1963], and Berger and Webster [1963]. About this time the paper of Holladay [1957] seems to have been discovered, and a number of authors set about the task of studying best interpolation problems. Early contributors to this development included deBoor [1962, 1963], Walsh, Ahlberg and Nilson [1962], Ahlberg and Nilson [1963], Ahlberg, Nilson, and Walsh [1964], and Schoenberg [1964a, b, c]. The history of the development of best interpolation problems will be traced in greater detail in a later monograph.

In addition to the papers mentioned above dealing with best interpolation by splines, there were also a few isolated papers written in the early 1960s that dealt with constructive properties of spaces of piecewise polynomials (with no references to the other literature). These include Schwerdfeger [1960, 1961], Kahane [1961], Stone [1961], Aumann [1963], Brudnyi and Gopengauz [1963], Ream [1961], Peterson [1962], and Lawson

[1964]. More information on the early history of splines can be found in the historical notes sections of later chapters.

Despite the feverish activity in splines during the past 20 years, there are relatively few books on the subject. The following works deal to various degrees with the theory and applications of splines: Ahlberg, Nilson, and Walsh [1967b], Sard and Weintraub [1971], Schoenberg [1973], Schultz [1973b], Böhmer [1974], Prenter [1975], Fisher and Jerome [1975], deBoor [1978], and Stechin and Subbotin [1978].

Programs for dealing with splines can be found in the works of Späth [1973] and deBoor [1978]. The book by Laurent [1972] also has several chapters on splines, and they are starting to be mentioned in some of the elementary books on numerical analysis and approximation theory. There have been several conferences devoted entirely (or heavily) to splines, and their proceedings provide useful source material. Here we mention Greville [1969a], Schoenberg [1969b], Meir and Sharma [1973], Böhmer, Meinardus, and Schempp [1974, 1976], and Karlin, Micchelli, Pinkus, and Schoenberg [1976]. The article by van Rooij and Scherer [1974] contains a fairly complete bibliography of the theory of splines up until January 1973.

2
PRELIMINARIES

In this chapter we establish some notation and develop a number of useful tools. While for the most part the material belongs to classical approximation theory and numerical analysis, we include several quite recent results. The discussion in this chapter is restricted entirely to real-valued functions defined on an interval. Functions of several variables are treated in Chapter 13.

§ 2.1. FUNCTION CLASSES

Given a closed finite interval $[a,b]$, we define the space of *bounded real-valued functions on* $I=[a,b]$ as

$$B[I]=\{f:f \text{ is a real-valued function on } I \text{ and} |f(x)|<\infty \text{ for all } x\in I\}. \tag{2.1}$$

This is a linear space of functions.

In most practical problems, the functions of interest possess some degree of smoothness. For example, we will often deal with the space of *continuous functions on* I,

$$C[I]=\{f:f \text{ is continuous at each } x \text{ in } I\}. \tag{2.2}$$

This is a normed linear space with norm

$$\|f\|_{C[I]}=\max_{x\in I}|f(x)|. \tag{2.3}$$

In fact, $C[a,b]$ is a Banach space with the norm (2.3). We shall make very little use of Banach spaces in this book, however.

We shall often have to deal with *derivatives* of functions. We introduce the standard notation

$$D_+ f(x) = \lim_{h \downarrow 0} \frac{f(x+h)-f(x)}{h} \tag{2.4}$$

and

$$D_- f(x) = \lim_{h \downarrow 0} \frac{f(x)-f(x-h)}{h}. \tag{2.5}$$

When these limits exist we call them the *right and left derivatives of f at x*, respectively. When both left and right derivatives exist at a point and are equal, then we write

$$Df(x) = D_- f(x) = D_+ f(x). \tag{2.6}$$

We say a function f is *differentiable* on the closed interval $[a,b]$ provided that $Df(x)$ exists for all $a<x<b$, and that $D_+ f(a)$ and $D_- f(b)$ exist. In this case it is standard practice to write

$$Df(x) = \begin{cases} D_+ f(x) & \text{if } x=a \\ D_- f(x) & \text{if } x=b. \end{cases} \tag{2.7}$$

We can now introduce several classes of smoother functions. If r is a positive integer, we write

$$C^r[I] = \{ f: \text{the } r^{\text{th}} \text{ derivative } D^r f \text{ belongs to } C[I] \} \tag{2.8}$$

for the space of *r-times continuously differentiable functions*. These spaces are increasingly smooth subspaces of $C[I]$ as r increases; that is,

$$\cdots C^2[I] \subseteq C^1[I] \subseteq C[I].$$

There are other smooth subspaces of $C[I]$ lying in between these spaces —for example, the space of *absolutely continuous* functions defined by

$$\begin{aligned} AC[I] = \{ f: &\text{for any } \varepsilon>0, \text{ there exists } \delta>0, \\ &\text{so that for all } n \text{ and all } a \leqslant \underline{t}_1 \leqslant \bar{t}_1 \leqslant \underline{t}_2 \\ &\leqslant \bar{t}_2 \cdots \leqslant \underline{t}_n \leqslant \bar{t}_n \leqslant b \ \text{ with } \sum_{i=1}^{n} |\bar{t}_i - \underline{t}_i| \\ &< \delta, \sum_{i=1}^{n} |f(\bar{t}_i) - f(\underline{t}_i)| < \varepsilon \} \end{aligned} \tag{2.9}$$

satisfies $C^1[I] \subseteq AC[I] \subseteq C[I]$.

In addition to the spaces mentioned above, we shall also make extensive use of the classical *Lebesgue spaces* defined by

$$L_p[I] = \{f : f \text{ is measurable on } I \text{ and } \|f\|_p < \infty\}, \tag{2.10}$$

where

$$\|f\|_{L_p[I]} = \|f\|_p = \left[\int_a^b |f(x)|^p dx\right]^{1/p}, \qquad 1 \leqslant p < \infty, \tag{2.11}$$

and

$$\|f\|_{L_\infty[I]} = \operatorname*{ess\,sup}_{x \in I} |f(x)|, \qquad p = \infty. \tag{2.12}$$

$L_p[I]$ is a normed linear space (in fact, a Banach space) for each $1 \leqslant p \leqslant \infty$.

It is also useful to deal with certain subspaces of the L_p spaces where the functions possess smooth derivatives. Given $1 \leqslant p \leqslant \infty$ and any positive integer r, we define

$$L_p^r[I] = \{f : D^{r-1}f \in AC[I] \quad \text{and } D^r f \in L_p[I]\}. \tag{2.13}$$

The space $L_p^r[I]$ is called a *Sobolev* space. It is a normed linear space (in fact, a Banach space) with norm

$$\|f\|_{L_p^r[I]} = \sum_{j=0}^{r} \|D^j f\|_{L_p[I]}. \tag{2.14}$$

These classes of smooth functions are nested as follows:

$$C^r[I] \subseteq L_\infty^r[I] \subseteq L_p^r[I] \subseteq L_1^r[I] \subseteq C^{r-1}[I],$$

for all $1 \leqslant p \leqslant \infty$ and all positive integers r.

We shall introduce other spaces of smooth functions in later chapters as we need them—see in particular Sections 2.8 and 6.5. Analogous spaces of functions of several variables are introduced in Section 13.2.

§ 2.2. TAYLOR EXPANSIONS AND THE GREEN'S FUNCTION

In this section we discuss the Taylor theorem and the Green's function associated with the differential operator D^m. Analogous results for general differential operators L are given in Section 10.2. Let

$$(x-y)_+^0 = \begin{cases} 1, & x \geqslant y \\ 0, & x < y \end{cases}, \tag{2.15}$$

and

$$(x-y)_+^{m-1} = \begin{cases} (x-y)^{m-1}, & x \geqslant y, \quad m > 1 \\ 0, & x < y \end{cases}. \tag{2.16}$$

The following theorem is the classical Taylor expansion of a smooth function in terms of a polynomial, with an explicit remainder term:

THEOREM 2.1. Taylor Expansion

Let $f \in L_1^m[a,b]$. Then for all $a \leqslant x \leqslant b$,

$$f(x) = \sum_{j=0}^{m-1} \frac{D^j f(a)(x-a)^j}{j!} + \int_a^b \frac{(x-y)_+^{m-1} D^m f(y)\,dy}{(m-1)!}. \tag{2.17}$$

Moreover, there exists $a \leqslant \xi_x \leqslant b$ such that

$$f(x) = \sum_{j=0}^{m-1} \frac{D^j f(a)(x-a)^j}{j!} + \frac{D^m f(\xi_x)(x-a)^m}{m!}. \tag{2.18}$$

Proof. Integrating by parts $m-1$ times, we obtain

$$\int_a^x \frac{(x-y)^{m-1}}{(m-1)!} D^m f(y)\,dy = \sum_{i=1}^{m-1} \frac{(x-y)^{m-i}}{(m-i)!} D^{m-i} f(y)\Big|_a^x + \int_a^x Df(y)\,dy.$$

Expansion (2.17) follows. Applying the mean-value theorem for integrals to the remainder in (2.17), we obtain (2.18). ∎

There is a dual version of this result whose proof is almost identical.

THEOREM 2.2. Dual Taylor Expansion

Let $f \in L_1^m[a,b]$. Then for all $a \leqslant y \leqslant b$,

$$f(y) = \sum_{j=0}^{m-1} \frac{(-1)^j D^j f(b)(b-y)^j}{j!} + \int_a^b \frac{(x-y)_+^{m-1}(-1)^m D^m f(x)\,dx}{(m-1)!}. \tag{2.19}$$

It is convenient to have a symbol for the kernel in the Taylor expansion. We write

$$g_m(x;y)=\frac{(x-y)_+^{m-1}}{(m-1)!}. \tag{2.20}$$

It is clear from the definition of $g_m(x;y)$ that for each fixed y it is infinitely often right differentiable with respect to x. Similarly, for each fixed x it is infinitely often left differentiable with respect to y. Writing D_x and D_y for these derivatives, we immediately have the following properties of $g_m(x;y)$:

$$D_x^i g_m(x;y)=g_{m-i}(x;y), \qquad 0\leqslant i\leqslant m-1 \tag{2.21}$$

$$D_x^m g_m(x;y)=(-1)^m D_y^m g_m(x;y)=0, \qquad \text{all } x\neq y \tag{2.22}$$

$$(-1)^i D_y^i g_m(x;y)=g_{m-i}(x;y), \qquad 0\leqslant i\leqslant m-1 \tag{2.23}$$

$$\begin{aligned} D_x^i g_m(x;y)|_{x=y} &= (-1)^i D_y^i g_m(x;y)|_{x=y} \\ &= \delta_{i,m-1}, \qquad i=0,1,\ldots,m-1. \end{aligned} \tag{2.24}$$

The following theorem shows that g_m is the *Green's function* associated with the differential operator D^m:

THEOREM 2.3.

Let $h\in L_1[a,b]$ and real numbers $f_0,\ldots,f_{m-1}$ be given. Then the function

$$f(x)=\sum_{j=0}^{m-1} f_j\frac{(x-a)^j}{j!}+\int_a^b g_m(x;y)h(y)\,dy \tag{2.25}$$

solves the *initial-value problem*

$$D^m f(x)=h(x), \qquad \text{almost everywhere in } [a,b] \tag{2.26}$$

$$D^i f(a)=f_i, \qquad i=0,1,\ldots,m-1. \tag{2.27}$$

Proof. In view of the properties of g_m, it is clear that f satisfies (2.27). On the other hand,

$$D^{m-1}f(x)=f_{m-1}+\int_a^x h(y)\,dy,$$

and (2.26) also follows. ■

In view of Theorem 2.3, it is of interest to define the operator D^{-m} by

$$D^{-m}f(x)=\int_a^b g_m(x;y)f(y)\,dy, \qquad f\in L_1[a,b]. \tag{2.28}$$

D^{-m} is a kind of inverse to the differential operator D^m in the sense that

$$D^m: L_p^m[I]\mapsto L_p[I]$$

$$D^{-m}: L_p[I]\mapsto L_p^m[I]$$

and

$$D^m D^{-m}f=f.$$

The Taylor expansion shows that there is a connection between values of f and its various derivatives up to order m. Thus it should not be surprising that the derivatives of f can be estimated in terms of the size of f and the size of its mth derivative. We have the following useful tool:

THEOREM 2.4

There exist constants $C_{m,1},\ldots,C_{m,m-1}$ (depending only on m and $[a,b]$) such that for $j=1,2,\ldots,m-1$,

$$\|D^jf\|_{C[a,b]}\leqslant C_{m,j}\left(\varepsilon^{-j}\|f\|_{C[a,b]}+\varepsilon^{m-j}\|D^mf\|_{C[a,b]}\right) \tag{2.29}$$

for any $f\in C^m[a,b]$ and any $0<\varepsilon<(b-a)/2$.

Proof. The proof proceeds by induction on m and j. We consider first the case $m=2$ and $j=1$. Choose $N=\min\{n:(b-a)/n<\varepsilon\}$.

Then

$$\varepsilon\geqslant h=\frac{(b-a)}{N}=\frac{(b-a)(N-1)}{(N-1)N}\geqslant\frac{(N-1)\varepsilon}{N}\geqslant\frac{\varepsilon}{2}.$$

Let $x_i=a+ih$, $i=0,1,\ldots,N$, and let $I_i=[x_i,x_{i+1}]$. For any ξ in the first third of I_i and η in the last third of I_i, there exists θ in (ξ,η) such that

$$|Df(\theta)|=\left|\frac{f(\eta)-f(\xi)}{\eta-\xi}\right|\leqslant\frac{6}{\varepsilon}|f(\xi)|+\frac{6}{\varepsilon}|f(\eta)|. \tag{2.30}$$

Then for all $x\in I_i$,

$$|Df(x)|\leqslant|Df(\theta)+\int_\theta^x D^2f(t)\,dt|\leqslant\frac{12}{\varepsilon}\|f\|_{C[I_i]}+\varepsilon\|D^2f\|_{C[I_i]}. \tag{2.31}$$

We have proved (2.29) for $m=2$ and $j=1$.

Now suppose the result has been established for $m-1$ and $1 \leqslant j \leqslant m-2$. We next establish it for m and $j=m-1$. Applying (2.31) to $D^{m-2}f$, we obtain

$$|D^{m-1}f(x)| \leqslant C_{2,1}\left(\frac{1}{\varepsilon}\|D^{m-2}f\|_{C[I_i]}+\varepsilon\|D^m f\|_{C[I_i]}\right). \tag{2.32}$$

By the induction hypothesis, we can write

$$\|D^{m-2}f\|_{C[I_i]} \leqslant C_{m-1,m-2}\left(\frac{1}{\delta^{m-2}}\|f\|_{C[I_i]}+\delta\|D^{m-1}f\|_{C[I_i]}\right). \tag{2.33}$$

Choosing $\delta=\varepsilon/(2C_{2,1}C_{m-1,m-2})$ and substituting (2.33) in (2.32), we obtain

$$\|D^{m-1}f\|_{C[I_i]} \leqslant \frac{2^{m-2}C_{2,1}^{m-1}C_{m-1,m-2}^{m-1}}{\varepsilon^{m-1}}\|f\|_{C[I_i]}+\frac{1}{2}\|D^{m-1}f\|_{C[I_i]}$$
$$+C_{2,1}\varepsilon\|D^m f\|_{C[I_i]}.$$

This establishes (2.29) for m and for $j=m-1$.

To complete the proof we now proceed by downward induction on j. Assume (2.29) holds for m and j as well as for all smaller m's. Then substituting it in the inequality

$$\|D^{j-1}f\|_{C[I_i]} \leqslant C_{j,j-1}\left(\frac{1}{\varepsilon^{j-1}}\|f\|_{C[I_i]}+\varepsilon\|D^j f\|_{C[I_i]}\right).$$

we obtain (2.29) for $j-1$. ■

We also need the following analog of Theorem 2.4 for functions in the Sobolev spaces $L_p^m[a,b]$, $1 \leqslant p < \infty$.

THEOREM 2.5

There exist constants $C_{m,1},\ldots,C_{m,m-1}$ (depending only on m and $[a,b]$) such that for $j=1,2,\ldots,m-1$,

$$\|D^j f\|_{L_p[a,b]} \leqslant C_{m,j}\left(\varepsilon^{-j}\|f\|_{L_p[a,b]}+\varepsilon^{m-j}\|D^m f\|_{L_p[a,b]}\right) \tag{2.34}$$

for all $f \in L_p^m[a,b]$ and any $0<\varepsilon<(b-a)/2$.

Proof. Since the proof is very much like that of Theorem 2.4 we can be sketchy. It will be enough to establish the result for $m=2$ and $j=1$ as the rest of the proof proceeds as before. By (2.30) and (2.31) we have

$$|Df(x)| \leqslant \frac{6}{\varepsilon}|f(\zeta)| + \frac{6}{\varepsilon}|f(\eta)| + \int_{I_i} |D^2f(t)|\,dt$$

for all $x \in I_i$. Now if we integrate this inequality with respect to ζ over the first third of I_i and with respect to η over the last third of I_i, we obtain

$$|Df(x)| \leqslant \frac{36}{\varepsilon^2}\int_{I_i} |f(t)|\,dt + \int_{I_i} |D^2f(t)|\,dt.$$

Applying Hölder's inequality to each of the integrals and using the discrete Hölder inequality (cf. Remark 2.1), we obtain

$$\|Df\|^p_{L_p[I_i]} \leqslant 2^{p-1}h^p\left\{\left(\frac{36}{\varepsilon^2}\right)^p \|f\|^p_{L_p[I_i]} + \|D^2f\|^p_{L_p[I_i]}\right\}.$$

Now summing over $i = 0, 1, \ldots, N-1$ and using an elementary inequality (see Remark 2.1), we obtain

$$\|Df\|^p_{L_p[I]} \leqslant 2^{p-1}\varepsilon^p\left(\frac{36}{\varepsilon^2}\|f\|_{L_p[I]} + \|D^2f\|_{L_p[I]}\right)^p.$$

Taking the pth root of both sides, we obtain (2.34) for this case. ∎

§ 2.3. MATRICES AND DETERMINANTS

We frequently have to deal with matrices and determinants formed from a given set of functions. In this section we introduce some convenient notation for such matrices and determinants, and give several useful tools for dealing with them.

We begin with some notation. Let $\{u_i\}_1^m$ be a set of functions defined on a set I, and let $t_1, \ldots, t_m$ be points in I such that

$$t_1 < t_2 \cdots < t_m.$$

Then we define the matrix associated with $\{u_i\}_1^m$ and $\{t_i\}_1^m$ by

$$M\begin{pmatrix} t_1, \ldots, t_m \\ u_1, \ldots, u_m \end{pmatrix} = \begin{bmatrix} u_1(t_1) & u_2(t_1) & \cdots & u_m(t_1) \\ u_1(t_2) & u_2(t_2) & \cdots & u_m(t_2) \\ \cdots & & & \\ u_1(t_m) & u_2(t_m) & \cdots & u_m(t_m) \end{bmatrix}. \tag{2.35}$$

We denote the determinant by

$$D\begin{pmatrix} t_1,\ldots,t_m \\ u_1,\ldots,u_m \end{pmatrix} = \det M\begin{pmatrix} t_1,\ldots,t_m \\ u_1,\ldots,u_m \end{pmatrix}. \tag{2.36}$$

To see how such matrices can arise, consider the following basic interpolation problem:

PROBLEM 2.6. Lagrange Interpolation

Given real numbers $y_1,\ldots,y_m$, find u in $\mathcal{U} = \text{span}\,\{u_i\}_1^m$ such that

$$u(t_j) = y_j, \qquad j = 1,2,\ldots,m.$$

Discussion. The problem is to determine coefficients $c_1,\ldots,c_m$ such that

$$\sum_{i=1}^{m} c_i u_i(t_j) = y_j, \qquad j = 1,2,\ldots,m.$$

This is a linear system of m equations for the m unknown coefficients, and can be written in matrix form as $M\mathbf{c} = \mathbf{y}$, where M is given in (2.35) and $\mathbf{c} = (c_1,\ldots,c_m)^T$, $\mathbf{y} = (y_1,\ldots,y_m)^T$. This system has a unique solution for arbitrary $\mathbf{y}$ precisely when the matrix M is nonsingular. This in turn is equivalent to the nonvanishing of the determinant D.

When M is nonsingular, we can find an explicit formula for the interpolating function. For each $i = 1,2,\ldots,m$, let L_i be the unique function in $\mathcal{U}$ such that

$$L_i(t_j) = \delta_{ij} = \begin{cases} 1, & i = j \\ 0, & i \neq j, \end{cases} \quad j = 1,2,\ldots,m. \tag{2.37}$$

The functions $L_1,\ldots,L_m$ are called the *Lagrange functions*. In terms of them the unique solution of Problem 2.6 is given by

$$u(x) = \sum_{i=1}^{m} y_i L_i(x). \qquad \blacksquare$$

It will also be useful to define matrices associated with a set of points $t_1 \leqslant t_2 \leqslant \cdots \leqslant t_m$ where some of the t's are equal to each other. In order to describe exactly where the equalities hold, we suppose that

$$t_1 \leqslant t_2 \leqslant \cdots \leqslant t_m = \overbrace{\tau_1,\ldots,\tau_1}^{l_1},\ldots,\overbrace{\tau_d,\ldots,\tau_d}^{l_d}, \tag{2.38}$$

where each τ_i is repeated exactly l_i times with $\Sigma_{i=1}^{d} l_i = m$. Then given any sufficiently differentiable functions $u_1, \ldots, u_m$, we define

$$M\begin{pmatrix} t_1, \ldots, t_m \\ u_1, \ldots, u_m \end{pmatrix} = \begin{bmatrix} u_1(\tau_1) & u_2(\tau_1) & \cdots & u_m(\tau_1) \\ Du_1(\tau_1) & Du_2(\tau_1) & \cdots & Du_m(\tau_1) \\ & \cdots & & \\ D^{l_1-1}u_1(\tau_1) & D^{l_1-1}u_2(\tau_1) & \cdots & D^{l_1-1}u_m(\tau_1) \\ & \cdots & & \\ u_1(\tau_d) & u_2(\tau_d) & \cdots & u_m(\tau_d) \\ Du_1(\tau_d) & Du_2(\tau_d) & \cdots & Du_m(\tau_d) \\ & \cdots & & \\ D^{l_d-1}u_1(\tau_d) & D^{l_d-1}u_2(\tau_d) & \cdots & D^{l_d-1}u_m(\tau_d) \end{bmatrix}. \tag{2.39}$$

We can give a more compact definition of this matrix if we introduce the integers

$$d_i = \max\{j : t_i = \cdots = t_{i-j}\}, \qquad i = 1, 2, \ldots, m.$$

Then

$$M\begin{pmatrix} t_1, \ldots, t_m \\ u_1, \ldots, u_m \end{pmatrix} = \left[D^{d_i} u_j(t_i) \right]_{i,j=1}^{m}.$$

This kind of matrix arises, for example, in the following interpolation problem:

PROBLEM 2.7. Hermite Interpolation

Given real numbers $\{v_{ij}\}_{j=1,i=1}^{l_i,d}$, find u in $\mathfrak{U} = \operatorname{span}\{u_i\}_1^m$ such that

$$D^{j-1}u(t_i) = v_{ij}, \qquad j = 1, 2, \ldots, l_i$$
$$i = 1, 2, \ldots, d.$$

Discussion. In this problem we are interpolating both function values and certain derivatives. The determination of the required coefficients again involves the solution of a linear system that can be written in matrix form, this time with the matrix M defined in (2.39). ■

The matrix M defined in (2.35) can be thought of as a function of the vector $\mathbf{t} = (t_1, \ldots, t_m)$ on the *open simplex*

$$T = \{\mathbf{t} = (t_1, \ldots, t_m) \in I^m : t_1 < t_2 < \cdots < t_m\}.$$

If the functions $u_1,\ldots,u_m$ are continuous on I^m then the matrix M varies continuously (and so does its determinant) as $\mathbf{t}$ runs over T. Similarly, if the functions $u_1,\ldots,u_m$ all belong to $C^m[I]$, then the matrix M defined in (2.39) defines a function on the *closed simplex*

$$\overline{T}=\{\mathbf{t}=(t_1,\ldots,t_m)\in I^m\colon t_1\leqslant t_2\leqslant\cdots\leqslant t_m\}.$$

On the other hand, on T the matrix M is *not* a continuous function, no matter how smooth the functions $u_1,\ldots,u_m$ may be. The following example makes this point clearer:

EXAMPLE 2.8

Suppose $t_1<t_2^\nu$ for all ν, and that $t_2^\nu\downarrow t_1$ as $\nu\to\infty$. Then

$$M\begin{pmatrix} t_1, & t_1 \\ u_1, & u_2 \end{pmatrix}=\begin{bmatrix} u_1(t_1) & u_2(t_1) \\ Du_1(t_1) & Du_2(t_1) \end{bmatrix}, \tag{2.40}$$

while

$$\lim_{\nu\to\infty} M\begin{pmatrix} t_1, & t_2^\nu \\ u_1, & u_2 \end{pmatrix}=\begin{bmatrix} u_1(t_1) & u_2(t_1) \\ u_1(t_1) & u_2(t_1) \end{bmatrix}. \tag{2.41}$$

Discussion. The matrices in (2.40) and (2.41) are in general different. For example, this is the case if $u_1(x)=1$ and $u_2(x)=x$. The matrix in (2.41) always has zero determinant (since two rows are equal), while the determinant of the matrix in (2.40) may well be nonzero. ■

The above discussion shows that the nonsingularity of a matrix for all $\mathbf{t}$ in the open simplex T does not imply anything about the nonsingularity of the matrix for $\mathbf{t}$ on the boundary of $\overline{T}$. Or saying it another way, nonsingularity for distinct t's does not imply anything about the case where some of the t's are equal to each other. On the other hand, it is true that if the determinant maintains one sign for all distinct t's, then it must have the same sign for all $t_1\leqslant\cdots\leqslant t_m$; that is, the sign must remain the same as we go to the boundary of $\overline{T}$. This is a useful fact, and we prove it in the following lemma:

LEMMA 2.9

Let $\{u_i\}_1^m$ be functions defined on an interval I such that

$$D\begin{pmatrix} x_1,\ldots, & x_m \\ u_1,\ldots, & u_m \end{pmatrix}\geqslant 0 \qquad \text{for all } x_1<x_2<\cdots<x_m \text{ in } I. \tag{2.42}$$

Given $\{t_i\}_1^m$ with repetitions as in (2.38), suppose that for some $\varepsilon > 0$

$$u_j \in C^{l_i - 1}[\tau_i, \tau_i + \varepsilon), \qquad \begin{matrix} j = 1,2,\ldots,m \\ i = 1,2,\ldots,d \end{matrix}. \tag{2.43}$$

Then

$$D\begin{pmatrix} t_1,\ldots,t_m \\ u_1,\ldots,u_m \end{pmatrix} \geqslant 0. \tag{2.44}$$

The same assertion is valid if we replace $[\tau_i, \tau_i + \varepsilon)$ by $(\tau_i - \varepsilon, \tau_i]$ in (2.43), all $i = 1,2,\ldots,m$.

Proof. Let e be the number of equal signs in the sequence $t_1 \leqslant t_2 \leqslant \cdots \leqslant t_m$. When $e = 0$, the t's are distinct, and (2.44) follows from the assumption (2.42). We now proceed by induction on e; assuming the result holds for $e - 1$, we try to prove it for e. Writing the t's with repetitions as in (2.38), suppose that τ_k is the first τ which is repeated; that is,

$$t_1 \leqslant t_2 \leqslant \cdots \leqslant t_m = \tau_1 < \cdots < \tau_{k-1} < \overbrace{\tau_k = \cdots = \tau_k}^{l_k} < \cdots < \overbrace{\tau_d = \cdots = \tau_d}^{l_d},$$

with $l_k > 1$. Then with sufficiently small δ, the slightly perturbed set

$$\tilde{t}_1 \leqslant \tilde{t}_2 \leqslant \cdots \leqslant \tilde{t}_m =$$

$$\tau_1 < \cdots < \tau_{k-1} < \overbrace{\tau_k = \cdots = \tau_k}^{l_k - 1} < \tau_k + \delta < \cdots < \overbrace{\tau_d = \cdots = \tau_d}^{l_d}$$

has only $e - 1$ equal signs. Thus by the inductive hypothesis,

$$0 \leqslant D\begin{pmatrix} \tilde{t}_1,\ldots,\tilde{t}_m \\ u_1,\ldots,u_m \end{pmatrix} = \begin{bmatrix} u_1(\tau_1) & \cdots & u_m(\tau_1) \\ \vdots & & \vdots \\ u_1(\tau_{k-1}) & \cdots & u_m(\tau_{k-1}) \\ u_1(\tau_k) & \cdots & u_m(\tau_k) \\ \vdots & & \vdots \\ D^{l_k-2}u_1(\tau_k) & \cdots & D^{l_k-2}u_m(\tau_k) \\ u_1(\tau_k+\delta) & \cdots & u_m(\tau_k+\delta) \\ \vdots & & \vdots \\ D^{l_d-1}u_1(\tau_d) & \cdots & D^{l_d-1}u_m(\tau_d) \end{bmatrix}.$$

Now for each $i=1,2,\ldots,m$, the Taylor expansion asserts the existence of $\tau_k \leqslant \xi_{ik} \leqslant \tau_k+\delta$ with

$$u_i(\tau_k+\delta)=\sum_{j=0}^{l_k-2}\frac{D^j u_i(\tau_k)\delta^j}{j!}+\frac{D^{l_k-1}u_i(\xi_{ik})\delta^{l_k-1}}{(l_k-1)!}.$$

Substituting these expansions in the determinant and simplifying, we obtain

$$0\leqslant\frac{\delta^{l_k-1}}{(l_k-1)!}\begin{bmatrix} u_1(\tau_1) & \cdots & u_m(\tau_1)\\ \vdots & & \vdots\\ u_1(\tau_{k-1}) & \cdots & u_m(\tau_{k-1})\\ u_1(\tau_k) & \cdots & u_m(\tau_k)\\ \vdots & & \vdots\\ D^{l_k-2}u_1(\tau_k) & \cdots & D^{l_k-2}u_m(\tau_k)\\ D^{l_k-1}u_1(\xi_{1k}) & \cdots & D^{l_k-1}u_m(\xi_{mk})\\ \vdots & & \vdots\\ D^{l_d-1}u_1(\tau_d) & \cdots & D^{l_d-1}u_m(\tau_d)\end{bmatrix}.$$

Since $\xi_{ik}\downarrow\tau_k$ as $\delta\downarrow 0$ for $i=1,2,\ldots,k$, this determinant approaches the one in (2.44), and we conclude that this latter determinant is also nonnegative. This completes the proof under the hypothesis (2.43).

If each of the $u_1,\ldots,u_m$ is sufficiently smooth to the left of each τ_i, then we use the perturbed set

$$\tilde{t}_1\leqslant\tilde{t}_2\leqslant\cdots\leqslant\tilde{t}_m=\tau_1<\cdots<\tau_{k-1}<\tau_k-\delta,\ \overbrace{\tau_k,\ldots,\tau_k}^{l_k-1},\ldots,\ \overbrace{\tau_d,\ldots,\tau_d}^{l_d}$$

and expand to the left. After rearranging the rows and taking account of the minus signs in the Taylor expansion, we obtain (2.44). ■

§ 2.4. SIGN CHANGES AND ZEROS

In this section we introduce some counting procedures for sign changes and for zeros of functions.

DEFINITION 2.10. Sign Changes of a Vector

Let $\mathbf{v}=(v_1,\ldots,v_n)$ be a vector of real numbers. We define the number of *strong sign changes of* $\mathbf{v}$ by

$$S^-(\mathbf{v}) = \begin{array}{l}\text{the number of sign changes in}\\ \text{the sequence } v_1,\ldots,v_n, \text{ where}\\ \text{zeros are ignored.}\end{array} \tag{2.45}$$

Similarly, we define the number of *weak sign changes of* $\mathbf{v}$ by

$$S^+(\mathbf{v}) = \begin{array}{l}\text{the maximum number of sign}\\ \text{changes in the sequence } v_1,\ldots,v_n,\\ \text{where each zero can be regarded}\\ \text{as either } +1 \text{ or } -1, \text{ whichever}\\ \text{makes the count largest.}\end{array} \tag{2.46}$$

It is clear that

$$S^-(\mathbf{v}) \leqslant S^+(\mathbf{v}) \qquad \text{for all } \mathbf{v}. \tag{2.47}$$

It is also relatively easy to show that for all $\mathbf{v}$,

$$S^+\big(v_1, -v_2,\ldots,(-1)^{r-1}v_r\big) + S^-(v_1,v_2,\ldots,v_r) \geqslant r-1, \tag{2.48}$$

and equality holds if all the v_i's are nonzero.

DEFINITION 2.11. Sign Changes of a Function

Let f be a bounded real-valued function on a subset I of the real-line $\mathbf{R}$. We call

$$S_I^-(f) = \sup_n \{S^-[f(t_1),\ldots,f(t_n)] : t_1 < t_2 < \cdots < t_n \in I\} \tag{2.49}$$

the number of *strong sign changes of f on I*. Similarly, we define the number of *weak sign changes of f on I* by

$$S_I^+(f) = \sup_n \{S^+[f(t_1),\ldots,f(t_n)] : t_1 < \cdots < t_n \in I\}. \tag{2.50}$$

It follows immediately from (2.47) that

$$S_I^-(f) \leqq S_I^+(f). \tag{2.51}$$

We illustrate $S_I^-(f)$ and $S_I^+(f)$ in Figure 3.

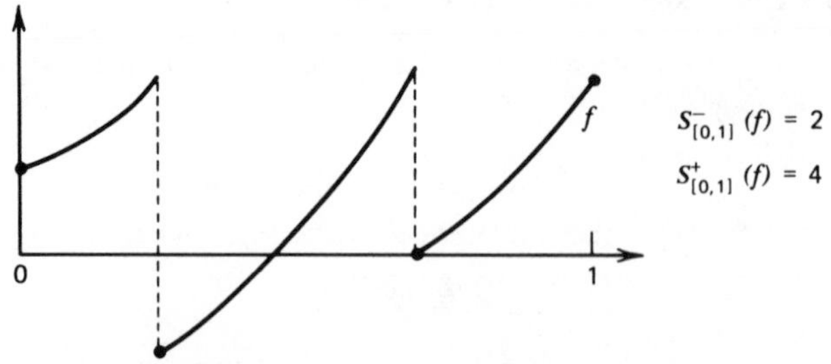

Figure 3. Strong and weak sign changes.

DEFINITION 2.12. Simple Zeros

Let $f \in B[I]$. We say that f has a zero at the point $\zeta \in I$ provided $f(\zeta)=0$. We shall write

$$Z_I(f) = \text{number of zeros of } f \text{ in the set } I. \tag{2.52}$$

If a function vanishes identically on a subinterval of I, then $Z_I(f)=\infty$. Thus, it is useful to make the following definition:

DEFINITION 2.13. Separated Zeros

Let $f \in B[I]$. We say that $t_1 < t_2 < \cdots < t_n \in I$ are separated zeros of f provided f has a zero at each point $t_1, \ldots, t_n$ and there exist

$$t_1 < y_1 < t_2 < y_2 < \cdots < t_{n-1} < y_{n-1} < t_n$$

in I with $f(y_i) \neq 0$, $i = 1, 2, \ldots, n-1$. We write

$$Z_I^{\text{sep}}(f) = \max\{n : f \text{ has } n \text{ separated zeros in } I\}. \tag{2.53}$$

Up to this point we have been working with arbitrary subsets of I of the real line. For the remainder of the section we restrict our attention to *intervals* (open, closed, or half closed). Our immediate goal is to introduce the idea of double zeros.

DEFINITION 2.14. Double Zeros

If $f \in B[I]$ is a function with a zero at a point t in the interior of I, then we say that *f changes sign at t* provided

$$f(t-\varepsilon)f(t+\varepsilon) < 0 \qquad \text{for all } \varepsilon > 0 \text{ sufficiently small.}$$

If f has a zero at a point t in (a,b) and does not change sign at this point, then we say f has a *double zero at t*. We write

$$Z_I^2(f) = \begin{array}{l}\text{number of zeros of } f\text{, count-}\\ \text{ing double zeros twice.}\end{array} \tag{2.54}$$

For continuous functions defined on an interval, there is a close connection between zeros and sign changes. We have the following:

THEOREM 2.15

Let $f \in C[I]$ where I is a subinterval of $\mathbf{R}$. Then

$$S_I^-(f) \leqslant Z_I^{\text{sep}}(f) \tag{2.55}$$

and

$$S_I^+(f) \leqslant Z_I^2(f). \tag{2.56}$$

Proof. If a continuous function on an interval is such that $f(t_1)\cdot f(t_2) < 0$, then f must vanish somewhere in between. ■

It is also possible to define zeros of higher multiplicity provided f has enough derivatives.

DEFINITION 2.16. Multiple Zeros

Suppose $f \in C^z[I]$ and $t \in I$. We say that f has a *zero of multiplicity z at the point t* provided

$$f(t) = Df(t) = \cdots D^{z-1}f(t) = 0 \neq D^z f(t). \tag{2.57}$$

As usual, if t is an endpoint of the interval I, then D is to be understood as either the left or right derivative. The behavior of a function in the vicinity of a multiple zero depends on whether that zero is odd or even. We have the following:

If f has an odd zero at t in (a,b), then f must change sign at the point t, (2.58)

and

if f has an even zero at t in (a,b), then f does not change sign at t. (2.59)

DEFINITION 2.17. Number of Multiple Zeros

Suppose $f \in C^r[I]$ where I is a subinterval of $\mathbf{R}$. Then we define the *number of zeros of f on I, counting multiplicities up to order r* by

$$Z_I^r(f) = \begin{array}{l}\text{number of zeros of } f \text{ on } I, \text{ count-} \\ \text{ing multiplicities as in (2.57) up} \\ \text{to } z = r.\end{array} \tag{2.60}$$

If $f \in C^\infty[I]$, we define

$$Z_I^*(f) = \begin{array}{l}\text{number of zeros of } f \text{ on } I, \text{ count-} \\ \text{ing multiplicities of all orders.}\end{array} \tag{2.61}$$

We conclude this section with a useful tool for dealing with zeros of functions. First we need a definition.

DEFINITION 2.18. Rolle's Points

If f is an absolutely continuous function on the interval (c,d), then we say that c is a *left Rolle's point of f* provided that

$$\text{either } f(c) = 0 \text{ or for every } \varepsilon > 0, \text{ there exists some}$$

$$c < t < c + \varepsilon \text{ with } f(t) \cdot Df(t) > 0.$$

Similarly, we say that d is a *right Rolle's point of f* provided

$$\text{either } f(d) = 0 \text{ or for every } \varepsilon > 0, \text{ there exists some}$$

$$d - \varepsilon < t < d \text{ with } f(t) \cdot Df(t) < 0.$$

Examples of left and right Rolle's points of a function are shown in Figure 4. Geometrically, c or d is a Rolle's point if the function f moves away from the axis as we move into the interval.

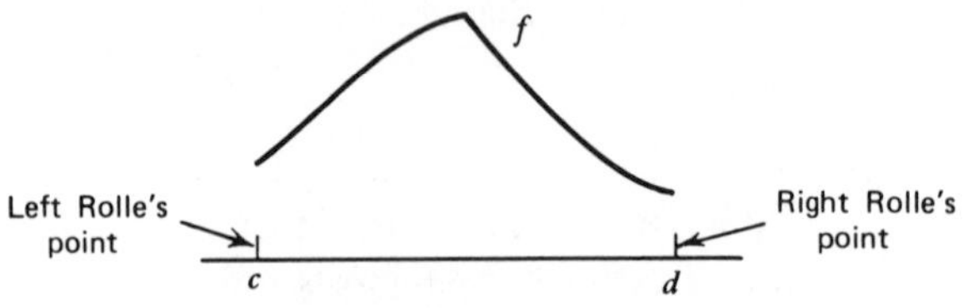

Figure 4. The extended Rolle's theorem.

THEOREM 2.19. Extended Rolle's Theorem

Suppose $f \in AC(c,d)$, and that c and d are left and right Rolle's points of f, respectively. Then Df has at least one sign change on the interval (c,d). If Df is continuous on (c,d), then it has at least one zero on this interval.

Proof. Since f is absolutely continuous, we can write $f(d)-f(c)=\int_c^d Df(t)\,dt$. If $f(c)=f(d)$, this implies either $Df(t)=0$ almost everywhere or it must change sign. This is the usual Rolle's theorem. Now suppose $f(c)<f(d)$. Then $Df(t)$ must be positive somewhere (in fact, on a set of positive measure) in (c,d). If $f(d)>0$, then since d is a right Rolle's point, $Df(t)<0$ for some t near d, and we have the desired sign change. If $f(c)<0$, then since c is a left Rolle's point, $Df(t)<0$ for some t near c, and again Df changes sign. The case where $f(c)>f(d)$ can be argued similarly. The last statement is obvious since a continuous function can only change sign by passing through zero. ■

§ 2.5. TCHEBYCHEFF SYSTEMS

In this section we explore the connection between properties of determinants formed from a set of functions $u_1,\ldots,u_m$ and the number of zeros or sign changes linear combinations of these functions can possess. Throughout the section we suppose that I is an (open, closed, or half-open) *interval* in $\mathbf{R}$.

First we must define some terminology.

DEFINITION 2.20. Tchebycheff System

A set $U=\{u_i\}_1^m$ of functions in $C[I]$ is called a *Tchebycheff* (T$-$) *system* provided

$$D\begin{pmatrix} t_1,\ldots,t_m \\ u_1,\ldots,u_m \end{pmatrix}>0 \qquad \text{for all } t_1<t_2<\cdots<t_m \text{ in } I. \tag{2.62}$$

The following theorem characterizes T-systems in terms of sign changes:

THEOREM 2.21

If $U=\{u_i\}_1^m$ is a T-system, then

$$Z_I\left(\sum_{i=1}^{m} c_i u_i\right) \leqq m-1 \qquad \text{for all real } c_1,\ldots,c_m, \text{ not all } 0, \tag{2.63}$$

where Z_I counts the number of simple zeros on I. Conversely, if $U=\{u_i\}_1^m$ is a set of functions in $C[I]$ such that (2.63) holds, then either U or the set $\tilde{U}=\{u_1,\dots,u_{m-1},-u_m\}$ is a T-system.

Proof. By the discussion of Problem 2.6, there can exist a nontrivial $\mathbf{c}=(c_1,\dots,c_m)$ such that $u=\sum_{i=1}^m c_i u_i$ vanishes at m points $t_1<\cdots<t_m$ if and only if t e determinant D in (2.62) is zero. Hence if U is a T-system, (2.63) follows. Conversely, if (2.63) holds, then D can never be zero. Since D is a continuous function of the t's, it must have one sign for all $t_1<\cdots<t_m$ in I. We conclude that either U or $\tilde{U}$ forms a T-system.

Tchebycheff systems play an important role in approximation theory, statistics, and other parts of analysis. There are many examples of T-systems. We give just two.

EXAMPLE 2.22

Let $U=\{1,x,\dots,x^{m-1}\}$.

Discussion. Since this set of functions spans the space of polynomials of order m, it is particularly important. We introduce the special notation

$$VM(t_1,\dots,t_m)=M\begin{pmatrix} t_1,\dots,t_m \\ 1,\dots,x^{m-1}\end{pmatrix} \tag{2.64}$$

and

$$V(t_1,\dots,t_m)=D\begin{pmatrix} t_1,\dots,t_m \\ 1,\dots,x^{m-1}\end{pmatrix}, \tag{2.65}$$

where VM is the *Vandermonde matrix* and V is the *Vandermonde determinant*. It is shown in Theorem 3.5 that V is positive for all choices of $t_1 \leqslant \cdots \leqslant t_m$, and thus it follows that U is a T-system. In fact, with some algebraic manipulation it can be shown that

$$V(t_1,\dots,t_m)=\prod_{1\leqslant i<j\leqslant m}(t_j-t_i) \tag{2.66}$$

whenever $t_1<t_2<\cdots<t_m$, while

$$V(t_1,\dots,t_m)=\prod_{1\leqslant i<j\leqslant d}(\tau_j-\tau_i)^{l_j l_i}\prod_{i=1}^{d}\prod_{\nu=1}^{l_i-1}\nu! \tag{2.67}$$

whenever $\quad t_1\leqslant t_2\leqslant\cdots\leqslant t_m=\overbrace{\tau_1,\dots,\tau_1}^{l_1}<\cdots<\overbrace{\tau_d,\dots,\tau_d}^{l_d}.$ ■

EXAMPLE 2.23

Let $U=\{1,\cos(x),\sin(x),\ldots,\cos(rx),\sin(rx)\}$.

Discussion. This set of functions is a T-system on the interval $I=[0,2\pi)$. It is not a T-system on any larger interval. See Karlin and Studden [1966], p. 180. ■

Theorem 2.21 shows that linear combinations of a T-system cannot have more than $m-1$ zeros. The following result shows that the same is true even if we count double zeros:

THEOREM 2.24

Let $U=\{u_i\}_1^m$ be a T-system on I. Then

$$Z_I^2\left(\sum_{i=1}^{m} c_i u_i\right) \leqslant m-1 \qquad \text{for all } c_1,\ldots,c_m \in \mathbf{R}, \text{ not all } 0.$$

Proof. Suppose $u \in \mathcal{U} = \operatorname{span}\{u_i\}_1^m$ is a nontrivial function with $Z_I^2(u) = m$, and the zeros of u are $t_1 < t_2 < \cdots < t_p$. If we add to this set $t_i - \varepsilon$ for the first double zero t_i, and the points $t_i + \varepsilon$ for all double zeros t_i, then for sufficiently small $\varepsilon > 0$ we obtain a set of points

$$x_1 < x_2 < \cdots < x_r$$

with $r \geqslant m+1$. Moreover, we see that

$$\eta(-1)^{i-1} u(x_i) \geqslant 0, \qquad i=1,2,\ldots,r,$$

where $\eta = +1$ or -1. Since $u \in \mathcal{U}$,

$$0 = D\begin{pmatrix} x_1,\ldots,x_{m+1} \\ u,u_1,\ldots,u_m \end{pmatrix} = \sum_{i=1}^{m+1} (-1)^{i-1} u(x_i) D\begin{pmatrix} x_1,\ldots,x_{i-1},x_{i+1},\ldots,x_{m+1} \\ u_1,\ldots,u_m \end{pmatrix}.$$

This implies $u(x_i)=0, i=1,2,\ldots,m+1$, which contradicts Theorem 2.21. ■

So far we have been working with *systems* of functions. In some applications it is more convenient to talk about *spaces*.

DEFINITION 2.25. Tchebycheff Space

An m-dimensional linear space $\mathcal{U}$ is called a *Tchebycheff* (T-) *space* provided it has a basis that is a T-system.

THEOREM 2.26

If $\mathfrak{U}$ is a T-space, then for any basis $U=\{u_i\}_1^m$, either U or $\tilde{U}=\{u_1,\ldots,u_{m-1},-u_m\}$ is a T-system.

Proof. Since $\mathfrak{U}$ is a T-space, there exists some basis $\{v_i\}_1^m$ which forms a T-system. Now if U is also a basis, then there must exist some nonsingular matrix A such that $(u_1,\ldots,u_m)^T=A(v_1,\ldots,v_m)^T$. But then

$$D\begin{pmatrix} t_1,\ldots,t_m \\ u_1,\ldots,u_m \end{pmatrix}=\det(A)\cdot D\begin{pmatrix} t_1,\ldots,t_m \\ v_1,\ldots,v_m \end{pmatrix}$$

$$\text{for all } t_1<t_2<\cdots<t_m \text{ in } I.$$

It follows that either U or $\tilde{U}$ is a T-system. ∎

In terms of this new terminology, Theorem 2.21 asserts that an m-dimensional linear space $\mathfrak{U}$ is a T-space if and only if $Z_I(u)\leqslant m-1$ for all nontrivial u in $\mathfrak{U}$.

We now introduce a subclass of T-systems with stronger determinantal properties.

DEFINITION 2.27 Complete Tchebycheff System

Suppose $U=\{u_i\}_1^m$ is such that $\{u_i\}_1^k$ is a T-system for each $k=1,2,\ldots,m$. Then we say that U is a *Complete Tchebycheff* (CT-) system.

The powers discussed in Example 2.22 provide an example of a CT-system. The following is an example of a T-system that is not a CT-system.

EXAMPLE 2.28

Let $U=\{\sin(x),\cos(x)\}$ on $[0,\pi)$.

Discussion. U is clearly a T-system on $[0,\pi)$. It is not a CT-system, however, as $\sin(x)$ is not positive throughout this interval. In fact, it is not possible to find any basis for $\mathfrak{U}=\text{span}(U)$ that forms a CT-system, since any linear combination of $\sin(x)$ and $\cos(x)$ always has a zero on $[0,\pi)$. ∎

The situation encountered in Example 2.28 cannot occur on open intervals, and we now have the following surprising result:

THEOREM 2.29

Suppose $U=\{u_i\}_1^m$ is a T-system on (a,b). Then there exist functions $V=\{v_i\}_1^m$ with $\text{span}(V)=\text{span}(U)$ such that V is a CT-system on (a,b).

There are several possible proofs of this remarkable result. We shall give a proof in the next section utilizing results on Weak Tchebycheff systems (see p. 41). It can be shown by example that Theorem 2.29 is true only for open intervals (see Remark 2.3). In the terminology of T-spaces, Theorem 2.29 asserts that every T-space on an open interval is a CT-space.

We now introduce a still stronger form of Tchebycheff system.

DEFINITION 2.30. Order Complete Tchebycheff System

A set of functions $\{u_i\}_1^m$ with the property that

$$\{u_{i_\nu}\}_{\nu=1}^k \qquad \text{is a T-system for all } 1 \leqslant i_1 < i_2 < \cdots < i_k \leqslant m \text{ and all } 1 \leqslant k \leqslant m$$

is called an *Order Complete Tchebycheff* (OCT-) *system.*

We shall show in Theorem 3.7 that the powers $1, x, \ldots, x^{m-1}$ form an OCT-system on any subinterval of $(0, \infty)$. The following theorem shows that OCT-systems have an important *variation-diminishing property*.

THEOREM 2.31. Descartes' Rule of Signs

Suppose $U = \{u_i\}_1^m$ is an OCT-system. Then

$$Z\left(\sum_{i=1}^m c_i u_i\right) \leqq S^-(c_1, \ldots, c_m) \qquad \text{for any real } c_1, \ldots, c_m, \text{ not all 0,} \tag{2.68}$$

where Z counts the number of simple zeros in $[a,b]$ and S^- counts the number of strong sign changes in the coefficients [cf. (2.45)].

Proof. Let $S^-(c) = d-1$. Then we can divide the coefficients into d sets,

$$\overbrace{c_1, \ldots, c_{n_2}}, \overbrace{c_{n_2+1}, \ldots, c_{n_3}}, \ldots, \overbrace{c_{n_d+1}, \ldots, c_m}, \tag{2.69}$$

such that all c_i's in any one group have the same sign, and at least one c_i in each group is nonzero. Now with $n_1 = 0$, $n_{d+1} = m$, define

$$v_j = \sum_{i=n_j+1}^{n_{j+1}} |c_i| u_i, \qquad j = 1, 2, \ldots, d. \tag{2.70}$$

We claim $\{v_j\}_1^d$ is a CT-system. Indeed, for any $1 \leqslant k \leqslant d$ and $t_1 < \cdots < t_k$ in $[a,b]$, we have

$$D\begin{pmatrix} t_1, \ldots, t_k \\ v_1, \ldots, v_k \end{pmatrix} = \sum_{i_1=n_1+1}^{n_2} \cdots \sum_{i_k=n_k+1}^{n_{k+1}} |c_{i_1}| \cdots |c_{i_k}| D\begin{pmatrix} t_1, \ldots, t_k \\ u_{i_1}, \ldots, u_{i_k} \end{pmatrix}, \tag{2.71}$$

which is positive since U is an OCT-system and at least one c_i is nonzero in each group. Let

$$\tilde{c}_j = (-1)^{j-1}\operatorname{sgn}(\text{first group of } c\text{'s}), \qquad j = 1, 2, \ldots, d. \tag{2.72}$$

Then clearly

$$\tilde{u} = \sum_{j=1}^{d} \tilde{c}_j v_j \equiv u = \sum_{i=1}^{m} c_i u_i, \tag{2.73}$$

and using Theorem 2.21 on the T-system $\{v_j\}_1^d$, we obtain

$$Z\left(\sum_{i=1}^{m} c_i u_i\right) = Z\left(\sum_{j=1}^{d} \tilde{c}_j v_j\right) \leqslant d - 1 = S^-(c_1, \ldots, c_m). \qquad \blacksquare$$

So far in this section we have been working exclusively with continuous functions. If we work with smoother functions, then we can deal with determinants with repeated t's.

DEFINITION 2.32. Extended Tchebycheff System

Let $U = \{u_i\}_1^m$ belong to $C^{m-1}[I]$, where I is a subinterval of $\mathbf{R}$. We call U an *Extended Tchebycheff* (ET-) *system* provided

$$D\begin{pmatrix} t_1, \ldots, t_m \\ u_1, \ldots, u_m \end{pmatrix} > 0 \qquad \text{for all } t_1 \leqslant t_2 \leqslant \cdots \leqslant t_m \text{ in } I. \tag{2.74}$$

The following result characterizes ET-systems in terms of zeros:

THEOREM 2.33

If $U = \{u_i\}_1^m$ is an ET-system on I, then

$$Z_I^{m-1}\left(\sum_{i=1}^{m} c_i u_i\right) \leqslant m - 1 \qquad \text{for all real } c_1, \ldots, c_m, \text{ not all } 0, \tag{2.75}$$

where Z^{m-1} counts the number of zeros with multiplicities up to order $m-1$ (cf. Definition 2.17). Conversely, if U is a set of linearly independent functions in $C^{m-1}[I]$ satisfying (2.75), then either U or $\tilde{U} = \{u_1, \ldots, u_{m-1}, -u_m\}$ is an ET-system.

Proof. Suppose U is an ET-system, and that some nontrivial $u=\sum_{i=1}^{m}c_iu_i$ is such that $Z_I^{m-1}(u)\geqslant m$. If $t_1\leqslant t_2\leqslant\cdots\leqslant t_m$ are m of these zeros, then this can happen only if the determinant D in (2.74) is zero. This contradicts the assumption that U is an ET-system. Conversely, if (2.75) holds, then D in (2.74) can never be zero. For $\mathbf{t}=(t_1,\ldots,t_m)$ in the open simplex $T=\{\mathbf{t}:t_1<t_2<\cdots<t_m\in I\}$, D is a continuous function of $\mathbf{t}$, hence it must have one sign. By Lemma 2.9, D also maintains the same sign as the t's are allowed to coalesce. We conclude that D has one sign for all $\mathbf{t}$, and thus either U or $\tilde{U}$ is an ET-system. ■

As with T-systems, it is useful to define stronger versions of ET-systems. We say that $U=\{u_i\}_1^m$ is an *Extended Complete Tchebycheff* (ECT-) *system* provided that $\{u_i\}_1^k$ is an ET-system for each $k=1,2,\ldots,m$. We shall study ECT-systems in greater detail in Section 9.1. If U has the property that

$$\{u_{i_\nu}\}_{\nu=1}^{k} \qquad \text{is an ET-system for all } 1\leqslant i_1<i_2<\cdots<i_k\leqslant m \text{ and all } k=1,2,\ldots,m,$$

then we say that U is an *Order Complete Extended Tchebycheff* (OCET-) *system*. For OCET-*systems* we have the following important *variation-diminishing property*. It is a strengthening of Theorem 2.31 in that multiple zeros are now counted.

THEOREM 2.34. Descartes' Rule of Signs

Suppose $U=\{u_i\}_1^m$ is an OCET-*system*. Then

$$Z_I^{m-1}\left(\sum_{i=1}^{m}c_iu_i\right)\leqslant S^-(c_1,\ldots,c_m) \qquad \text{for any real } c_1,\ldots,c_m, \text{ not all 0.} \tag{2.76}$$

Proof. Since the proof is very similar to that of Theorem 2.31, we can be brief. Let $S^-(\mathbf{c})=d-1$. Then dividing the $c_1,\ldots,c_m$ into groups as in (2.69), let $V=\{v_i\}_1^d$ be as defined in (2.70). Since U is an OCET-system, the expansion (2.71) implies that V is an ET-system (in fact, an ECT-system). Then with $\tilde{c}_1,\ldots,\tilde{c}_d$ and $\tilde{u}$ as in (2.72) and (2.73), Theorem 2.33 applied to the ET-system V implies

$$Z_I^{m-1}\left(\sum_{i=1}^{m}c_iu_i\right)=Z_I^{m-1}\left(\sum_{j=1}^{d}\tilde{c}_jv_j\right)\leqq d-1=S^-(c_1,\ldots,c_m). \qquad ■$$

§ 2.6. WEAK TCHEBYCHEFF SYSTEMS

We shall see in Chapter 4 that spaces of spline functions are not generally T-spaces. Thus it is useful to introduce a weaker form of T-system capable of encompassing splines.

DEFINITION 2.35. Weak Tchebycheff System

Let $u_1,\ldots,u_m$ be a set of bounded real-valued functions defined on a set $I \subseteq R$. We say that the $\{u_i\}_1^m$ form a *Weak Tchebycheff* (WT-) *system* provided they are linearly independent, and

$$D\begin{pmatrix} t_1,\ldots,t_m \\ u_1,\ldots,u_m \end{pmatrix} \geqslant 0 \qquad \text{for all } t_1<t_2<\cdots<t_m \text{ in } I. \tag{2.77}$$

In contrast with the definition of T-systems in the previous section, here we have deliberately allowed arbitrary subsets I of the real-line $\mathbf{R}$, and we have not required the functions $\{u_i\}_1^m$ to be continuous. We begin with some examples of WT-systems.

EXAMPLE 2.36

Let $U=\{u_i\}_1^m$ be a T-system on an interval I. Then U is a WT-system on any subset J of I.

Discussion. In fact, it is clear that for any set of $t_1<\cdots<t_m$ in J, the determinant D in (2.77) is actually positive. ■

The method of Example 2.36 is one way of constructing a wide variety of WT-systems on arbitrary point sets. Another method is illustrated in the following example:

EXAMPLE 2.37

Let $U=\{u_i\}_1^m$ be a T-system on an interval I. Suppose we define functions $u_1,\ldots,u_m$ by

$$\tilde{u}_i(x)=\begin{cases} u_i(x), & x\in I\setminus J \\ 0, & x\in J, \end{cases}$$

$i=1,2,\ldots,m$, where J is any subset of I. Then if $\tilde{U}=\{\tilde{u}_i\}_1^m$ has dimension m, it is a WT-system on J.

Discussion. In this case D will be positive for all $t_1<\cdots<t_m$ in $I\setminus J$. If any one of the t's lies in J, then $D=0$. The functions in U can be quite wild—for example, we can take $I=\mathbf{R}$ and $J=$ the set of rational numbers. ■

We shall see in Chapter 4 that various spline spaces have bases that form WT-systems. The following example involves the spline x_+^0 defined in (2.15):

EXAMPLE 2.38

Let $U=\{1,x_+^0\}$ on $I=[-1,1]$. Then U forms a WT-system.

Discussion. We may check directly that $D \geqslant 0$ for all $-1 \leqslant t_1 < t_2 \leqslant 1$. ∎

In the previous section we saw that the T-systems could be characterized in terms of zeros. The following theorem gives a characterization of WT-systems in terms of sign changes:

THEOREM 2.39

Let $U=\{u_i\}_1^m$ be a linearly independent set of m functions defined on a set I. If U is a WT-system, then

$$S_I^-\left(\sum_{i=1}^m c_i u_i\right) \leqq m-1, \qquad \text{any real } c_1,\ldots,c_m, \text{ not all } 0. \tag{2.78}$$

Conversely, if (2.78) holds, then either U or $\tilde{U}=\{u_1,\ldots,u_{m-1},-u_m\}$ is a WT-system on I.

Proof. We first show that if U is a WT-system, then (2.78) must hold. Suppose it does not. Then there exists $u \in \mathfrak{U} = \operatorname{span}(U)$ and $T=\{t_1<t_2<\cdots<t_{m+1}\}$ with

$$(-1)^i u(t_i) > 0, i=1,2,\ldots,m+1. \tag{2.79}$$

CASE 1. Suppose rank $U|T=m$, where, in general,

$$\operatorname{rank} U|T = \operatorname{rank}\{[u_1(t),\ldots,u_m(t)] : t\in T\}.$$

Then since u is a linear combination of $u_1,\ldots,u_m$,

$$0 = D\begin{pmatrix} t_1,\ldots,t_{m+1} \\ u_1,\ldots,u_m,u \end{pmatrix} = \sum_{i=1}^{m+1} (-1)^i u(t_i) D_i, \tag{2.80}$$

where

$$D_i = D\begin{pmatrix} t_1,\ldots,t_{i-1},t_{i+1},\ldots,t_{m+1} \\ u_1,\ldots,u_m \end{pmatrix}, \qquad i=1,2,\ldots,m+1.$$

Since U is a WT-system, each of the D_i is nonnegative, and (2.80) implies they are all zero. But this contradicts the assumption that rank $U|T=m$.

CASE 2. Suppose rank $U|T=k<m$. We now show that there is a new set of points $\tilde{T}$ and a new $\tilde{u}\in\mathfrak{U}$ such that $(-1)^i\tilde{u}(\tilde{t}_i)>0$, $i=1,2,\ldots,m+1$ and rank $U|\tilde{T}=k+1$. This process could then be continued until Case 1 applies. To construct $\tilde{u}$ and $\tilde{T}$, first note that since rank $U|I=m$, there must exist some point $t\in I$ such that rank $U|\hat{T}=k+1$, where $\hat{T}=T\cup\{t\}$. Suppose $t_v<t<t_{v+1}$, where we set $t_0=-\infty$ and $t_{m+2}=\infty$ for convenience. Suppose t_μ is the first point to the right of t such that rank $U|\tilde{T}=k+1$, where $\tilde{T}=\hat{T}\setminus\{t_\mu\}$. (If none exists to the right, then there must be one to the left, and a similar argument carries through.) $\tilde{T}$ is our new set. Now by elementary linear algebra (see Remark 2.6) there exists a function l_t in $\mathfrak{U}$ that vanishes at all points of T and with $l_t(t)=1$. Similarly, for each $i=v+1,\ldots,\mu-1$ there exists a function $l_i\in\mathfrak{U}$ that vanishes at all points of $\hat{T}$, except at t_i where it has the value 1. But then for sufficiently small $\varepsilon>0$ we easily check that the function

$$\tilde{u}=\varepsilon u+(-1)^{v+1}l_t+\sum_{i=\nu+1}^{\mu-1}(-1)^{i+1}l_i$$

has alternating signs on the sequence of points making up $\tilde{T}$.

We now establish the converse. Suppose that neither U nor $\tilde{U}$ is a WT-system. This means that for some $X=\{x_1<\cdots<x_m\}$ and $Y=\{y_1<\cdots<y_m\}$ in I,

$$D\binom{X}{U}D\binom{Y}{U}:=D\binom{x_1,\ldots,x_m}{u_1,\ldots,u_m}D\binom{y_1,\ldots,y_m}{u_1,\ldots,u_m}<0.$$

The fact that both of these determinants are nonzero implies that we can replace the rows of $D\binom{X}{U}$ by rows taken from $D\binom{Y}{U}$ one at a time, always keeping the determinant nonzero. But $D\binom{X}{U}$ and $D\binom{Y}{U}$ have opposite signs, and we conclude that at some point the removal of a row and replacement by another (in its natural place) causes a switch in sign of the determinant. It follows that for some $Z=\{z_1<\cdots<z_m\}$, there is a $\tilde{Z}$ obtained from Z by removing z_k and replacing it by z such that $D\binom{Z}{L}D\binom{\tilde{Z}}{L}<0$.

Let $L_1,\ldots,L_m$ be the Lagrange functions in $\mathfrak{U}$ corresponding to the points $z_1,\ldots,z_m$ [cf. (2.37)]. Then for some nonzero constant α (cf. proof of Theorem 2.26)

$$D\binom{Z}{L}=\alpha D\binom{Z}{U} \quad \text{and } D\binom{\tilde{Z}}{L}=\alpha D\binom{\tilde{Z}}{U},$$

hence $D\binom{Z}{L}D\binom{\tilde{Z}}{L}<0$. Since $D\binom{Z}{L}=1$, we conclude $D\binom{\tilde{Z}}{L}<0$. Assuming that $0\leqslant j\leqslant m$ is such that $z_j<z<z_{j+1}$, expanding $D\binom{\tilde{Z}}{L}$ out we deduce that

$$\operatorname{sgn} L_k(z)=\begin{cases}(-1)^{k-j}, & \text{if } 0\leqslant j\leqslant k-2\\ -1, & \text{if } k-1\leqslant j\leqslant k.\\ (-1)^{k-j-1}, & \text{if } k+1\leqslant j\leqslant m\end{cases}$$

Using the Lagrange functions $L_1,\dots,L_m$, we can construct $g\in\mathfrak{U}$ such that

$$g(z_i)=\operatorname{sgn} L_k(z)\begin{cases}(-1)^{j-i+1}, & i=1,2,\dots,j\\ (-1)^{j-i}, & i=j+1,\dots,m.\end{cases}$$

Now for $\varepsilon>0$ sufficiently small, the function $u=L_k+\varepsilon g$ satisfies (2.79) with $T=\{z_1,\dots,z_j,z,z_{j+1},\dots,z_m\}$. ■

In our next theorem we explore what happens when we restrict a WT-system on a set I to a subset J of I.

THEOREM 2.40

Let $U=\{u_i\}_1^m$ be a WT-system on the set I, and let J be a subset of I. Suppose

$$\mathfrak{U}_J=\operatorname{span}\{u_i|_J\}_{i=1}^m$$

is of dimension l. Then there exists a set $V=\{v_j\}_1^l$ spanning $\mathfrak{U}_J$ such that V is a WT-system on J.

Proof. Let $v_1,\dots,v_l$ be a basis for $\mathfrak{U}_J$, and suppose $v_{l+1},\dots,v_m$ are chosen such that $\{v_i\}_1^m$ span $\mathfrak{U}$. By subtracting linear combinations of the $v_1,\dots,v_l$ if necessary, we can arrange that $v_i(x)=0$ for all $x\in J, i=l+1,\dots,m$. By the linear independence of $\{u_i\}_1^m$ on I, there exist $t_{l+1},\dots,t_m$ in $I\setminus J$ such that

$$D\begin{pmatrix}t_{l+1},\dots,t_m\\ v_{l+1},\dots,v_m\end{pmatrix}\neq 0.$$

Since $v_{l+1},\dots,v_m$ vanish on J,

$$D\begin{pmatrix}t_1,\dots,t_l\\ v_1,\dots,v_l\end{pmatrix}D\begin{pmatrix}t_{l+1},\dots,t_m\\ v_{l+1},\dots,v_m\end{pmatrix}=D\begin{pmatrix}t_1,\dots,t_m\\ v_1,\dots,v_m\end{pmatrix}$$

for all $t_1 < t_2 < \cdots < t_l$ in J. We conclude that the first determinant on the left is of one sign for all such $t_1, \ldots, t_l$, hence either V or $\tilde{V} = \{v_1, \ldots, v_{l-1}, -v_l\}$ is a WT-system. ■

If $\{u_i\}_1^k$ is a WT-system on I for each $1 \leqslant k \leqslant m$, then we say that $U = \{u_i\}_1^m$ is a *Complete Weak Tchebycheff* (CWT-) *system* on I. The following result for WT-systems should be compared with Theorem 2.29 on T-systems:

THEOREM 2.41

Let $U = \{u_i\}_1^m$ be a WT-system on the set I. Then there exists a CWT-system $V = \{v_i\}_1^m$ spanning the same space as U.

Proof. By the linear independence of $u_1, \ldots, u_m$ on I, there exist points $t_1, \ldots, t_m$ in I such that $D\begin{pmatrix} t_1, \ldots, t_m \\ u_1, \ldots, u_m \end{pmatrix} \neq 0$. Let $L_1, \ldots, L_m$ be the Lagrange functions in $\mathfrak{U} = \text{span}(U)$ corresponding to these points, and set $U_{m-1} = \{L_i\}_1^{m-1}$. It is obvious that U_{m-1} has dimension $m-1$. By Theorem 2.40, we conclude that either U_{m-1} or $\tilde{U}_{m-1} = \{L_1, \ldots, L_{m-2}, -L_{m-1}\}$ is a WT-system on $J = I \backslash \{t_m\}$. This in turn implies by Theorem 2.39 that

$$S_J^-\left(\sum_{i=1}^{m-1} c_i L_i\right) \leqslant m-2 \qquad \text{for all } c_1, \ldots, c_{m-1}, \text{ not all } 0.$$

But since each of the functions $L_1, \ldots, L_{m-1}$ vanishes at t_m, this implies

$$S_I^-\left(\sum_{i=1}^{m-1} c_i L_i\right) \leqslant m-2 \qquad \text{for all } c_1, \ldots, c_{m-1}, \text{ not all } 0.$$

Using Theorem 2.39 again, it follows that either U_{m-1} or $\tilde{U}_{m-1}$ is a WT-system of $m-1$ functions on I. This process can be continued to construct WT-systems U_i of i functions for $i = m-2, \ldots, 1$. ■

If $\mathfrak{U}$ is an m-dimensional linear space defined on a set I, we say that $\mathfrak{U}$ is a *Weak Tchebycheff* (WT-) *space* provided it has a basis that is a WT-system. We define *Complete Weak Tchebycheff* (CWT-) *spaces* similarly. It is clear that the analog of Theorem 2.26 holds here; that is, if $\mathfrak{U}$ is a WT-space and $U = \{u_i\}_1^m$ is a basis for it, then either U or $\tilde{U} = \{u_1, \ldots, u_{m-1}, -u_m\}$ must be a WT-system. Theorem 2.39 asserts that an m-dimensional linear space $\mathfrak{U}$ is a WT-space if and only if $S_I^-(u) \leqslant m-1$ for all nontrivial u in $\mathfrak{U}$. Theorem 2.41 asserts that every WT-space $\mathfrak{U}$ is automatically a CWT-space.

As an application of Theorem 2.41, we now establish Theorem 2.29 in which it was asserted that a T-space on an open interval I is a CT-space.

Proof of Theorem 2.29. If $U=\{u_i\}_1^m$ is a T-system on the open interval I, then it is automatically a WT-system. We conclude from Theorem 2.41 that there exist functions $V=\{v_i\}_1^{m-1}$ in $\mathcal{U}=\text{span}(U)$ such that V forms a WT-system on I. We claim V must be a T-system of $m-1$ functions on I. Let $v\in\mathcal{V}$. Since $\mathcal{V}\subseteq\mathcal{U}$, v can have at most $m-1$ zeros. If it has $m-1$ zeros $t_1<t_2<\cdots<t_{m-1}$, then it must change sign at each zero (otherwise some would be double zeros, contradicting Theorem 2.24). Now let $x_0<t_1<x_1<t_2<x_2<\cdots<t_{m-1}<x_m$. Then $(-1)^i v(x_0)v(x_i)>0$ for $i=0,1,\dots,m$, which implies $S_I^-(v)=m-1$. This is impossible since V is a WT-system. We conclude that $Z_I(v)\leqslant m-2$ for any $v\in\mathcal{V}$, and thus (since the signs of the determinants formed from the $v_1,\dots,v_{m-1}$ are plus) V forms a T-system of $m-1$ functions.

We call $U=\{u_i\}_1^m$ an *Order Complete Weak Tchebycheff* (OCWT-) *system* on the set I provided

$$\{u_{i_\nu}\}_{\nu=1}^{k} \quad \text{is a } WT\text{-system on } I \text{ for all } 1\leqslant i_1<i_2<\cdots<i_k\leqslant m \text{ and all } 1\leqslant k\leqslant m.$$

The following theorem gives an important variation-diminishing property of OCWT-systems.

THEOREM 2.42

Let $U=\{u_i\}_1^m$ be an OCWT-system on the set I. Then

$$S_I^-\left(\sum_{i=1}^m c_i u_i\right)\leqslant S^-(c_1,\dots,c_m), \qquad \text{all real } c_1,\dots,c_m, \text{ not all } 0. \quad (2.81)$$

Proof. The proof is very similar to the proof of Theorem 2.31. Suppose $S^-(c)=d$. Then we can divide the set of coefficients as in (2.69). Let $v_1,\dots,v_d$ be as defined in (2.70). We claim that $V=\{v_i\}_1^d$ is a CWT-system. Indeed, by the expansion (2.71) and the OCWT-property of U, we see that determinants formed from $v_1,\dots,v_d$ are always nonnegative. It also follows from (2.71) that for some $t_1<\cdots<t_d$, the determinant is positive, hence $v_1,\dots,v_d$ are linearly independent. Now with $\{\tilde{c}_j\}_1^d$ as in (2.72), Theorem 2.39 applied to the WT-system V implies

$$S_I^-\left(\sum_{i=1}^m c_i u_i\right)=S_I^-\left(\sum_{j=1}^d \tilde{c}_j v_j\right)\leqslant d-1=S^-(c_1,\dots,c_m). \qquad \blacksquare$$

For us, the most important example of an OCWT-system will be provided by the B-splines discussed in Chapter 4. Theorem 2.42 will then yield a valuable variation-diminishing property of B-spline expansions.

Up to this point we have concentrated on sign changes of functions in WT-spaces. In the remainder of this section we examine zero properties. As shown in Example 2.37, it can happen that all of the functions in a WT-space $\mathfrak{U}$ vanish at the same (possibly infinite) set of points. In order to get any meaningful results, we shall have to exclude such points.

DEFINITION 2.43. Essential Points

Let $\mathfrak{U}$ be an m-dimensional linear space defined on a subset I of $\mathbf{R}$. We say that $t \in I$ is an *essential point of I relative to* $\mathfrak{U}$ provided there exists $u \in \mathfrak{U}$ with $u(t) \neq 0$.

We shall need the following simple lemma:

LEMMA 2.44

Suppose $\mathfrak{U}$ is an m-dimensional linear space defined on I, and suppose $x_1 < x_2 < \cdots < x_n$ are essential points of I (relative to $\mathfrak{U}$). Then there exists $u \in \mathfrak{U}$ such that $u(x_i) \neq 0$, $i = 1, 2, \ldots, n$.

Proof. We construct $g_1, \ldots, g_n$ consecutively with $g_i(x_j) \neq 0$, $j = 1, 2, \ldots, i$. Since x_1 is essential, it is clear that there exists $g_1 \in \mathfrak{U}$ with $g_1(x_1) \neq 0$. Suppose now that g_{i-1} has been constructed. If $g_{i-1}(x_i) \neq 0$, then we may take $g_i = g_{i-1}$. If $g_{i-1}(x_i) = 0$, let h be such that $h(x_i) \neq 0$. Then for $\varepsilon \neq 0$ sufficiently small, we may take $g_i = g_{i-1} + \varepsilon h$. ■

Note that in Lemma 2.44 there is no restriction on n. It can be arbitrarily large. In the following theorem we give a bound on the number of zeros a function in a WT-space can have at essential points. We call such zeros *essential zeros*.

THEOREM 2.45

Let $\mathfrak{U}$ be a WT-space of dimension $m > 1$. Then

$$Z_I^{\text{sep, ess}}(u) \leqslant m \qquad \text{for all nontrivial } u \in \mathfrak{U}, \tag{2.82}$$

where Z counts essential separated zeros (cf. Definition 2.13). Moreover, if $u \in \mathfrak{U}$ has m essential separated zeros $x_1 < x_2 < \cdots < x_m$, then

$$u(x) = 0 \qquad \text{for all } x \in I \text{ with } x < x_1 \text{ or } x > x_m.$$

Proof. We prove the second statement first. Suppose $u \in \mathfrak{U}$ has m separated essential zeros $x_1 < x_2 < \cdots < x_m$. Let $x_i < y_i < x_{i+1}$, $i = 1, 2, \ldots, m-1$

be such that $u(y_i)\neq 0$. Suppose now that u is not zero for some $y_m > x_m$. We shall show that this leads to a contradiction. By Lemma 2.44 there exists $v\in\mathfrak{U}$ with $v(x_i)\neq 0$, $i=1,\dots,m$. But then [cf. (2.48)],

$$S^-[v(x_1),u(y_1),\dots,v(x_m),u(y_m)]$$
$$+S^+[v(x_1),-u(y_1),\dots,v(x_m),-u(y_m)]\geqslant 2m-1.$$

Since all of the components are nonzero, we conclude that either

$$S^-[v(x_1),u(y_1),\dots,v(x_m),u(y_m)]\geqslant m \tag{2.83}$$

or

$$S^-[v(x_1),-u(y_1),\dots,v(x_m),-u(y_m)]\geqslant m. \tag{2.84}$$

Now choose $\varepsilon>0$ such that

$$\varepsilon \max_{1\leqslant i\leqslant m}|v(x_i)|<\min_{1\leqslant i\leqslant m}|u(y_i)|.$$

Then, if (2.83) holds, we see that $S^-(u+\varepsilon v)\geqslant m$. Similarly, if (2.84) holds, then $S^-(u-\varepsilon v)\geqslant m$. In either case this contradicts the assumption that $\mathfrak{U}$ is a WT-space. If u is nonzero for a point $y_0<x_1$, we reach a similar contradiction.

The bound (2.82) now follows. Indeed, if u had $m+1$ separated zeros, then it would have m separated zeros $x_1<\cdots<x_m$ and a point $y_{m+1}>x_m$ where u is not zero. This is impossible by what we have already shown. ■

The following example shows that the bound (2.82) cannot be improved:

EXAMPLE 2.46

Let $U=\{1,\sin(x),\cos(x)\}$ and $I=[0,2\pi]$.

Discussion. U is a WT-system on $[0,2\pi]$. The function $\sin(x)$ has three separated essential zeros, namely the points $0,\pi$, and 2π. ■

The following theorem shows that if we work on sets with only essential points, then the zeros of functions in a WT-space must occur in separated "intervals:"

THEOREM 2.47

Let $\mathfrak{U}$ be a WT-space of dimension m on the set I. Suppose that all points of I are essential with respect to $\mathfrak{U}$. Then for every $u\in\mathfrak{U}$ there exist intervals $I_1,\ldots,I_l$ and points $y_1<\cdots<y_{l-1}$ with $l\leqslant m$ such that

$$I_1<y_1<I_2<y_2<\cdots<y_{l-1}<I_l$$

$$u(y_i)\neq 0,\qquad i=1,2,\ldots,l-1$$

$$\mathfrak{Z}(u)=\{x\in I: u(x)=0\}=\bigcup_{i=1}^{l}(I\cap I_i).$$

Proof. Let $l=\max\{j$: there exist $z_1<y_1<z_2<\cdots<y_{j-1}<z_j$ with $u(z_i)=0$, $i=1,2,\ldots,j$ and $u(y_i)\neq 0$, $i=1,2,\ldots,j-1\}$. By Theorem 2.45, $l\leqslant m$. Let $\mathfrak{Z}_i=\{y_{i-1}<x<y_i: u(x)=0\}$. If z and $\tilde{z}$ are two points in $\mathfrak{Z}_i$, then all points in I between z and $\tilde{z}$ must also be in $\mathfrak{Z}_i$, for otherwise l would not be maximal. We conclude that each of the $\mathfrak{Z}_i$ has the form $I\cap I_i$, where I_i is some subinterval of $\mathbf{R}$. ■

When the set I in Theorem 2.47 is an interval, then the "intervals" making up the zero set of u are ordinary subintervals of $\mathbf{R}$. Example 2.38 shows an example of a WT-space with functions that have zeros on intervals. The splines discussed in Chapter 4 provide further examples.

We have remarked earlier that every T-system is automatically a WT-system. In the following theorem we explore to what extent the reverse is true:

THEOREM 2.48

Suppose $\mathfrak{U}$ is a WT-space of continuous functions on the open interval $I=(a,b)$. Suppose that all the points of I are essential with respect to $\mathfrak{U}$. Finally, suppose

$$Z_I(u)<\infty \qquad \text{for all nontrivial } u\in\mathfrak{U}. \tag{2.85}$$

Then $\mathfrak{U}$ is a T-space on I.

Proof. We shall show that for any nontrivial $u\in\mathfrak{U}$, $Z_I(u)\leqslant m-1$. Suppose to the contrary that some such u has $Z_I(u)\geqslant m$. Since u only has finitely many zeros, they must be distinct, separated zeros. Let x_1 be the smallest, and x_r the largest. Then by Theorem 2.45 we conclude that $u(x)=0$ for $x<x_1$ and $x>x_r$. This is a contradiction of (2.85), and the theorem is proved. ■

Theorem 2.48 is clearly also valid if I is a half-open interval. Example 2.46 shows that it does not hold, however, if I is a closed interval. In view of Theorem 2.47, the hypothesis (2.85) will be satisfied whenever $\mathfrak{U}$ does not contain any functions that vanish on intervals. The connection between WT- and T-systems will be discussed further in Section 11.4.

In this section we have given direct proofs for all of our results on WT-systems. Some of the results can be obtained from the analogous results on T-systems by smoothing techniques—see Remarks 2.4 and 2.5.

§ 2.7. DIVIDED DIFFERENCES

Although the basic properties of divided differences are developed in most books on numerical analysis, some of their finer properties (and in particular the case of repeated t's) are often not fully discussed. Thus in this section we give a complete treatment.

Divided differences can be defined in several (equivalent) ways. We shall define them as quotients of determinants. The advantages of this approach are: (1) it permits a rapid and clean derivation of the properties of divided differences, and (2) it can also be used to discuss some important kinds of generalized divided differences. Here we deal only with the classical divided differences; for generalizations, see §9.1 and Remark 2.7.

DEFINITION 2.49. Divided Differences

Given points $t_1,\ldots,t_{r+1}$ and a function f, we define its rth *order divided difference* over the points $t_1,\ldots,t_{r+1}$ by

$$[t_1,\ldots,t_{r+1}]f=\frac{D\begin{pmatrix} t_1,\ldots,t_{r+1} \\ 1,x,\ldots,x^{r-1},f\end{pmatrix}}{D\begin{pmatrix} t_1,\ldots,t_{r+1} \\ 1,x,\ldots,x^{r}\end{pmatrix}}. \tag{2.86}$$

In this definition we have tacitly assumed that the t's are in increasing order (in order for the determinants to make sense). But the definition makes sense, in general, if we agree that

$$[t_1,\ldots,t_{r+1}]f=[\tilde{t}_1,\ldots,\tilde{t}_{r+1}]f,$$

where $\tilde{t}_1 \leqslant \tilde{t}_2 \leqslant \cdots \leqslant \tilde{t}_{r+1}$ consists of the points $\{t_i\}_1^{r+1}$ in their natural order.

When the t's are distinct, then $[t_1,\ldots,t_{r+1}]f$ is defined for any function that has finite values at these points. When some of the t's occur more than once, then the value of the determinant in the numerator of (2.86) depends on certain derivatives of f, and the corresponding divided difference only makes sense for functions f that possess the required derivatives.

The following theorem gives some basic properties of divided differences:

THEOREM 2.50

If $t_1,\ldots,t_{r+1}$ are pairwise distinct, then

$$[t_1,\ldots,t_{r+1}]f=\sum_{i=1}^{r+1}\frac{f(t_i)}{\omega'(t_i)}=\sum_{i=1}^{r+1}\frac{f(t_i)}{\prod_{\substack{j=1\\ j\neq i}}^{r+1}(t_i-t_j)} \tag{2.87}$$

where

$$\omega(t)=(t-t_1)(t-t_2)\cdots(t-t_{r+1}).$$

More generally, if

$$t_1,\ldots,t_{r+1}=\overbrace{\tau_1,\ldots,\tau_1}^{l_1},\ldots,\overbrace{\tau_d,\ldots,\tau_d}^{l_d} \tag{2.88}$$

with $\tau_1<\tau_2<\cdots<\tau_d$, then

$$[t_1,\ldots,t_{r+1}]f=\sum_{i=1}^{d}\sum_{j=1}^{l_i}\alpha_{ij}D^{j-1}f(\tau_i), \tag{2.89}$$

where

$$\alpha_{i,l_i}\neq 0,\qquad i=1,2,\ldots,d.$$

Thus the rth divided difference is a linear functional defined on all sufficiently smooth functions. Moreover, if f and g agree on the point set $\{t_i\}_1^{r+1}$ in the sense that

$$D^{j-1}f(\tau_i)=D^{j-1}g(\tau_i),\qquad \begin{array}{l} j=1,2,\ldots,l_i,\\ i=1,2,\ldots,d\end{array} \tag{2.90}$$

then $[t_1,\ldots,t_{r+1}]f=[t_1,\ldots,t_{r+1}]g$.

Proof. The expansion (2.87) follows directly from the definition and the explicit formulae (2.66) for the vanderMonde determinant. To prove (2.89) we need only expand the determinant in the numerator of (2.86) with respect to its last column. The denominator is always positive since it is a vanderMonde. The fact that the coefficients α_{il_i} of the highest derivatives of f at each τ_i in (2.89) are nonzero follows from

$$\alpha_{il_i} = \frac{V\left(\overbrace{\tau_1,\dots,\tau_1}^{l_1},\dots,\overbrace{\tau_i,\dots,\tau_i}^{l_i-1},\dots,\overbrace{\tau_d,\dots,\tau_d}^{l_d}\right)}{V(t_1,\dots,t_{r+1})}.$$

It is clear from (2.89) that the divided difference is linear; that is, $[\cdot](f+g)=[\cdot]f+[\cdot]g$ and $[\cdot](\alpha f)=\alpha[\cdot]f$. The fact that $[\cdot]f=[\cdot]g$ when f and g agree on the t's is also clear from (2.89). ■

In our next theorem we give several other important properties of divided differences. Of particular importance is the fact that divided differences can be computed recursively.

THEOREM 2.51

Given any points $t_1,\dots,t_{r+1}$ (not necessarily in order) and any sufficiently smooth function f,

$$[t_1,\dots,t_{r+1}]f = \frac{[t_2,\dots,t_{r+1}]f-[t_1,\dots,t_r]f}{t_{r+1}-t_1} \tag{2.91}$$

provided $t_1 \neq t_{r+1}$. If $t_1=t_2=\cdots=t_{r+1}$, then

$$[t_1,\dots,t_{r+1}]f = \frac{D^r f(t_1)}{r!}. \tag{2.92}$$

In general, if $f \in C^r[a,b]$, where $a=\min_{1\leqslant i\leqslant r+1} t_i$ and $b=\max_{1\leqslant i\leqslant r+1} t_i$, then

$$[t_1,\dots,t_{r+1}]f = \frac{D^r f(\theta)}{r!}, \qquad \text{some } a\leqslant\theta\leqslant b. \tag{2.93}$$

Concerning the divided differences of powers of x, we have

$$[t_1,\dots,t_{r+1}]x^j = \begin{cases} 0, & j=0,1,\dots,r-1 \\ \rho_{j-r}(t_1,\dots,t_{r+1}), & j=r,r+1,\dots \end{cases}, \tag{2.94}$$

where $\rho_0(t_1,\dots,t_{r+1})=1$ and

$$\rho_l(t_1,\dots,t_{r+1}) = \sum_{1\leqslant i_1\leqslant i_2\leqslant\cdots\leqslant i_l\leqslant r+1} (t_{i_1}t_{i_2}\cdots t_{i_l}). \tag{2.95}$$

The sum in (2.95) is taken over all choices of l integers, allowing repetitions. For example, $\rho_1(t_1,\ldots,t_{r+1})=t_1+\cdots+t_{r+1}$, while $\rho_2(t_1,t_2)=t_1^2+t_1t_2+t_2^2$. The number of terms in the sum is $(r+l)!/(r!l!)$.

Proof. We begin with the first half of (2.94). By the definition of the divided difference, if we take $f(x)=x^j$ with $0\leqslant j\leqslant r-1$, then two columns of the determinant in the numerator of (2.86) will be the same, and the value will be zero. If we take $f(x)=x^r$, then the numerator and denominator are identical, and the value of the divided difference is 1.

We prove the remaining assertions first in the case that $t_1=t_2=\cdots=t_{r+1}$. In this case, the denominator in (2.86) is

$$V(t_1,\ldots,t_{r+1})=\prod_{\nu=1}^{r}\nu!,$$

while the numerator is

$$D^rf(t_1)\,V(t_1,\ldots,t_r)=D^rf(t_1)\prod_{\nu=1}^{r-1}\nu!,$$

and (2.92) follows. Moreover, for $j\geqslant r$,

$$[t_1,\ldots,t_{r+1}]x^j=D^rx^j|_{x=t_1}=\frac{j!t_1^{j-r}}{(j-r)!r!}=\rho_{j-r}(t_1,\ldots,t_1),$$

which completes the proof of (2.94) in this case.

Suppose now that $t_1\neq t_{r+1}$. We establish the theorem by induction on r. For $r=1$ it is easily checked. Suppose it holds for divided differences of order $r-1$. Our first task is to prove (2.91). Consider the linear functional

$$\lambda f=[t_1,\ldots,t_{r+1}]f-\frac{[t_2,\ldots,t_{r+1}]f-[t_1,\ldots,t_r]f}{(t_{r+1}-t_1)}$$

It is clear that $\lambda x^i=0$, $i=0,1,\ldots,r-1$ by (2.94). Moreover, using (2.94) for r points and with $j=r$ we have

$$\lambda x^r=1-\frac{(t_2+\cdots+t_{r+1})-(t_1+\cdots+t_r)}{(t_{r+1}-t_1)}=0.$$

On the other hand, λf has the form of the sum in (2.89). Thus the set of linear equations $\lambda x^i=0$, $i=0,1,\ldots,r$ provides a nonsingular system for the coefficients in the expansion of λ, and we conclude $\lambda=0$. This proves (2.91).

By the inductive hypothesis and the definition of the ρ's, we have

$$[t_1,\ldots,t_{r+1}]x^j = \frac{[t_2,\ldots,t_{r+1}]x^j - [t_1,\ldots,t_r]x^j}{(t_{r+1}-t_1)}$$

$$= \frac{\rho_{j-r+1}(t_2,\ldots,t_{r+1}) - \rho_{j-r+1}(t_1,\ldots,t_r)}{(t_{r+1}-t_1)}$$

$$= \rho_{j-r}(t_1,\ldots,t_{r+1}),$$

and (2.94) is also established for all r. Finally, property (2.93) follows directly from the Peano representation (established in Theorem 4.23):

$$[t_1,\ldots,t_{r+1}]f = \int_{t_1}^{t_{r+1}} \frac{Q(x)D^r f(x)\,dx}{(r-1)!},$$

where Q is a nonnegative function with $\int_{t_1}^{t_{r+1}} Q(x)\,dx = 1/r$. ■

The recursion relation (2.91) permits us to compute the divided difference by generating the triangular array

$$\begin{array}{llll}
[t_1]f & [t_2]f \quad \cdots & [t_r]f & [t_{r+1}]f \\
[t_1,t_2]f & & & [t_r,t_{r+1}]f \\
\cdots & & & \ddots \\
[t_1,\ldots,t_r]f & [t_2,\ldots,t_{r+1}]f & & \\
[t_1,\ldots,t_{r+1}]f. & & &
\end{array}$$

We emphasize once again that the recursion formula (2.91) does not require the t's to be in order. For example, we have

$$[\tau_1,\tau_2,\tau_2,\tau_3]f = \frac{[\tau_2,\tau_2,\tau_3]f - [\tau_1,\tau_2,\tau_2]f}{(\tau_3-\tau_1)}$$

$$= [\tau_2,\tau_1,\tau_2,\tau_3]f = \frac{[\tau_1,\tau_2,\tau_3]f - [\tau_2,\tau_1,\tau_2]f}{(\tau_3-\tau_2)},$$

assuming $\tau_1 \neq \tau_3 \neq \tau_2$.

Our next result is a form of *Leibniz rule* for differencing the product of two functions.

THEOREM 2.52

For any $t_1, t_2, \ldots, t_{r+1}$ and appropriately smooth functions f and g,

$$[t_1, \ldots, t_{r+1}]fg = \sum_{i=1}^{r+1} [t_1, \ldots, t_i]f \cdot [t_i, \ldots, t_{r+1}]g. \tag{2.96}$$

Proof. We proceed by induction on r. For $r=0$ the result is trivial. Suppose it holds for $r-1$. If $t_1 = t_{r+1}$, (2.96) is the *usual Leibniz rule*

$$\frac{D^r fg}{r!} = \sum_{i=1}^{r+1} \frac{D^{i-1}f(t_1)\,D^{r+1-i}g(t_1)}{(i-1)!(r+1-i)!}$$

$$= \frac{1}{r!} \sum_{i=0}^{r} \binom{r}{i} D^i f(t_1)\,D^{r-i}g(t_1). \tag{2.97}$$

We suppose now that $t_1 \neq t_{r+1}$. Then, using (2.91) and the Leibniz rule for r points, we obtain

$$[t_1, \ldots, t_{r+1}]fg = \frac{[t_2, \ldots, t_{r+1}]fg - [t_1, \ldots, t_r]fg}{(t_{r+1} - t_1)}$$

$$= \frac{\sum_{i=2}^{r+1} [t_2, \ldots, t_i]f \cdot [t_i, \ldots, t_{r+1}]g - \sum_{i=1}^{r} [t_1, \ldots, t_i]f \cdot [t_i, \ldots, t_r]g}{(t_{r+1} - t_1)}$$

If we add and subtract the term

$$\sum_{i=2}^{r+1} [t_1, \ldots, t_{i-1}]f \cdot [t_i, \ldots, t_{r+1}]g = \sum_{i=1}^{r} [t_1, \ldots, t_i]f \cdot [t_{i+1}, \ldots, t_r]g$$

and combine like terms, we obtain

$$[t_1, \ldots, t_{r+1}]fg = \Bigg(\sum_{i=2}^{r+1} [t_1, \ldots, t_i]f \cdot [t_i, \ldots, t_{r+1}]g \cdot (t_i - t_1)$$

$$+ \sum_{i=1}^{r} [t_1, \ldots, t_i]f \cdot [t_i, \ldots, t_{r+1}]g \cdot (t_{r+1} - t_i) \Bigg) \Big/ (t_{r+1} - t_1)$$

$$= \sum_{i=1}^{r+1} [t_1, \ldots, t_i]f \cdot [t_i, \ldots, t_{r+1}]g.$$

■

The following theorem shows that for smooth functions, the divided difference over an arbitrary set of points with repetitions is the limit of divided differences over distinct points. In particular, we see that for any $f \in C^r[I]$, $[t_1,\ldots,t_{r+1}]f$ is a continuous function on the *closed simplex*

$$\bar{T}=\{\mathbf{t}=(t_1,\ldots,t_{r+1})\in I^{r+1}: \quad t_1 \leqslant t_2 \leqslant \cdots \leqslant t_{r+1}\}.$$

THEOREM 2.53

Let $t_{1,\varepsilon},\ldots,t_{r+1,\varepsilon}$ be a sequence of points with $t_{i,\varepsilon}\to t_i$ as $\varepsilon\to 0$, $i=1,2,\ldots,r+1$. Then for all sufficiently smooth f,

$$[t_{1,\varepsilon},\ldots,t_{r+1,\varepsilon}]f \to [t_1,\ldots,t_{r+1}]f, \text{ as } \varepsilon\to 0. \tag{2.98}$$

Proof. It suffices to consider the case where just one t moves with ε. For convenience of notation, let $u_1=1,\ldots,u_{r+1}=x^r$. Suppose $t_1 \leqslant \cdots \leqslant t_i < t_{i+1} = \cdots = t_{i+l} < t_{i+l+1} \leqslant \cdots \leqslant t_{r+1}$, and that $t_{i+l,\varepsilon}\downarrow t_{i+l}$. Then using Taylor's expansion as in the proof of Lemma 2.9, we obtain

$$[t_1,\ldots,t_{i+l-1},t_{i+l,\varepsilon},t_{i+l+1},\ldots,t_{r+1}]f$$

$$= \frac{\begin{vmatrix} \cdots & & & \cdots \\ u_1(t_{i+1}) & \cdots & u_r(t_{i+1}) & f(t_{i+1}) \\ \cdots & & & \\ D^{l-2}u_1(t_{i+1}) & \cdots & D^{l-2}u_r(t_{i+1}) & D^{l-2}f(t_{i+1}) \\ D^{l-1}u_1(\xi_1) & \cdots & D^{l-1}u_r(\xi_r) & D^{l-1}f(\eta) \\ \cdots & & & \cdots \end{vmatrix}}{\begin{vmatrix} \cdots & & \cdots \\ u_1(t_{i+1}) & \cdots & u_{r+1}(t_{i+1}) \\ \cdots & & \\ D^{l-2}u_1(t_{i+1}) & \cdots & D^{l-2}u_{r+1}(t_{i+1}) \\ D^{l-1}u_1(\xi_1) & \cdots & D^{l-1}u_{r+1}(\xi_{r+1}) \\ \cdots & & \cdots \end{vmatrix}},$$

where $\xi_1,\ldots,\xi_{r+1}$ and η all go to t_{i+1} as $\varepsilon\to 0$. Taking the limit, we obtain (2.98). ■

At times it is useful to have more specific information about how the divided difference changes as we move one or more of the t's. Our next two theorems give explicit formulae for the partial derivatives of divided differences with respect to the t's.

THEOREM 2.54

Let $t_1 \leqslant t_2 \leqslant \cdots \leqslant t_{r+1}$ be given. Fix $1 \leqslant j \leqslant r+1$, and suppose that t_j is not equal to any other t. Then for any sufficiently differentiable f,

$$\frac{\partial}{\partial t_j}[t_1,\ldots,t_{r+1}]f=[t_1,\ldots,t_{j-1},t_j,t_j\, t_{j+1},\ldots,t_{r+1}]f. \tag{2.99}$$

Proof. By the recursion formula (2.91) and the fact that the order of the t's can be rearranged without affecting the value of the divided difference,

$$\begin{aligned}
&[t_1,\ldots,t_{j-1},t_j+\varepsilon,\ldots,t_{r+1}]f-[t_1,\ldots,t_{r+1}]f\\
&\quad=[t_1,\ldots,t_{j-1},t_{j+1},\ldots,t_{r+1},t_j+\varepsilon]f-[t_j,t_1,\ldots,t_{j-1},t_{j+1},\ldots,t_{r+1}]f\\
&\quad=\varepsilon[t_j,t_1,\ldots,t_{j-1},t_{j+1},\ldots,t_{r+1},t_j+\varepsilon]f\\
&\quad=\varepsilon[t_1,\ldots,t_j,t_j+\varepsilon,\ldots,t_{r+1}]f.
\end{aligned}$$

Dividing by ε and taking the limit as $\varepsilon \to 0$, we obtain (2.99). ■

Theorem 2.54 deals with the case where a single t, unequal to any other t, is moved. Given a set of t's with

$$t_1,\ldots,t_{r+1}=\overbrace{\tau_1,\ldots,\tau_1}^{l_1} \leqslant \cdots \leqslant \overbrace{\tau_d,\ldots,\tau_d}^{l_d}, \tag{2.100}$$

we now consider the case where the entire block $\tau_i,\ldots,\tau_i$ is moved together.

THEOREM 2.55

Let $\{t_i\}_1^{r+1}$ be as in (2.100). If $\tau_i<\tau_{i+1}$, then

$$\frac{\partial_+}{\partial \tau_i}[t_1,\ldots,t_{r+1}]f=l_i\begin{bmatrix} l_1 & l_i+1 & l_d\\ \tau_1,\ldots, & \tau_i,\ldots, & \tau_d\end{bmatrix}f, \tag{2.101}$$

where $\partial_+/\partial\tau_i$ denotes the right derivative. If $\tau_{i-1}<\tau_i$, then the same expression holds for the left derivative.

Proof. Suppose $\tau_i<\tau_{i+1}$ and we wish to move τ_i to the right. Then for any $\varepsilon>0$,

$$\begin{aligned}
&\begin{bmatrix} l_1 & l_i & l_d\\ \tau_1,\ldots, & \tau_i+\varepsilon,\ldots, & \tau_d\end{bmatrix}f-\begin{bmatrix} l_1 & l_i & l_d\\ \tau_1,\ldots, & \tau_i,\ldots, & \tau_d\end{bmatrix}f\\
&\quad=\sum_{r=1}^{l_i}\left(\begin{bmatrix} l_1 & l_i-r+1 & r-1 & l_d\\ \tau_1,\ldots, & \tau_i+\varepsilon, & \tau_i,\ldots, & \tau_d\end{bmatrix}f\right.\\
&\qquad\left.-\begin{bmatrix} l_1 & l_i-r & r & l_d\\ \tau_1,\ldots, & \tau_i+\varepsilon, & \tau_i,\ldots, & \tau_d\end{bmatrix}f\right).
\end{aligned}$$

Now the same argument used in Theorem 2.54 can be applied to each individual term, and after dividing by ε and taking the limit as $\varepsilon \downarrow 0$, we obtain (2.101). ■

It is often useful to be able to estimate the size of divided differences in terms of lower-order divided differences. We have the following:

THEOREM 2.56

Let $0 \leqslant i \leqslant r-1$ be fixed integers, and suppose

$$\gamma_j = \min_{1 \leqslant \nu \leqslant r+1-j} |t_{\nu+j} - t_\nu| > 0, \qquad j = i+1, \ldots, r.$$

Then

$$|[t_1, \ldots, t_{r+1}]f| \leqslant \sum_{\nu=0}^{r-i} \frac{\binom{r-i}{\nu}[t_{\nu+1}, \ldots, t_{\nu+i+1}]f}{\gamma_{i+1} \cdots \gamma_r}, \tag{2.102}$$

where $\binom{n}{m}$ is the usual binomial coefficient (see Remark 2.8).

Proof. Since

$$|[t_1, \ldots, t_{r+1}]f| \leqslant \frac{|[t_2, \ldots, t_{r+1}]f| + |[t_1, \ldots, t_r]f|}{(t_{r+1} - t_1)},$$

the result follows for $i = r-1$. Now we proceed by induction on i. Suppose (2.102) holds for i; we now prove it for $i-1$. We have

$$|[t_1, \ldots, t_{r+1}]f| \leqslant \sum_{\nu=0}^{r-i} \frac{\binom{r-i}{\nu} \dfrac{|[t_{\nu+2}, \ldots, t_{\nu+i+1}]f| + |[t_{\nu+1}, \ldots, t_{\nu+i}]f|}{(t_{\nu+i+1} - t_{\nu+1})}}{\gamma_{i+1} \cdots \gamma_r}.$$

Since $|t_{\nu+i+1} - t_{\nu+1}| \geqslant \gamma_i$, combining the above sums (and using a simple identity for binomial coefficients—see Remark 2.8) yields the assertion for $i-1$. ■

The special case of divided differences over equally spaced points is of particular importance.

THEOREM 2.57

Given $h > 0$ and a positive integer r, we define the rth *forward difference of* f at t by

$$\Delta_h^r f(t) = r!\,h^r[t, t+h, \ldots, t+rh]f. \tag{2.103}$$

It has the following properties:

$$\Delta_h^r f(t) = \sum_{i=0}^{r} (-1)^{r-i} \binom{r}{i} f(t+ih) \tag{2.104}$$

$$\Delta_h^r x^i = r!\, h^r \delta_{i,r}, \qquad i = 0, 1, \ldots, r \tag{2.105}$$

$$|\Delta_h^r f(t)| \leqslant 2^r \|f\|_\infty. \tag{2.106}$$

Moreover, if $f \in L_1^r[t, t+rh]$, then

$$\Delta_h^r f(t) = h^r \int_0^{rh} \frac{D^r f(t+u) N^r(u/h)\, du}{h}, \tag{2.107}$$

where

$$0 \leqslant N^r(x) \leqslant 1, \qquad \int_0^r N^r(u)\, du = 1, \qquad \|N^r\|_{L_q[0,r]} \leqslant 1, \qquad \text{all } 1 \leqslant q \leqslant \infty.$$

For such f we may also write

$$\Delta_h^r f(t) = \int_0^h \cdots \int_0^h D^r f(t+s_1+\cdots+s_r)\, ds_1 \cdots ds_r. \tag{2.108}$$

If $f \in L_p^r[a, b+rh]$, then

$$\|\Delta_h^r f\|_{L_p[a,b]} \leqslant h^r \|D^r f\|_{L_p[a,b+rh]}. \tag{2.109}$$

Finally, if $f \in C^r[t, t+rh]$, then

$$\Delta_h^r f(t) = h^r D^r f(\theta), \qquad \text{for some} \quad t \leqslant \theta \leqslant t+rh. \tag{2.110}$$

Proof. The expansion (2.103) follows from the general expansion (2.87) with $t_i = t+(i-1)h, i = 1, 2, \ldots, r+1$. Then (2.105) follows from (2.94). The estimate (2.106) is a consequence of (2.104) and the fact that $\Sigma_{i=0}^r \binom{r}{i} = 2^r$ (see Remark 2.8). The representation (2.107) and the properties of N^r are discussed in Section 4.4. The alternate integral relation (2.108) can be established by induction using the fact that $\Delta_h^r f(t) = \Delta_h \Delta_h^{r-1} f(t)$. We obtain the inequality (2.109) by applying the Minkowski inequality (see Remark 2.2) to the integral representation (2.107). Finally, (2.110) also follows from (2.107) with the help of the mean-value theorem. ■

§ 2.8. MODULI OF SMOOTHNESS

In order to establish sharp theorems on how well a given function can be approximated, it is useful to have some precise measure of the smoothness of the function. In this section we discuss various moduli of smoothness.

DEFINITION 2.58. Modulus of Smoothness

Given $1 \leqslant p \leqslant \infty$, a positive integer r, and $0<t\leqslant (b-a)/r$, the function

$$\omega_r(f;t)_p = \omega_r(f;t)_{L_p[a,b]} = \sup_{0<h<t} \|\Delta_h^r f\|_{L_p[I_{rh}]} \tag{2.111}$$

is called the rth *modulus of smoothness of f in L_p*. Here $\Delta_h^r f$ is the rth forward difference [see (2.104)], and $I_{rh}=[a,b-rh]$.

When $r=1$, the rth modulus of smoothness is traditionally referred to as the *modulus of continuity*. The following theorem gives several basic properties of the modulus of smoothness:

THEOREM 2.59

Fix $r \geqslant 1$. Then

$$\omega_r(f;t_1+t_2)_p \leqslant \omega_r(f;t_1)_p + \omega_r(f;t_2)_p; \tag{2.112}$$

$$\omega_r(f_1+f_2;t)_p \leqslant \omega_r(f_1;t)_p + \omega_r(f_2;t)_p; \tag{2.113}$$

$$\omega_r(f;t)_p \leqslant \omega_r(f;\tilde{t})_p, \qquad 0<t<\tilde{t}; \tag{2.114}$$

$$\omega_r(f;kt)_p \leqslant k^r \omega_r(f;t)_p, \qquad \text{any positive integer } k; \tag{2.115}$$

$$\omega_r(f;\lambda t)_p \leqslant \lceil \lambda \rceil^r \omega_r(f;t)_p, \qquad \text{all } \lambda>0, \text{ where } \lceil\lambda\rceil = \min\{\text{integers } i : i \geqslant \lambda\}; \tag{2.116}$$

$$\omega_r(f;t)_p \leqslant 2^j \omega_{r-j}(f;t)_p, \qquad 0<j; \tag{2.117}$$

$$\omega_r(f;t)_p \leqslant 2^r \|f\|_p \tag{2.118}$$

$$\omega_r(f;t)_p \leqslant t^j \omega_{r-j}(D^j f;t)_p, \qquad \text{if } f \in L_p^j[a,b], \quad 1 \leqslant p < \infty \tag{2.119}$$

$$\text{if } f \in C^j[a,b], \quad p=\infty;$$

$$\omega_r(f;t)_p \leqslant t^r \|D^r f\|_p, \qquad \text{if } f \in L_p^r[a,b], \quad 1 \leqslant p < \infty \tag{2.120}$$

$$\text{if } f \in C^r[a,b], \quad p=\infty;$$

$$\omega_r(f;t)_p \to 0 \quad \text{as } t\to 0, \qquad \text{if } f \in L_p[a,b], \quad 1 \leqslant p < \infty \tag{2.121}$$

$$\text{if } f \in C[a,b], \quad p=\infty;$$

$$\text{if } \lim_{t\downarrow 0} t^{-r}\omega_r(f;t)_p = 0, \qquad \text{then } f \in \mathcal{P}_r; \tag{2.122}$$

$$\omega_r(f;t)_q \leqslant |I|^{1/q-1/p} \omega_r(f;t)_p, \qquad \text{all } 1 \leqslant q \leqslant p \leqslant \infty. \tag{2.123}$$

Discussion. These properties follow more or less directly from the definition. For details, see Timan [1963], Lorentz [1966], and Johnen [1972]. Some of the properties can also be derived from the analogous properties of the K-functional and its equivalence with the modulus of smoothness, as shown in the following section. ■

Our next theorem gives a useful formula for the modulus of smoothness of a function that is obtained by a change of variable from another function.

THEOREM 2.60

Suppose $y=\alpha x+\beta$ maps the interval $[a,b]$ onto the interval $[c,d]$, and that $x=\gamma y+\delta$ is the inverse mapping taking $[c,d]$ onto $[a,b]$. Suppose f is defined on $[a,b]$, and that $F(y)=f(\gamma y+\delta)$. Then

$$\omega_r(F;t)_{L_p[c,d]}=\alpha^{1/p}\omega_r(f;\alpha^{-1}t)_{L_p[a,b]}. \tag{2.124}$$

Proof. We have $\gamma=\alpha^{-1}$ and

$$\omega_r(F;t)_{L_p[c,d]}=\sup_{0<h<t}\left[\int_c^{d-rh}\left|\sum_{i=0}^{r}(-1)^{r-i}\binom{r}{i}F(y+ih)\right|^p dy\right]^{1/p}$$

$$=\sup_{0<h<t}\left[\int_c^{d-rh}\left|\sum_{i=0}^{r}(-1)^{r-i}\binom{r}{i}f(\gamma(y+ih)+\delta)\right|^p dy\right]^{1/p}$$

$$=\sup_{0<h<t}\left[\int_a^{b-rh\gamma}\left|\sum_{i=0}^{r}(-1)^{r-i}\binom{r}{i}f(x+i\gamma h)\right|^p \alpha\, dx\right]^{1/p}$$

$$=\sup_{0<h<t}\left[\int_a^{b-r\gamma h}\left|\sum_{i=0}^{r}(-1)^{r-i}\binom{r}{i}f(x+i\gamma h)\right|^p dx\right]^{1/p}\alpha^{1/p}$$

$$=\alpha^{1/p}\omega_r(f;\gamma t)_{L_p[a,b]}.$$ ■

In (2.119) we have given a simple estimate of $\omega_r(f;t)_p$ in terms of lower-order moduli of smoothness. In the following theorem we give estimates in the opposite direction:

THEOREM 2.61. Marchaud Inequalities

Let $1 \leqslant p \leqslant \infty$, and suppose r is a positive integer. Then there exist positive constants C_1 and δ (depending only on a, b, r, and p) such that for all $0<t<\delta$,

$$\omega_j(f;t)_p \leqslant C_1 t^j \left[\|f\|_p + \int_t^{\delta} u^{-1-j} \omega_r(f;u)_p \, du \right], \qquad j=1,2,\ldots,r-1. \tag{2.125}$$

On the other hand, if for some $\delta > 0$,

$$\int_0^{\delta} u^{-1-j} \omega_r(f;u)_p \, du < \infty, \tag{2.126}$$

then $f \in L_p^j[a,b]$, and

$$\omega_{r-j}(D^j f;t)_p \leqslant C_1 \int_0^t u^{-1-j} \omega_r(f;u)_p \, du, \tag{2.127}$$

all $0<t<\delta$, where C_1 is a constant (depending only on a, b, r, and p). If (2.126) holds for $p=\infty$, then $f \in C^j[a,b]$, and (2.127) also holds for $p=\infty$.

Discussion. For a detailed direct proof of these results, see Johnen [1972]. These assertions also follow from the corresponding results on the K-functional—cf. Theorems 2.71 and 2.72. ■

The second part of Theorem 2.61 is particularly interesting since it asserts that when $\omega_r(f;u)$ goes to zero fast enough with respect to u, then f must have a certain number of smooth derivatives.

We close this section by introducing some interesting smooth spaces of functions that are characterized in terms of their moduli of smoothness. First, given $r \geqslant 0$, $1 \leqslant p \leqslant \infty$, and $0<\alpha \leqslant 1$, let

$$\mathrm{Lip}_p^{r,\alpha}[a,b] = \{ f \in L_p[a,b] : \omega_{r+1}(f;t)_p = \mathcal{O}(t^{r+\alpha}) \} \tag{2.128}$$

(where $\mathcal{O}$ is the standard "big oh" see Remark 2.9). When $r=0$ and $p=\infty$, the classical *Lipschitz space* is defined by

$$\mathrm{Lip}^{\alpha}[a,b] = \{ f \in C[a,b] : \quad \text{there exists } M < \infty \text{ with}$$
$$|f(x+h) - f(x)| \leqslant M h^{\alpha} \text{ for all } a \leqslant x \leqslant x+h \leqslant b \}.$$

The following application of Theorem 2.61 shows that the Lipschitz spaces (2.128) lie between the classical smooth spaces in the sense that

$$L_p^{r+1}[a,b] \subseteq \mathrm{Lip}_p^{r,\alpha}[a,b] \subseteq L_p^r[a,b] \qquad \text{if } 1 \leqslant p < \infty$$

and

$$C^{r+1}[a,b] \subseteq \mathrm{Lip}_\infty^{r,\alpha}[a,b] \subseteq C^r[a,b] \qquad \text{if } p = \infty.$$

THEOREM 2.62

For $1 \leqslant p < \infty$ and $0 < \alpha < 1$,

$$\mathrm{Lip}_p^{r,\alpha}[a,b] = \{ f \in L_p^r[a,b] : D^r f \in \mathrm{Lip}_p^\alpha[a,b] \}.$$

Similarly, if $p = \infty$, then

$$\mathrm{Lip}_\infty^{r,\alpha}[a,b] = \{ f \in C^r[a,b] : D^r f \in \mathrm{Lip}^\alpha[a,b] \}.$$

Proof. Let $1 \leqslant p < \infty$. Then $D^r f \in \mathrm{Lip}_p^\alpha$ coupled with (2.119) implies

$$\omega_{r+1}(f;t)_p \leqslant t^r \omega_1(D^r f;t)_p \leqslant C t^{r+\alpha},$$

and it follows that $f \in \mathrm{Lip}_p^{r,\alpha}[a,b]$. Conversely, if $f \in \mathrm{Lip}_p^{r,\alpha}$, then

$$\int_0^\delta u^{-r-1} \omega_{r+1}(f;u)_p \, du < \infty, \qquad \text{some } \delta > 0,$$

and by Theorem 2.61, we conclude that $f \in L_p^r[a,b]$ and

$$\omega_1(D^r f;t)_p \leqslant C \int_0^t u^{-1-r} u^{r+\alpha} \, du = c t^\alpha;$$

that is, $D^r f \in \mathrm{Lip}_p^\alpha[a,b]$. The proof for $p = \infty$ is similar. ■

Theorem 2.62 does not cover the case where $\alpha = 0$, and indeed, the spaces $\mathrm{Lip}_p^{r,0}[a,b]$ have a somewhat different behavior. $\mathrm{Lip}_\infty^{1,0}[a,b]$ is the classical *Zygmund space*, and thus we introduce the notation

$$Z_p^r[a,b] = \{ f \in L_p[a,b] : \omega_{r+2}(f;t)_p = \mathcal{O}(t^{r+1}) \}. \qquad (2.129)$$

THEOREM 2.63

Let $1 \leqslant p < \infty$. Then

$$Z_p^r[a,b] = \{ f \in L_p^r[a,b] \colon D^r f \in Z_p^0[a,b] \}.$$

Similarly, if $p = \infty$, then

$$Z_\infty^r[a,b] = \{ f \in C^r[a,b] \colon D^r f \in Z_\infty^0[a,b] \}.$$

Proof. The proof follows the same pattern as that of Theorem 2.62, using (2.119) and Theorem 2.61. ■

The spaces $\mathrm{Lip}_\infty^{r;\alpha}[a,b]$ and $Z_\infty^r[a,b]$ are examples of what are called *Besov spaces*—see Section 6.5.

§ 2.9. THE *K*-FUNCTIONAL

In this section we introduce an expression called the *K-functional* which measures the smoothness of a function in terms of how well it can be approximated by smooth functions. This will provide an alternate way of characterizing smooth function classes (which as it turns out is equivalent to using the moduli of smoothness defined in the previous section). We also include several applications of the K functional which will be useful later on.

DEFINITION 2.64. The *K*-Functional

Suppose $1 \leqslant p < \infty$, $t > 0$, and that r is a positive integer. For $f \in L_p[I]$ we define

$$K_{r,p}(t)f = \inf_{g \in L_p^r[I]} (\|f-g\|_p + t^r \|D^r g\|_p). \tag{2.130}$$

If $p = \infty$ and $f \in C[I]$, we define

$$K_{r,\infty}(t)f = \inf_{g \in C^r[I]} (\|f-g\|_\infty + t^r \|D^r g\|_\infty). \tag{2.131}$$

We call $K_{r,p}(t)$ the *K-functional of Peetre*.

It is clear that $K_{r,p}(t)$ is a nonlinear functional defined on the space $L_p[I]$ if $1 \leqslant p < \infty$, and on $C[I]$ if $p = \infty$. It is a measure of how well the function f can be approximated by smoother functions (in $L_p^r[I]$ or $C^r[I]$) while maintaining a control on the size of the rth derivative of the approximant.

The parameter t controls the balance between the size of the derivative and the error in the approximation. The following theorem lists some elementary properties of the K-functional:

THEOREM 2.65

Let $1 \leqslant p \leqslant \infty$ and $r > 0$. Then

$$K_{r,p}(t_1 + t_2)f \leqslant 2^{r-1}\left[K_{r,p}(t_1)f + K_{r,p}(t_2)f\right]; \tag{2.132}$$

$$K_{r,p}(t)(f_1 + f_2) \leqslant K_{r,p}(t)f_1 + K_{r,p}(t)f_2; \tag{2.133}$$

$$K_{r,p}(t)f \leqslant K_{r,p}(\tilde{t})f \qquad \text{if } t \leqslant \tilde{t}; \tag{2.134}$$

$$K_{r,p}(t)f \leqslant \|f\|_p; \tag{2.135}$$

$$K_{r,p}(t)f \leqslant t^r \|D^r f\|_p \quad \begin{array}{ll} \text{if } f \in L_p^r[a,b], & 1 \leqslant p < \infty \\ \text{if } f \in C^r[a,b], & p = \infty; \end{array} \tag{2.136}$$

$$\lim_{t \to 0} K_{r,p}(t)f = 0 \quad \begin{array}{ll} \text{if } f \in L_p[a,b], & 1 \leqslant p < \infty \\ \text{if } f \in C[a,b], & p = \infty; \end{array} \tag{2.137}$$

$$\text{if } \lim_{t \to 0} t^{-r} K_{r,p}(t)f = 0, \qquad \text{then} \quad f \in \mathcal{P}_r. \tag{2.138}$$

Proof. Most of these properties follow directly from the definition. For details; see Butzer and Berens [1967] or Johnen and Scherer [1977]. ■

Some further properties of the K-functional are established in Theorems 2.71 and 2.72 below. We now turn to the task of establishing the equivalence of the K-functional with the modulus of smoothness introduced in the previous section. An essential tool in proving this equivalence is the following lemma which shows how to construct a smooth approximation to a given function.

LEMMA 2.66

Let $1 \leqslant p < \infty$, $t > 0$, and suppose r is a positive integer. Let $J = [c,d]$ and $\tilde{J} = [c, d + r^2 t]$. Then for all $f \in L_p[\tilde{J}]$ there exists $g \in L_p^r[J]$ such that

$$\|f - g\|_{L_p[J]} \leqslant \int_0^r \|\Delta_{tu}^r f\|_{L_p[J]}\, du \leqslant r^{r+1} \omega_r(f; t)_{L_p[\tilde{J}]} \tag{2.139}$$

and

$$t^r \|D^r g\|_{L_p[J]} \leqslant (2^r - 1) \max_{1 \leqslant i \leqslant r} \|\Delta_{it}^r f\|_{L_p[J]} \leqslant r^r (2^r - 1) \omega_r(f; t)_{L_p[\tilde{J}]}. \tag{2.140}$$

Similarly, if $\tilde{J}=[c-r^2t,d]$, then the same inequalities hold with Δ_{tu}^r and Δ_{kt}^r replaced by Δ_{-tu}^r and Δ_{-kt}^r, respectively. Similar results hold for $f\in C[\tilde{J}]$ with $p=\infty$.

Proof. We prove only the case of $1\leqslant p<\infty$. The proof for $p=\infty$ is similar. Given $f\in L_p[\tilde{J}]$, we need to construct an approximation $g\in L_p^r[J]$. To this end we take the so-called *Steklov average of f*:

$$g(x)=\int_0^{rt}\left[f(x)+(-1)^{r+1}\Delta_u^r f(x)\right]\frac{N^r(u/t)\,du}{t}$$

$$=\sum_{i=1}^{r}(-1)^{i+1}\binom{r}{i}\int_0^{rt}f(x+iu)\frac{N^r(u/t)\,du}{t}, \qquad (2.141)$$

where N^r is the function that appears in the representation (2.107) of the rth forward difference. Let $F=D^{-r}f$ [cf. (2.28)]. Then by (2.107), for all $1\leqslant i\leqslant r$,

$$\int_0^{rt}f(x+iu)\frac{N^r(u/t)\,du}{t}=\int_0^{irt}f(x+u)\frac{N^r(u/it)\,du}{it}=(it)^{-r}\Delta_{it}^r F(x).$$

It follows from (2.141) that

$$g(x)=\sum_{i=1}^{r}(-1)^{i+1}\binom{r}{i}(it)^{-r}\Delta_{it}^r F(x), \qquad (2.142)$$

and thus that $g\in L_p^r[J]$. Now by the properties of N^r,

$$\|f-g\|_{L_p[J]}\leqslant\int_0^r\|\Delta_{tu}^r f\|_{L_p[J]}N^r(u)\,du\leqslant\int_0^r\|\Delta_{tu}^r f\|_{L_p[J]}\,du\leqslant r^{r+1}\omega_r(f;t)_{L_p[\tilde{J}]}.$$

Here we have used the Minkowski inequality—cf. Remark 2.2. We have proved (2.139).

If we differentiate (2.142) r times, we obtain

$$D^r g(x)=\sum_{i=1}^{r}(-1)^{i+1}\binom{r}{i}(it)^{-r}\Delta_{it}^r f(x).$$

Since $\Sigma_{i=1}^r\binom{r}{i}=2^r-1$ and

$$\max_{1\leqslant i\leqslant r}\|\Delta_{it}^r f\|_{L_p[J]}\leqslant r^r\omega_r(f;t)_{L_p[\tilde{J}]},$$

the inequality (2.140) also follows. ∎

The following theorem clarifies the close relationship between the K-functional and the modulus of smoothness:

THEOREM 2.67

Let $I=[a,b]$ and $1 \leqslant p < \infty$. Then there exist constants C_1 and C_2 (depending only on r, p, and I) such that

$$C_1\omega_r(f;t)_p \leqslant K_{r,p}(t)f \leqslant C_2\omega_r(f;t)_p \tag{2.143}$$

for all $f \in L_p[I]$. A similar assertion holds with $p=\infty$ if $f \in C[a,b]$.

Proof. We consider only the case of $1 \leqslant p < \infty$; the case of $p=\infty$ is similar. Let $g \in L_p^r[I]$, where we write $I=[a,b]$. Then by the properties (2.106) and (2.109) of divided differences,

$$\begin{aligned}\|\Delta_h^r f\|_{L_p[I_{rh}]} &\leqslant \|\Delta_h^r(f-g)\|_{L_p[I_{rh}]} + \|\Delta_h^r g\|_{L_p[I_{rh}]}\\ &\leqslant 2^r\big(\|f-g\|_{L_p[I_{rh}]} + h^r\|D^r g\|_{L_p[I_{rh}]}\big),\end{aligned}$$

where $I_{rh}=[a,b-rh]$. Taking the infimum over all g on the right-hand side, and then the supremum over all $0<h \leqslant t$, we obtain the left-hand side of (2.143) with $C_1 = 1/2^r$.

The proof of the upper inequality in (2.143) is somewhat more complicated. Let $\eta_i = a + i(b-a)/3$, $i=0,1,\dots,3$, and define $J_1=[\eta_0,\eta_2]$, $J_2=[\eta_1,\eta_3]$, and $J_3=[\eta_1,\eta_2]$. Let g_1 and g_2 be the functions associated with f and the intervals J_1 and J_2 as in Lemma 2.66. We now blend g_1 and g_2 together to form a function g defined on I which approximates f well and does not have too large an rth derivative.

To this end, let $\psi \in L_\infty^r[I]$ be any function such that $0 \leqslant \psi(t) \leqslant 1$ on I, $\psi(t)=0$ on $[\eta_0,\eta_1]$, $\psi(t)=1$ on $[\eta_2,\eta_3]$, and $\|D^i\psi\|_{L_\infty[I]} \leqslant C_8 < \infty$, $i=0,1,\dots,r$. Such a function is constructed in Theorem 4.37. Then

$$g=(1-\psi)g_1+\psi g_2 = g_1+\psi(g_2-g_1)$$

belongs to $L_p^r[I]$. We now estimate $\|f-g\|$ and $\|D^r g\|$.

By Lemma 2.66,

$$\|f-g\|_{L_p[I]} \leqslant \|f-g_1\|_{L_p[J_1]} + \|f-g_2\|_{L_p[J_2]} \leqslant C_7\omega_r(f;t)_{L_p[I]}.$$

Using the Leibniz rule and the properties of ψ, we obtain

$$\|D^r g\|_{L_p[J_3]} \leqslant C_6\Big(\|D^r g_1\|_{L_p[J_3]} + \max_{0 \leqslant i \leqslant r}\|D^i g_2 - D^i g_1\|_{L_p[J_3]}\Big).$$

Now, estimating the intermediate derivative (cf. Theorem 2.5) gives

$$\|D^r g\|_{L_p[J_3]} \leqslant C_5(2\|D^r g_1\|_{L_p[J_3]} + \|D^r g_2\|_{L_p[J_3]} + \|g_1 - g_2\|_{L_p[J_3]})$$

$$\leqslant 2C_5\Big(\|D^r g_1\|_{L_p[J_3]} + \|D^r g_2\|_{L_p[J_3]} + \max_{1 \leqslant i \leqslant 2} \|f - g_i\|_{L_p[J_3]}\Big)$$

$$\leqslant 2C_5\Big(\|D^r g_1\|_{L_p[J_1]} + \|D^r g_2\|_{L_p[J_2]} + \max_{1 \leqslant i \leqslant 2} \|f - g_i\|_{L_p[J_i]}\Big),$$

where we have also used the fact that $J_3 \subseteq J_1$ and $J_3 \subseteq J_2$. The same estimate for $D^r g$ holds on J_1 and J_2 since g reduces to g_1 and g_2, respectively, on these intervals. We conclude that

$$t^r\|D^r g\|_{L_p[I]} \leqslant C_4 \max_{1 \leqslant i \leqslant 2} \left(\|f - g_i\|_{L_p[J_i]} + \|D^r g_i\|_{L_p[J_i]}\right)$$

$$\leqslant C_3 \omega_r(f; t)_{L_p[I]}.$$

Combining this with our previous estimate for $\|f - g\|$, we see that

$$K_{r,p}(t)f \leqslant \|f - g\|_{L_p[I]} + t^r\|D^r g\|_{L_p[I]} \leqslant C_2 \omega_r(f; t)_{L_p[I]}. \qquad \blacksquare$$

The remainder of this section is devoted to some applications of the K-functional. Our first result shows that in obtaining error bounds for various approximation processes, it is often sufficient to establish the bounds for smooth functions only, and the desired bounds for less-smooth function classes will follow automatically.

THEOREM 2.68

Let $1 \leqslant p < \infty$, and suppose r is a positive integer. Let $\mathbb{S}$ be a set of functions in $L_p[I]$ such that for each $g \in L_p^r[I]$ there exists an element $s_g \in \mathbb{S}$ satisfying

$$\|g - s_g\|_p \leqslant C_0 + C_1 t^r \|D^r g\|_p, \qquad \text{some } t > 0, \tag{2.144}$$

with C_0 and C_1 constants depending only on r. Then there exists a constant C_2 (depending only on r, p, and I) such that for each $f \in L_p[I]$ there exists $s_f \in \mathbb{S}$ satisfying

$$\|f - s_f\|_p \leqslant C_0 + C_2 \omega_r(f; t)_p. \tag{2.145}$$

For $p = \infty$, (2.145) holds for all $f \in C[I]$ provided (2.144) holds for all $g \in C^r[I]$.

Proof. We consider the case of $1 \leqslant p < \infty$. Let $f \in L_p[I]$. Then for any $g \in L_p^r[I]$,

$$\|f - s_g\|_p \leqslant \|f - g\|_p + \|g - s_g\|_p \leqslant C_0 + \max(C_1, 1)(\|g - f\|_p + t^r \|D^r g\|_p).$$

Since the K-functional is defined as an infimum, if we vary g in $L_p^r[I]$, we can find some $g^* \in L_p^r[I]$ so that this inequality becomes

$$\|f - s_{g^*}\|_p \leqslant C_0 + 2\max(1, C_1) K_{r,p}(t) f.$$

We may now use the upper bound on K in terms of ω_r in Theorem 2.67 to complete the proof. The proof for $f \in C[I]$ is similar. ∎

The following variant of Theorem 2.68 shows that the same idea can be applied to linear approximation processes:

THEOREM 2.69

Let $1 \leqslant p < \infty$, and suppose r is a positive integer. Let L be a bounded linear operator mapping $L_p[I]$ into itself. Suppose that for $t > 0$,

$$\|g - Lg\|_p \leqslant C_0 + C_1 t^r \|D^r g\|_p, \qquad \text{all } g \in L_p^r[I]. \tag{2.146}$$

Then there exists a constant C_2 (depending on r, p, I, and $\|L\|$) such that

$$\|f - Lf\|_p \leqslant C_0 + C_2 \omega_r(f; t)_p, \qquad \text{all } f \in L_p[I]. \tag{2.147}$$

For $p = \infty$, (2.147) holds for all $f \in C[I]$ provided L is a bounded linear mapping of $C[I]$ into itself satisfying (2.146).

Proof. We treat only the case of $1 \leqslant p < \infty$. For any $f \in L_p[I]$ and $g \in L_p^r[I]$,

$$\begin{aligned}\|f - Lf\|_p &\leqslant \|f - g\|_p + \|g - Lg\|_p + \|Lf - Lg\|_p \\ &\leqslant C_0 + \max(1, C_1)(1 + \|L\|)(\|f - g\|_p + t^r \|D^r g\|_p).\end{aligned}$$

The rest of the proof is exactly as in Theorem 2.68. ∎

In a number of applications it is useful to be able to *extend* a function from an interval J to a larger interval I. This is trivial if we do not care how smooth the resulting function is—we can just define the extension to be zero outside of J. When f has a certain number of derivatives on J, it is also easy to extend it to I to be a function with the same number of derivatives—for example, to extend the function beyond the right endpoint of J we may tack on a polynomial that agrees with the Taylor expansion of f at that point. On the other hand, if we want to extend a function while *preserving some moduli of smoothness*, the problem is no longer quite so simple. We have the following elegant application of the K-functional to the solution of this extension problem:

THEOREM 2.70. Whitney Extension Theorem

Let $J=[a,b]\subseteq I=[c,d]$. Suppose $1\leqslant p<\infty$ and $r\geqslant 1$. Then there exists a linear operator T mapping $L_p[J]$ into $L_p[I]$ such that Tf extends f from J to I [i.e., $Tf(x)=f(x)$, all $x\in J$] and

$$\omega_r(Tf;t)_{L_p[I]}\leqslant C_1\omega_r(f;t)_{L_p[J]}. \tag{2.148}$$

Moreover, if $f\in L_p^r[J]$, then $Tf\in L_p^r[I]$, and

$$\|D^rTf\|_{L_p[I]}\leqslant C_2\|D^rf\|_{L_p[J]}. \tag{2.149}$$

The constants C_1 and C_2 depend only on r and the ratio $|I|/|J|$. A similar result holds for continuous functions with $p=\infty$.

Proof. We give the proof only for $1\leqslant p<\infty$. We consider first the case where $J=[0,1]$ and $I=[-1,1]$. Given $f\in L_p[J]$, let

$$Tf(x)=\begin{cases} f(x), & 0\leqslant x\leqslant 1\\ \sum_{i=0}^{r} c_i f(-2^{-i}x), & -1\leqslant x\leqslant 0,\end{cases} \tag{2.150}$$

where $c_0,\ldots,c_r$ are chosen as the solution of the system

$$\sum_{i=0}^{r} c_i(-2^{-i})^j=1, \qquad j=0,1,\ldots,r. \tag{2.151}$$

Clearly T defines a linear extension operator mapping functions defined on J into functions defined on I. The condition (2.151) assures that $TP=P$ for all polynomials $P\in\mathcal{P}_r$. This in turn assures that for any $g\in L_p^r[J]$, its Taylor expansion about zero for $x\geqslant 0$ agrees with the Taylor expansion of Tg about zero for $x\leqslant 0$. This means that Tg has continuous derivatives up to order $r-1$ at zero, hence $Tg\in L_p^r[I]$. For any such g we also have

$$\|D^rTg\|_{L_p[I]}\leqslant C_2\|D^rg\|_{L_p[J]},$$

where C_2 depends only on r. This proves (2.149).

Now for any $f \in L_p[J]$ and $g \in L_p^r[J]$,

$$\|\Delta_h^r Tf\|_{L_p[I]} \leqslant \|\Delta_h^r T(f-g)\|_{L_p[I]} + \|\Delta_h^r Tg\|_{L_p[I]}$$

$$\leqslant 2^r \|T(f-g)\|_{L_p[I]} + h^r \|D^r Tg\|_{L_p[I]}$$

$$\leqslant C_3(\|f-g\|_{L_p[J]} + h^r \|D^r g\|_{L_p[J]}).$$

Taking the infimum over all $g \in L_p^r[J]$ on the right-hand side and the supremum over all $h \leqslant t$ on the left-hand side, we obtain

$$\omega_r(f;t)_{L_p[I]} \leqslant C_3 K_{r,p}(t)_J f.$$

Applying Theorem 2.67 to estimate K in terms of ω_r, we obtain (2.148), at least for $J=[0,1]$ and $I=[-1,1]$. A similar argument shows that f can be extended from J to $[0,2]$. It remains to prove the general case. We accomplish this with a change of variables. Given $f \in L_p[a,b]$, let $F(y)=f[a+(b-a)y]$. Then $F \in L_p[0,1]$, and by property (2.124) of moduli of smoothness,

$$\omega_r(F;t)_{L_p[0,1]} = \left(\frac{1}{b-a}\right)^{1/p} \omega_r[f;(b-a)t]_{L_p[a,b]}.$$

Let TF be the extension of F to $[-1,2]$, and define

$$Tf(y) = TF\left[\frac{(3y-d-2c)}{(d-c)}\right].$$

Clearly $Tf \in L_p^m[c,d]$, and

$$\omega_r(Tf;t)_{L_p[c,d]} = \left(\frac{d-c}{3}\right)^{1/p} \omega_r\left[TF;\frac{3t}{(d-c)}\right]_{L_p[-1,2]}.$$

Now, using property (2.124) of moduli of smoothness, we obtain

$$\omega_r(Tf;t)_{L_p[c,d]} \leqslant C_1 \lfloor (b-a) \rfloor \left(\frac{1}{b-a}\right)^{1/p} \left\lfloor \frac{3}{d-c} \right\rfloor \left(\frac{d-c}{3}\right)^{1/p} \omega_r(f;t)_{L_p[a,b]}.$$

It is now clear that the constant depends on $|I|/|J|$. A similar proof works for $f \in C[a,b]$. ■

We conclude this section by proving two of the more difficult properties of the K-functional—certain so-called Marchaud-type inequalities. In view of the equivalence of the K-functional with ω_r, these results establish the classical Marchaud inequalities for the modulus of smoothness (cf. Theorem 2.61).

THEOREM 2.71. Marchaud Inequality

Let $1 \leqslant p < \infty$ and $0 < t \leqslant 1$. Then for every $f \in L_p[a,b]$,

$$K_{j,p}(t)f \leqslant C_1 t^j \left[\|f\|_p + \int_t^1 s^{-1-j} K_{r,p}(s) f\, ds \right], \tag{2.152}$$

$j = 1, 2, \ldots, r-1$. Here C_1 is a constant depending only on j, r, and $[a,b]$. The same inequalities hold for $p = \infty$ if $f \in C[a,b]$.

Proof. We consider the case of $1 \leqslant p < \infty$. Let $f \in L_p[a,b]$. For each $h \geqslant 0$, let $g_h \in L_p^r[a,b]$ be such that

$$\|f - g_h\|_p + h^r \|D^r g_h\|_p \leqslant 2K_{r,p}(h)f. \tag{2.153}$$

Given $0 < t \leqslant 1$, let n be such that $1/2 < 2^n t \leqslant 1$, and define $\varphi_k = g_{2^k t}$. Then $g_t = \varphi_0$ can be written as the telescoping series

$$\varphi_0 = \sum_{k=0}^{n-1} (\varphi_k - \varphi_{k+1}) + \varphi_n.$$

Fix $1 \leqslant j \leqslant r-1$. Then applying Theorem 2.5 to estimate the jth derivative of each term (via the 0th and rth derivatives), we obtain

$$t^j \|D^j g_t\|_p \leqslant C_3 \sum_{k=0}^{n-1} 2^{-jk} \left[\|\varphi_k - \varphi_{k+1}\|_p + (2^k t)^r \|D^r(\varphi_k - \varphi_{k+1})\|_p \right]$$
$$+ 2^{-nj} \left[\|\varphi_n\|_p + (2^n t)^r \|D^r \varphi_n\|_p \right].$$

By adding and subtracting f inside the norm of the first term on the right and applying the triangle inequality we obtain

$$t^j \|D^j g_t\|_p \leqslant C_2 \sum_{k=0}^{n} 2^{-jk} \left[\|f - \varphi_k\|_p + (2^k t)^r \|D^r \varphi_k\|_p \right] + 2^{-nj} \|f\|_p$$
$$\leqslant C_2 \sum_{k=0}^{n} 2^{-jk} K_{r,p}(2^k t) f + 2^{-nj} \|f\|_p.$$

The second inequality follows from (2.153). Since (2.153) also asserts that $\|f-g_t\|_p \leqslant 2K_{r,p}(t)f$,

$$\begin{aligned} K_{j,p}(t)f &\leqslant \|f-g_t\|_p + t^j\|D^j g_t\|_p \\ &\leqslant C_2\left[2^{-nj}\|f\|_p + t^j\sum_{k=0}^{n}(2^k t)^{-j}K_{r,p}(2^k t)\right] \\ &\leqslant C_2\left[2^{-nj}\|f\|_p + t^j\int_{t/2}^{2^n t} s^{-1-j}K_{r,p}(s)\,ds\right] \\ &\leqslant C_2 t^j\left[\|f\|_p + \int_{t/2}^{1} s^{-1-j}K_{r,p}(s)f\,ds\right]. \end{aligned}$$

The result follows since

$$\int_{t/2}^{1} s^{-1-j}K_{r,p}(s)f\,ds \leqslant 2^j\int_t^1 s^{-1-j}K_{r,p}(s)f\,ds. \qquad \blacksquare$$

Our next Marchaud-type inequality for the K-functional is a kind of inverse approximation theorem. It asserts that if f can be approximated sufficiently well by smooth functions (in the sense that the K-functional is small), then f itself must be smooth (in particular, it must possess an appropriate number of smooth derivatives).

THEOREM 2.72

Suppose $1 \leqslant p < \infty$, and that r is a positive integer. If $f \in L_p[a,b]$ is such that

$$\int_0^1 s^{-1-j}K_{r,p}(s)f\,ds < \infty, \tag{2.154}$$

for some $1 \leqslant j \leqslant r-1$, then $f \in L_p^j[a,b]$ with

$$\|D^j f\|_p \leqslant C_1\left[\|f\|_p + \int_0^1 s^{-1-j}K_{r,p}(s)f\,ds\right] \tag{2.155}$$

and

$$K_{r-j,p}(t)D^j f \leqslant C_1\int_0^t s^{-1-j}K_{r,p}(s)f\,ds. \tag{2.156}$$

Here C_1 is a constant that depends only on j, r, p, and $[a,b]$. If $f \in C[a,b]$ and (2.154) holds with $p=\infty$, then we conclude that $f \in C^j[a,b]$, and (2.155) and (2.156) are also valid with $p=\infty$.

Proof. We consider only the case of $1 \leqslant p < \infty$. Given $f \in L_p[a,b]$, let g_h be defined as in (2.153). By the same techniques used to prove Theorem 2.71, we obtain

$$\sum_{k=0}^{\infty} \|D^j(\varphi_k - \varphi_{k+1})\|_p \leqslant C_4 \sum_{k=0}^{\infty} (2^{-k}t)^{-j} K_{r,p}(2^{-k-1}t)f$$

$$\leqslant C_3 \int_0^t s^{-1-j} K_{r,p}(s) f\, ds. \tag{2.157}$$

We conclude that $\sum_{k=0}^{\infty}(\varphi_k - \varphi_{k+1})$ converges in $L_p^j[a,b]$, hence φ_k must also have a limit element in $L_p^j[a,b]$. But by (2.153),

$$\|f - \varphi_k\|_p \leqslant 2K_{r,p}(2^{-k}t)f,$$

which goes to 0 as $k \to \infty$ since $K_{r,p}(t)$ is monotone increasing in t while the integral (2.154) is finite. We conclude that φ_k converges to f, and thus that $f \in L_p^j[a,b]$.

To establish the inequality (2.155), we use (2.157) for $t = 1$, giving

$$\|D^j f\|_p \leqslant \|D^j g_1\|_p + C_4 \sum_{k=0}^{\infty} 2^{-kj} K_{r,p}\left[2^{-(k+1)}\right] f.$$

Taking account of the monotonicity of $K_{r,p}$ and using (2.153), we obtain (2.155).

To get (2.156), we note that

$$K_{r-j,p}(t) D^j f \leqslant \|D^j(f - g_t)\|_p + t^{r-j} \|D^{r-j} D^j g\|_p$$

$$\leqslant \|D^j(f - g_t)\|_p + t^{r-j} \|D^r g\|_p$$

$$\leqslant \sum_{k=0}^{\infty} \|D^j(\varphi_k - \varphi_{k+1})\|_p + t^{r-j} \|D^r g_t\|_p.$$

Using (2.153) and (2.157) again, we get

$$K_{r-j,p}(t) D^j f \leqslant C_2 \left[\int_0^t s^{-1-j} K_{r,p}(s) f\, ds + t^{-j} K_{r,p}(t) f \right],$$

which immediately implies (2.156). ∎

§ 2.10. *n*-WIDTHS

In this section we briefly review the theory of n-widths. The idea is to obtain estimates on the maximum rate of convergence one can expect in using finite dimensional linear spaces to approximate given classes of smooth functions. The results are useful in deciding when we should be satisfied with the approximation power of particular approximating spaces.

To motivate the definition of n-width, consider approximation of functions in the space $L_p^\sigma[a,b]$ by polynomials. Given a function $f \in L_p^\sigma[a,b]$, we define the distance of f to $\mathcal{P}_n$ by

$$d(f, \mathcal{P}_n)_{L_p[a,b]} = \inf_{g \in \mathcal{P}_n} \|f-g\|_{L_p[a,b]}.$$

Jackson's theorem (see Theorem 3.12) asserts that

$$d(f, \mathcal{P}_n)_{L_p[a,b]} \leqslant C_1\left(\frac{1}{n}\right)^\sigma \|D^\sigma f\|_{L_p[a,b]}, \qquad \text{all } f \in L_p^\sigma[a,b]. \quad (2.158)$$

This bound can be arbitrarily large for functions f in $L_p^\sigma[a,b]$. Thus we introduce the class

$$UL_p^\sigma[a,b] = \left\{ f \in L_p^\sigma[a,b] : \|D^\sigma f\|_{L_p[a,b]} \leqslant 1 \right\}, \quad (2.159)$$

and consider the distance of the entire set UL_p^σ from $\mathcal{P}_n$ defined by

$$d(UL_p^\sigma, \mathcal{P}_n)_{L_p[a,b]} = \sup_{f \in UL_p^\sigma} d(f, \mathcal{P}_n)_{L_p[a,b]}.$$

The result (2.158) implies

$$d(UL_p^\sigma, \mathcal{P}_n)_{L_p[a,b]} \leqslant C_1\left(\frac{1}{n}\right)^\sigma.$$

This says that the worst function in UL_p^σ can still be approximated to order $(1/n)^\sigma$ by polynomials of order n.

In view of the above discussion, we may now ask the following question: is it possible to get a better order of convergence if we use some other n-dimensional linear space rather than the polynomials $\mathcal{P}_n$? To answer this question, we define the quantity

$$d_n(UL_p^\sigma, L_p) = \inf_{X_n} d(UL_p^\sigma, X_n), \quad (2.160)$$

where the infimum is taken over all n-dimensional subspaces X_n of $L_p[a,b]$. If $d_n \geqslant C_2(1/n)^\sigma$, then we can conclude that the polynomials do as well (in order of approximation) as any other n-dimensional linear spaces. We shall see later that this is indeed the case for UL_p^σ. The quantity $d_n(UL_p^\sigma, L_p)$ is called the *n-width of UL_p^σ in L_p*.

The idea of n-width of a set can be cast in a more general setting.

DEFINITION 2.73. n-Width

Let X be a normed linear space with norm $\|\cdot\|_X$. Given an n-dimensional linear subspace $X_n \subseteq X$, we define the *distance of $f \in X$ from X_n* as

$$d(f, X_n)_X = \inf_{g \in X_n} \|f - g\|_X. \tag{2.161}$$

Given a subset $A \subseteq X$, we define the *distance of A from X_n* to be

$$d(A, X_n)_X = \sup_{f \in A} d(f, X_n)_X. \tag{2.162}$$

Finally, we define the *n-width of A in X* as

$$d_n(A, X) = \inf_{X_n} d(A, X_n)_X, \tag{2.163}$$

where the infimum is taken over all n-dimensional subspaces X_n contained in X.

The n-width $d_n(A, X)$ is a measure of how hard it is to approximate functions in A using arbitrary n-dimensional linear subspaces of X. In a way, it measures how nasty A is. We are primarily interested in the asymptotic behavior of $d_n(A, X)$ as $n \to \infty$. We shall write

$$d_n(A, X) \approx \eta_n \tag{2.164}$$

provided η_n is a sequence such that for some constants $0 \leqslant C_1 \leqslant C_2 < \infty$,

$$C_1 \eta_n \leqslant d_n(A, X) \leqslant C_2 \eta_n \qquad \text{as } n \to \infty.$$

The importance of knowing the asymptotic behavior of $d_n(A, X)$ is that it may help to decide if a given sequence of linear subspaces X_n does a good job of approximating functions in A. We introduce the following definitions:

DEFINITION 2.74. Optimal Sequences

If X_n is a sequence of n-dimensional linear subspaces of X with

$$d(A, X_n)_X \approx d_n(A, X),$$

then we say that X_n is an *asymptotically optimal sequence*. If we are lucky enough to find a sequence of n-dimensional linear subspaces X_n of X with

$$d(A,X_n)=d_n(A,X),$$

then we say that X_n is an *optimal sequence*.

While asymptotically optimal sequences are not very hard to find, optimal sequences are somewhat rarer. Thus it is interesting to note that for several common smooth spaces of functions, certain spaces of splines turn out to be optimal.

We turn now to the question of calculating the behavior of $d_n(A,X)$ for a particular A and X. To this end, we observe that if X_n is any particular subspace of X of dimension n, then

$$d_n(A,X) \leqslant d(A,X_n)_X. \tag{2.165}$$

On the other hand, if F_n is a sequence of functions in A, then

$$\inf_{X_n} d(F_n,X_n)_X \leqslant d_n(A,X). \tag{2.166}$$

We may use (2.165) to obtain upper bounds on d_n, and (2.166) to obtain lower bounds. The latter can often be used in conjunction with the following simple lemma.

LEMMA 2.75

Let X be a normed linear space and $X_n=\text{span}\,\{\phi_i\}_1^n$ be an n-dimensional linear subspace of X. Suppose λ is a bounded linear functional on X such that

$$\lambda\varphi_i=0, \qquad i=1,2,\ldots,n. \tag{2.167}$$

Then for all $f\in X$,

$$d(f,X_n)_X \geqslant \frac{|\lambda f|}{\|\lambda\|}. \tag{2.168}$$

Proof. For any $\{a_i\}_1^n$,

$$\|f-\sum_{i=1}^n a_i\varphi_i\|_X \geqslant \frac{\left|\lambda\left(f-\sum_{i=1}^n a_i\varphi_i\right)\right|}{\|\lambda\|}=\frac{|\lambda f|}{\|\lambda\|}. \qquad \blacksquare$$

To illustrate these ideas, we now compute the n-width of the set $UL_p^\sigma[a,b]$ defined in (2.159) as a subset of $L_p[a,b]$.

THEOREM 2.76

Let $1 \leqslant p < \infty$, and $UL_p^\sigma[a,b]$ be as defined in (2.159). Then

$$d_n(UL_p^\sigma, L_p) \approx \left(\frac{1}{n}\right)^\sigma. \tag{2.169}$$

Proof. The upper bound on d_n follows from (2.165) with $X_n = \mathcal{P}_n$ (cf. Example 3.14). We derive the lower bound using Lemma 2.75. Given $X_n = \text{span}\,\{\varphi_j\}_1^n$, define λ on $L_p[a,b]$ by

$$\lambda f = \sum_{i=0}^{n} c_i \int_{\eta_i}^{\eta_{i+1}} f(t)\,dt,$$

where $\eta_i = a + ih$, $i = 0, 1, \ldots, n+1$ and $h = (b-a)/(n+1)$, and where $\{c_i\}_0^n$ are chosen so that

$$\lambda \varphi_j = \sum_{i=0}^{n} c_i \int_{\eta_i}^{\eta_{i+1}} \varphi_j(t)\,dt = 0, \qquad j = 1, 2, \ldots, n.$$

We may normalize the c's so that $\Sigma_{i=0}^n |c_i|^{p'} = 1$, where $1/p + 1/p' = 1$. Then

$$|\lambda f| \leqslant \int_a^b |f(t)| \cdot \left| \sum_{i=0}^{n} c_i \chi_i(t) \right| dt \leqslant \|f\|_p \left\| \sum_{i=0}^{n} c_i \chi_i \right\|_{p'},$$

where χ_i is the characteristic function of $I_i = (\eta_i, \eta_{i+1})$. Since

$$\left\| \sum_{i=0}^{n} c_i \chi_i \right\|_{p'} = \left(\sum_{i=0}^{n} |c_i|^{p'} \int_{I_i} dt \right)^{1/p'} = h^{1-1/p},$$

we conclude that $\|\lambda\| \leqslant h^{1-1/p}$.

We now construct a nasty function in UL_p^σ. In Theorem 4.34 we construct (a perfect spline) $B = B_{\sigma+1}$ such that

$$B \in C^{\sigma-1}[-1,1]$$

$$D^j B(-1) = D^j B(1) = 0, \qquad j = 0, 1, \ldots, \sigma - 1$$

$$\int_{-1}^{1} B(x)\,dx = 1$$

$$|D^\sigma B(x)| = 2^{\sigma-1}\sigma!, \qquad \text{all } -1 \leqslant x \leqslant 1.$$

Let

$$F(x)=\sum_{i=0}^{n}\frac{|c_i|^{p'/p}}{\sigma!2^{2\sigma-1}}\operatorname{sgn}(c_i)B\left[\frac{2(x-\bar{\eta}_i)}{h}\right]h^{\sigma-1/p},$$

where $\bar{\eta}_i=(\eta_i+\eta_{i+1})/2$, $i=0,1,\ldots,n$. Then clearly $F\in L_p^\sigma[a,b]$, and

$$\|D^\sigma F\|_p^p=\sum_{i=0}^{n}|c_i|^{p'}h^{-1}\int_{\eta_i}^{\eta_{i+1}}dt=1.$$

On the other hand,

$$|\lambda F|=\sum_{i=0}^{n}\frac{|c_i|^{1+p'/p}}{\sigma!2^{2\sigma-1}}h^{\sigma-1/p}\cdot h=\frac{h^{\sigma+1-1/p}}{\sigma!2^{2\sigma-1}}.$$

By Lemma 2.75, we conclude

$$d_n(UL_p^\sigma,L_p)\geqslant\frac{h^{\sigma+1-1/p}}{\sigma!2^{2\sigma-1}h^{1-1/p}}=\frac{h^\sigma}{\sigma!2^{2\sigma-1}}\geqslant C_2\left(\frac{1}{n}\right)^\sigma. \qquad ■$$

Theorem 2.76 shows that the sequence of polynomial spaces $\mathcal{P}_1,\mathcal{P}_2,\ldots$ is an *asymptotically optimal sequence* of approximation spaces for $UL_p^\sigma[a,b]$ in $L_p[a,b]$. The problem of determining the n-width of $UL_p^\sigma[a,b]$ in $L_q[a,b]$ for $p\neq q$ is quite delicate. We quote the following result:

THEOREM 2.77

Let $\sigma\geqslant 1$. Then

$$d_n(UL_p^\sigma[a,b],L_q[a,b])\approx\begin{cases}\left(\frac{1}{n}\right)^{\sigma+1/2-1/p}, & 1\leqslant p\leqslant 2<q\leqslant\infty\\ \left(\frac{1}{n}\right)^{\sigma}, & 2\leqslant p<q\leqslant\infty\\ \left(\frac{1}{n}\right)^{\sigma+1/q-1/p}, & 1\leqslant p<q\leqslant 2\\ \left(\frac{1}{n}\right)^{\sigma}, & 1\leqslant q\leqslant p\leqslant\infty\end{cases}. \tag{2.170}$$

Discussion. The proof of this theorem has engaged the talents of a number of researchers over quite a period of years. Kolmogorov [1936] first established the case of $p=q=2$. The cases $(p,q)=(1,2)$ and (∞,∞) were treated by Stechkin [1954]. A number of authors have contributed to the case of $p \geqslant q$, including Lorentz [1960], Tihomirov [1960, 1969]. Babadshanov and Tihomirov [1967], Makavoz [1972], Solomiak and Tihomirov [1967], Korneichuk [1974], and Scholz [1974]. The case of $p<q$ is more difficult. For $1 \leqslant p<q \leqslant 2$, it was done by Ismagilov [1974]. Gluskin [1974] and Mayorov [1975] focussed on the $(1,\infty)$ case. For the remaining cases, and a historical discussion, see Kashin [1977a, b]. For a discussion of the role of splines as optimal subspaces, see Dahmen, deBoor, and DeVore [1980]. ■

§ 2.11. PERIODIC FUNCTIONS

Given $a<b$, we define the space of *continuous periodic functions* on $I=[a,b]$ by

$$\mathring{C}[a,b]=\{f \in C[a,b]: f(b)=f(a)\}. \tag{2.171}$$

Similarly, we define the space of *m-times continuously differentiable periodic functions* by

$$\mathring{C}^m[a,b]=\{f \in C^m[a,b]: D^jf(a)=D^jf(b), \quad j=0,1,\ldots,m\}. \tag{2.172}$$

We define the *periodic Sobolev space* by

$$\mathring{L}_p^m[a,b]=L_p^m[a,b] \cap \mathring{C}^{m-1}[a,b]. \tag{2.173}$$

These spaces are all normed linear (in fact, Banach) spaces with the usual norms defined in Section 2.1.

At times it is convenient to think of periodic functions as being defined on a circle obtained by bending the interval $[a,b]$ such that the endpoint b is joined to the endpoint a. Periodic functions have a natural extension from $[a,b]$ to all of $\mathbf{R}$—for example, to extend f from $[a,b]$ to $[a,b+(b-a)]$, we may take

$$f(x)=f[x-(b-a)], \qquad b \leqslant x \leqslant b+(b-a). \tag{2.174}$$

If f belongs to one of the spaces defined above, then its extension belongs to the analogous space on $[a,b+(b-a)]$.

To measure the smoothness of periodic functions we introduce the *periodic modulus of smoothness*:

$$\mathring{\omega}_r(f;t)_p=\sup_{0<h \leqslant t} \|\Delta_h^r f\|_{L_p[a,b]}. \tag{2.175}$$

This definition assumes that f has been extended periodically from $[a,b]$ to $[a,b+(b-a)]$. It differs from (2.111) in the nonperiodic case only in that the p-norm is taken over all $[a,b]$ rather than over a shortened interval.

It is a simple matter to show that the periodic modulus of smoothness has exactly the same properties as the modulus of smoothness discussed in Section 2.8. We can also introduce the K-functional corresponding to our periodic spaces. We define

$$\mathring{K}_{m,p}(t)f = \inf_{g \in \mathring{L}_p^m[a,b]} \left(\|f-g\|_p + t^m \|D^m g\|_p \right). \tag{2.176}$$

The fact that this functional is equivalent to the modulus of smoothness is proved in almost exactly the same way as the proof of Theorem 2.67 in the nonperiodic case. We also note that the important applications of the K-functional to establishing approximation theorems incorporated in Theorems 2.68 and 2.69 also have immediate periodic analogs.

The n-widths of various spaces of periodic functions can be computed by the same techniques illustrated in Section 2.10 for the nonperiodic case. In particular, we note that the n-widths

$$d_n\left(U\mathring{L}_p^\sigma[a,b], L_q[a,b]\right)$$

have exactly the same behavior as in the nonperiodic case—cf. (2.170).

§ 2.12. HISTORICAL NOTES

Section 2.1

The spaces and notation introduced here belong to classical analysis. Sobolev spaces (especially of multidimensional functions) play an important role in the theory of differential equations. For a detailed treatment, see Adams [1975]. Theorem 2.5 which connects the norms of various derivatives of a function is proved by Adams [1975], page 71, for $1 \leq p < \infty$ (his statement looks a little different, but reduces to ours if we set his $\varepsilon = t^{1/m}$). See also Friedman [1969], page 19.

Section 2.2

Taylor's expansion and the use of the Green's function to solve initial value problems belong to elementary calculus and the theory of ordinary differential equations, respectively.

Section 2.3

The matrices and determinants discussed here arise frequently in approximation theory and numerical analysis. We have followed Karlin and Studden [1966] and Karlin [1968] in our choice of notation. The ideas inherent in Lemma 2.9 belong to the standard bag of tricks for this area—see for example, Karlin and Studden [1966], page 8.

Section 2.4

In defining sign changes and zeros we have followed Karlin and Studden [1966] and Karlin [1968] where further results and more references can be found. Rolle's theorem is a part of calculus. The extended version given in Theorem 2.19 is taken from Schumaker [1976b].

Section 2.5

Tchebycheff systems and their various refinements play an important role in many parts of analysis as well as in probability and statistics. For an extensive treatment, see Karlin and Studden [1966] and Karlin [1968]. These books deal only with functions that belong to $C(I)$, where I is an interval. For a discussion of discontinuous T-systems on general sets see Zielke [1979]. We have followed classical notation for the most part. Order Complete Tchebycheff systems are referred to as Descartes systems in the literature—we have introduced the adjective "OC" as we want to use it with ET- and WT-systems also. In approximation theory the T- and ET-spaces are sometimes referred to as Haar and Markov spaces, respectively.

Section 2.6

In this section we have given a fairly complete treatment of the basic properties of WT-systems. Although WT-systems were mentioned in the book by Karlin and Studden [1966], they have not received too much attention until recently. A number of results on WT-systems can also be interpreted as results in the theory of total positivity—see Karlin [1968].

One of the earliest papers on WT-systems *per se* was by Jones and Karlowitz [1970], where the basic characterization Theorem 2.39 was established for a WT-system of continuous functions on an interval. Theorem 2.39 was later established for general WT-systems using a Bernstein polynomial smoothing approach (cf. Remark 2.5). The direct proof given here is credited to Zielke [1979].

Theorems 2.40 and 2.41 can be found in the article by Stockenberg [1977b] with different proof. Theorem 2.41 was also established (for the case of continuous WT-systems on an interval) independently by Sommer and Strauss [1977], using the smoothing technique outlined in Remark 2.4. The variation-diminishing property for OCWT-systems in Theorem 2.42 also follows from general results on totally positive matrices—see Karlin [1968].

Zero properties of WT-systems were developed by Bartelt [1975], Stockenberg [1977a], and Sommer and Strauss [1977]. The latter authors dealt only with continuous WT-systems. We have followed Stockenberg; however, our proof of Theorem 2.45 is much simpler than the original.

There is an important moment theory for T-systems (cf. Karlin and Studden [1966]). Much of this theory can also be carried over to WT-systems—see Micchelli and Pinkus [1977].

Section 2.7

In most numerical analysis texts divided differences are introduced via the recursion relation (2.91)—this is one of the reasons why the case of multiple t's is often ignored. The definition via quotients of determinants is credited to Popoviciu [1959]. Still another approach is to define them as coefficients of certain interpolating polynomials (cf. deBoor [1978] or Conte and deBoor [1972]).

We have not included every possible known result about divided differences. On the other hand, a number of the results presented here are not easily located in the classical literature—for example, the Leibniz rule, the explicit formulae for derivatives, and the exact expressions for the differences of x^j. For additional information, consult Richardson [1954], Milne-Thompson [1960], Conte and deBoor [1972], Isaacson and Keller [1966], and deBoor [1978].

Section 2.8

For more information on moduli of smoothness, see Timan [1963] or Lorentz [1966] and the references therein. Johnen [1972] has recently given a unified treatment including the case of several variables.

Section 2.9

The K-functional was introduced by Peetre [1963, 1964], and it has proved to be an important tool in several parts of analysis. For its role in approximation theory as well as development of its basic theory, see Butzer and Berens [1967]. Our definition (2.130) is at minor variance with the standard one in that we have used t^r in place of t as the multiplier of $\|D^r g\|_p$. Our rationale for doing this is that now the parameter t in the K-functional and the parameter t in the modulus of smoothness play the same role. Our discussion of the connection between the moduli of smoothness and the K-functional follows the work of DeVore [1976], where the history of this equivalence can be found. Our proofs of the Marchaud inequalities for the K-functional in Theorems 2.71 and 2.72 follow the work of Johnen and Scherer [1977]. The elegant applications of the K-functional in Theorems 2.68 to 2.70 are taken from the article of DeVore [1976].

Section 2.10

The theory of n-widths was initiated by Kolmogorov. It has not received much attention in the general approximation theory books with the notable exception of Lorentz [1966], which we have followed.

Theorem 2.76 is, of course, contained in Theorem 2.77. We have given a complete proof to illustrate the techniques required to get lower bounds (and to provide a nice application of the perfect B-spline). The n-widths of the unit balls in Besov spaces (cf. § 6.5) have also been calculated (see e.g., Scholz [1974]).

§ 2.13. REMARKS

Remark 2.1

The discrete Hölder inequality asserts that for all nonnegative $a_1,\ldots,a_n$,

$$\left(\sum_{i=1}^{n} a_i\right)^p \leqslant n^{p-1}\sum_{i=1}^{n} a_i^p, \qquad 1\leqslant p<\infty.$$

Conversely, it is known that

$$\sum_{i=1}^{n} a_i^p \leqslant \left(\sum_{i=1}^{n} a_i\right)^p.$$

For these and other related inequalities, see Hardy, Littlewood, and Pólya [1959] or Beckenbach and Bellman [1961].

Remark 2.2

The Minkowski inequality for functions of two variables asserts that

$$\left(\int_c^d\left|\int_a^b f(x,y)\,dx\right|^p dy\right)^{1/p} \leqslant \int_a^b\left(\int_c^d |f(x,y)|^p\,dy\right)^{1/p} dx,$$

with equality holding precisely when $f(x,y)=\varphi(x)\psi(y)$. See Hardy, Littlewood, and Pólya [1959], p. 148, or Beckenbach and Bellman [1961], p. 22.

Remark 2.3

The question of when a T-space is a CT space is somewhat delicate. The fact that this is so on open intervals is proved in the articles by Nemeth [1969] and by Zielke [1973]. The failure of this assertion for other types of intervals was demonstrated by examples in the works of Volkov [1958], Nemeth [1966], and Zielke [1975]. See Zielke [1979] for more details.

Remark 2.4

We have given direct proofs of all results on WT-systems. For WT-systems of continuous functions on an interval, it is possible to derive many of their properties from analogous results on T-systems using an important smoothing idea. The idea is to smooth the functions $\{u_i\}_1^m$ into a T-system $\{u_{i,\varepsilon}\}_1^m$. One approach to doing this is to define

$$u_{i,\varepsilon}(t)=\int_a^b u_i(x)L_\varepsilon(t;x)\,dx, \qquad i=1,2,\ldots,m,$$

where

$$L_\varepsilon(t;x)=\frac{1}{\varepsilon\sqrt{2\pi}}\exp\left(-\frac{1}{2}\left(\frac{t-x}{\varepsilon}\right)^2\right), \qquad \text{all } x,t.$$

It can be shown easily that $u_{i,\varepsilon}\to u_i$ uniformly on $[a',b']$ as $\varepsilon\downarrow 0$ for any $a<a'<b'<b$. Moreover, using a basic composition formula (see Karlin [1968], p. 16) and the strict total positivity of the kernel L_ε (see Karlin [1968], p. 99), it follows that $\{u_{i,\varepsilon}\}_1^n$ is a T-system.

Remark 2.5

There is an alternate way to smooth WT-systems in $C[a,b]$ into T-systems as a result of the work done by Bastien and Dubuc [1976]. It utilizes the classical Bernstein polynomials defined by

$$B_nf(x)=\sum_{i=0}^{n}\binom{n}{i}f\left(a+\frac{i(b-a)}{n}\right)\left(\frac{x-a}{b-a}\right)^i\left(\frac{b-x}{b-a}\right)^{n-i}.$$

For an extensive treatment of Bernstein polynomials, see Lorentz [1953]. It is known that

$$\|f-B_nf\|_{C[a,b]} \leq \frac{5}{4}\omega\left[f; \frac{(b-a)}{\sqrt{n}}\right].$$

Now to smooth a WT-system $\{u_i\}_1^m$, we may take

$$u_{i,n}(x) = B_n u_i(x), \qquad i=1,2,\ldots,m.$$

As $n\to\infty$ these functions converge uniformly to the u_i on $[a,b]$. Moreover, by the variation-diminishing property

$$Z_{(a,b)}(B_nf) \leq S^-_{(a,b)}(B_nf) \leq S^-_{[a,b]}(f)$$

of Bernstein polynomials (cf. Pólya and Schoenberg [1958]), it is easily argued that $\{u_{i,n}\}_1^m$ forms a T-system on (a, b) for n sufficiently large. This approach can also be used to derive some properties of WT-systems on arbitrary sets I.

Remark 2.6

The following is a standard result from linear algebra, but for convenience we state it explicitly here:

LEMMA 2.78

Let M be a (not necessarily square) matrix such that rank ($M\backslash i$th row of M)$<$rank (M). Then there exists $\mathbf{c}$ such that $M\mathbf{c}=\mathbf{r}$, where $\mathbf{r}$ has all 0 components except for the ith which is 1.

Proof. By the assumptions, it follows that the rank of the augmented matrix $[M|\mathbf{r}]$ is the same as the rank of M. Thus is follows that $\mathbf{r}$ must be a linear combination of the columns of M. ■

Remark 2.7

If $\{u_i\}_1^m$ and $\{u_i\}_1^{m+1}$ are T-systems on $[a,b]$, then given any function defined on $[a,b]$, we can define its *generalized divided difference with respect to* $\{u_i\}_1^{m+1}$ by

$$[t_1,\ldots,t_{m+1}]f = \frac{D\begin{pmatrix} t_1,\ldots,t_{m+1} \\ u_1,\ldots,u_m,f \end{pmatrix}}{D\begin{pmatrix} t_1,\ldots,t_{m+1} \\ u_1,\ldots,u_{m+1} \end{pmatrix}}, \qquad a \leq t_1 < \cdots < t_{m+1} \leq b.$$

Such generalized divided differences were introduced by Popoviciu [1959]. It is clear that many of the basic properties of ordinary divided differences carry over. If $\{u_i\}_1^{m+1}$ is actually a CT-system, then it is even possible to give recursion relations—cf. Mühlbach [1973]. When $\{u_i\}_1^{m+1}$ is an ECT-system, we may allow repeated t's. Such divided differences are studied in more detail in Section 9.1.

Remark 2.8

The classical Binomial Theorem states that for all real numbers x and y,

$$(x+y)^r = \sum_{i=0}^{r} \binom{r}{i} x^i y^{r-i},$$

where $\binom{r}{i} = \mathrm{r}!/(\mathrm{r}-i)!i!$ are called the *binomial coefficients*. They satisfy the recursion

$$\binom{r}{i} + \binom{r}{i-1} = \binom{r+1}{i}.$$

Taking $x=y=1$ in the binomial theorem, we obtain $2^r = \sum_{i=0}^{r} \binom{r}{i}$.

Remark 2.9

If for some constant C,

$$f(t) \leqslant Cg(t) \qquad \text{as } t \to 0,$$

it is standard practice to write

$$f(t) = \mathcal{O}[g(t)].$$

If $f(t)/g(t) \to 0$ as $t \to 0$, then we write

$$f(t) = \mathrm{o}[g(t)].$$

The symbols $\mathcal{O}$ and o are called "big oh" and "little oh," respectively.

3
POLYNOMIALS

The space of polynomials $\mathcal{P}_m$ has played an important role in approximation theory and numerical analysis for many years. In Section 1.2 we have listed a number of properties of $\mathcal{P}_m$ that help account for the usefulness of this space. In this chapter we develop some of these properties in greater detail. We concentrate on polynomials in one variable only—for several variables, see Section 13.3. It is not our purpose to provide a comprehensive treatment of polynomials, but rather to provide background material and to illustrate a number of techniques to be used later.

§ 3.1. BASIC PROPERTIES

Throughout this chapter we are interested in the space

$$\mathcal{P}_m = \left\{ p(x) = \sum_{i=1}^{m} c_i x^{i-1}, \qquad c_1, \ldots, c_m, x \text{ real} \right\} \tag{3.1}$$

of real-valued *polynomials of order m* with real coefficients. We begin by showing that $\mathcal{P}_m$ is a finite dimensional linear space with a convenient basis.

THEOREM 3.1

$\mathcal{P}_m$ is a linear subspace of $C^\infty(\mathbf{R})$. Moreover, given any real number a, the functions $1, x-a, \ldots, (x-a)^{m-1}$ form a basis for $\mathcal{P}_m$.

Proof. It is clear from the definition that each $p \in \mathcal{P}_m$ is infinitely often differentiable on $\mathbf{R}$. Since $\alpha p + \beta q \in \mathcal{P}_m$ for all $p, q \in \mathcal{P}_m$ and all $\alpha, \beta \in \mathbf{R}$, it follows that $\mathcal{P}_m$ is a linear subspace of $C^\infty(\mathbf{R})$. Since each of the functions $1, \ldots, (x-a)^{m-1}$ is clearly in $\mathcal{P}_m$, to show that they form a basis we need only prove that they are linearly independent. Suppose $p(x)=\sum_{i=1}^{m} c_i(x-a)^{i-1}\equiv 0$. Then for any b, all derivatives of p must vanish at b; that is,

$$\begin{bmatrix} p(b) \\ Dp(b) \\ \vdots \\ D^{m-1}p(b) \end{bmatrix}$$

$$= \begin{bmatrix} 1 & b-a & (b-a)^2 & \cdots & (b-a)^{m-1} \\ 0 & 1 & 2(b-a) & \cdots & (m-1)(b-a)^{m-2} \\ \vdots & & & & \vdots \\ 0 & 0 & 0 & \cdots & (m-1)! \end{bmatrix} \begin{bmatrix} c_1 \\ c_2 \\ \vdots \\ c_m \end{bmatrix} = \begin{bmatrix} 0 \\ 0 \\ \vdots \\ 0 \end{bmatrix}.$$

This is a homogeneous system of m-equations which is clearly nonsingular, and it follows that $c_1 = c_2 = \cdots = c_m = 0$. ■

The practical significance of Theorem 3.1 is that once having chosen a basis for $\mathcal{P}_m$, each polynomial will have a unique set of coefficients associated with it. This formally establishes the fact that polynomials can be stored on a digital computer. The following well-known algorithm shows that polynomials are easily evaluated, and thus that $\mathcal{P}_m$ satisfies both properties (1.4) and (1.5) of *computer compatibility*.

ALGORITHM 3.2 Horner's Scheme to Compute $p(x) = \sum_{i=1}^{m} c_i(x-a)^{i-1}$

1. $u \leftarrow x - a$;
2. $p \leftarrow c_m$;
3. For $i \leftarrow m-1$ step -1 until 1 do $p \leftarrow u * p + c_i$.

Discussion. The fact that the final value of p will be $p(x)$ follows from the observation that $p(x)$ can be written in nested form as

$$p(x) = c_1 + u\{c_2 + u[c_3 + \cdots + u(c_m)]\cdots\}.$$

It is clear that this algorithm requires just $m-1$ multiplications and m additions and/or subtractions. ■

It follows from the definition that the derivative and indefinite integral of a polynomial are again polynomials. In particular, if $p(x)=\sum_{i=1}^{m} c_i(x-a)^{i-1}$, then

$$Dp(x)=\sum_{i=1}^{m-1} ic_{i+1}(x-a)^{i-1}$$

while

$$D_a^{-1}p(x)=\int_a^x p(t)\,dt=\sum_{i=2}^{m+1}\frac{c_{i-1}}{(i-1)}(x-a)^{i-1}.$$

The coefficients of Dp and $D_a^{-1}p$ are easily computed from the coefficients of p. Once we have them, we can evaluate Dp or $D_a^{-1}p$ at any given x by Horner's scheme. The derivatives of p can also be computed directly by synthetic division—see Remark 3.1.

We close this section by stating two important Markov-type inequalities.

THEOREM 3.3

Let $g\in\mathcal{P}_m$. Then

$$\|Dg\|_{L_\infty[a,b]}\leqslant\frac{2(m+1)^2}{(b-a)}\|g\|_{L_\infty[a,b]}. \tag{3.2}$$

Moreover, for all $1\leqslant p\leqslant q\leqslant\infty$,

$$\|g\|_{L_q[a,b]}\leqslant\left[\frac{2(p+1)}{(b-a)}(m+1)^2\right]^{1/p-1/q}\|g\|_{L_p[a,b]}. \tag{3.3}$$

Discussion. See Timan [1963], pp. 218 and 236. ■

§ 3.2. ZEROS AND DETERMINANTS

One of the most important properties of the space $\mathcal{P}_m$ is the following bound on the number of zeros a nontrivial polynomial in $\mathcal{P}_m$ can have.

THEOREM 3.4

Given p, let $Z^*(p)$ denote the number of zeros of p on the real line, counting multiplicities as in (2.61). Then

$$Z^*(p)\leqslant m-1 \qquad \text{for all nontrivial } p\in\mathcal{P}_m. \tag{3.4}$$

Proof. For $m=1$ the result is clear since a nonzero constant can never vanish. We now proceed by induction. Suppose the assertion is valid for $m-1$. If $p \in \mathcal{P}_m$ and $Z^*(p) \geq m$, then by Rolle's Theorem 2.19 the polynomial $Dp \in \mathcal{P}_{m-1}$ has at least $m-1$ zeros. By the inductive hypothesis, this is possible only if Dp is identically zero. But then p is a constant, and since it vanishes at least one point, it must itself be zero. ■

We can now establish that $\mathcal{P}_m$ is an ET-space.

THEOREM 3.5

The set of functions $u_1 = 1$, $u_2(x) = x, \ldots, u_m(x) = x^{m-1}$ forms an ET-system on **R**. In other words, the VanderMonde determinant

$$V(t_1, t_2, \ldots, t_m) = D\begin{pmatrix} t_1, \ldots, t_m \\ u_1, \ldots, u_m \end{pmatrix}$$

is positive for all $t_1 \leq t_2 \leq \cdots \leq t_m$.

Proof. By Theorem 2.33, either $\{u_1, \ldots, u_m\}$ or $\{u_1, \ldots, u_{m-1}, -u_m\}$ must form an ET-system; that is, $V(t_1, \ldots, t_m)$ must have one strict sign. Thus to complete the proof, it suffices to show that V is actually positive for some choice of t's. But by examination,

$$V(0,0,\ldots,0) = \prod_{i=1}^{m-1} i! > 0. \quad ■$$

There are several other proofs of the positivity of the VanderMonde determinants. In fact, explicit formulae for their values can be derived [cf. (2.66)–(2.67)].

An immediate corollary of Theorem 3.5 is the fact that the Hermite Interpolation Problem 2.7 can be solved using polynomials. In particular, we have the following:

THEOREM 3.6. Hermite Interpolation

Let $\tau_1 < \tau_2 < \cdots < \tau_d$ and positive integers $l_1, l_2, \ldots, l_d$ be prescribed with $\sum_{i=1}^{d} l_i = m$. Then for any given set of real numbers $\{z_{ij}\}_{j=1, i=1}^{l_i, d}$ there exists a unique $p \in \mathcal{P}_m$ with

$$D^{j-1}p(\tau) = z_{ij}, \qquad j = 1, 2, \ldots, l_i$$
$$i = 1, 2, \ldots, d.$$

Proof. If we write $p(x)=\sum_{i=1}^{m} c_i x^{i-1}$, then the desired coefficients can be computed from the linear system $M\mathbf{c}=\mathbf{z}$ whose determinant is $V(t_1,\ldots,t_m)$ with

$$t_1 \leqslant \cdots \leqslant t_m = \overbrace{\tau_1,\ldots,\tau_1}^{l_1},\ldots,\overbrace{\tau_d,\ldots,\tau_d}^{l_d}$$

(cf. the discussion of Problem 2.7). For other ways of finding p numerically, see the classical numerical analysis books. Using divided differences, it is possible to give an exact expression for the difference between f and p. We have

$$[t_1,\ldots,t_m,x]f=\frac{D\begin{pmatrix} t_1,\ldots,t_m,x \\ 1,\ldots,t^{m-1},f \end{pmatrix}}{D\begin{pmatrix} t_1,\ldots,t_m,x \\ 1,t,\ldots,t^m \end{pmatrix}}.$$

Expanding both numerator and denominator out, we see that

$$f(x)-p(x)=(x-t_1)\cdots(x-t_m)[t_1,\ldots,t_m,x]f. \qquad (3.5) \quad \blacksquare$$

In Theorem 3.5 we have shown that the functions $u_1=1,\ldots,u_m=x^{m-1}$ form an ET-system on **R**. Since this is true for all m, we have actually shown that they form an ECT-system. The following theorem shows that if we restrict ourselves to the interval $(0,\infty)$, this set of functions forms an OCET-system:

THEOREM 3.7

Let $u_i=x^{i-1}$, $i=1,2,\ldots,m$. Then for all $1\leqslant p\leqslant m$,

$$D\begin{pmatrix} t_1,\ldots,t_p \\ u_{i_1},\ldots,u_{i_p} \end{pmatrix}>0 \qquad \text{for all} \qquad \begin{matrix} 0<t_1\leqslant t_2\leqslant & \cdots \leqslant t_p \\ 1\leqslant i_1<i_2< & \cdots <i_p\leqslant m \end{matrix}. \qquad (3.6)$$

Proof. Let G be the number of gaps in the sequence of integers $i_1,\ldots,i_p$. We proceed by induction on G and p. If $G=0$ then by inspection we have

$$D\begin{pmatrix} 0,\ldots,0 \\ u_1,\ldots,u_i \end{pmatrix}=\prod_{j=1}^{i-1} j!. \qquad (3.7)$$

But then for all $1<i<i+p\leqslant m$ and all $0<t_1\leqslant t_2\leqslant\cdots\leqslant t_p$,

$$D\begin{pmatrix} t_1,\ldots,t_p \\ u_{i+1},\ldots,u_{i+p}\end{pmatrix}=D\begin{pmatrix} 0,\ldots,0 \\ u_1,\ldots,u_i\end{pmatrix}D\begin{pmatrix} t_1,\ldots,t_p \\ u_{i+1},\ldots,u_{i+p}\end{pmatrix}\Big/\prod_{j=1}^{i-1} j!$$

$$=D\begin{pmatrix} 0,\ldots,0,t_1,\ldots,t_p \\ u_1,\ldots,u_i,u_{i+1},\ldots,u_{i+p}\end{pmatrix}\Big/\prod_{j=1}^{i-1} j!$$

$$=V(0,\ldots,0,t_1,\ldots,t_p)\Big/\prod_{j=1}^{i-1} j!>0.$$

This proves the result for $G=0$ and all p.

Suppose now that (3.6) has been established for $p-1$ and for sequences of t's with, at most, $G-1$ gaps. Consider a sequence of t's with G gaps. For expediency, in the remainder of the proof we shorten the notation for the determinant D in (3.6) to $D\begin{pmatrix} 1,\ldots,p \\ i_1,\ldots,i_p\end{pmatrix}$. Then by a basic identity for determinants (see Remark 3.2),

$$D\begin{pmatrix} 1,\ldots,p-1 \\ i_2,\ldots,i_{p-1},\nu\end{pmatrix}D\begin{pmatrix} 1,\ldots,p \\ i_i,\ldots,i_p\end{pmatrix}=D\begin{pmatrix} 1,\ldots,p-1 \\ i_2,\ldots,i_p\end{pmatrix}D\begin{pmatrix} 1,\ldots,p \\ i_1,\ldots,i_{p-1},\nu\end{pmatrix}$$

$$-D\begin{pmatrix} 1,\ldots,p-1 \\ i_1,\ldots,i_{p-1}\end{pmatrix}D\begin{pmatrix} 1,\ldots,p \\ i_2,\ldots,i_p,\nu\end{pmatrix},$$

where ν is any integer in the set $\{1,2,\ldots,m\}\backslash\{i_1,\ldots,i_p\}$. The determinant D in (3.7) is the second on the left. Each of the other determinants is either of order $p-1$ or is formed from a sequence of t's with, at most, $G-1$ gaps, and thus are all nonzero. We conclude that D is also nonzero.

To determine the sign of $D\begin{pmatrix} 1,\ldots,p \\ i_1,\ldots,i_p\end{pmatrix}$, suppose ν lies between the integers i_j and i_{j+1}. Then, taking account of the number of interchanges necessary to put $i_2,\ldots,i_{p-1},\nu$ into natural order, we see that the first determinant on the left has sign $(-1)^{p-j-1}$. A similar consideration shows that the second one on the right has the same sign, while the last has the opposite sign. Since the first and third determinants on the right are positive, (3.7) follows for sequences with G gaps. ■

Theorem 3.7 is not true if t's are permitted to take nonpositive values. For example, the determinant $D\begin{pmatrix} t \\ u_2\end{pmatrix}=t$ can be negative, zero, or positive, depending on t.

§ 3.3. VARIATION DIMINISHING PROPERTIES

In this section we make some observations about how the shape of a polynomial can be related to the sign structure of its set of coefficients. The simplest manifestation of this phenomenon is the obvious fact that a polynomial with all positive coefficients is positive throughout $(0,\infty)$. Our first theorem is a generalization of this observation.

THEOREM 3.8. Descartes' Rule of Signs

Let $Z^*_{(0,\infty)}$ count the number of zeros on $(0,\infty)$ with multiplicities, and let S^- count strong sign changes [see (2.45)]. Then

$$Z^*_{(0,\infty)}\left(\sum_{i=1}^{m} c_i t^{i-1}\right) \leqslant S^-(c_1, c_2, \ldots, c_m), \tag{3.8}$$

for all $c_1, \ldots, c_m$, not all 0.

Proof. Since $1, x, \ldots, x^{m-1}$ form an OCET-system on $(0,\infty)$ by Theorem 3.7, the assertion (3.8) follows immediately from Theorem 2.34. ■

Theorem 3.8 gives bounds on the number of zeros a polynomial can have on the interval $(0,\infty)$. The following result deals with the case of an arbitrary interval (a,b):

THEOREM 3.9. Budan-Fourier

Let p be a nontrivial polynomial in $\mathcal{P}_m$. Then

$$\begin{aligned} Z^*_{(a,b)}(p) \leqslant S^-\left[p(a), Dp(a), \ldots, D^{m-1}p(a)\right] \\ - S^-\left[p(b), Dp(b), \ldots, D^{m-1}p(b)\right]. \end{aligned} \tag{3.9}$$

If we assume that p is of *exact order* (i.e., its $m-1$st derivative is a nonzero constant), then we can state this in the slightly stronger form,

$$\begin{aligned} Z^*_{(a,b)}(p) \leqslant m-1-S^+\left[p(a), -Dp(a), \ldots, (-1)^{m-1}D^{m-1}p(a)\right] \\ - S^+\left[p(b), Dp(b), \ldots, D^{m-1}p(b)\right]. \end{aligned} \tag{3.10}$$

Proof. We prove the stronger version first. For $m=1$ it is trivial. We now proceed by induction on m. Suppose $m>1$ and that (3.9) has been established for polynomials of order $m-1$. For expediency, we introduce the notation

$$A_j = S^+\left[(-1)^j D^j p(a),\ldots,(-1)^{m-1} D^{m-1} p(a)\right]$$

$$B_j = S^+\left[D^j p(b),\ldots,D^{m-1} p(b)\right]$$

for $j=0,1,\ldots,m-1$. Let $\alpha = A_0 - A_1$ and $\beta = B_0 - B_1$.

Clearly, α and β can take on only the values 0 or 1. We claim that $\alpha=1$ is only possible if a is a left Rolle's point (cf. Definition 2.18) for p. If $p(a)=0$, then a is automatically a Rolle's point. Say $p(a)>0$. Then A_0 must have the pattern

$$\left(+1,\overbrace{0,\ldots,0}^{r},(-1)^{r+1},\ldots\right) \qquad \text{for some} \quad 0 \leqslant r.$$

This implies $D^{r+1}p(a)>0$, and thus

$$Dp(t) = \int_a^t \cdots \int_a^{\zeta_{r-1}} D^{r+1} p > 0 \qquad \text{for} \quad a < t < a+\varepsilon$$

if $\varepsilon > 0$ is sufficiently small. This proves that a is a left Rolle's point in this case. If $p(a)<0$, the proof is similar. The same kind of argument may be given to show that $\beta=1$ is possible only if b is a right Rolle's point for p.

We now claim that

$$Z^*(p) \leqslant Z^*(Dp) + 1 - \alpha - \beta. \tag{3.11}$$

To prove this, we consider two cases:

1. $Z^*(p)=0$. Then (3.11) surely holds; if not, both α and β are 1. But if they are both 1, then by the extended Rolle's Theorem 2.19, Dp has a zero between the Rolle's points a and b.

2. $Z^*(p)>0$. Suppose p has zeros $z_1 \leqslant z_2 \leqslant \cdots \leqslant z_k$. Then Dp has at least $k-1$ zeros in $[z_1, z_k]$. In addition, if $\alpha=1$, then Dp has a zero in (a, z_1), while if $\beta=1$, then Dp has another zero in (z_k, b). Thus for any combination of values of α and β, (3.11) holds.

Now combining (3.11) with the induction hypothesis, we obtain

$$\begin{aligned} Z^*(p) &\leqslant Z^*(Dp) + 1 - \alpha - \beta \leqslant m - 2 + 1 - \alpha - \beta - A_1 - B_1 \\ &\leqslant m - 1 - A_0 - B_0, \end{aligned}$$

which is (3.10). The bound (3.9) in terms of strong sign changes follows from the identity (2.48). ∎

The following example shows that the assumption that p be of exact order in the second part of Theorem 3.9 cannot be dispensed with.

EXAMPLE 3.10

Let $p(t) = t$.

Discussion. If we consider p to be a polynomial of order $m=3$, then (3.10) would assert

$$Z^*_{(-1,1)}(p) \leqslant 2 - S^+(-,-,0) - S^+(+,+,0) = 0,$$

which, of course, is incorrect since p has a zero at 0. ■

The Budan-Fourier theorem can be used to give a short proof of Descartes' rule of signs for polynomials. Indeed, suppose we choose $a=0$ and b very large in (3.9). Then since $c_i = D^{i-1}p(0)/(i-1)!$, $i=1,2,\ldots,m$,

$$S^-\left[p(0), Dp(0), \ldots, D^{m-1}p(0)\right] = S^-(c_1, \ldots, c_m).$$

On the other hand, if b is sufficiently large, then $p(b), Dp(b), \ldots, D^{m-1}p(b)$ all have the same sign, and the second term in (3.9) is zero. Thus (3.9) reduces to the statement (3.8) of Descarte's rule of signs.

§ 3.4. APPROXIMATION POWER OF POLYNOMIALS

In this section we justify our claim that polynomials are capable of approximating smooth functions well. There is a very extensive body of material on this subject, and only a few core results will be presented here. We start with one of the seminal results in the area, proved in 1885.

THEOREM 3.11. Weierstrass Approximation Theorem

Let $\varepsilon > 0$. Then for any $f \in C[a,b]$ there exists a polynomial p (depending on f and ε) such that $\|f-p\|_\infty < \varepsilon$.

Discussion. We shall prove this in Theorem 3.13 below. For a direct proof, see any book on approximation theory. ■

Weierstrass' approximation theorem asserts that every continuous function on a closed interval can be approximated uniformly to any prescribed accuracy by a polynomial. It should be emphasized, however, that the theorem does not say anything about the order of the polynomial. Indeed, if the function f is rather wild, then it will generally take an extremely high-degree polynomial to approximate it well. Considerable work has gone into the question of relating the smoothness of a function to how well it can be approximated by a polynomial of *given* order. Of particular interest is the *rate* at which the error approaches zero as the order of the polynomial is increased. The following theorem gives precise information on this connection in terms of the modulus of smoothness of the function:

THEOREM 3.12. Jackson's Theorem

Let $1 \leqslant p < \infty$ and $1 \leqslant \sigma \leqslant m$. Then there is a constant C_1 (depending only on p and $[a,b]$) such that for every $f \in L_p[a,b]$ there exists a polynomial $P_f \in \mathcal{P}_m$ with

$$\|f - P_f\|_p \leqslant C_1 \omega_\sigma\left(f; \frac{(b-a)}{2m}\right)_p. \tag{3.12}$$

The same result holds with $p = \infty$ if $f \in C[a,b]$.

Proof. We prove the result only for $1 \leqslant p < \infty$, as the case of $p = \infty$ is similar. We assume at first that $I = [a,b] = [-1,1]$; the general result will be obtained later by a change of variables. We shall construct the polynomial P_f associated with f explicitly. As a first step toward the construction, let T be the operator extending $f \in L_p^\sigma[I]$ to Tf in $L_p^\sigma[-4,4]$. By Theorem 2.70, T has the property

$$\omega_\sigma(Tf;t)_{L_p[-4,4]} \leqslant C_9 \omega_\sigma(f;t)_{L_p[-1,1]}.$$

Using Legendre polynomials, it is possible to construct polynomials $V_m \in \mathcal{P}_m$ with $V_m(x) \geqslant 0$ and

$$\int_{-1/\sigma}^{1/\sigma} V_m(t)\,dt = 1 \tag{3.13}$$

$$\int_{-3}^{3} |t|^\sigma V_m(t)\,dt \leqslant C_8 m^{-\sigma}, \tag{3.14}$$

where C_8 depends only on σ. We outline this construction in Remark 3.3.

In terms of the polynomials V_m, we now define a linear operator L_m mapping $L_p[I]$ into $\mathcal{P}_m$ by

$$L_m f(x) = \int_{-2}^{2} Tf(y)\varphi_m(y-x)\,dy,$$

where

$$\varphi_m(u) = -\sum_{k=1}^{\sigma} (-1)^k \binom{\sigma}{k} k^{-1} V_m\left(\frac{u}{k}\right).$$

Our first task is to establish a bound on $\|f-L_mf\|_p$. For all $1 \leqslant k \leqslant \sigma$ and $-1 \leqslant x \leqslant 1$, we have

$$\frac{(-2-x)}{k} \leqslant \frac{-1}{\sigma} \leqslant \frac{1}{\sigma} \leqslant \frac{(2-x)}{k}.$$

Thus we can write

$$\int_{-2}^{2} Tf(y)\frac{V_m\left[\frac{(y-x)}{k}\right]dy}{k} = \int_{(-2-x)/k}^{(2-x)/k} Tf(x+ky)V_m(y)\,dy$$

$$= \int_{-1/\sigma}^{1/\sigma} Tf(x+ky)V_m(y)\,dy + R_k(x),$$

where

$$R_k(x) = \int_{J_{x,k}} Tf(x+ky)V_m(y)\,dy$$

and

$$J_{x,k} = \left[\frac{(-2-x)}{k}, \frac{(2-x)}{k}\right] \Big\backslash \left[-\frac{1}{\sigma}, \frac{1}{\sigma}\right].$$

It follows that

$$L_mf(x) = \int_{-1/\sigma}^{1/\sigma} \left[f(x) + (-1)^{\sigma+1}\Delta_y^{\sigma} Tf(x)\right] V_m(y)\,dy$$

$$+ \sum_{k=1}^{\sigma} (-1)^{k+1}\binom{\sigma}{k} R_k(x). \qquad (3.15)$$

Let $L_m^{(1)}f$ and $L_m^{(2)}f$ be the first and second terms of this expansion, respectively.

We work on the second term. Let

$$J = \left\{y: \frac{1}{\sigma} \leqslant |y| \leqslant 3\right\} \subseteq [-3,3].$$

Since $\sigma|y|\geqslant 1$ for all $y\in J$, using (3.14) and Minkowski's inequality (see Remark 2.2), we obtain

$$\|R_k\|_{L_p[I]}=\|\int_{J_{x,k}} Tf(x+ky)V_m(y)\,dy\|_{L_p[I]}$$

$$\leqslant \|Tf\|_{L_p[-4,4]}\int_J V_m(y)\,dy\leqslant \|f\|_{L_p[I]}\int_{-3}^{3}(\sigma|y|)^\sigma V_m(y)\,dy$$

$$\leqslant C_8 m^{-\sigma}\|f\|_{L_p[I]}.$$

This proves that

$$\|L_m^{(2)}f\|_{L_p[I]}\leqslant 2^\sigma C_8 m^{-\sigma}\|f\|_{L_p[I]}. \tag{3.16}$$

We turn now to the first term in (3.15). Assume for a moment that $f\in L_p^\sigma[I]$. Then by Theorem 2.70,

$$\|D^\sigma Tf\|_{L_p[-2,2]}\leqslant \|D^\sigma Tf\|_{L_p[-4,4]}\leqslant C_7\|D^\sigma Tf\|_{L_p[I]}.$$

By (3.13),

$$f(x)=\int_{-1/\sigma}^{1/\sigma} f(y)V_m(y)\,dy,$$

and using (2.109) to estimate the σth forward difference, we obtain

$$\|f-L_m^{(1)}\|_{L_p[I]}\leqslant \int_{-1/\sigma}^{1/\sigma}\|\Delta_y^\sigma Tf\|_{L_p[I]}V_m(y)\,dy$$

$$\leqslant \|D^\sigma Tf\|_{L_p[-2,2]}\int_{-3}^{3}|y|^\sigma V_m(y)\,dy\leqslant C_6 m^{-\sigma}\|D^\sigma f\|_{L_p[I]}. \tag{3.17}$$

Combining the estimates (3.16) and (3.17), we have

$$\|f-L_mf\|_{L_p[I]}\leqslant C_5\big[m^{-\sigma}\|f\|_{L_p[I]}+m^{-\sigma}\|D^\sigma f\|_{L_p[I]}\big]$$

for all $f\in L_p^\sigma[I]$. Applying Theorem 2.69, we find that

$$\|f-L_mf\|_{L_p[I]}\leqslant C_4\big[m^{-\sigma}\|f\|_{L_p[I]}+\omega_\sigma(f;m^{-1})_{L_p[I]}\big] \tag{3.18}$$

for all $f\in L_p[I]$.

Statement (3.18) is very nearly the statement of the theorem. It remains to remove the extra term involving $\|f\|_p$ on the right-hand side. By Theorem 3.18 below, there exists a polynomial $q \in \mathcal{P}_\sigma$ such that

$$\|f-q\|_{L_p[I]} \leqslant C_3 \omega_\sigma(f;2)_{L_p[I]} = C_3(2m)^\sigma \omega_\sigma(f;m^{-1})_{L_p[I]}.$$

Now set

$$P_f = L_m(f-q)+q.$$

Then since $\omega_\sigma(f-q;t)=\omega_\sigma(f;t)$ (as $q \in \mathcal{P}_\sigma$), (3.18) implies

$$\|f-P_f\|_{L_p[I]} \leqslant \|(f-q)-L_m(f-q)\|_{L_p[I]}$$

$$\leqslant C_3 m^{-\sigma}\|f-q\|_{L_p[I]} + \omega_\sigma(f-q;m^{-1})_{L_p[I]} \leqslant C_2\omega_\sigma(f;m^{-1})_{L_p[I]}. \tag{3.19}$$

The proof is complete for $I=[-1,1]$.

To get the result for general $[a,b]$, we make a change of variables. Given $f \in L_p[a,b]$, let $F(v)=f([v(b-a)+(b+a)]/2)$. Then $F \in L_p[-1,1]$, and there exists P_F satisfying (3.19). We now define

$$P_f(x) = P_F\left(\frac{2(x-a)-(b-a)}{b-a}\right).$$

This is again a polynomial in $\mathcal{P}_m$, and using Theorem 2.60 on change of variable in a moduli of smoothness, we obtain

$$\|f-P_f\|_{L_p[a,b]} = \|F-P_F\|_{L_p[-1,1]} \leqslant C_2\omega_\sigma(F;m^{-1})_{L_p[-1,1]}$$

$$\leqslant C_2\left(\frac{b-a}{2}\right)^{-1/p}\omega_\sigma\left(f;\frac{b-a}{2m}\right)_{L_p[a,b]}. \quad \blacksquare$$

Although the proof of Theorem 3.12 is mathematically constructive, it is clear that it does not immediately translate into a numerical method for producing polynomial approximations. Indeed, it is not an easy task to design a numerical algorithm realizing the order of approximation guaranteed by Jackson's theorem. Interpolating polynomials certainly do not do the trick (see Section 3.6)—neither do the Bernstein polynomials (see Remark 3.4).

The following simple corollary of Jackson's theorem provides a proof of the Weierstrass approximation theorem.

THEOREM 3.13

Let $1 \leqslant p < \infty$. Then for every $f \in L_p[a,b]$ there exists a sequence of polynomials $P_m f \in \mathcal{P}_m$ with

$$\|f - P_m f\|_{L_p[a,b]} \to 0 \qquad \text{as} \quad m \to \infty.$$

Thus the sequence of spaces $\mathcal{P}_1, \mathcal{P}_2, \ldots$ is dense in $L_p[a,b]$. The same result holds for $p = \infty$ if $f \in C[a,b]$.

Proof. By Jackson's theorem there exist $P_m f \in \mathcal{P}_m$ such that

$$\|f - P_m f\|_p \leqslant C_1 \omega_1\left(f; \frac{b-a}{2m}\right)_p.$$

But as $m \to \infty$ the modulus of continuity goes to zero [cf. (2.121)]. ∎

To emphasize the fact that Jackson's theorem provides *rates* of convergence, we give two simple examples.

EXAMPLE 3.14

If $1 \leqslant p < \infty$, then for all $f \in L_p^\sigma[a,b]$,

$$d(f, \mathcal{P}_m)_p \leqslant C_1 \left(\frac{1}{m}\right)^\sigma \|D^\sigma f\|_{L_p[a,b]}.$$

The same holds with $p = \infty$ if $f \in C[a,b]$.

Discussion. This estimate follows from (3.12) and property (2.120) of the modulus of smoothness. Here the rate of convergence is $(1/m)^\sigma$. By Theorem 2.76, we observe that a better order of approximation cannot be achieved with any other finite dimensional linear spaces. More precisely,

$$d(UL_p^\sigma, \mathcal{P}_m)_p \approx d_m(UL_p^\sigma, L_p) \approx \left(\frac{1}{m}\right)^\sigma$$

so that $\mathcal{P}_1, \mathcal{P}_2, \ldots$ is an asymptotically optimal sequence of spaces for approximating UL_p^σ in L_p. ∎

EXAMPLE 3.15

Let

$$\Lambda_\omega^\sigma = \{f \in C^\sigma[a,b] : \omega(D^\sigma f; t) \leqslant C\omega(t), C < \infty\}, \tag{3.20}$$

where ω is a given continuous function on $[0,b-a]$ with $\omega(0)=0$ and $\omega(t_1+t_2)\leqslant\omega(t_1)+\omega(t_2)$. Then

$$d(f,\mathcal{P}_m)_p\leqslant C_1\left(\frac{1}{m}\right)^{\sigma}\omega\left(\frac{1}{m}\right). \tag{3.21}$$

Discussion. It is known (cf. Lorentz [1966]) that $d_m(\Lambda_\omega^\sigma[a,b],C[a,b])\geqslant C_2\omega(1/m)$, hence

$$d_m(\Lambda_\omega^\sigma[a,b],C[a,b])\approx\left(\frac{1}{m}\right)^{\sigma}\omega\left(\frac{1}{m}\right).$$

Thus the sequence of polynomials $\mathcal{P}_1,\mathcal{P}_2,\ldots$ again provides an asymptotically optimal sequence of approximating spaces. The best known example of Λ_ω^σ is the space $\mathrm{Lip}^{\sigma,\alpha}[a,b]$, where $\omega(t)=t^\alpha$ [cf. (2.128)]. ■

§ 3.5. WHITNEY-TYPE THEOREMS

In the previous section we gave estimates of how well smooth functions can be approximated by polynomials of order m. Our bounds involved the quantity $1/m$. In this section we give a different kind of estimate for $d(f,\mathcal{P}_m)_p$ in which the *length of the interval*, rather than the order of the polynomial, plays the key role. To illustrate the kind of result we are looking for, we begin with two simple examples.

THEOREM 3.16

For all $f\in C[a,b]$,

$$d(f,\mathcal{P}_1)_\infty\leqslant\frac{1}{2}\omega(f;b-a).$$

Proof. To obtain this estimate, we define $q\in\mathcal{P}_1$ by $q=[\max_{a\leqslant x\leqslant b}f(x)-\min_{a\leqslant x\leqslant b}f(x)]/2$. Then for any $a\leqslant x\leqslant b$,

$$|f(x)-q(x)|\leqslant\frac{1}{2}\omega(f;b-a).$$ ■

THEOREM 3.17

For all $f\in C[a,b]$,

$$d(f,\mathcal{P}_2)_\infty\leqslant\omega_2(f;b-a).$$

Proof. Let q be the polynomial in $\mathcal{P}_2$ that interpolates to f at the points a and b. Let

$$M = \|f-q\|_\infty = \max_{a \leqslant x \leqslant b} \delta(x), \qquad \delta(x) = |f(x) - q(x)|.$$

Suppose δ assumes its maximum at η. If η is in the first half of the interval, then with $h = \eta - a$,

$$|f(\eta-h) - 2f(\eta) + f(\eta+h)| = |\delta(\eta-h) - 2\delta(\eta) + \delta(\eta+h)| \geqslant M.$$

This implies that $M \leqslant \omega_2(f;h) \leqslant \omega_2(f;b-a)$. A similar argument can be used when η falls in the second half of the interval $[a,b]$. ■

The following general result of this type is called a Whitney-type theorem:

THEOREM 3.18

Let $1 \leqslant p < \infty$. Then there is a constant C_1 (depending only on p and σ) such that for every $f \in L_p[a,b]$ there exists a polynomial $q \in \mathcal{P}_\sigma$ with

$$\|f-q\|_{L_p[a,b]} \leqslant C_1 \omega_\sigma(f; b-a)_{L_p[a,b]}. \tag{3.22}$$

A similar result holds with $p = \infty$ for all $f \in C[a,b]$.

Proof. We consider only the case of $1 \leqslant p < \infty$. It suffices to establish the result for $I = [a,b] = [-1,1]$, as the general result follows from a change of variables as in the proof of Theorem 3.12. Now, given $f \in L_p[I]$, by Lemma 2.66 and the Whitney Extension Theorem 2.70, we can find $g \in L_p^\sigma[I]$ such that

$$\|f-g\|_{L_p[I]} \leqslant C_4 \omega_\sigma(f;1)_{L_p[I]}$$

and

$$\|D^\sigma g\|_{L_p[I]} \leqslant C_3 \omega_\sigma(f;1)_{L_p[I]}.$$

Now let

$$q(x) = \sum_{j=0}^{\sigma-1} \frac{D^j g(0) x^j}{j!}$$

be the Taylor expansion of g about zero. Then

$$|D^i(g-q)(x)| \leqslant \int_0^x |D^{i+1}(g-q)(u)|\,du \leqslant \|D^{i+1}(g-q)\|_{L_p[I]},$$

and so

$$\|D^i(g-q)\|_{L_p[I]} \leqslant 2^{1/p}\|D^{i+1}(g-q)\|_{L_p[I]}, \qquad i=0,1,\ldots,\sigma-1.$$

This implies

$$\|g-q\|_{L_p[I]} \leqslant 2^{\sigma/p}\|D^\sigma g\|_{L_p[I]} \leqslant C_2\omega_\sigma(f;1)_{L_p[I]},$$

and we conclude that

$$\|f-q\|_{L_p[I]} \leqslant \|f-g\|_{L_p[I]}+\|g-q\|_{L_p[I]} \leqslant C_1\omega_\sigma(f;1)_{L_p[I]}. \qquad \blacksquare$$

Theorem 3.18 was used in the proof of Jackson's theorem on polynomial approximation power in the previous section. In Section 6.4 we will need the following improved version of Theorem 3.18 in which a polynomial is constructed that approximates as in (3.22), and whose derivatives approximate the derivatives of f *simultaneously*.

THEOREM 3.19

Let $1 \leqslant \sigma \leqslant m$. Then there is a constant C_1 (depending only on m) such that for all $f \in C^{\sigma-1}[a,b]$ there exists a polynomial $p_f \in \mathcal{P}_m$ with

$$\|D^j(f-p_f)\|_{L_q[a,b]} \leqslant C_1(b-a)^{\sigma-j-1+1/q}\omega_{m-\sigma+1}(D^{\sigma-1}f;b-a)_{C[a,b]}, \tag{3.23}$$

for $j=0,1,\ldots,\sigma-1$ and $1 \leqslant q \leqslant \infty$.

Proof. Since $D^{\sigma-1}f \in C[a,b]$, by Theorem 3.18 there exists a polynomial $g \in \mathcal{P}_{m-\sigma+1}$ such that

$$\|D^{\sigma-1}f-g\|_\infty \leqslant C_1\omega_{m-\sigma+1}(D^{\sigma-1}f;b-a)_\infty. \tag{3.24}$$

Now define

$$p_f(x) = \sum_{j=0}^{\sigma-2}\frac{D^jf(a)(x-a)^j}{j!} + \int_a^b \frac{(x-y)_+^{\sigma-2}g(y)\,dy}{(\sigma-2)!}.$$

Since $g \in \mathcal{P}_{m-\sigma+1}$, $p_f \in \mathcal{P}_m$ and (cf. Theorem 2.3) $D^{\sigma-1}p_f = g$, while $D^jp_f(a) = D^jf(a)$, $j=0,1,\ldots,\sigma-2$. Thus

$$|D^j(p_f-f)(x)| \leqslant \int_a^b |D^{j+1}(f-p_f)(y)|\,dy \leqslant (b-a)\|D^{j+1}(f-p_f)\|_\infty. \tag{3.25}$$

for $j=0,1,\ldots,\sigma-2$. This implies that

$$\|D^j(f-p_f)\|_\infty \leqslant (b-a)^{\sigma-j-1}\|D^{\sigma-1}(f-p_f)\|_\infty$$

$$=(b-a)^{\sigma-j-1}\|D^{\sigma-1}f-g\|_\infty.$$

Coupling this with (3.24), we obtain (3.23) for $q=\infty$. To obtain it for $1\leqslant q<\infty$, we simply integrate the qth power of (3.23) over $[a,b]$ and then take the qth root of both sides. ■

There is also a version of Theorem 3.19 for L_p^σ functions.

THEOREM 3.20

Let $1\leqslant p\leqslant\infty$ and $1\leqslant\sigma\leqslant m$. Then there is a constant C_1 (depending only on m and p) such that for every $f\in L_p^\sigma[a,b]$ there exists a polynomial $p_f\in\mathcal{P}_m$ with

$$\|D^j(f-p_f)\|_{L_q[a,b]}\leqslant C_1(b-a)^{\sigma-j+1/q-1/p}\omega_{m-\sigma}(D^\sigma f;b-a)_{L_p[a,b]} \quad (3.26)$$

for $j=0,1,\ldots,\sigma-1$ and all $1\leqslant q\leqslant\infty$.

Proof. Since $D^\sigma f\in L_p[a,b]$, by Theorem 3.18 we can find $g\in\mathcal{P}_{m-\sigma}$ such that

$$\|D^\sigma f-g\|_p\leqslant C_1\omega_{m-\sigma}(D^\sigma f;b-a)_p. \quad (3.27)$$

We now define

$$p_f(x)=\sum_{j=0}^{\sigma-1}\frac{D^jf(a)(x-a)^j}{j!}+\int_a^b\frac{(x-y)_+^{\sigma-1}g(y)\,dy}{(\sigma-1)!}.$$

Then $p_f\in\mathcal{P}_m$ and $D^\sigma p_f=g$, while $D^jp_f(a)=D^jf(a)$, $j=0,1,\ldots,\sigma-1$. Applying Hölder's inequality to the integral in (3.25), we obtain

$$|D^j(f-p_f)(x)|\leqslant(b-a)^{1-1/q}\|D^{j+1}(f-p_f)\|_q.$$

Integrating this over the interval $[a,b]$, we obtain

$$\|D^j(f-p_f)\|_q\leqslant(b-a)\|D^{j+1}(f-p_f)\|_q$$

for $j=0,1,\ldots,\sigma-2$. A similar argument with Hölder's inequality shows that

$$\|D^{\sigma-1}(f-p_f)\|_q \leqslant (b-a)^{1-1/p+1/q}\|D^{\sigma}(f-p_f)\|_p.$$

Combining these estimates with (3.27), we obtain (3.26). ■

§ 3.6. THE INFLEXIBILITY OF POLYNOMIALS

In Section 1.2 we asserted that the main defect of polynomials for approximation purposes is their relative inflexibility. In this section we illustrate how this inflexibility manifests itself, and give some indication of what the underlying causes are.

We begin with an example. Suppose we want to approximate the function $f(x)=1/(1+x^2)$ on the interval $[-5,5]$ by polynomials. One natural approach is to choose m points in the interval, and to interpolate at these points. Suppose we take equally spaced points; that is,

$$t_i=-5+10\frac{(i-1)}{(m-1)}, \qquad i=1,2,\ldots,m. \tag{3.28}$$

Then by Theorem 3.6 there exists a unique polynomial $L_m f$ in $\mathcal{P}_m$ which interpolates f at the points $\{t_i\}_1^m$. We graph f and $L_m f$ for $m=5$ and $m=15$ in Figure 5.

Figure 5 shows that the polynomial $L_{15}f$ does a much better job than $L_5 f$ in the middle part of the interval $[-5,5]$, but it is much worse at the ends. It seems reasonable to hope, however, that if m is increased (so that $L_m f$ and f match at more and more points), $L_m f$ will approximate f well

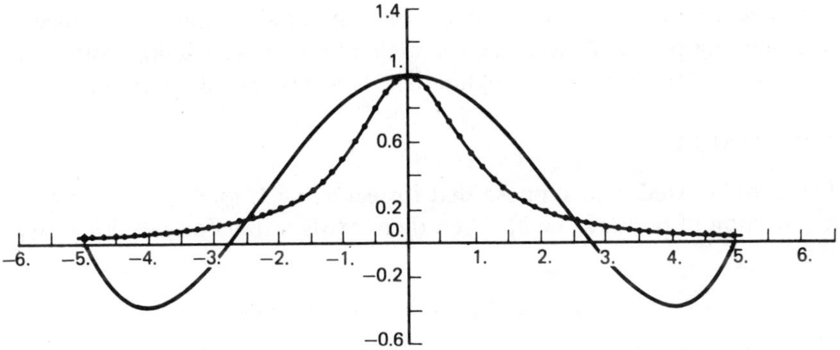

Figure 5a. Lagrange interpolation of $1/(1+x^2)$ at five equidistant points on $[-5,5]$.

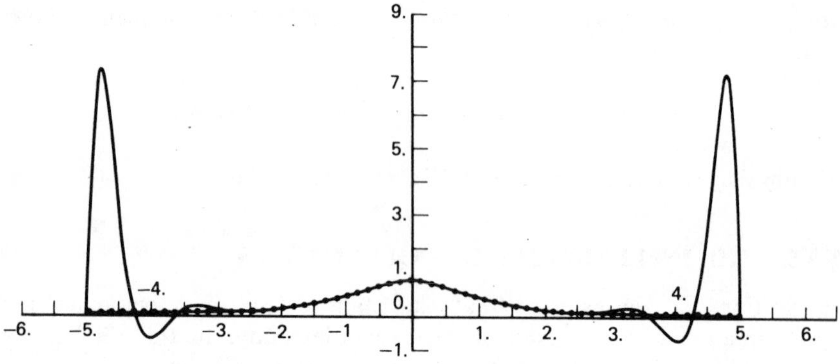

Figure 5b. Lagrange interpolation of $1/(1+x^2)$ at 15 equidistant points on $[-5,5]$.

throughout $[-5,5]$. The following theorem shows that this does not happen:

THEOREM 3.21. Runge

Let $f(x)=1/(1+x^2)$ on $[-5,5]$, and let $L_m f$ be the polynomial of order m that interpolates f at m equally spaced points as in (3.28). Then

$$|f(x)-L_m f(x)|\to\infty \quad \text{as} \quad m\to\infty \quad \text{for} \quad |x|>3.64\,.$$

Discussion. See Runge [1901] (or e.g., Isaacson and Keller [1966], p. 275). The nonconvergence here is certainly not due to a lack of smoothness of f since f is clearly infinitely often differentiable on $\mathbb{R}$. ■

Theorem 3.21 shows that sequences of polynomials interpolating at equally spaced points may not converge. The following theorem shows that convergence cannot be guaranteed for any predetermined sequence of interpolating points. (There are some choices of interpolating points that do quite well for low-order polynomials, however. See Remark 3.5.)

THEOREM 3.22

Let $[a,b]$ be fixed, and suppose that for each $m\geqslant 1$, $t_{m1}<t_{m2}<\cdots<t_{mm}$ is a collection of points in $[a,b]$. Then there exists a function $f\in C[a,b]$ such that

$$\|f-L_m f\|_\infty\to\infty \quad \text{as} \quad m\to\infty,$$

where $L_m f$ is the unique polynomial of order m interpolating f at $t_{m1},\ldots,t_{mm}$.

Discussion. The original proof of this result is credited to Faber [1914]. For a modern proof using the uniform boundedness principle, see Cheney [1966], page 215. While this result shows that we cannot make interpolating polynomials converge to arbitrary continuous functions no matter how we select the interpolation points, it is possible to establish convergence results on special sets of interpolation points for smoother classes of functions. ■

The nonconvergence phenomenon in Theorems 3.21 and 3.22 can be regarded as a manifestation of the inflexibility of polynomials. If polynomials are forced to follow a curve in one interval, they may respond by oscillating wildly elsewhere. This tendency to oscillate becomes increasingly pronounced as the order of the polynomial is increased. For low orders (say at most 5 or 10) it may be acceptable. Unfortunately, in order to achieve suitable accuracy we usually have to use fairly high-order polynomials (cf. Jackson's Theorem 3.12).

One way of explaining the tendency of polynomials to oscillate is the following observation: The coefficients of the derivative of a high-order polynomial will be much larger, in general, than the coefficients of the original polynomial. Indeed, the coefficient of x^{m-2} in Dp is $(m-1)a_m$, where a_m is the coefficient of x^{m-1} in p. This implies the likelihood of steep derivatives.

Another factor in the inflexibility of polynomials is deeply rooted in one of their most conspicuous properties, heralded earlier as a virtue: polynomials are smooth. In fact, polynomials are too smooth. As a function of a complex variable, they are analytic, which means that their values everywhere in the complex plane are determined by their value in any arbitrarily small set. We may establish an assertion of this type on $\mathbf{R}$ without reference to complex variables as follows:

THEOREM 3.23

Let $p \in \mathcal{P}_m$. Then $p(x)$ is determined for all $x \in \mathbf{R}$ by its values in any interval (c,d), no matter how small.

Proof. Given any points $c < t_1 < t_2 < \cdots < t_m < d$, p is uniquely determined by its values at these m points. ■

§ 3.7. HISTORICAL NOTES

Section 3.1

The material here is classical, and can be found in most books on approximation theory or numerical analysis.

Section 3.2

The results on zeros, VanderMonde matrices, and Hermite interpolation can be found in practically any book on numerical analysis. The result of Theorem 3.7 on subdeterminants of the VanderMonde is probably less well known in such circles than it should be. The polynomials are utilized as prime examples of various forms of T-systems in the books by Karlin and Studden [1966] and Karlin [1968]. The method of proof of Theorem 3.7 follows ideas used by Karlin [1968].

Section 3.3

Descartes' rule of signs for polynomials can be proved in a variety of ways. For example, a different proof is given by Karlin [1968], page 317. The Budan-Fourier Theorem 3.9 also has a number of different proofs. Karlin [1968], page 316, gives a proof based on Descarte's rule of signs. The inductive method presented here was developed for splines; see Schumaker [1976c]. Still another proof can be found in the article by Melkman [1974a].

Section 3.4

Jackson-type theorems have a long and rich history. It began with the proof of Theorem 3.12 for $p=\infty$ and $\sigma=1$ by Jackson in 1913. His proof relied on converting the problem to one involving approximation of periodic functions by trigonometric polynomials. Later, several other authors gave direct constructive proofs for the case of $p=\infty$ and for all $\sigma \geqslant 1$. The theorem for $p<\infty$ seems to be of more recent vintage. Potapov [1956, 1961] considered the case of $\sigma=1$, $1 \leqslant p<\infty$. Bak and Newman [1972] also did this case independently. The complete result as presented here is credited to DeVore [1976], and we have followed his proof. For another proof, see Oswald [1978]. Recently there has also been considerable interest in bounds for approximation of monotone functions by monotone polynomials; see, for example, DeVore [1977a] and references therein.

Section 3.5

The first Whitney-type approximation theorem was given by Whitney [1957], where the case of $p=\infty$ and $f \in C[a,b]$ were considered. His proof delivers estimates for the size of the constants. The result for $1 \leqslant p<\infty$ is credited to DeVore [1976], and we have followed his proof; see also Brudnyi [1964].

Section 3.6

The observation that interpolating polynomials do not always converge to the function being interpolated was made early by Meray [1884, 1896]. The example given here is taken from the work of Runge [1901]. These results led to a quest for conditions on the interpolating points or on the functions to be interpolated which suffice to guarantee convergence. A host of both positive and negative results are known, and the problem remains an active research area.

§ 3.8. REMARKS

Remark 3.1

The following generalization of Horner's scheme computes the derivatives of a polynomial at a given point.

ALGORITHM 3.24. Synthetic Division—To Compute $D^j p(t)/j!$, $j=0,1,\ldots,d-1$ for a Given Polynomial $p(x)=\sum_{i=1}^{m} c_i (x-a)^{i-1}$)

1. $u \leftarrow t-a$;
2. For $j \leftarrow 1$ step 1 until d
   For $i \leftarrow m-1$ step -1 until j
   $c(i) \leftarrow c(i+1)\cdot u + c(i)$.

Discussion. Upon completion of this algorithm, we have $c(i) = D^{i-1}p(t)/(i-1)!$, $i=1,2,\ldots,d$. If it is run with $d=m-1$, it produces all derivatives of p at t. For a proof, see the standard numerical analysis texts (for example, Conte and deBoor [1972]). ∎

Remark 3.2

Let $A=(A_{ij})_{i=1,j=1}^{r,\ r+1}$ be any r by $r+1$ matrix of numbers. Given any

$$1 \leqslant i_1 < \cdots < i_p \leqslant r \quad \text{and} \quad 1 \leqslant j_1 < \cdots < j_p \leqslant r+1,$$

we define an *associated minor of* A by

$$A\begin{pmatrix} i_1,\ldots,i_p \\ j_1,\ldots,j_p \end{pmatrix} = \det\left(A_{i_\nu j_\mu}\right)_{\nu=1,\mu=1}^{p\quad p}.$$

It is shown in the book of Karlin [1968], page 8, that

$$A\begin{pmatrix} 2,\dots,r-1,r+1 \\ 1,\dots,r-1 \end{pmatrix} A\begin{pmatrix} 1,\dots,r \\ 1,\dots,r \end{pmatrix} = A\begin{pmatrix} 2,\dots,r \\ 1,\dots,r-1 \end{pmatrix} A\begin{pmatrix} 1,\dots,r-1,r+1 \\ 1,\dots,r \end{pmatrix}$$
$$-A\begin{pmatrix} 1,\dots,r-1 \\ 1,\dots,r-1 \end{pmatrix} A\begin{pmatrix} 2,\dots,r+1 \\ 1,\dots,r \end{pmatrix}.$$

If A is $r+1$ by r instead, then the same result holds with the rows and columns interchanged.

Remark 3.3

We sketch here the construction of the polynomials V_m used in the proof of Jackson's Theorem 3.12. Let $n=4k$, and let P_{2k} be the Legendre polynomial of degree $2k$ defined on $[-1,1]$ (see e.g., Szegö [1939]). Then P_{2k} has $2k$ zeros

$$-1 < -x_k^{(k)} < \cdots < -x_1^{(k)} < 0 < x_1^{(k)} < \cdots < x_k^{(k)} < 1.$$

Given $1 \leqslant \sigma \leqslant k$, define

$$W_n(x) = c_n \left[\frac{p_{2k}(x)}{(x-x_1^{(k)})(x+x_1^{(k)})\cdots(x-x_{(k)}^{(k)})(x+x_{(k)}^{(k)})} \right]^2,$$

where c_n is chosen so that $\int_{-1}^{1} w_n(x)\,dx = 1$. Clearly $W_n(x) \geqslant 0$. Using the Gauss-quadrature rule and some properties on the location of the zeros of the Legendre polynomial, it can be shown that

$$\int_{-1}^{1} |x|^\sigma W_n(x)\,dx \leqslant C_1 n^{-\sigma},$$

with a constant C_1 dependent only on σ. For arbitrary $n \geqslant 4$, we define $W_n = W_{\lfloor n/4 \rfloor}$, where as usual $\lfloor n/4 \rfloor$ denotes the biggest integer in $n/4$. The desired V_n are obtained after a change of variable from the interval $[-1,1]$ to $[a,b]$. For more details, see DeVore [1968, 1976].

Remark 3.4

By the error bound in Remark 2.5, the Bernstein polynomials introduced there can be used to provide a proof of Weierstrass' Approximation Theorem 3.11. However, since B_m is a positive linear operator, there are saturation results (cf. Lorentz [1966] or DeVore [1972]) which assert that no matter how smooth a function may be, $\|f - B_m f\|$ cannot go to zero faster than $1/m$ unless $f \in \mathcal{P}_m$ (i.e., f is a polynomial itself). Thus the Bernstein polynomials do not produce good enough approximations to be used to prove Jackson's theorem.

Remark 3.5

Despite the negative nature of Theorem 3.22, there is at least one set of interpolation points that produces relatively good approximations for polynomials of low order. For example, it is known that for $m \leqslant 20$, interpolation at the zeros of the Tchebycheff polynomial, given by

$$t_j = \frac{\left[a+b-(a-b)\cos((2j-1)\pi/2m)\right]}{2}, \qquad j=1,2,\ldots,m,$$

produces a polynomial whose deviation from f (in the uniform norm) is no more than four times as great as the error obtained by using the best approximating polynomial. For more on this point, see deBoor [1978].

4

POLYNOMIAL SPLINES

In Chapter 1 we discussed the need for convenient classes of functions for approximation purposes. Our discussion of polynomials and their properties led us to the conclusion that spaces of smooth piecewise polynomials should be useful for approximation purposes. In this chapter we present the basic properties of such spaces.

§ 4.1. BASIC PROPERTIES

Let $[a,b]$ be a finite closed interval, and let

$$\Delta=\{x_i\}_1^k \qquad \text{with } a=x_0<x_1<\cdots<x_k<x_{k+1}=b$$

be a partition of it into k subintervals

$$I_i=[x_i,x_{i+1}), \qquad i=0,1,\ldots,k-1 \text{ and } I_k=[x_k,x_{k+1}].$$

Let m be a positive integer, and let $\mathfrak{M}=(m_1,\ldots,m_k)$ be a vector of integers with $1\leqslant m_i\leqslant m$, $i=1,2,\ldots,k$.

DEFINITION 4.1. Polynomial Splines

We call the space

$$\begin{aligned}\mathcal{S}(\mathcal{P}_m;\mathfrak{M};\Delta)=\{&s\text{: there exist polynomials } s_0,\ldots,s_k \\ &\text{in } \mathcal{P}_m \text{ such that } s(x)=s_i(x) \text{ for } x\in I_i,\ i=0,1,\ldots,k, \\ &\text{and } D^js_{i-1}(x_i)=D^js_i(x_i) \text{ for } j=0,1,\ldots,m-1-m_i, \\ &i=1,\ldots,k\}\end{aligned} \tag{4.1}$$

the *space of polynomial splines of order m with knots $x_1,\ldots,x_k$ of multiplicities $m_1,\ldots,m_k$.*

We call $\mathfrak{M}$ the *multiplicity vector*. It controls the nature of the space $\mathcal{S}(\mathcal{P}_m;\mathfrak{M};\Delta)$ by controlling the smoothness of the splines at the knots. If

$m_i = m$, we interpret the definition to mean that the two polynomial pieces s_{i-1} and s_i in the intervals adjoining the knot x_i are unrelated to each other (and thus there may be a jump discontinuity at x_i). If $m_i < m$, then we have forced these two polynomial pieces to tie together smoothly in the sense that the spline s and its first $m-1-m_i$ derivatives are all continuous across the knot.

To illustrate the nature of the spline space $\mathcal{S}(\mathcal{P}_m; \mathfrak{M}; \Delta)$ as we vary $\mathfrak{M}$, we give two examples.

EXAMPLE 4.2

If $\mathfrak{M} = (m, m, \ldots, m)$, then

$$\mathcal{S}(\mathcal{P}_m; \mathfrak{M}; \Delta) = \mathcal{P}\mathcal{P}_m(\Delta),$$

where $\mathcal{P}\mathcal{P}_m(\Delta)$ is the space of piecewise polynomials (cf. Definition 1.1).

Discussion. This space contains functions with possible jump discontinuities at the knots. It is the least smooth of the spline spaces. ■

EXAMPLE 4.3

Let $\mathfrak{M} = (1, 1, \ldots, 1)$. Then

$$\mathcal{S}(\mathcal{P}_m; \mathfrak{M}; \Delta) = \mathcal{S}_m(\Delta),$$

where $\mathcal{S}_m(\Delta)$ is the space of splines of order m with *simple knots* (compare Definition 1.2).

Discussion. This space of splines is a subset of $C^{m-2}[a,b]$. It is the smoothest space of piecewise polynomials of order m with genuine knots at the points $x_1, \ldots, x_k$. (If we tried to make the pieces join together any smoother, the knots would disappear.) ■

In most practical applications, the natural setting for approximation problems is a closed interval $[a,b]$, and thus we have elected to define splines on such intervals. On the other hand, every spline has a natural extension to the whole real line. Indeed, if $s \in \mathcal{S}(\mathcal{P}_m; \mathfrak{M}; \Delta)$, then we define

$$s(x) = \begin{cases} s_0(x), & x < a \\ s_k(x), & x > b, \end{cases} \tag{4.2}$$

where s_0 and s_k are the polynomials defining s in the intervals I_0 and I_k, respectively [cf. (4.1)].

We now show that $\mathcal{S}(\mathcal{P}_m;\mathfrak{M};\Delta)$ is a finite dimensional linear space, and we give a basis for it. First we identify the dimension of $\mathcal{S}$.

THEOREM 4.4

Let $K=\sum_{i=1}^k m_i$. Then

$$\mathcal{S}(\mathcal{P}_m;\mathfrak{M};\Delta) \qquad \text{is a linear space of dimension } m+K.$$

Proof. It is clear from the definition of $\mathcal{S}$ that it is a linear space. Let s be a typical element in $\mathcal{S}$, and suppose that $s_0,s_1,\dots,s_k$ are the associated polynomial pieces on the intervals $I_0,I_1,\dots,I_k$, respectively. We can write each such polynomial in the form

$$s_i(x)=\sum_{j=1}^m c_{ij}\frac{x^{j-1}}{(j-1)!}.$$

For each $i=0,1,\dots,k$, let $\mathbf{c}_i=(c_{i1},\dots,c_{im})^T$. Then the continuity conditions on s can be written as the following linear system of equations:

$$A\mathbf{c}=\begin{bmatrix} A_1 & -A_1 & & & & \\ & A_2 & -A_2 & & & \\ & & A_3 & -A_3 & & \\ & & & \ddots & & \\ & & & & A_k & -A_k \end{bmatrix}\begin{bmatrix}\mathbf{c}_0\\ \mathbf{c}_1\\ \vdots \\ \mathbf{c}_k\end{bmatrix}=\mathbf{0},$$

where

$$A_i=\begin{bmatrix} 1 & x_i & x_i^2/2 & & \cdots & x_i^{m-1}/(m-1)! \\ 0 & 1 & x_i & & \cdots & x_i^{m-2}/(m-2)! \\ \cdots & & & & & \\ 0 & 0 & \cdots & 1 & \cdots & x_i^{m_i}/m_i! \end{bmatrix}.$$

The matrix A_i is of size $m-m_i$ by m. It is clear that the matrix A has rank equal to the number of its rows, namely $\sum_{i=1}^k(m-m_i)$. Since A is a transformation from the $m(k+1)$ dimensional vector space $\mathbf{R}^{m(k+1)}$ into $\mathbf{R}^{\sum_1^k(m-m_i)}$, it follows that the linear space of all vectors $\mathbf{c}$ which satisfy $A\mathbf{c}=\mathbf{0}$ is of dimension

$$m(k+1)-\sum_{i=1}^k(m-m_i)=m+\sum_{i=1}^k m_i=m+K.$$

This is, of course, also the dimension of $\mathcal{S}$. ■

In the proof of Theorem 4.4 we have counted the number of free parameters in the piecewise polynomial representation of elements in $\mathcal{S}$, and subtracted the number of conditions on these polynomial pieces to obtain the dimension. The introduction of the matrices A_i was just to check that all the conditions are independent of each other.

Now that we know the dimension of $\mathcal{S}(\mathcal{P}_m;\mathfrak{M};\Delta)$, we may construct a basis for it. Since obviously

$$\mathcal{P}_m \subseteq \mathcal{S}(\mathcal{P}_m;\mathfrak{M};\Delta),$$

any basis for $\mathcal{S}$ must include at least a basis for $\mathcal{P}_m$. We may take the functions $1, x-a, \ldots, (x-a)^{m-1}$. To find an additional K basis elements, we recall the notation

$$(x-y)_+^j = (x-y)^j (x-y)_+^0, \qquad j>0,$$

where

$$(x-y)_+^0 = \begin{cases} 0, & x<y \\ 1, & x \geqslant y. \end{cases}$$

(We have already encountered these functions in the Taylor expansion, and as Green's functions for initial-value problems; see Section 2.2.)

THEOREM 4.5

A basis for $\mathcal{S}(\mathcal{P}_m;\mathfrak{M};\Delta)$ is given by

$$\left\{\rho_{i,j}(x) = (x-x_i)_+^{m-j}\right\}_{j=1,i=0}^{m_i \quad k}, \tag{4.3}$$

where $x_0 = a$ and $m_0 = m$.

Proof. By definition, $(x-x_i)_+^j$ is identically 0 for $x<x_i$, and is a polynomial of degree j for $x \geqslant x_i$. Since

$$D^\nu (x-x_i)_+^j \big|_{x=x_i} = 0, \qquad \nu = 0,1,\ldots,j-1,$$

it follows that each of the functions in (4.3) at least belongs to $\mathcal{S}$. Since there are precisely $m+K$ of them, it remains only to check that they are linearly independent.

Suppose that for some set of coefficients $\{c_{ij}\}$, the spline

$$s(x) = \sum_{i=0}^{k} \sum_{j=1}^{m_i} c_{ij} (x-x_i)_+^j \tag{4.4}$$

is identically zero on the interval $[a,b]$. Then for $x \in I_0$ we have $\sum_{j=1}^{m} c_{0j}(x - x_0)^{m-j} = 0$, and by the linear independence of these functions (cf. Theorem 3.1), $c_{01} = \cdots = c_{0m} = 0$. But then for $x \in I_1$, $s \equiv 0$ implies $\sum_{j=1}^{m_1} c_{1j}(x - x_1)^{m-j} = 0$. These are again linearly independent polynomials, so these coefficients must also be zero. This process can be continued moving one interval to the right at a time to show that all of the c's are zero, and the desired linear independence is established. ■

Theorem 4.5 shows that every spline $s \in \mathcal{S}(\mathcal{P}_m; \mathfrak{M}; \Delta)$ has *a unique* representation in the form of (4.4). We call the basis in (4.3) a *one-sided basis* for $\mathcal{S}$. While this basis is useful for theoretical purposes, it is not well suited for numerical applications. For example, to evaluate $s(x)$ for x near the right end of the interval $[a,b]$, it is necessary to evaluate all of the basis elements and compute the entire sum. In the next section we shall show that it is possible to define a more local and more symmetric basis for $\mathcal{S}$, which is very important for numerical applications.

§ 4.2. CONSTRUCTION OF A LOCAL BASIS

To motivate what we want to do in this section, consider the following example:

EXAMPLE 4.6

Let $[a,b] = [0,5]$ and $\Delta = \{1,2,3,4\}$. Find a basis for $\mathcal{S}_2(\Delta)$.

Discussion. The basis of one-sided functions for this space is given by the functions (see Figure 6)

$$1,\ x,\ (x-1)_+,\ (x-2)_+,\ (x-3)_+,\ (x-4)_+. \tag{4.5}$$

On the other hand, it is clear that the "hat" functions $B_1, \ldots, B_6$ shown in Figure 6 are also linearly independent members of $\mathcal{S}_2(\Delta)$, hence they also form a basis. Each of these functions is nonzero only on a relatively small set, and the basis has considerably more symmetry. ■

Given a function f, we define its *support* to be the set

$$\text{support}(f) = \text{closure}\{x : f(x) \neq 0\}. \tag{4.6}$$

Example 4.6 suggests that it may be possible to construct a basis for $\mathcal{S}(\mathcal{P}_m; \mathfrak{M}; \Delta)$ consisting of splines with relatively small supports. Clearly, any such basis must be constructed from linear combinations of the one-sided basis elements in Theorem 4.5. The following lemma deals with

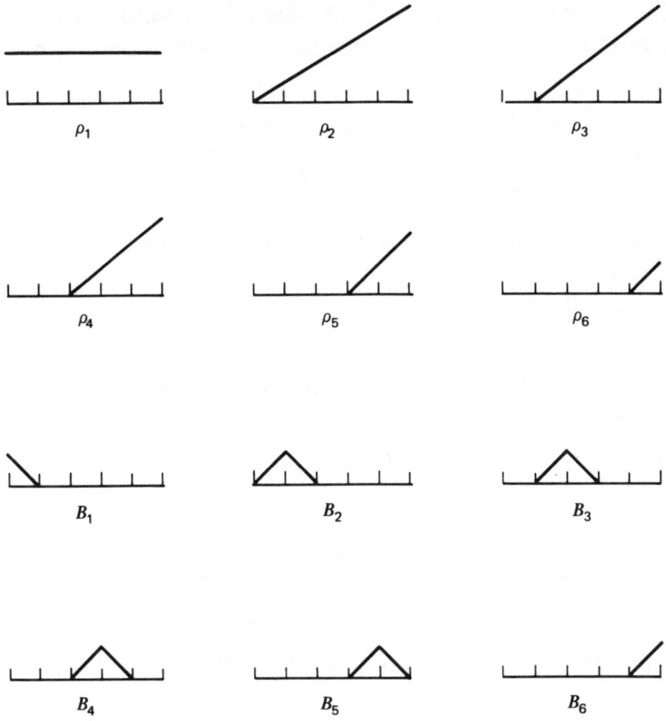

Figure 6. Local versus one-sided bases for $\mathbb{S}_2(\Delta)$.

the question of when it is possible for a linear combination of such functions to vanish outside of a finite interval:

LEMMA 4.7

Let $\tau_1 < \tau_2 < \cdots < \tau_d$ and $1 \leqslant l_i \leqslant m$, $i = 1, 2, \ldots, d$ be given. Then if $\sum_{i=1}^d l_i > m$, there exist $\{\alpha_{ij}\}_{i=1,j=1}^{d \;\; l_i}$, not all 0, so that

$$B(x) = \sum_{i=1}^{d} \sum_{j=1}^{l_i} \alpha_{ij} \frac{(x-\tau_i)_+^{m-j}}{(m-j)!}$$

satisfies

$$B(x) = 0 \qquad \text{for } x < \tau_1 \text{ and } x > \tau_d.$$

On the other hand, if $\sum_{i=1}^d l_i \leqslant m$, then no such nontrivial B exists.

Proof. By the one-sided nature of the plus functions, it is clear that for any $\{\alpha_{ij}\}$, $B(x)=0$ for $x<\tau_1$. If $B(x)$ is to be zero for $x>\tau_d$, then we must require

$$B(x)=\sum_{i=1}^{d}\sum_{j=1}^{l_i}\alpha_{ij}\sum_{\nu=0}^{m-j}\frac{x^\nu}{\nu!}\frac{(-\tau_i)^{m-j-\nu}}{(m-j-\nu)!}=0, \qquad \text{all } x>\tau_d$$

(where we have used the binomial theorem to expand $(x-\tau_i)^{m-j}$). Setting

$$\gamma_{ij\nu}=\begin{cases}\dfrac{(-\tau_i)^{m-j-\nu}}{(m-j-\nu)!}, & \nu=0,\ldots,m-j\\ 0, & \nu=m-j+1,\ldots,m-1,\end{cases}$$

we can rewrite this set of equations as

$$\sum_{i=1}^{d}\sum_{j=1}^{l_i}\alpha_{ij}\sum_{\nu=0}^{m-1}\frac{x^\nu}{\nu!}\gamma_{ij\nu}=\sum_{\nu=0}^{m-1}\frac{x^\nu}{\nu!}\sum_{i=1}^{d}\sum_{j=1}^{l_i}\alpha_{ij}\gamma_{ij\nu}=0.$$

Since the powers $1,x,\ldots,x^{m-1}$ are linearly independent, this is equivalent to

$$\sum_{i=1}^{d}\sum_{j=1}^{l_i}\alpha_{ij}\gamma_{ij\nu}=0, \qquad \nu=0,1,\ldots,m-1. \tag{4.7}$$

This is a homogeneous system of m equations for the $\sum_{i=1}^{d}l_i$ coefficients, and thus it always has a nontrivial solution if $\sum_{i=1}^{d}l_i>m$.

It is of interest to examine the system (4.7) in more detail. Writing the equations in the order $\nu=m-1,\ldots,0$, we can write them in matrix form as

$$[M_1\cdots M_d]\begin{bmatrix}\alpha_{11}\\ \vdots\\ \alpha_{1l_1}\\ \vdots\\ \alpha_{d1}\\ \vdots\\ \alpha_{dl_d}\end{bmatrix}=0$$

where for $i=1,2,\ldots,d$,

$$M_i = \begin{bmatrix} 1 & 0 & \cdots & 0 \\ -\tau_i & 1 & & \\ & & \ddots & \\ & & & 1 \\ \frac{(-\tau_i)^{m-2}}{(m-2)!} & & & \\ \frac{(-\tau_i)^{m-1}}{(m-1)!} & \frac{(-\tau_i)^{m-2}}{(m-2)!} & \cdots & \frac{(-\tau_i)^{m-l_i}}{(m-l_i)!} \end{bmatrix}$$

Suppose now that $\sum_{i=1}^{d} l_i \leqslant m$. Then the square matrix in the first $\sum_{i=1}^{d} l_i$ rows of this matrix is a nonzero multiple of the VanderMonde matrix, and so it is of full rank. It follows that in this case the only solution to (4.7) is the trivial one with all α's equal to zero. ■

Lemma 4.7 shows that in order to construct a linear combination of the plus functions which vanishes outside of a finite interval, we must have $\sum_{i=1}^{d} l_i \geqslant m+1$. We now consider the case where this sum is exactly $m+1$ in more detail. In this case we have one more unknown than equations, and by a result of linear algebra, $B(x)$ must be given by

$$B(x) = C_1 \det \begin{bmatrix} 1 \quad 0 & \cdots & 0 & \cdots & 1 \quad 0 & \cdots & 0 \\ \tau_1 \quad 1 & \cdots & 0 & \cdots & \tau_d \quad 1 & \cdots & 0 \\ \frac{\tau_1^{l_1-1}}{(l_1-1)!} & \cdots & 1 & & \frac{\tau_d^{l_d-1}}{(l_d-1)!} & \cdots & 1 \\ \vdots & & & & \vdots & & \\ \frac{\tau_1^{m-1}}{(m-1)!} & & & & \frac{\tau_d^{m-1}}{(m-1)!} & & \\ \frac{(x-\tau_1)_+^{m-1}}{(m-1)!} & \cdots & \frac{(x-\tau_1)^{m-l_1}}{(m-l_1)!} & \cdots & \frac{(x-\tau_d)_+^{m-1}}{(m-1)!} & \cdots & \frac{(x-\tau_d)_+^{m-l_d}}{(m-l_d)!} \end{bmatrix}$$

where C_1 is a nonzero constant. If we compare this with the definition of the divided difference [see (2.86)], we see that (except for a constant multiplier) $B(x)$ is the $m+1$st divided difference of the function $(x-y)_+^{m-1}$ taken over the points

$$\overbrace{\tau_1,\ldots,\tau_1}^{l_1} \ldots \overbrace{\tau_d,\ldots,\tau_d}^{l_d}.$$

This observation suggests that in order to construct local support basis elements for the space $\mathcal{S}(\mathcal{P}_m;\mathfrak{M};\Delta)$, we should work with divided differences of the function $(x-y)_+^{m-1}$. We do exactly that in Theorem 4.9 below. First we state another definition.

DEFINITION 4.8. Extended Partition

Let $a<x_1<x_2<\cdots<x_k<b$ and $1\leqslant m_i\leqslant m$, $i=1,2,\ldots,k$ be given. Suppose

$$y_1\leqslant y_2\leqslant\cdots\leqslant y_{2m+K}$$

is such that

$$y_1\leqslant\cdots\leqslant y_m\leqslant a,\qquad b\leqslant y_{m+K+1}\leqslant\cdots\leqslant y_{2m+K} \tag{4.8}$$

and

$$y_{m+1}\leqslant\cdots\leqslant y_{m+K}=\overbrace{x_1,\ldots,x_1}^{m_1},\ldots,\overbrace{x_k,\ldots,x_k}^{m_k}.$$

Then we call $\tilde{\Delta}=\{y_i\}_1^{2m+K}$ an *extended partition associated with* $\mathcal{S}(\mathcal{P}_m;\mathfrak{M};\Delta)$.

We note that the points $\{y_i\}_{i=m+1}^{m+K}$ in an extended partition $\tilde{\Delta}$ associated with $\mathcal{S}(\mathcal{P}_m;\mathfrak{M};\Delta)$ are uniquely determined. The first and last m points in $\tilde{\Delta}$ can be chosen arbitrarily, subject to (4.8).

THEOREM 4.9

Let $\tilde{\Delta}=\{y_i\}_1^{2m+K}$ be an extended partition associated with $\mathcal{S}(\mathcal{P}_m;\mathfrak{M};\Delta)$, and suppose $b<y_{2m+K}$. For $i=1,2,\ldots,m+K$, let

$$B_i(x)=(-1)^m(y_{i+m}-y_i)[y_i,\ldots,y_{i+m}](x-y)_+^{m-1},\qquad a\leqslant x\leqslant b. \tag{4.9}$$

Then $\{B_i\}_1^{m+K}$ form a basis for $\mathcal{S}(\mathcal{P}_m;\mathfrak{M};\Delta)$ with

$$B_i(x)=0\qquad\text{for } x\notin[y_i,y_{i+m}] \tag{4.10}$$

and

$$B_i(x)>0\qquad\text{for } x\in(y_i,y_{i+m}). \tag{4.11}$$

Moreover,

$$\sum_{i=1}^{m+K}B_i(x)=1\qquad\text{for all } a\leqslant x\leqslant b, \tag{4.12}$$

and

$$S^{-}\left(\sum_{i=1}^{m+K} c_i B_i\right) \leqslant S^{-}(c_1,\ldots,c_{m+K}), \qquad \text{any } c_1,\ldots,c_{m+K} \text{ not all 0.}$$

(4.13)

Discussion. The fact that $B_1,\ldots,B_{m+K}$ are splines in the space $\mathcal{S}(\mathcal{P}_m;\mathfrak{M};\Delta)$ is shown in Theorem 4.14, while their linear independence follows from Theorem 4.18. The properties (4.10) and (4.11) are discussed in Theorem 4.17, while (4.12) is dealt with in Theorem 4.20. The variation-diminishing property (4.13) follows from Theorem 2.42 and the fact (shown in Theorem 4.65 below) that $\{B_i\}_1^{m+K}$ is an Order-Complete Weak Tchebycheff system. ■

The $B_1,\ldots,B_{m+K}$ defined in Theorem 4.9 are called *B-splines.* We discuss them in great detail in the following section. The use of B-splines in numerical computation is described in Chapter 5. We have included the factor $(-1)^m$ in the definition of the B-splines in order to make them positive (where they are nonzero). The factor $(y_{i+m}-y_i)$ is a normalization factor designed to produce the identity (4.12) which asserts that the B-splines form a *partition of unity*. Property (4.13) is called the *variation-diminishing property* of B-splines.

The following corollary of Theorem 4.9 deals with the case where the extended partition $\tilde{\Delta}$ is chosen with

$$b = y_{m+K+1} = \cdots = y_{2m+K}. \tag{4.14}$$

This is fairly common practice in applications.

COROLLARY 4.10

Suppose $\tilde{\Delta}$ is an extended partition associated with $\mathcal{S}(\mathcal{P}_m;\mathfrak{M};\Delta)$ such that (4.14) holds. Define $\{B_i\}_1^{2m+K}$ as in (4.9) with the exception of B_{m+K}, whose value at b we alter to

$$B_{m+K}(b) = \lim_{x\uparrow b} B_{m+K}(x). \tag{4.15}$$

Then $\{B_i\}_1^{m+K}$ form a basis for $\mathcal{S}(\mathcal{P}_m;\mathfrak{M};\Delta)$, and properties (4.10) to (4.13) hold.

Proof. The proof is identical except that we must now use (4.15) to assure that B_{m+K} lies in $\mathcal{S}$. Indeed, if (4.14) holds, then by the properties of the function $(x-y)_+^{m-1}$, the divided difference defining $B_{m+K}(x)$ in (4.9) vanishes at b, and does not reduce to a polynomial in $\mathcal{P}_m$ in the last interval I_k. This point is further clarified in the next example. ■

EXAMPLE 4.11

Construct the B-spline basis for the space $\mathcal{S}_2(\Delta)$ defined in Example 4.6.

Discussion. We choose the extended partition $y_1=y_2=0$, $y_3=1$, $y_4=2$, $y_5=3$, $y_6=4$, and $y_7=y_8=5$. Then we may apply Corollary 4.10. The resulting B-splines are shown in Figure 6. The last B-spline is given by

$$B_6(x)=\begin{cases}(-1)^2[4,5,5](x-y)_+=(x-4)_+-(x-5)_+-(x-5)_+^0, & 0\leqslant x<5\\ \lim\limits_{x\uparrow 5} B_6(x)=1, & x=5,\end{cases}$$

whereas $[4,5,5](5-y)_+=0$. ■

§ 4.3. B-SPLINES

In the previous section we have seen that divided differences of the Green's function $(x-y)_+^{m-1}$ turn out to be piecewise polynomials which are useful for constructing local bases for spline spaces. In this section we examine such objects, called B-splines, in more detail. The discussion is independent of Section 4.2.

We begin with the definition of the objects of interest to us in this section.

DEFINITION 4.12. B-Splines

Let

$$\cdots \leqslant y_{-1} \leqslant y_0 \leqslant y_1 \leqslant y_2 \leqslant \cdots$$

be a sequence of real numbers. Given integers i and $m>0$, we define

$$Q_i^m(x)=\begin{cases}(-1)^m[y_i,\ldots,y_{i+m}](x-y)_+^{m-1}, & \text{if } y_i<y_{i+m}\\ 0, & \text{otherwise}\end{cases} \tag{4.16}$$

for all real x. We call Q_i^m the mth *order B-spline associated with the knots* $y_i,\ldots,y_{i+m}$.

For $m=1$, the B-spline associated with $y_i<y_{i+1}$ is particularly simple. It is the piecewise constant function

$$Q_i^1(x)=\begin{cases}\dfrac{1}{(y_{i+1}-y_i)}, & y_i\leqslant x<y_{i+1}\\ 0, & \text{otherwise.}\end{cases} \tag{4.17}$$

We can also give explicit formulae for Q_i^m in case either y_i or y_{i+m} is a knot of multiplicity m.

THEOREM 4.13

Suppose $y_i<y_{i+1}=\cdots=y_{i+m}$. Then

$$Q_i^m(x)=\begin{cases}\dfrac{(x-y_i)^{m-1}}{(y_{i+m}-y_i)^m}, & y_i\leqslant x<y_{i+m}\\ 0, & \text{otherwise.}\end{cases} \tag{4.18}$$

Similarly, if $y_i=\cdots=y_{i+m-1}<y_{i+m}$, then

$$Q_i^m(x)=\begin{cases}\dfrac{(y_{i+m}-x)^{m-1}}{(y_{i+m}-y_i)^m}, & y_i\leqslant x<y_{i+m}\\ 0, & \text{otherwise.}\end{cases} \tag{4.19}$$

Proof. In these cases the determinants in the numerator and denominator of the divided difference can be computed explicitly [cf. (2.66)–(2.67)]. ■

We now discuss the basic structure of $Q_i^m(x)$ in general. The following theorem shows that it is a polynomial spline of order m with knots at $y_i,\ldots,y_{i+m}$:

THEOREM 4.14

Let $y_i<y_{i+m}$, and suppose

$$y_i\leqslant\cdots\leqslant y_{i+m}=\overbrace{\tau_1,\ldots,\tau_1}^{l_1},\ldots,\overbrace{\tau_d,\ldots,\tau_d}^{l_d}.$$

Then

$$Q_i^m(x)=\sum_{j=1}^{d}\sum_{k=1}^{l_j}\alpha_{jk}(x-\tau_j)_+^{m-k} \tag{4.20}$$

with $\alpha_{jl_j} \neq 0$, $j=1,2,\ldots,d$. Moreover,

$$D_-^k Q_i^m(\tau_j) = D_+^k Q_i^m(\tau_j), \qquad k=0,1,\ldots,m-l_j-1, \qquad j=1,2,\ldots,d. \tag{4.21}$$

Thus Q_i^m is a polynomial spline of order m with knots at $\tau_1,\ldots,\tau_d$ of multiplicities $l_1,\ldots,l_d$.

Proof. The expansion (4.20) follows from the expansion (2.89) for divided differences. The continuity properties (4.21) hold for the individual plus functions, and thus also for Q_i^m. ■

Our next result will be useful in delineating the structure of Q_i^m for a general set of $y_i,\ldots,y_{i+m}$. It is also of crucial importance in numerical computations involving B-splines.

THEOREM 4.15

Let $m \geqslant 2$, and suppose $y_i < y_{i+m}$. Then for all $x \in \mathbf{R}$,

$$Q_i^m(x) = \frac{(x-y_i)Q_i^{m-1}(x) + (y_{i+m}-x)Q_{i+1}^{m-1}(x)}{(y_{i+m}-y_i)}. \tag{4.22}$$

Proof. For $y_i < y_{i+1} = \cdots = y_{i+m}$ or $y_i = \cdots = y_{i+m-1} < y_{i+m}$, the result follows from Theorem 4.13. Thus we may assume $y_{i+1} < y_{i+m}$ and $y_i < y_{i+m-1}$. Observe that $(x-y)_+^{m-1} = (x-y)_+^{m-2}(x-y)$. Applying Leibniz's rule for the divided difference of a product (see Theorem 2.52), we obtain

$$(-1)^m[y_i,\ldots,y_{i+m}](x-y)_+^{m-1} = (-1)^m[y_i,y_{i+1}](x-y)$$
$$\cdot[y_{i+1},\ldots,y_{i+m}](x-y)_+^{m-2} + (x-y_i)(-1)^m[y_i,\ldots,y_{i+m}](x-y)_+^{m-2}.$$

Substituting

$$(-1)^m[y_i,\ldots,y_{i+m}](x-y)_+^{m-2}$$
$$= \frac{(-1)^{m-1}}{(y_{i+m}-y_i)}\left\{[y_i,\ldots,y_{i+m-1}](x-y)_+^{m-2} - [y_{i+1},\ldots,y_{i+m}](x-y)_+^{m-2}\right\}$$

and rearranging, we obtain (4.22). ■

Theorem 4.15 provides a *recursion relation* whereby B-splines of order m can be related to B-splines of order $m-1$. The following result shows that the derivative of a B-spline of order m can also be written in terms of two B-splines of lower order:

THEOREM 4.16

Let $y_i < y_{i+m}$, and suppose D_+ is the right derivative operator. Then

$$D_+ Q_i^m(x) = (m-1)\frac{\left[Q_i^{m-1}(x) - Q_{i+1}^{m-1}(x)\right]}{(y_{i+m}-y_i)}. \tag{4.23}$$

Proof. If either y_i or y_{i+m} has multiplicity m, then the result follows directly from Theorem 4.13. If not, then

$$D_+ Q_i^m(x) = (-1)^m [y_i,\ldots,y_{i+m}] D_+(x-y)_+^{m-1}$$

$$= (-1)^m [y_i,\ldots,y_{i+m}](m-1)(x-y)_+^{m-2}$$

$$= (-1)^m (m-1)\frac{\left\{[y_{i+1},\ldots,y_{i+m}](x-y)_+^{m-2} - [y_i,\ldots,y_{i+m-1}](x-y)_+^{m-2}\right\}}{(y_{i+m}-y_i)}$$

$$= (m-1)\frac{\left[Q_i^{m-1}(x) - Q_{i+1}^{m-1}(x)\right]}{(y_{i+m}-y_i)}.$$

■

We can now say considerably more about the shape of Q_i^m.

THEOREM 4.17

Let $m > 1$, and suppose $y_i < y_{i+m}$. Then

$$Q_i^m(x) > 0 \qquad \text{for } y_i < x < y_{i+m} \tag{4.24}$$

and

$$Q_i^m(x) = 0 \qquad \text{for } x < y_i \text{ and } y_{i+m} < x. \tag{4.25}$$

At the endpoints of the interval (y_i, y_{i+m}) we have

$$\begin{aligned}(-1)^{k+m-\alpha_i} D_+^k Q_i^m(y_i) &= 0, \qquad k = 0,1,\ldots,m-1-\alpha_i,\\ &> 0, \qquad k = m-\alpha_i,\ldots,m-1\end{aligned} \tag{4.26}$$

and

$$(-1)^{m-\beta_{i+m}} D_-^k Q_i^m(y_{i+m}) = 0, \qquad k = 0, 1, \ldots, m-1-\beta_{i+m}$$
$$> 0, \qquad k = m - \beta_{i+m}, \ldots, m-1, \qquad (4.27)$$

where

$$\alpha_i = \max\{j : y_i = \cdots = y_{i+j-1}\}$$
$$\beta_{i+m} = \max\{j : y_{i+m} = \cdots = y_{i+m-j+1}\}.$$

(α_i tells how many of the points $y_i \leqslant \cdots \leqslant y_{i+m}$ are equal to y_i, while β_{i+m} tells how many of them are equal to y_{i+m}.)

Proof. To prove (4.24), we proceed by induction on m. For $m=1$ it is clear from (4.17). Assume now that it holds for order $m-1$. Then for $y_i < x < y_{i+m}$, both factors $(x-y_i)$ and $(y_{i+m}-x)$ in the recursion (4.22) are positive. Moreover, both $Q_i^{m-1}(x)$ and $Q_{i+1}^{m-1}(x)$ are nonnegative, and at least one of them is positive. It follows that $Q_i^m(x) > 0$.

By the definition of the plus functions, it is clear that $Q_i^m(x) = 0$ for $x < y_i$. On the other hand, for $x > y_{i+m}$, $Q_i^m(x)$ is the mth divided difference of the polynomial $(x-y)^{m-1}$, and thus is also zero. The vanishing of the indicated derivatives at y_i and y_{i+m} now follows from the fact that Q_i^m is zero outside of (y_i, y_{i+m}), coupled with the continuity properties (4.21) of these derivatives.

It remains to establish the assertions about the remaining derivatives in (4.26) and (4.27). If either $\alpha_i = m$ or $\beta_{i+m} = m$, we may verify the signs of these derivatives directly. If both α_i and β_{i+m} are less than m, we may proceed by induction on m. The case of $m=2$ is easy to check. Now suppose (4.26) holds for splines of order $m-1$. Then coupling

$$(-1)^{k+m-\alpha_i} D_+^{k-1} Q_i^{m-1}(y_i) > 0 \qquad \text{and} \quad (-1)^{k+m-1-\alpha_i} D_+^{k-1} Q_{i+1}^{m-1}(y_i) \geqslant 0$$

with formula (4.23), we obtain (4.26) for $k = m-\alpha_i, \ldots, m-1$. The proof of (4.27) is similar. ■

Theorems 4.13 to 4.17 provide rather detailed information on the shape of the B-spline $Q_i^m(x)$. It will be shown in Theorem 4.57 that

$$Z_{(y_i, y_{i+m})}(D_+^j Q_i^m) \leqslant j, \qquad j = 1, \ldots, m-1.$$

In fact, if $D_+^j Q_i^m$ is continuous, then it has exactly j zeros in (y_i, y_{i+m}). To give a feeling for what the B-splines look like, we have graphed some typical B-splines for various choices of m and knot locations in Figure 7.

Our next theorem deals with the linear independence of B-splines.

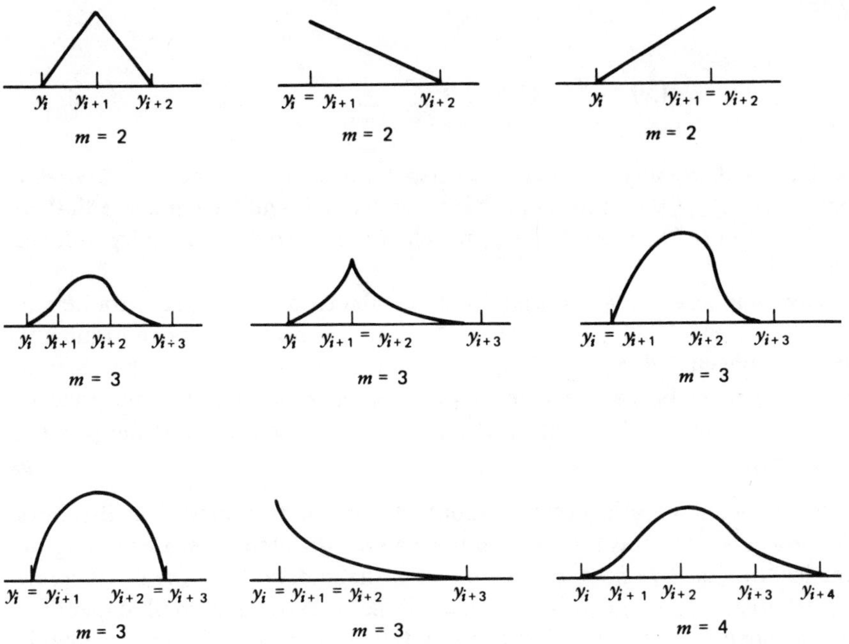

Figure 7. Shapes of some B-splines.

THEOREM 4.18

Let $y_l < y_{l+1}$. Then the B-splines $\{Q_i^m\}_{i=l+1-m}^{l}$ span $\mathcal{P}_m$ on $[y_l, y_{l+1})$. More generally, if $l < r$ and $y_{r-1} < y_r$, then

$$\{Q_i^m\}_{i=l-m+1}^{r-1} \qquad \text{are linearly independent on } [y_l, y_r).$$

Proof. Restricted to the interval $I_l = [y_l, y_{l+1})$, each of the splines $\{Q_i^m\}_{l+1-m}^{l}$ is in $\mathcal{P}_m$. Hence to establish the first assertion, we need only show that these functions are linearly independent on I_l. Suppose now that

$$s(x) = \sum_{i=l+1-m}^{l} c_i Q_i^m(x) = 0 \qquad \text{for all } x \in I_l. \tag{4.28}$$

Suppose not all c_i are zero, and let c_p be the first nonzero one. Suppose

$$y_p, \ldots, y_l = \overbrace{\tau_1, \ldots, \tau_1}^{l_1}, \ldots, \overbrace{\tau_d, \ldots, \tau_d}^{l_d}.$$

Then

$$\tilde{s}(x)=\sum_{i=p}^{l} c_i Q_i^m(x)=\sum_{j=1}^{d}\sum_{k=1}^{l_j}\alpha_{jk}(x-\tau_j)_+^{m-k}$$

with $\alpha_{1l_1}\neq 0$. We also observe that $\tilde{s}(x)=0$ for $x<y_p=\tau_1$ and for $x>y_l=\tau_d$. But since $\sum_{i=1}^{d} l_i \leqslant m$, this contradicts Lemma 4.7, and we conclude that all of the c's in (4.28) must be zero, which is the desired linear independence assertion.

Now suppose $[y_l,y_r)$ consists of more than one nontrivial subinterval, and that $s(x)=\sum_{i=l-m+1}^{r-1} c_i Q_i^m(x)\equiv 0$ on it. Then by the above argument, the coefficients of all the B-splines with support including the interval $[y_\nu,y_{\nu+1})$ must be zero whenever $[y_\nu,y_{\nu+1})$ is a nontrivial subinterval of $[y_l,y_r)$. But this implies that all coefficients must be zero, and the proof is complete. ■

So far we have said nothing about the *size* of B-splines. The B-splines Q_i^m introduced in this section can have wildly different sizes depending on the location of the knots. For example, in the interval $[y_i,y_{i+1})$ the B-spline $Q_i^1(x)=1/(y_{i+1}-y_i)$ can be extremely large or extremely small, depending on the spacing of the y's. For computational purposes it is not acceptable to deal with functions that are too small or too large. This suggests that we should introduce some *normalization* of the B-splines.

DEFINITION 4.19. Normalized B-Splines

Let

$$N_i^m(x)=(y_{i+m}-y_i)Q_i^m(x), \tag{4.29}$$

where Q_i^m is the B-spline defined in Definition 4.12. We call N_i^m the *normalized B-spline* associated with the knots $y_i,\ldots,y_{i+m}$.

For $m=1$, the normalized B-spline associated with $y_i<y_{i+1}$ is given by

$$N_i^1(x)=\begin{cases}1, & y_i\leqslant x<y_{i+1}\\ 0, & \text{otherwise.}\end{cases} \tag{4.30}$$

It follows from our next theorem that for all $m\geqslant 1$,

$$0\leqslant N_i^m(x)\leqslant 1, \qquad \text{all } x\in\mathbf{R}. \tag{4.31}$$

THEOREM 4.20

The B-splines form a *partition of unity*; that is,

$$\sum_{i=j+1-m}^{j} N_i^m(x)=1 \qquad \text{for all } y_j \leqslant x < y_{j+1}. \tag{4.32}$$

Proof. By (4.30) the assertion is trivial for $m=1$. We now proceed by induction. Assume the result is correct for splines of order $m-1$. Then by the recursion relation (4.22)

$$\sum_{i=j+1-m}^{j} N_i^m(x)= \sum_{i=j+1-m}^{j} \left[(x-y_i)Q_i^{m-1}(x)+(y_{i+m}-x)Q_{i+1}^{m-1}(x)\right]$$

$$= \sum_{i=j+2-m}^{j} (x-y_i+y_{i+m-1}-x)Q_i^{m-1}(x)= \sum_{i=j+2-m}^{j} N_i^{m-1}(x)=1. \quad \blacksquare$$

The same kind of argument used in Theorem 4.20 can also be used to derive explicit expansions for important polynomials in terms of B-splines.

THEOREM 4.21

Let $l \leqslant r$ and $y_l < y_{r+1}$. Then for any $y \in \mathbf{R}$,

$$(y-x)^{m-1}= \sum_{i=l+1-m}^{r} \varphi_{i,m}(y)N_i^m(x), \qquad \text{all } y_l \leqslant x < y_{r+1}, \tag{4.33}$$

where

$$\varphi_{i,m}(y)= \prod_{\nu=1}^{m-1} (y-y_{i+\nu}).$$

Moreover, for $j=1,2,\ldots,m$,

$$x^{j-1}= \sum_{i=l+1-m}^{r} \xi_i^{(j)}N_i^m(x), \qquad \text{all } y_l \leqslant x < y_{r+1}, \tag{4.34}$$

where

$$\xi_i^{(j)}=(-1)^{j-1}\frac{(j-1)!}{(m-1)!}D^{m-j}\varphi_{i,m}(0), \qquad i=l+1-m,\ldots,r.$$

Proof. We proceed by induction on m. For $m=1$ the assertions follow immediately from (4.30), and the fact that $\varphi_{i,1}(y)=\xi_i^{(1)}=1$ for all i. Assuming the result holds for $m-1$, we now prove it for m. Using the recursion relation, we have

$$\sum_{i=l+1-m}^{r} \varphi_{i,m}(y)N_i^m(x)=\sum_{i=l+1-m}^{r} \varphi_{i,m}(y)\big[(x-y_i)Q_i^{m-1}(x)$$
$$+(y_{i+m}-x)Q_{i+1}^{m-1}(x)\big].$$

Rearranging this sum and using the fact that $Q_{l+1-m}^{m-1}(x)=Q_{r+1}^{m-1}(x)=0$ for x in the interval $[y_l,y_{r+1})$, we obtain

$$\sum_{i=l+1-m}^{r} Q_i^{m-1}(x)\big[(x-y_i)\varphi_{i,m}(y)+(y_{i+m-1}-x)\varphi_{i-1,m}(y)\big].$$

Now the quantity in brackets can be rewritten as

$$\varphi_{i,m-1}(y)\big[(x-y_i)(y-y_{i+m-1})+(y_{i+m-1}-x)(y-y_i)\big]$$
$$=\varphi_{i,m-1}(y)(y-x)(y_{i+m-1}-y_i).$$

We conclude that

$$\sum_{i=l+1-m}^{r} \varphi_{i,m}(y)N_i^m(x)=(y-x)\sum_{i=l+1-m}^{r} \varphi_{i,m-1}(y)N_i^{m-1}(x)$$
$$=(y-x)(y-x)^{m-2}=(y-x)^{m-1}.$$

The identity (4.34) follows if we differentiate (4.33) $m-j$ times with respect to y and evaluate it at $y=0$. ■

The coefficients $\xi_i^{(j)}$ in Theorem 4.21 can also be written in terms of the classical symmetric functions (see Remark 4.1) as

$$\xi_i^{(j)}=\frac{\operatorname{symm}_{j-1}(y_{i+1},\dots,y_{i+m-1})}{\binom{m-1}{j-1}}. \tag{4.35}$$

For reference, we note that

$$\xi_i^{(1)}=1 \quad \text{and} \quad \xi_i^{(2)}=\frac{(y_{i+1}+\cdots+y_{i+m-1})}{(m-1)}. \tag{4.36}$$

At times it is useful to have bounds on the size of derivatives of B-splines. Even for the normalized splines, such bounds depend heavily on the spacing of the knots. For example, the linear B-spline

$$N_i^2(x)=\begin{cases}\dfrac{(x-y_i)}{(y_{i+1}-y_i)}, & y_i \leqslant x < y_{i+1}\\[2ex] \dfrac{(y_{i+2}-x)}{(y_{i+2}-y_{i+1})}, & y_{i+1}\leqslant x<y_{i+2}\end{cases}$$

has derivative

$$D_+N_i^2(x)=\begin{cases}\dfrac{1}{(y_{i+1}-y_i)}, & y_i \leqslant x < y_{i+1}\\[2ex] \dfrac{1}{(y_{i+2}-y_{i+1})}, & y_{i+1}\leqslant x<y_{i+2}.\end{cases}$$

This becomes arbitrarily large if the knots are close together. The following theorem gives bounds on the size of the derivatives in terms of the knot spacing.

THEOREM 4.22

Let $N_i^m(x)$ be the normalized B-spline defined over the knots $y_i \leqslant \cdots \leqslant y_{i+m}$. Suppose l and x are such that $y_l \leqslant x < y_{l+1}$, and define

$$\Delta_{i,l,j}=\min\{(y_{\nu+j}-y_\nu): y_i \leqslant y_\nu \leqslant y_l < y_{l+1} \leqslant y_{\nu+j} \leqslant y_{i+m}\}$$

for $j=1,2,\ldots,m$. Suppose $\sigma>0$, and that $\Delta_{i,l,m-\sigma+1}>0$. Then

$$|D_+^\sigma N_i^m(x)| \leqslant \frac{\Gamma_{m\sigma}}{\Delta_{i,l,m-1}\cdots\Delta_{i,l,m-\sigma}}, \tag{4.37}$$

where

$$\Gamma_{m\sigma}=\frac{(m-1)!}{(m-\sigma-1)!}\binom{\sigma}{\lfloor \frac{\sigma}{2}\rfloor} \leqslant 2^\sigma \frac{(m-1)!}{(m-\sigma-1)!},$$

and as usual, $\lfloor \sigma/2 \rfloor = \max\{j: j \leqslant \sigma/2\}$.

Proof. By the definition of $N_i^m(x)$,

$$D_+^\sigma N_i^m(x)=\frac{(-1)^m(m-1)!}{(m-\sigma-1)!}(y_{i+m}-y_i)[y_i,\ldots,y_{i+m}](x-y)_+^{m-\sigma-1}.$$

Repeatedly using the relation (2.91) to express the divided difference in terms of lower-order ones, we obtain

$$|D_+^\sigma N_i^m(x)|$$

$$\leqslant \frac{(m-1)!(y_{i+m}-y_i)}{(m-\sigma-1)!} \frac{\sum_{\nu=0}^{\sigma} \binom{\sigma}{\nu} |[y_{\nu+i},\ldots,y_{\nu+i+m-\sigma}](x-y)_+^{m-\sigma-1}|}{(\Delta_{i,l,m} \cdots \Delta_{i,l,m-\sigma+1})}$$

$$\leqslant \frac{(m-1)!}{(m-\sigma-1)!} \left(\binom{\sigma}{\lfloor \frac{\sigma}{2} \rfloor} \right) \sum_{\nu=0}^{\sigma} \frac{N_{\nu+i}^{m-\sigma}(x)}{(\Delta_{i,l,m-1} \cdots \Delta_{i,l,m-\sigma})}.$$

The result follows since the B-splines sum to 1. ■

Our next theorem establishes the so-called *Peano representation* for divided differences. It also gives values for the moments of the B-spline Q_i^m.

THEOREM 4.23

Fix $0 \leqslant j \leqslant m-1$. Then for all f in the space $L_1^{m-j}[y_i, y_{i+m}]$,

$$[y_i,\ldots,y_{i+m}]f = \int_{y_i}^{y_{i+m}} \frac{(-1)^j D_+^j Q_i^m(x) D^{m-j} f(x)\,dx}{(m-1)!}. \tag{4.38}$$

Moreover,

$$\int_{y_i}^{y_{i+m}} (-1)^j D_+^j Q_i^m(x) x^\nu\,dx = \begin{cases} 0, & \nu = 0,1,\ldots,j-1 \\ \dfrac{\nu!(m-1)!}{(m+\nu-j)!} \rho_{\nu-j}(y_i,\ldots,y_{i+m}), & \nu = j,\ldots,m-1. \end{cases} \tag{4.39}$$

where ρ is the function defined in (2.95). In particular, the first two moments of Q_i^m are given by

$$\int_{y_i}^{y_{i+m}} Q_i^m(x)\,dx = \frac{1}{m} \tag{4.40}$$

and

$$\int_{y_i}^{y_{i+m}} x Q_i^m(x)\,dx = \frac{y_i + \cdots + y_{i+m}}{m(m+1)}. \tag{4.41}$$

Proof. By the dual Taylor expansion (2.19), if $f \in L_1^{m-j}[y_i, y_{i+m}]$, we can write

$$f(y) = \sum_{k=0}^{m-j-1} \frac{(-1)^k D^k f(y_{i+m})(y_{i+m}-y)^k}{k!}$$

$$+ \int_{y_i}^{y_{i+m}} \frac{(-1)^{m-j} D_x^j (x-y)_+^{m-1} D^{m-j} f(x)\,dx}{(m-1)!}.$$

Applying the divided difference operator $[y_i, \ldots, y_{i+m}]$ to both sides, we obtain (4.38). To establish (4.39), take $f(x) = x^{m-j+\nu}$ and recall formula (2.94) for its divided difference. ■

Before proving our next result, we need to introduce another B-spline $\tilde{Q}_i^m$. It is closely related to Q_i^m, and, in fact, it is identical except at m-tuple knots.

LEMMA 4.24

Given $y_i < y_{i+m}$, let

$$\tilde{Q}_i^m(x) = [y_i, \ldots, y_{i+m}](y-x)_+^{m-1}, \qquad \text{all } x \in \mathbf{R}. \tag{4.42}$$

Then

$$\tilde{Q}_i^m(x) = Q_i^m(x) \qquad \text{for all } x \in \mathbf{R} \setminus J_i^m,$$

where $J_i^m = \{m\text{-tuple knots of } Q_i^m\}$ (cf. Definition 4.12).

Proof. It is easily checked that

$$(x-y)_+^{m-1} - (-1)^m (y-x)_+^{m-1} = (x-y)^{m-1}.$$

Applying the divided difference over $[y_i, \ldots, y_{i+m}]$, we see that

$$[y_i, \ldots, y_{i+m}](x-y)_+^{m-1} - (-1)^m [y_i, \ldots, y_{i+m}](y-x)_+^{m-1} = 0$$

since the mth order divided difference of the polynomial $(y-x)^{m-1}$ is 0. ■

The reason for the difference between $\tilde{Q}_i^m$ and Q_i^m is that $\tilde{Q}_i^m$ is left continuous while Q_i^m is right continuous. This makes no difference, of course, except at an m-tuple knot.

As another application of the representation (4.38) of divided differences, we have the following interesting result connecting inner products of B-splines with divided differences.

THEOREM 4.25

Let $y_i < y_{i+m}$ and $y_j < y_{j+n}$. Then

$$\int_{-\infty}^{\infty} Q_i^m(x) Q_j^n(x)\,dx = \frac{(-1)^m (n-1)!(m-1)!}{(m+n-1)!} \cdot [y_i,\ldots,y_{i+m}]_x [y_j,\ldots,y_{j+n}]_y (y-x)_+^{m+n-1}. \tag{4.43}$$

(The subscripts x and y on the divided difference symbols indicate which variables they operate on.)

Proof. Let $f(x) = [y_j,\ldots,y_{j+n}](y-x)^{m+n-1}$. Then $(-1)^m D^m f(x) = (m+n-1)!\tilde{Q}_j^n(x)/(n-1)!$. Substituting this in (4.38) and using the fact that $\tilde{Q}_j^n(x)$ and $Q_j^n(x)$ are equal (except perhaps at one point), we obtain (4.43). ∎

Inner-products of B-splines play an important role in several applications. It should be emphasized that although (4.43) is an important theoretical tool, it is not necessarily a good way to compute inner products. See Section 5.4.

Our next theorem deals with the question of what happens to a B-spline when we make a small perturbation in the location of its knots. The result will be a useful tool later on.

THEOREM 4.26

Let $y_i \leqslant \cdots \leqslant y_{i+m}$, and suppose $y_i^{(\nu)} \leqslant \cdots \leqslant y_{i+m}^{(\nu)}$ is a sequence of points with

$$y_j^{(\nu)} \to y_j, \qquad j = i,\ldots,i+m \text{ as } \nu \to \infty.$$

Let Q_i^m and $Q_{i,\nu}^m$ be the mth order B-spline associated with these knot sets, respectively. Then for all $k = 0,1,\ldots,m-1$,

$$D_+^k Q_{i,\nu}^m(x) \to D_+^k Q_i^m(x), \qquad \text{all } x \in \mathbf{R} \backslash J_i^k$$

where

$$J_i^k = \{y_j : y_j \text{ is a knot of } Q_i^m \text{ of multiplicity } m-k \text{ or more}\}.$$

The convergence is uniform on any closed set excluding J_i^k.

Proof. We consider the case $k=0$ first. If $y_i = y_{i+m}$, then $Q_i^m(x)=0$ for all x. Since $Q_{i,\nu}^m(x)=0$ outside of $[y_i^{(\nu)}, y_{i+m}^{(\nu)}]$, which is shrinking to the single point y_i, the assertion follows in this case.

Suppose now that $y_i < y_{i+m}$. For $m=1$ the result is obvious. We proceed by induction on m. Using the recursion relation (4.22) on both Q_i^m and $Q_{i,\nu}^m$, we obtain

$$|Q_i^m(x) - Q_{i,\nu}^m(x)|$$

$$\leqslant \frac{|Q_i^{m-1}(x)(x-y_i)(y_{i+m}^{(\nu)}-y_i^{(\nu)}) - Q_{i,\nu}^{m-1}(x)(x-y_i^{(\nu)})(y_{i+m}-y_i)|}{(y_{i+m}-y_i)(y_{i+m}^{(\nu)}-y_i^{(\nu)})}$$

$$+\frac{|Q_{i+1}^{m-1}(x)(y_{i+m}-x)(y_{i+m}^{(\nu)}-y_i^{(\nu)}) - Q_{i+1,\nu}^{m-1}(x)(y_{i+m}^{(\nu)}-x)(y_{i+m}-y_i)|}{(y_{i+m}-y_i)(y_{i+m}^{(\nu)}-y_i^{(\nu)})}.$$

The expressions on the right go to zero as $\nu\to\infty$. They go to zero uniformly on any closed interval excluding J_i^0.

For $k>0$ our assertion follows from the fact that the derivatives of a B-spline can be written in terms of lower-order B-splines (cf. Theorem 4.16). ■

Two examples of the convergence of B-splines defined on perturbed sets of knots are shown in Figure 8. The second example shows clearly that there is no convergence at an m-tuple knot of Q_i^m.

In Theorem 4.26 we have shown the continuity of B-splines as functions of their knots. In some applications it is useful to have more—namely, the actual derivatives with respect to the knots. The following theorem gives explicit expressions for such derivatives, where they exist.

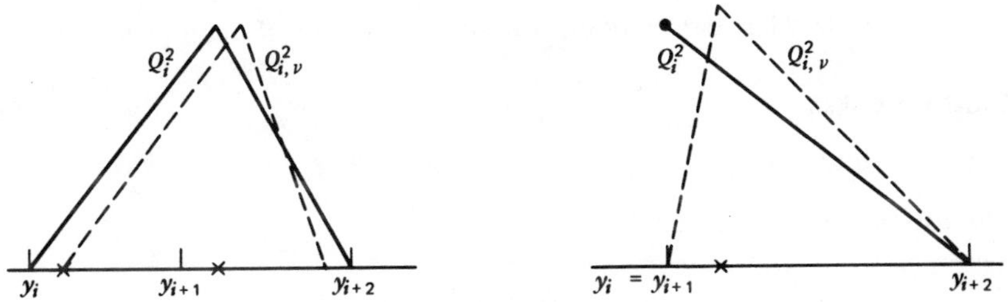

Figure 8. B-Splines with perturbed knots.

THEOREM 4.27

Suppose

$$y_i \leqslant y_{i+1} \leqslant \cdots \leqslant y_{i+m} = \overbrace{\tau_1}^{l_1} \leqslant \overbrace{\tau_2}^{l_2} \leqslant \cdots \leqslant \overbrace{\tau_d}^{l_d}.$$

Fix $1 \leqslant j \leqslant d$, and suppose $l_j \leqslant m-2$. Then

$$\frac{\partial}{\partial \tau_j} N_i(x)$$

$$= \begin{cases} (-1)^{m-1}\begin{bmatrix} l_1+1, l_2 \cdots l_d - 1 \\ \tau_1, \tau_2, \ldots, \tau_d \end{bmatrix}(x-y)_+^{m-1}, & \text{if } j=1 \text{ and } l_1=1, \\ (-1)^m \begin{bmatrix} l_1 - 1 \cdots l_{d-1}, l_d+1 \\ \tau_1 \cdots \tau_{d-1}, \tau_d \end{bmatrix}(x-y)_+^{m-1}, & \text{if } j=d \text{ and } l_d=1, \\ l_j(y_{i+m}-y_i)(-1)^m \begin{bmatrix} l_1 \cdots l_j+1 \cdots l_d \\ \tau_1, \ldots, \tau_j, \ldots, \tau_d \end{bmatrix}(x-y)_+^{m-1}, & \text{otherwise.} \end{cases} \tag{4.44}$$

Here $\partial/\partial\tau_j$ is to be interpreted as a right derivative if $\tau_j = \tau_{j-1}$ and as a left derivative if $\tau_j = \tau_{j+1}$. The same formulas are valid when l_j is $m-1$ or m for all x excluding $x = y_j$.

Proof. If $j=1$ and $l_1=1$, then

$$(-1)^m \frac{\partial}{\partial \tau_1} N_i(x) = \frac{\partial}{\partial \tau_1}\begin{bmatrix} l_2 \cdots l_d \\ \tau_2, \ldots, \tau_d \end{bmatrix}(x-y)_+^{m-1} - \frac{\partial}{\partial \tau_1}\begin{bmatrix} l_1 \cdots l_d - 1 \\ \tau_1, \ldots, \tau_d \end{bmatrix}(x-y)_+^{m-1}.$$

The first term is zero, and the second term can be computed using Theorem 2.55 on the derivatives of divided differences. The case where $j=d$ and $l_d=1$ is similar. In all other cases we may apply Theorem 2.55 directly. ■

To illustrate what can happen, we give a number of examples.

EXAMPLE 4.28

Let $m=1$ and $y_1 < y_2$.

Discussion. In this case we have

$$\frac{\partial}{\partial y_1} N_1(x) = [y_1, y_1](x-y)_+^0 = \begin{cases} 0 & x \neq y_1 \\ \text{not defined}, & x = y_1. \end{cases}$$ ■

EXAMPLE 4.29

Let $m=2$ and $y_1<y_2<y_3$.

Discussion. In this case we have

$$\frac{\partial}{\partial y_1} N_1(x)=\begin{cases} -[y_1,y_1,y_2](x-y)_+^1, & x\neq y_1 \\ \text{not defined}, & x=y_1. \end{cases}$$

In Figure 9*a* we show N_1 and its derivative. We also graph the B-spline N_1^ε corresponding to a slight perturbation of y_1 to illustrate the direction in which N_1 is changing. Taking the derivative with respect to y_2, we obtain

$$\frac{\partial}{\partial y_2} N_1(x)=\begin{cases} (y_3-y_1)[y_1,y_2,y_2,y_3](x-y)_+^1, & x\neq y_2 \\ \text{not defined}, & x=y_2. \end{cases}$$

See Figure 9*b*. ■

EXAMPLE 4.30

Let $m=3$ and $y_1<y_2<y_3<y_4$.

Discussion. Here we compute

$$\frac{\partial}{\partial y_2} N_1(x)=-[y_2,y_2,y_3,y_4](x-y)_+^2+[y_1,y_2,y_2,y_3](x-y)_+^2,$$

valid for all x. We illustrate this derivative in Figure 9*c*. ■

EXAMPLE 4.31

Let $m=3$ and $y_1<y_2=y_3<y_4$.

Discussion. We write $y_1\leqslant\cdots\leqslant y_4=\tau_1<\tau_2=\tau_2<\tau_3$. Then the derivative of N_1 with respect to τ_2 is given by

$$\frac{\partial}{\partial \tau_2} N_1(x)=\begin{cases} -2[y_2,y_2,y_3,y_4](x-y)_+^2+2[y_1,y_2,y_2,y_3](x-y)_+^2, & x\neq y_2 \\ \text{not defined}, & x=y_2 \end{cases}.$$

See Figure 9*d*. ■

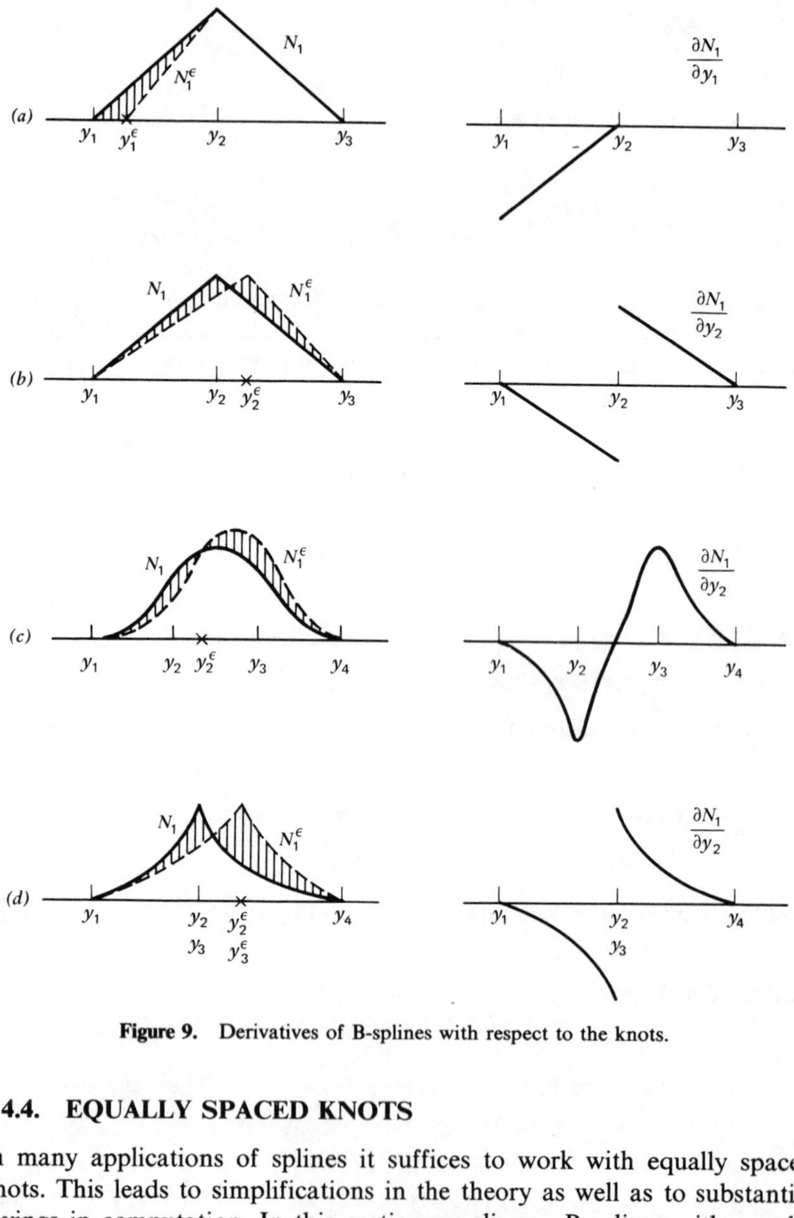

Figure 9. Derivatives of B-splines with respect to the knots.

§ 4.4. EQUALLY SPACED KNOTS

In many applications of splines it suffices to work with equally spaced knots. This leads to simplifications in the theory as well as to substantial savings in computation. In this section we discuss B-splines with equally spaced knots.

We say that a set of knots $\cdots y_i, y_{i+1}, \ldots$ is *uniform with spacing h* provided

$$y_{i+1} - y_i = h \qquad \text{for all } i. \tag{4.45}$$

For uniformly spaced knots it turns out that any B-spline can be obtained from one basic B-spline by translation and scaling. Let

$$Q^m(x)=\frac{(-1)^m\Delta^m(x-y)_+^{m-1}}{m!}=\sum_{i=0}^{m}\frac{(-1)^i\binom{m}{i}(x-i)_+^{m-1}}{m!}. \tag{4.46}$$

This is the usual B-spline associated with the simple knots $0,1,\ldots,m$. It belongs to $C^{m-2}(-\infty,\infty)$. Associated with Q^m, we also introduce the normalized version

$$N^m(x)=mQ^m(x). \tag{4.47}$$

The following theorem shows that any B-spline associated with uniformly spaced knots can be obtained from Q^m or N^m by a translation (and possible scaling):

THEOREM 4.32

Suppose $y_i,\ldots,y_{i+m}$ are uniformly spaced with spacing h. Then

$$Q_i^m(x)=\frac{1}{h}Q^m\left(\frac{x-y_i}{h}\right) \tag{4.48}$$

and

$$N_i^m(x)=N^m\left(\frac{x-y_i}{h}\right). \tag{4.49}$$

Proof. For equally spaced y's the divided difference in the definition of Q_i^m in (4.16) becomes the forward difference operator, and we have

$$Q_i^m(x)=\frac{(-1)^m\Delta_h^m(x-y)_+^{m-1}}{h^m m!}=\sum_{i=0}^{m}\frac{(-1)^i\binom{m}{i}(x-y_i)_+^{m-1}}{h^m m!}$$

$$=\frac{Q^m\left(\frac{x-y_i}{h}\right)}{h}.$$

Assertion (4.49) follows since $N_i^m(x)=mhQ_i^m(x)$. ■

Concerning the size of the normalized spline N^m, we note that

$$\|N^m\|_{L_1[0,m]}=\|N^m\|_{L_\infty[0,m]}=1 \tag{4.50}$$

and thus

$$\|N^m\|_{L_q[0,m]} \leqslant 1 \qquad \text{for all } 1 \leqslant q \leqslant \infty. \tag{4.51}$$

For convenient reference we give the explicit formulae for the polynomial pieces of N^m for $m=2,3,4$ in Table 1. The normalized B-splines N^3 and N^4 are shown in Figure 10, along with their values at the knots.

Table 1. The B-Splines $N^m(x)$ for $m=2,3,4$

$N^2(x)=$	$x,$	$0 \leqslant x \leqslant 1$
	$(2-x),$	$1 \leqslant x \leqslant 2$
$N^3(x)=$	$x^2/2,$	$0 \leqslant x \leqslant 1$
	$(-2x^2+6x-3)/2,$	$1 \leqslant x \leqslant 2$
	$(3-x)^2/2,$	$2 \leqslant x \leqslant 3;$
$N^4(x)=$	$x^3/6,$	$0 \leqslant x \leqslant 1$
	$(-3x^3+12x^2-12x+4)/6,$	$1 \leqslant x \leqslant 2$
	$N^4(4-x),$	$2 \leqslant x \leqslant 4.$

Several of the formulae involving B-splines in the previous section can be simplified in the case of equally spaced knots. For example, the basic recursion formula (4.22) (which we now need only for Q^m) reads

$$Q^m(x) = \frac{xQ^{m-1}(x) + (m-x)Q^{m-1}(x-1)}{m}, \tag{4.52}$$

or in terms of the normalized B-spline N^m,

$$N^m(x) = xQ^{m-1}(x) + (m-x)Q^{m-1}(x-1). \tag{4.53}$$

Similarly, formula (4.23) for the derivative of the B-spline can now be written as

$$D_+N^m(x) = N^{m-1}(x) - N^{m-1}(x-1). \tag{4.54}$$

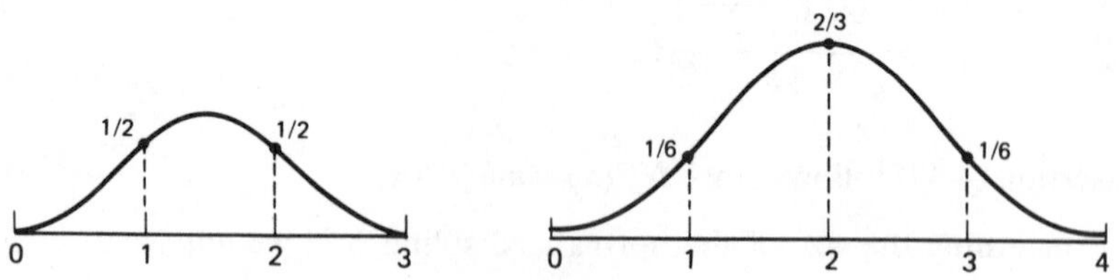

Figure 10. The B-splines N^3 and N^4.

For uniformly spaced knots the Peano representation formula (4.38) for divided differences becomes a representation for forward differences:

$$\Delta^m f(0) = \int_0^m N^m(x) D^m f(x)\,dx \tag{4.55}$$

for all $f \in L_1^m[0,m]$. By a change of variable, we obtain

$$\Delta_h^m f(t) = h^{m-1} \int_0^{mh} N^m\left(\frac{x}{h}\right) D^m f(x+t)\,dx \tag{4.56}$$

for all $h > 0$, all $t > 0$, and all $f \in L_1^m[t, t+mh]$.

The inner products of B-splines on uniform partitions can be computed explicitly. In particular, define

$$I_j^{m,n} = \int_0^m N^m(x) N^n(x+j)\,dx, \qquad j = 0, 1, \ldots, m-1. \tag{4.57}$$

Then if $\{N_i^m\}$ are the B-splines associated with a uniform partition of spacing h, then for all i and j,

$$\int_0^{mh} N_i^m(x) N_{i+j}^n(x)\,dx = h I_j^{m,n}, \qquad j = 0, 1, \ldots, m-1. \tag{4.58}$$

We give the values of $I_1^{m,m}, \ldots, I_{m-1}^{m,m}$ in Table 2 for the most commonly used values of m.

Table 2. Inner Products of B-Splines

	j				
m	0	1	2	3	4
1	1	0	0	0	0
2	$\frac{4}{6}$	$\frac{1}{6}$	0	0	0
3	$\frac{66}{120}$	$\frac{26}{120}$	$\frac{1}{120}$	0	0
4	$\frac{2{,}416}{5{,}040}$	$\frac{1{,}191}{5{,}040}$	$\frac{120}{5{,}040}$	$\frac{1}{5{,}040}$	0
5	$\frac{156{,}190}{362{,}880}$	$\frac{88{,}234}{362{,}880}$	$\frac{14{,}608}{362{,}880}$	$\frac{502}{362{,}880}$	$\frac{1}{362{,}880}$

For our next theorem we need to introduce a slightly translated version of the B-spline N^m defined in (4.47):

$$M^m(x)=N^m\left(x+\frac{m}{2}\right), \qquad \text{all } x\in\mathbf{R}. \tag{4.59}$$

This spline is symmetric about the origin and has support on $[-m/2, m/2]$. For m even it has simple knots at the integers, while for m odd the knots are at the midpoints between the integers. The following theorem shows that M^m can be defined by a convolution process:

THEOREM 4.33

For all $1\leqslant i\leqslant m-1$,

$$M^m(x)=M^i*M^{m-i}(x)=\int_{-\infty}^{\infty}M^i(x-y)M^{m-i}(y)\,dy. \tag{4.60}$$

Moreover,

$$M^m(x)=\frac{1}{2\pi}\int_{-\infty}^{\infty}\psi_m(u)e^{iux}\,du, \tag{4.61}$$

where

$$\psi_m(u)=\left(\frac{\sin(u/2)}{u/2}\right)^m. \tag{4.62}$$

Proof. To prove (4.60) it suffices to prove that

$$M^m(x)=M^{m-1}*M^1(x)=M^1*M^{m-1}(x). \tag{4.63}$$

This we prove by induction on m. The case of $m=2$ is easy. Suppose now that it holds for $m-1$. Then substituting the explicit expansion of $M^{m-1}(x)$ in terms of plus functions [cf. (4.46) and (4.48)], we obtain

$$M^{m-1}*M^1(x)=\sum_{j=0}^{m-1}\frac{(-1)^j\binom{m-1}{j}}{(m-2)!}\int_{x-1/2}^{x+1/2}\left(y+\frac{m-1}{2}-j\right)_+^{m-2}dy$$

$$=\sum_{j=0}^{m-1}\frac{(-1)^j\binom{m-1}{j}}{(m-1)!}\left[\left(x+\frac{m}{2}-j\right)_+^{m-1}-\left(x+\frac{m}{2}-j-1\right)_+^{m-1}\right]$$

$$=\sum_{j=0}^{m}\frac{(-1)^j}{(m-1)!}\left[\binom{m-1}{j}+\binom{m-1}{j-1}\right]\left(x+\frac{m}{2}-j\right)_+^{m-1},$$

provided we interpret $\binom{m-1}{-1}=\binom{m-1}{m}=0$. Now, combining the binomial coefficients (cf. Remark 2.8), we see that this sum is precisely the expansion of $M^m(x)$.

To prove (4.61), we take the *Fourier transform* of M^m:

$$\widehat{M^m}(u)=\int_{-\infty}^{\infty} M^m(x)e^{-iux}\,dx.$$

For $m=1$ we have

$$\widehat{M^1}(u)=\frac{\sin(u/2)}{u/2}, \tag{4.64}$$

by direct evaluation of the integral. Now since $M^m=M^1*\cdots*M^1$ is the convolution of m copies of M^1, it follows that

$$\widehat{M^m}(u)=\left[\widehat{M^1}(u)\right]^m=\psi_m(u). \tag{4.65}$$

Now formula (4.61) is just the inverse Fourier transformation. ■

§ 4.5. THE PERFECT B-SPLINE

In this section we introduce a special B-spline with some particularly nice properties.

THEOREM 4.34. The Perfect B-Spline

Let

$$y_i=\cos\left(\frac{m-i}{m}\right)\pi, \qquad i=0,1,\ldots,m, \tag{4.66}$$

and let

$$B_m^*(x)=m(-1)^m[y_0,y_1,\ldots,y_m](x-y)_+^{m-1}. \tag{4.67}$$

We call B^* the mth order *perfect B-spline*. It has the properties

$$\int_{-1}^{1} B_m^*(x)\,dx=1 \tag{4.68}$$

and

$$|D_+^{m-1}B_m^*(x)|=2^{m-2}(m-1)!, \qquad \text{all } -1\leqslant x\leqslant 1. \tag{4.69}$$

Proof. The property (4.68) follows from (4.40) and our choice of normalization. To prove (4.69), we use the fact (see Remark 4.3) that with $y_0,\ldots,y_m$ as in (4.66),

$$(-1)^m[y_0,y_1,\ldots,y_m]f=\frac{2^{m-2}}{m}\big[f(y_0)-2f(y_1)+2f(y_2)$$
$$+\cdots+(-1)^{m-1}2f(y_{m-1})+(-1)^m f(y_m)\big]. \tag{4.70}$$

It follows that

$$D_+^{m-1}B_m^*(x)=2^{m-2}(m-1)!\big[(x-y_0)_+^0-2(x-y_1)_+^0$$
$$+\cdots+(-1)^m(x-y_m)_+^0\big],$$

and thus that

$$D_+^{m-1}B_m^*(x)=(-1)^i 2^{m-2}(m-1)! \qquad \text{for } y_i\leqslant x<y_{i+1},\ i=0,1,\ldots,m-1. \tag{4.71}$$ ■

The B-spline B_m^* has support on $[-1,1]$. It is called *perfect* because its $m-1$st derivative is of constant absolute value. For convenient reference we list the formulae for the polynomial pieces of the perfect B-splines of order $m=2$ and $m=3$.

EXAMPLE 4.35

For $m=2$ the perfect B-spline is given by

$$B_2^*(x)=\begin{cases} x+1, & -1\leqslant x\leqslant 0\\ 1-x, & 0\leqslant x\leqslant 1.\end{cases} \tag{4.72}$$

For $m=3$ the perfect B-spline is given by

$$B_3^*(x)=\begin{cases} 2(x+1)^2, & -1\leqslant x\leqslant -1/2\\ 1-2x^2, & -1/2\leqslant x\leqslant 1/2\\ 2(1-x)^2, & 1/2\leqslant x\leqslant 1.\end{cases} \tag{4.73}$$

Discussion. For higher m the knots continue to be located symmetrically about the origin, and thus B_m^* is always symmetric about zero. ■

It is also of interest to have bounds on the intermediate derivatives of the perfect B-spline. Theorem 2.4 yields such bounds, but without good estimates for the constants. We have the following more precise result:

THEOREM 4.36

For $0 \leqslant j \leqslant m-2$,

$$\|D_+^j B_m^*(x)\|_{C[-1,1]} \leqslant \frac{2^{j+1}(m-1)!}{(m-j-2)!}. \tag{4.74}$$

Proof. See deBoor [1976c]. ■

The bounds (4.74) are not sharp. For $1 \leqslant j \leqslant \lfloor m/2 \rfloor$ they can be improved by noting that $B_m^* = m/2 \cdot N_m^*$ and applying Theorem 4.22.

One important application of the perfect B-splines is to the construction of transition functions that smoothly connect one function with another. The following theorem deals with a *transition function* connecting the function 0 with the function 1 (see Figure 11), which is optimal in a certain sense.

THEOREM 4.37

Let

$$g(x) = \begin{cases} 0, & x < -1 \\ \int_{-1}^{x} B_m^*(t)\,dt, & -1 \leqslant x < 1 \\ 1, & 1 \leqslant x. \end{cases} \tag{4.75}$$

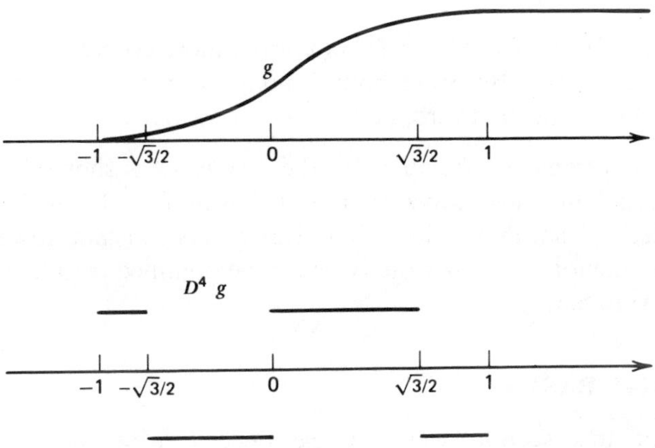

Figure 11. The optimal transition function for $m=4$.

Then $g \in L_\infty^m[\mathbf{R}]$. Moreover, it is the unique solution of the minimization problem

$$\underset{f \in U}{\text{minimize}} \|D^m f\|_{L_\infty[\mathbf{R}]}, \tag{4.76}$$

where

$$U = \left\{ f \in L_\infty^m[\mathbf{R}] : f(x) \equiv 0 \text{ for } x \leqslant -1 \text{ and } f(x) \equiv 1 \text{ for } x \geqslant 1 \right\}.$$

Proof. In view of (4.68), it is clear that g belongs to U. To prove that it provides a minimum in (4.76), we note that by integration by parts,

$$\int_{-1}^{1} D^m f(x) DT_m(x)\,dx = (-1)^{m-1} 2^{m-1} m! \tag{4.77}$$

for any $f \in U$, where T_m is the mth Tchebycheff polynomial of the first kind (see Remark 4.2). Now suppose $f \in U$ is such that $\|D^m f\|_\infty \leqslant \|D^m g\|_\infty$. Then with $\delta = g - f$, the fact that $D^m \delta$ has the same sign as DT_m everywhere on $(-1,1)$ (note that the zeros of DT_m are precisely at the $y_1, \ldots, y_{m-1}$), we have

$$0 \leqslant \int_{-1}^{1} D^m \delta(x) DT_m(x)\,dx \leqslant \int_{-1}^{1} D^m g(x) DT_m(x)\,dx$$
$$- \int_{-1}^{1} D^m f(x) DT_m(x)\,dx \leqslant 0.$$

Here we have used the fact that

$$|D^m g(x)| = 2^{m-2}(m-1)!, \qquad \text{all } -1 \leqslant x \leqslant 1$$

as well as (4.77). It follows that $D^m \delta(x) = 0$ almost everywhere on $(-1,1)$, and thus that $\delta \in \mathcal{P}_m$. But since both f and g are in U, δ and its first $m-1$ derivatives must vanish at zero, and we conclude that $\delta = 0$. ∎

The optimal transition function for the case $m = 4$ is shown in Figure 11 along with its fourth derivative. It is clear that on $(-1,1)$ the function g is also a perfect spline; that is, its mth derivative has constant absolute value. Transition functions for other intervals can be obtained from g by a simple change of variables.

§ 4.6. DUAL BASES

Throughout this section we shall be dealing with the spline space $\mathcal{S}(\mathcal{P}_m; \mathfrak{M}; \Delta)$ and its normalized B-spline basis $\{N_i^m\}_1^n$, $n = m + K$. A set of linear functionals $\{\lambda_j\}_1^n$ defined on $\mathbb{S}$ is called a *dual basis* provided

$$\lambda_j N_i^m = \delta_{ij} = \begin{cases} 1, & \text{if } i=j \\ 0, & \text{otherwise.} \end{cases} \tag{4.78}$$

The usefulness of a dual basis is implicit in the fact that

$$\text{if } s = \sum_{i=1}^{n} c_i N_i^m, \qquad \text{then } \lambda_j s = c_j, \quad j=1,2,\ldots,n. \tag{4.79}$$

For example, we may use a dual basis to examine the connection between the size of a spline and the size of its B-spline coefficients.

THEOREM 4.38

Let $s = \sum_{i=1}^n c_i N_i^m$. Then

$$\|s\|_{L_\infty(R)} \leqslant \|\mathbf{c}\|_\infty = \max_{1 \leqslant i \leqslant n} |c_i|. \tag{4.80}$$

Conversely, if $\{\lambda_j\}_1^n$ is a dual basis for $\{N_i^m\}_1^n$, then

$$\|\mathbf{c}\|_\infty \leqslant \max_{1 \leqslant j \leqslant n} \|\lambda_j\| \|s\|_{L_\infty(\mathbf{R})}, \tag{4.81}$$

where

$$\|\lambda_j\| = \sup_{\substack{s \in \mathbb{S} \\ s \neq 0}} |\lambda_j s| / \|s\|_{L_\infty(\mathbf{R})}.$$

Proof. The assertion (4.80) follows directly from the fact that the sum of the absolute values of the B-splines is 1. The inequality (4.81) follows from (4.79). ■

Before giving a construction of a dual basis for general m, we first give two examples to show that for $m=1$ and for $m=2$ the construction is simple.

EXAMPLE 4.39

Find a dual basis for $\mathbb{S} = \text{span}\{N_1^1, \ldots, N_n^1\}$.

Discussion. We construct two different dual bases. Clearly, one possibility is to take

$$\lambda_j s = s(y_j), \qquad j=1,2,\ldots,n. \tag{4.82}$$

For this dual basis we have $\|\lambda_j\|=1, j=1,2,\ldots,n.$

We may also construct a dual basis using local integrals. Let

$$\lambda_j s=\int_{y_j}^{y_{j+1}} \frac{s(t)\,dt}{(y_{j+1}-y_j)}, \qquad j=1,2,\ldots,n. \tag{4.83}$$

Again, $\|\lambda_j\| \leqslant 1, j=1,2,\ldots,n.$ ■

EXAMPLE 4.40

Construct a dual basis for $\mathbb{S}=\{N_1^2,\ldots,N_n^2\}$.

Discussion. Again, there are several possibilities. For example, using point-evaluation functionals, we can define

$$\lambda_j s=\begin{cases} s(y_{j+1}), & \text{if } y_j<y_{j+1}<y_{j+2} \\ s(y_{j+1}^{+}), & \text{if } y_j=y_{j+1}<y_{j+2} \\ s(y_{j+1}^{-}), & \text{if } y_j<y_{j+1}=y_{j+2}, \end{cases} \tag{4.84}$$

for $j=1,2,\ldots,n$. This basis satisfies $\|\lambda_j\|=1, j=1,2,\ldots,n.$

To get a different dual basis, we may use local integrals. Choose $0<\varepsilon<1$, and for $j=1,2,\ldots,n$, define

$$\lambda_j s=\int_{y_j}^{y_{j+2}} s(t)\varphi_j(t)\,dt, \tag{4.85}$$

where

$$\varphi_j(t)=\begin{cases} \dfrac{(1+\varepsilon)}{\varepsilon h_{j+1}}, & y_{j+1}\leqslant t<y_{j+1}+\varepsilon h_{j+1} \\ \dfrac{-\varepsilon}{(1-\varepsilon)h_{j+1}}, & y_{j+1}+\varepsilon h_{j+1}\leqslant t<y_{j+2} \\ 0, & \text{otherwise,} \end{cases}$$

provided $h_{j+1}=(y_{j+2}-y_{j+1})>0$, and

$$\varphi_j(t)=\begin{cases} \dfrac{-\varepsilon}{(1-\varepsilon)h_j}, & y_j\leqslant t<y_j+(1-\varepsilon)h_j \\ \dfrac{(1+\varepsilon)}{\varepsilon h_j}, & y_j+(1-\varepsilon)h_j\leqslant t<y_{j+1} \\ 0, & \text{otherwise,} \end{cases}$$

provided $h_j=(y_{j+1}-y_j)>0$.

It is easily checked that the linear functionals (4.85) form a dual basis. Moreover,

$$|\lambda_j s| \leqslant \|s\|_\infty \int |\varphi_j(t)|\,dt \leqslant (1+2\varepsilon)\|s\|_\infty,$$

so $\|\lambda_j\| \leqslant 1+2\varepsilon, j=1,2,\ldots,n.$ ■

In the following theorem we construct a dual basis for $\{N_i^m\}_1^n$ using local integrals:

THEOREM 4.41

Let $y_1 \leqslant \cdots \leqslant y_{n+m}$ be such that $y_i < y_{i+m}$, $i=1,2,\ldots,n$. Let $N_1^m,\ldots,N_n^m$ be the associated normalized B-splines. Then there is a dual set of linear functionals $\lambda_1,\ldots,\lambda_n$ with

$$|\lambda_j f| \leqslant (2m+1)9^{m-1} h_j^{-1/p} \|f\|_{L_p[\tilde{I}_j]}, \qquad 1 \leqslant p \leqslant \infty, \tag{4.86}$$

where $\tilde{I}_j = (y_j, y_{j+m})$ and $h_j = y_{j+m} - y_j$, $j=1,2,\ldots,n.$

Proof. For each $j=1,2,\ldots,n$, let

$$G_j(x) = g\left(\frac{2x-y_j-y_{j+m}}{y_{j+m}-y_j}\right),$$

where g is the transition function defined in Theorem 4.37. By the properties of g, we have

$$\begin{aligned} G_j(x) &= 0, && x \leqslant y_j \\ 0 \leqslant G_j(x) &\leqslant 1, && y_j \leqslant x \leqslant y_{j+m} \\ G_j(x) &= 1, && x \geqslant y_{j+m}. \end{aligned}$$

Moreover, with $h_j = (y_{j+m} - y_j)$, we have

$$\|D^m G_j\|_\infty = \left(\frac{4}{h_j}\right)^m \frac{(m-1)!}{4} \tag{4.87}$$

and

$$\|D^{m-k} G_j\|_\infty \leqslant \left(\frac{4}{h_j}\right)^{m-k} \frac{(m-1)!}{(k-1)!}, \qquad k=1,2,\ldots,m. \tag{4.88}$$

For each j, let

$$\psi_j(x) = G_j(x)\varphi_j(x),$$

where

$$\varphi_j(x) = \frac{(x-y_{j+1})\cdots(x-y_{j+m-1})}{(m-1)!}.$$

We define the dual basis by

$$\lambda_j s = \int_{y_j}^{y_{j+m}} s(x) D^m \psi_j(x)\,dx, \qquad j=1,2,\ldots,n. \tag{4.89}$$

We first check that this is indeed a dual basis. By the representation (4.38) of divided differences, we have

$$\lambda_j N_i^m = \int_{y_j}^{y_{j+m}} N_i^m(x) D^m \psi_j(x)\,dx = (m-1)!(y_{i+m}-y_i)$$

$$\cdot [y_i,\ldots,y_{i+m}]\psi_j. \tag{4.90}$$

For $i>j$ this is zero since ψ_j agrees with the polynomial φ_j on the points $y_i,\ldots,y_{i+m}$, and its divided difference is zero. For $i<j$ we again get zero since now ψ_j agrees with the function 0 on $y_i,\ldots,y_{i+m}$. Finally, for $j=i$, ψ_j agrees with the polynomial $(t-y_j)\varphi_j(t)/(y_{j+m}-y_j)$ on $y_i,\ldots,y_{i+m}$. This polynomial is of degree m, and its mth divided difference is equal to $1/(m-1)!(y_{i+m}-y_i)$. It follows that $\lambda_i N_i^m = 1$.

Now we estimate $\|\lambda_j\|_\infty$. For all $f \in L_\infty(\mathbf{R})$ we have

$$|\lambda_j f| \leq \|f\|_{L_p(\tilde{I}_j)} \|D^m \psi_j\|_{L_{p'}(\tilde{I}_j)}, \qquad \frac{1}{p}+\frac{1}{p'}=1. \tag{4.91}$$

We may easily check that

$$\|D^k \varphi_j\|_\infty \leq \frac{h_j^{m-1-k}}{(m-k-1)!}, \qquad k=0,1,\ldots,m-1, \tag{4.92}$$

while $\|D^m\varphi_j\|_\infty = 0$. Thus using Leibniz's rule (2.97) together with the estimates in (4.88) and (4.92), we obtain

$$h_j\|D^m\psi_j\|_\infty \leqslant h_j \sum_{k=0}^{m} \binom{m}{k} \|D^k\varphi_j\|_\infty \|D^{m-k}G_j\|_\infty$$

$$\leqslant \left(\frac{4}{h_j}\right)^m \frac{(m-1)!\,h_j^m}{4(m-1)!} + \sum_{k=1}^{m-1} \binom{m}{k} \left(\frac{4}{h_j}\right)^{m-k} \frac{(m-1)!\,h_j^{m-k}}{(k-1)!(m-k-1)!}$$

$$\leqslant 4^{m-1} + \sum_{k=1}^{m-1} \binom{m}{k} \binom{m-2}{k-1} 4^{m-k}(m-1)$$

$$\leqslant 4^{m-1} + \sum_{k=1}^{m-1} \binom{m}{k} 2^{m-k} \cdot \sum_{k=1}^{m-1} \binom{m-2}{k-1} 2^{m-k}(m-1)$$

$$\leqslant 4^{m-1} + 2(m-1)\cdot 3^m \cdot 3^{m-2} \leqslant (2m-1)9^{m-1}. \qquad \blacksquare$$

It is of some interest to determine to what extent the bounds on the linear functionals in Theorem 4.41 can be improved with another choice of dual basis. Thus given m and $\Delta = \{y_1 \leqslant \cdots \leqslant y_{n+m}\}$, we define

$$D(m,\Delta) = \inf_{\{\lambda_i\}_1^n} \left\{ \max_{1 \leqslant i \leqslant n} \|\lambda_i\| : \{\lambda_i\}_1^n \text{ form a dual basis to } \{N_i\}_1^n \right\} \tag{4.93}$$

and

$$D(m) = \sup_{\Delta} \{ D(m,\Delta) : y_i < y_{i+m}, \quad i = 1,2,\ldots,n \}. \tag{4.94}$$

A related question is how small can we make the constant in the inequality (4.81). We define the best possible constant as

$$\tilde{D}(m,\Delta) = \sup \frac{\|\mathbf{c}\|_\infty}{\left\| \sum_{i=1}^{n} c_i N_i^m \right\|_\infty}, \tag{4.95}$$

and set

$$\tilde{D}(m) = \sup_{\Delta} \{ \tilde{D}(m,\Delta) : y_i < y_{i+m}, \quad i = 1,2,\ldots,n \}. \tag{4.96}$$

By (4.81) we note that

$$\tilde{D}(m,\Delta) \leqslant D(m,\Delta) \qquad \text{and } \tilde{D}(m) \leqslant D(m). \tag{4.97}$$

The following theorem gives some information on the size of these constants:

THEOREM 4.42

Let

$$d_m = \frac{\binom{2m-3}{m-2}}{\binom{m-2}{\lfloor (m-2)/2 \rfloor}}. \tag{4.98}$$

Then

$$d_m \leqslant \tilde{D}(m) \leqslant D(m) \leqslant (2m-1)\cdot 9^{m-1}. \tag{4.99}$$

The values of d_m for $m=1,2,\dots,10$ are shown in Table 3. In general,

$$\frac{m-1}{m}\cdot 2^{m-3/2} \leqslant d_m \leqslant \frac{m}{m-1}\cdot 2^{m-3/2}. \tag{4.100}$$

Table 3. The Constants in Theorem 4.42

m	2	3	4	5	6	7	8	9	10
d_m	1	3	5	$11\frac{2}{3}$	21	$46\frac{1}{5}$	$85\frac{4}{5}$	$183\frac{6}{7}$	$347\frac{2}{7}$

Proof. The upper bound on $D(m)$ in (4.99) follows from Theorem 4.41 since the constant $(2m-1)9^{m-1}$ derived there was independent of Δ. For the lower bound on $\tilde{D}(m)$, we choose a specific Δ and find a spline s so that the ratio $\|\mathbf{c}\|_\infty/\|s\|_\infty \geqq d_m$. Choose $\Delta=\{-1=y_1=\cdots=y_m, y_{m+1}=\cdots=y_{2m}=1\}$. The associated normalized B-splines are

$$N_i^m(x) = 2^{1-m}\binom{m-1}{i-1}(1-x)^{m-i}(1+x)^{i-1}, \qquad i=1,2,\dots,m. \tag{4.101}$$

We now construct a linear combination of these B-splines with large coefficients, but with norm equal to 1. Consider the Tchebycheff polynomial (see Remark 4.2):

$$T_{m-1}(x) = \cos[(m-1)\arccos(x)].$$

It has norm $\|T_{m-1}\|_\infty = 1$. Moreover, by differentiating the associated Rodrigues formula

$$T_{m-1}(x) = \frac{(-1)^{m-1}(1-x^2)^{1/2}D^{m-1}(1-x^2)^{m-3/2}}{1\cdot 3\cdot 5\cdot \cdots \cdot (2m-3)},$$

we find that the B-spline expansion of T_{m-1} is

$$T_{m-1}(x) = (-1)^{m-1}N_1^m(x) + \sum_{i=2}^{m} (-1)^{m-i}\frac{\binom{2m-3}{2i-3}}{\binom{m-2}{i-2}}N_i^m(x). \tag{4.102}$$

The largest coefficient is precisely d_m. We have established the lower bound in (4.99).

It remains to prove the estimates (4.100). To this end we use Wallis' inequality:

$$\frac{2^{2n}}{\sqrt{(n+1/2)\pi}} \leqslant \binom{2n}{n} \leqslant \frac{2^{2n}}{\sqrt{n\pi}}, \qquad \text{all } n \geqslant 1, \tag{4.103}$$

which also implies the related inequality

$$\frac{2^{2n-1}}{\sqrt{(n+1/2)\pi}} \leqslant \binom{2n-1}{n} \leqslant \frac{1}{2}\binom{2n}{n} \leqslant \frac{2^{2n-1}}{\sqrt{n\pi}}. \tag{4.104}$$

We conclude that for m even,

$$\sqrt{\frac{m-2}{m-1/2}}\; 2^{m-3/2} \leqslant d_m \leqslant 2^{m-3/2},$$

while for m odd,

$$\sqrt{\frac{m-1}{m-1/2}} \cdot 2^{m-3/2} \leqslant d_m \leqslant \sqrt{\frac{m}{m-1}} \cdot 2^{m-3/2}.$$

Using the facts that $d_2 = 1$, $d_3 = 3$, and $\sqrt{(m-2)/(m-1/2)} \geqq (m-1)/m$ for $m \geqq 4$, we obtain (4.100). ■

Theorems 4.42 and 4.41 together show that the quantities $D(m)$ and $\tilde{D}(m)$ have an order of growth that lies between 2^{m-1} and 9^{m-1}. In any case, this is a substantial growth as m increases. With a careful construction of a dual linear basis, it can be shown that $d_m = D(m)$ for $m = 1, \ldots, 10$ (cf. Examples 4.39 and 4.40 where $m = 1, 2$ are handled). It is conjectured that equality holds for all m.

The constants $D(m)$ defined in (4.94) can also be used to give an estimate of how independent the various B-splines are from each other. This gives some measure of how well conditioned the B-spline basis is.

THEOREM 4.43

For all $i = 1, 2, \ldots, n$,

$$d\left(N_i^m, \operatorname{span}\{N_j^m\}_{j=1, j\neq i}^n\right)_\infty$$
$$\geqq \frac{1}{\tilde{D}(m,\Delta)} \geqslant \frac{1}{D(m,\Delta)} \geqslant \frac{1}{D(m)} \geqslant \frac{1}{(2m-1)9^{m-1}}, \tag{4.105}$$

where, in general, $d(f, Y)_X$ denotes the distance of f to the linear subspace Y in X; that is,

$$d(f, Y)_X = \inf_{g \in Y} \|f - g\|_X. \tag{4.106}$$

Proof. For any constant B such that $\|c\|_\infty \leqslant B \|\Sigma_{i=1}^n c_i N_i^m\|_\infty$, we have

$$d\left(N_i^m, \operatorname{span}\{N_j^m\}_{j=1, j\neq i}^n\right) = \inf\left\{\left\|\sum_{j=1}^n c_j N_j^m\right\| : c_i = 1\right\}$$
$$\geqslant \frac{1}{B} \inf\{\|\mathbf{c}\|_\infty ; c_i = 1\} \geqslant \frac{1}{B}.$$

Since $\tilde{D}(m,\Delta)$ is the inf of such constants, the first inequality in (4.105) follows. The others follow from (4.94), (4.97), and (4.99). ∎

If we restrict attention to equally spaced knots, then the constant $\tilde{D}(m,\Delta)$ defined in (4.95) is, in fact, independent of the spacing, and thus independent of Δ. Thus we define

$$D_m^* = \tilde{D}(m,\Delta), \qquad \Delta \text{ any uniform partition.} \tag{4.107}$$

THEOREM 4.44

For all $m>1$,

$$\left(\frac{\pi}{2}\right)^{m-2} \leqslant D_m^* \leqslant \left(\frac{\pi}{2}\right)^m. \tag{4.108}$$

Proof. It suffices to work with unit spaced knots. We begin by showing that D_m^* is bounded from below by $\frac{1}{2}(\pi/2)^{m-2}$. As in the proof of Theorem 4.42, we obtain the lower bound by constructing a spline with $\|s\|_\infty$ small and with $\|\mathbf{c}\|_\infty=1$. We take s to be the restriction to the interval $[a,b]=[m,n+1]$ of the spline

$$E_m(x)=\sum_{j=-\infty}^{\infty}(-1)^j N^m(x-j), \qquad \text{all } x\in\mathbf{R}. \tag{4.109}$$

(Note that even though this is defined as an infinite sum, $E_m(x)$ makes sense for every x since at most m of the N's have value at any given x.) Now we observe that

$$E_m\left(i+\frac{m}{2}\right)=\sum_{j=-\infty}^{\infty}(-1)^j N^m\left(i+\frac{m}{2}-j\right)=\sum_{j=-\infty}^{\infty}(-1)^j M^m(i-j)$$

$$=(-1)^i\sum_{j=-\infty}^{\infty}(-1)^j M^m(j).$$

To evaluate this sum, we observe that with ψ_m as in (4.62), $\hat{\psi}_m=\widehat{\widehat{M^m}}=M^m$, and thus the function

$$\varphi_m(x)=\sum_{j=-\infty}^{\infty}\psi_m(x+2\pi j) \tag{4.110}$$

has a Fourier series given by

$$\varphi_m(x)=\sum_{\nu=-\infty}^{\infty}M^m(\nu)e^{i\nu x}.$$

If we evaluate this at $x=\pi$, we obtain

$$\varphi_m(\pi)=\sum_{j=-\infty}^{\infty}(-1)^j M^m(j),$$

that is,

$$E_m\left(i+\frac{m}{2}\right)=(-1)^i\varphi_m(\pi). \tag{4.111}$$

On the other hand, by (4.54)

$$D_+E_m(x)=2\sum_{j=-\infty}^{\infty}(-1)^jN^{m-1}(x-j)=2E_{m-1}(x).$$

From this we conclude by induction that for all i

$$E_m(x) \quad \text{is monotone on the intervals } \left(i+\frac{m}{2},i+1+\frac{m}{2}\right)$$

$$E_m\left(i+\frac{m+1}{2}\right)=0$$

$$E_m\left(x-i-\frac{m+1}{2}\right)=-E_m\left(i+\frac{m+1}{2}-x\right) \quad \text{for } i+\frac{m}{2}\leqslant x\leqslant i+1+\frac{m}{2}$$

$$E_m\left(x-i-\frac{m}{2}\right)=E_m\left(i+\frac{m}{2}-x\right) \quad \text{for } i+\frac{m-1}{2}\leqslant x\leqslant i+\frac{m+1}{2}. \tag{4.112}$$

(Compare Figure 12 where the splines E_m are shown for $m=2,3$.) Property (4.111) together with (4.112) implies

$$\|E_m\|_\infty=\varphi_m(\pi).$$

By some classical manipulations (see Schoenberg [1969a], pp. 177, 180.

$$\varphi_m(\pi)=\sum_{j=1}^{\infty}\frac{(-1)^{m(j-1)}}{(2j-1)^m}\bigg/\frac{1}{2}\left(\frac{\pi}{2}\right)^m\leqslant 1\bigg/\frac{1}{2}\left(\frac{\pi}{2}\right)^{m-2}.$$

This shows that $D_m^*\geqslant 1/\varphi_m(\pi)\geqslant\frac{1}{2}\left(\frac{\pi}{2}\right)^{m-2}$.

Figure 12. Euler splines [cf. (4.114)].

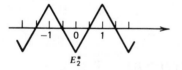

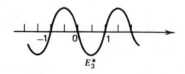

To complete the proof of (4.108), we now show that $D_m^* \leqslant \frac{1}{2}(\pi/2)^{m-2}$. Let $s(x)=\sum_{j=-\infty}^{\infty} c_j N^m(x-j)$. Then

$$s_\nu = s\left(\nu+\frac{m}{2}\right) = \sum_{j=-\infty}^{\infty} c_j M^m(\nu-j).$$

This is a discrete convolution transform whose inverse can be found (see Schoenberg [1972]):

$$c_j = \sum_{\nu=-\infty}^{\infty} \omega_\nu^m s_{\nu-j},$$

where ω_ν^m are the coefficients in the Fourier series expansion of $1/\varphi_m(x)$; that is,

$$\frac{1}{\varphi_m(x)} = \sum_{\nu=-\infty}^{\infty} \omega_\nu^m e^{i\nu x}. \tag{4.113}$$

It is also shown by Schoenberg [1969a, pp. 177, 182]. that $(-1)^\nu \omega_\nu^m > 0$, and that

$$\|\mathbf{c}\|_\infty \leqslant \sum_{-\infty}^{\infty} |\omega_\nu^m| \sup_\nu |s_\nu| \leqslant \sum_{\nu=-\infty}^{\infty} |\omega_\nu^m| \|s\|_\infty.$$

But by (4.113) with $x=\pi$,

$$\sum_{\nu=-\infty}^{\infty} |\omega_\nu^m| = \sum_{\nu=-\infty}^{\infty} (-1)^\nu \omega_\nu^m = \frac{1}{\varphi_m(\pi)}.$$

This implies $D_m^* \leqslant 1/\varphi_m(\pi)$, and the theorem is proved. ■

The spline

$$E_m^*(x) = \frac{E_m(x)}{\varphi_m(\pi)} \tag{4.114}$$

has norm 1 and interpolates the values $(-1)^i$ at the points $i+m/2$, all i. It is called the *Euler spline*.

§ 4.7 ZERO PROPERTIES

In Chapter 3 we saw that it was quite useful to have bounds on the number of zeros a polynomial can have. In this section we give similar bounds for polynomial splines. The results of this section will be used to show that spline spaces are Weak Tchebycheff spaces, and to examine important determinants formed from B-splines.

Our approach to zeros of polynomial splines will be similar to that used for polynomials. In particular, we intend to establish our bounds by induction, working with derivatives and an appropriate form of Rolle's theorem. To be useful, such a Rolle's theorem should assert that if a spline s has z zeros, then its derivative should have at least $z-1$ zeros. To formulate a precise theorem of this kind, we need to agree on what we mean by the derivative of a spline, and on how to count zeros.

The task of defining a zero count is complicated by the fact that (1) splines and their derivatives may have jumps at the knots, and (2) splines may vanish identically on intervals. (Even if we look at splines that do not vanish on intervals, as soon as we take their derivative, zero intervals may appear.)

DEFINITION 4.45. Isolated Zero

Let $x_1 < x_2 < \cdots < x_k$ and $1 \leqslant m_i \leqslant m$, $i = 1,2,\ldots,k$. Given a spline $s \in \mathbb{S}(\mathcal{P}_m; \mathfrak{M}; \Delta)$, we define the multiplicity of a zero at a point $t \in \mathbf{R}$ as follows:

Isolated Zero at t. Suppose that s does not vanish identically on any interval containing t, and that $s(t-) = D_-s(t) = \cdots = D_-^{l-1}s(t) = 0 \neq D_-^l s(t)$, while $s(t+) = D_+s(t) = \cdots = D_+^{r-1}s(t) = 0 \neq D_+^r s(t)$. Then we say that s has an isolated zero at t of multiplicity (4.115)

$$z = \begin{cases} \alpha+1, & \text{if } \alpha \text{ is even and } s \text{ changes sign at } t \\ \alpha+1, & \text{if } \alpha \text{ is odd and } s \text{ does not change sign at } t \\ \alpha, & \text{otherwise,} \end{cases}$$

where $\alpha = \max(l, r)$.

Some observations are in order. First, in this definition we have considered s to be defined on all of $\mathbf{R}$ (cf. the discussion about extending splines to all of $\mathbf{R}$ in Section 4.1). The meaning of "s changes sign" in (4.115) is the usual one—cf. Definition 2.14. If s jumps through zero at the point t (i.e., if $l = r = 0$ and s changes sign at t), then the point t counts as a zero of multiplicity 1. In general, odd-order zeros are associated with a change in sign, while even-order ones are associated with no change. We now define interval zeros.

DEFINITION 4.46. Interval Zeros

Let $s \in \mathcal{S}(\mathcal{P}_m; \mathfrak{M}; \Delta)$. We define interval zeros of s as follows:

Left-End Interval. Suppose that $s(x)=0$ for $-\infty < x < x_p$, while $s(y) \neq 0$ for some $x_p < y < x_{p+1}$. Then we say that $(-\infty, x_p)$ is a zero of s of multiplicity (4.116)

$$z = m + \sum_{i=1}^{p-1} m_i.$$

Interior Interval. Suppose that $s(x)=0$ for $x_p < x < x_q$ and does not vanish identically on any larger interval containing (x_p, x_q). Then we say that (x_p, x_q) is a zero interval of s of multiplicity (4.117)

$$z = \begin{cases} \alpha+1, & \text{if } \alpha \text{ is even and } s \text{ changes sign} \\ \alpha+1, & \text{if } \alpha \text{ is odd and } s \text{ does not change sign} \\ \alpha, & \text{otherwise,} \end{cases}$$

where $\alpha = m + \sum_{i=p+1}^{q-1} m_i$.

Right-End Interval. Suppose that $s(x)=0$ for $x_q < x < \infty$, while $s(y) \neq 0$ for some $x_{q-1} < y < x_q$. Then we say that (x_q, ∞) is a zero interval of s of multiplicity (4.118)

$$z = m + \sum_{q+1}^{k} m_i.$$

Concerning Definition 4.46, we note that if s vanishes on an interval, then the endpoints of that interval must be either $-\infty$, ∞, or a knot. The meaning of "s changes sign" in (4.117) is that for every $\varepsilon > 0$ there exist $x_p - \varepsilon < t_1 < x_p < x_q < t_2 < x_q + \varepsilon$, with $s(t_1)s(t_2) < 0$. As with isolated zeros, a spline changes sign across an odd interval zero, but it does not change sign across an even one.

Definition 4.46 allows zero intervals to be counted with multiplicity greater than m. The exact count depends on where the knots of the spline are located. Figures 13 and 14 illustrate some of the possible types of zeros that can occur for linear and quadratic splines, respectively.

We are now ready to define our zero counting procedure.

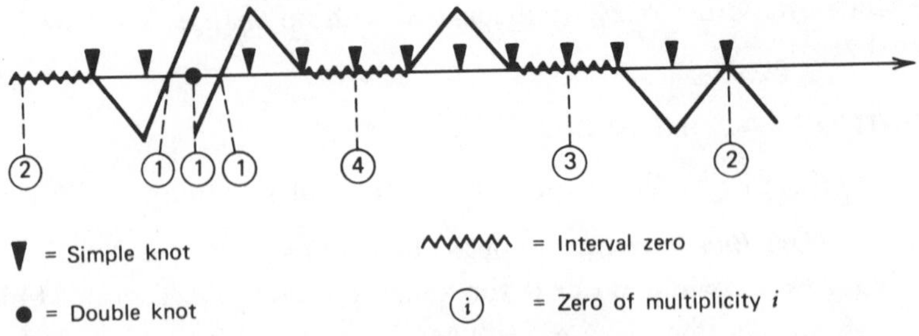

Figure 13. Zeros of a linear spline.

DEFINITION 4.47

Given a spline $s \in \mathcal{S}(\mathcal{P}_m; \mathcal{M}; \Delta)$, let $T_1, T_2, \ldots, T_d$ be points or intervals where s has zeros of multiplicities $z(T_1), \ldots, z(T_d)$, counting as in Definitions 4.45 and 4.46. We call

$$Z^{\mathcal{S}}(s) = \sum_{i=1}^{d} z(T_i) \tag{4.119}$$

the *number of zeros of s* on **R**, *relative to* $\mathcal{S} = \mathcal{S}(\mathcal{P}_m; \mathcal{M}; \Delta)$.

It is clear that the count $Z^{\mathcal{S}}(s)$ depends on $\mathcal{S}$. Since a given spline s can belong to more than one space $\mathcal{S}$, the way we count zeros of s will depend

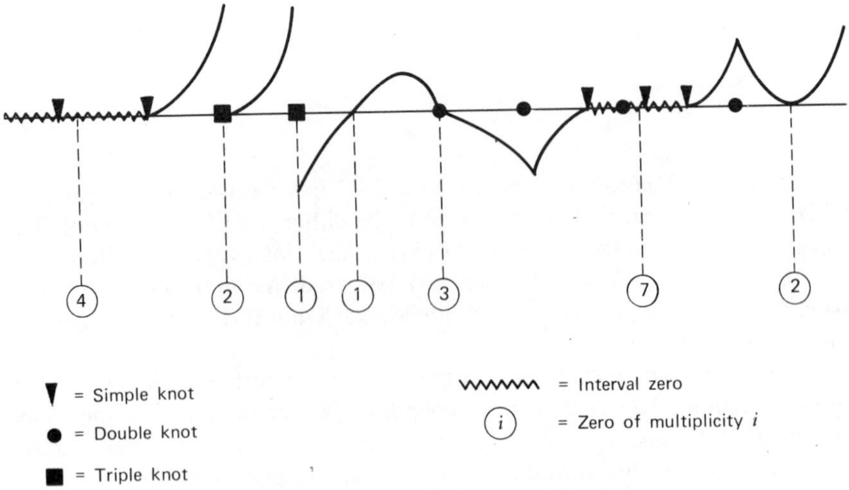

Figure 14. Zeros of a quadratic spline (m).

on which space we are counting in. The following example makes this point clearer:

EXAMPLE 4.48

Consider the spline s shown in Figure 13.

Discussion. We may consider s to belong to the space $\mathcal{S}(\mathcal{P}_m;\mathfrak{M};\Delta)$, with $m=2$, $\Delta=\{1,2,\ldots,13\}$, and $\mathfrak{M}=(1,1,2,1,1,\ldots,1)$. Relative to this space we have $Z^{\mathcal{S}}(s)=14$. On the other hand, we may also regard s as being a member of the spline space $\mathcal{S}(\mathcal{P}_m;\tilde{\mathfrak{M}};\Delta)$, with $m=3$, Δ as before, and $\tilde{\mathfrak{M}}=(2,2,3,2,2,\ldots,2)$. Relative to this space we have $\tilde{Z}^{\mathcal{S}}(s)=17$, since the interval $(-\infty,1)$ now has multiplicity 3, while the interval $(9,11)$ now has multiplicity 5. As still another example, we can also regard s as belonging to $\hat{\mathcal{S}}=\mathcal{S}(\mathcal{P}_m;\hat{\mathfrak{M}};\hat{\Delta})$, with $m=2$, $\hat{\Delta}=\{1,2,3,4,5,7,8,9,11,12,13\}$, and $\hat{\mathfrak{M}}=(1,1,2,1,\ldots,1)$. Here we have discarded the unused knots at 6 and 10. Now $\hat{Z}^{\mathcal{S}}(s)=12$. An infinitude of other counts can be obtained by considering s to be in spline spaces of higher order, or with additional unused knots. ■

While the dependence of $Z^{\mathcal{S}}(s)$ on the space $\mathcal{S}$ may seem unnatural, in practice we shall usually be working with one fixed space $\mathcal{S}(\mathcal{P}_m;\mathfrak{M};\Delta)$, and we will not have to worry about other counts.

Before stating Rolle's theorem for polynomial splines, we need to say something about derivatives. Because of their piecewise nature, polynomial splines do not have derivatives of arbitrary order at all points. It is clear from Definition 4.1, however, that there is no problem if we work with right derivatives.

THEOREM 4.49

Let $s\in\mathcal{S}(\mathcal{P}_m;\mathfrak{M};\Delta)$. Then $D_+s(x)$ exists for all x and is a right continuous function. Moreover,

$$D_+s\in\mathcal{S}(\mathcal{P}_{m-1};\mathfrak{M}';\Delta), \tag{4.120}$$

where

$$\mathfrak{M}'=(m_1',\ldots,m_k'), \qquad m_i'=\min(m-1,m_i),\ i=1,2,\ldots,k. \tag{4.121}$$

Proof. The fact that s is a piecewise polynomial assures that $D_+s(x)$ exists for all $x\notin\Delta$, while at the knots both left and right derivatives exist. The continuity of the right derivative follows from the fact that s is defined to be a polynomial on the left closed interval $[x_i,x_{i+1})$ for each $i=1,2,\ldots,k$. Clearly D_+s is a piecewise polynomial of order $m-1$. But if

$s, D_+s, \ldots, D_+^r s$ are all continuous across a knot x_i, then so are $D_+s, \ldots, D_+^{r-1}D_+s$, and (4.121) follows. ■

THEOREM 4.50. Rolle's Theorem For Splines

Suppose $s \in \mathbb{S}(\mathcal{P}_m; \mathfrak{M}; \Delta)$ and that s is continuous. Then

$$Z_{[a,b]}^{\mathfrak{D}\mathbb{S}}(D_+s) \geqslant Z_{[a,b]}^{\mathbb{S}}(s) - 1, \tag{4.122}$$

where $\mathfrak{D}\mathbb{S} = \mathbb{S}(\mathcal{P}_{m-1}; \mathfrak{M}'; \Delta)$ with $\mathfrak{M}'$ as in (4.121).

Proof. By Theorem 4.49, D_+s belongs to the space $\mathfrak{D}\mathbb{S}$. Now, if s has a z-tuple zero at the point t (relative to $\mathbb{S}$), then we claim that D_+s has a $z-1$ tuple zero at the same point or on the same interval (relative to $\mathfrak{D}\mathbb{S}$). For example, for isolated zeros we have the following situation:

α	s changes sign	$z_t(s)$	$\alpha-1$	Ds changes sign	$z_t(Ds)$
Even	Yes	$\alpha+1$	Odd	No	α
Even	No	α	Odd	Yes	$\alpha-1$
Odd	Yes	α	Even	No	$\alpha-1$
Odd	No	$\alpha+1$	Even	Yes	α

A similar situation holds for interval zeros. In addition to the zeros that D_+s inherits from s, we observe that by the extended Rolle's Theorem 2.19, between any two zeros of s, the spline D_+s must have a sign change. (Recall that we are assuming s is continuous, and thus it is absolutely continuous since it is a piecewise polynomial.) Assuming that there are a total of d points and intervals $T_1, \ldots, T_d$, where s has zeros of multiplicities $z_1, \ldots, z_d$ with $S_{[a,b]}^{\mathbb{S}}(s) = \sum_{i=1}^{d} z_i$, we find that

$$Z_{[a,b]}^{\mathfrak{D}\mathbb{S}}(D_+s) \geqslant \sum_{i=1}^{d} (z_i - 1) + d - 1 = Z_{[a,b]}^{\mathbb{S}}(s) - 1. \quad ■$$

Rolle's Theorem 4.50 for splines has been proved only for splines that are continuous (i.e., with no jumps). The following lemma will be useful in smoothing out splines with jump discontinuities:

LEMMA 4.51

Let s be a spline of order m with an m-tuple knot at ξ. Given any $\delta > 0$, there exists a spline s_δ of order m with a simple knot at $\xi - \delta$ and an $m-1$ tuple knot at ξ so that

$$s_\delta(x) = s(x) \qquad \text{for all } x \notin (\xi - \delta, \xi). \tag{4.123}$$

Moreover, if p_L and p_R are the polynomial pieces of s to the left and right of ξ, respectively, then for δ sufficiently small,

$$p_L(x) \leqslant s_\delta(x) \leqslant p_R(x), \qquad \xi - \delta \leqslant x \leqslant \xi. \tag{4.124}$$

Similarly, there exists a spline $\tilde{s}_\delta$ of order m with an $m-1$ tuple knot at ξ and a simple knot at $\xi+\delta$ such that $\tilde{s}$ agrees with s outside $(\xi, \xi+\delta)$, and (4.124) holds on $(\xi, \xi+\delta)$.

Proof. We discuss the construction of s_δ; the construction of $\tilde{s}_\delta$ is similar. Suppose

$$s(x) = p_L(x) + \sum_{i=0}^{m-1} c_i \frac{(x-\xi)_+^i}{i!}$$

for x in a neighborhood of ξ. Say $c_0 > 0$, then for any $\delta > 0$,

$$s_\delta(x) = p_L(x) + c_0 \frac{(x-\xi+\delta)_+^{m-1}}{(m-1)!\,\delta^{m-1}} + \sum_{i=1}^{m-1} \left[c_i - \frac{\delta^{-i} c_0}{(m-i-1)!} \right] \frac{(x-\xi)_+^i}{i!}$$

is an mth order spline with a simple knot at $\xi - \delta$ and an $m-1$ tuple knot at ξ. Clearly (4.123) holds for $x \leqslant \xi - \delta$. For $x \geqslant \xi - \delta$ we have

$$\frac{(x-\xi+\delta)_+^{m-1}}{(m-1)!\,\delta^{m-1}} = \sum_{i=0}^{m-1} \frac{(x-\xi)^i \delta^{-i}}{i!\,(m-i-1)!}.$$

Substituting this in $s_\delta(x)$, we see that (4.123) also holds for $x \geqslant \xi$.

To prove assertion (4.124), suppose for concreteness that $c_0 > 0$. Then clearly $s_\delta(x) \geqslant s(x) = p_L(x)$ for $\xi - \delta \leqslant x \leqslant \xi$. On the other hand, $s_\delta(\xi) = p_R(\xi)$, while $D_+ s_\delta(\xi-) \to \infty$ as $\delta \downarrow 0$. This implies $s_\delta(x) \leqslant p_R(x)$ for $\xi - \delta \leqslant x \leqslant \xi$ if δ is sufficiently small. ■

Figure 15 illustrates some of the typical cases arising in Lemma 4.51. The following lemma makes use of this method for splitting multiple knots to show that given any spline s, there is a continuous spline s_δ with the same number of knots, which is a perturbation of s, and which has the same number of zeros as s.

LEMMA 4.52

Let $s \in \mathcal{S}(\mathcal{P}_m; \mathfrak{M}; \Delta)$ with $K = \sum_{i=1}^k m_i$ knots. Then for all $\delta > 0$ sufficiently small, there exist $\tilde{\Delta}$ and $\tilde{\mathfrak{M}}$ with $K = \sum_{i=1}^{\tilde{k}} \tilde{m}_i$ and $\tilde{m}_i < m$, $i = 1, 2, \ldots, \tilde{k}$, and a spline s_δ in $\mathcal{S}(\mathcal{P}_m; \tilde{\mathfrak{M}}; \tilde{\Delta})$ such that s and s_δ are identical except in small

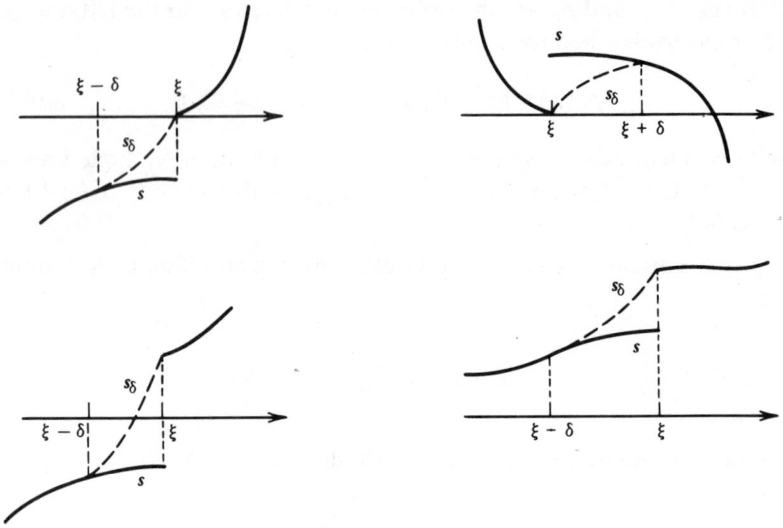

Figure 15. Splitting m-tuple knots.

intervals of length δ near the m-tuple knots of s, and such that

$$Z^{\tilde{\mathbb{S}}}_{[a,b]}(s_\delta) = Z^{\mathbb{S}}_{[a,b]}(s).$$

Proof. We may use Lemma 4.51 to split each m-tuple knot of s into a simple and an $m-1$ tuple knot. We need to exercise a little care at knots that are isolated zeros of s. If $s(\xi+)=0$, we should split the knot by moving one knot to $\xi-\delta$. If $s(\xi-)=0$, one knot should be split off and moved to $\xi+\delta$. Figure 15 shows some typical situations. It is easily checked that if we observe this rule, then s and s_δ have the same sign change properties, and the same multiplicity of zero at ξ. If s vanishes on an interval (x_i, x_j), then s_δ will vanish at least on an interval of the form $(x_i+\delta, x_j-\delta)$. Since s_δ will change sign exactly when s does, the multiplicity of the interval zeros of s and s_δ will also be the same. ■

Our main theorem concerning zeros of splines is the following:

THEOREM 4.53

For all $s \in \mathbb{S}(\mathcal{P}_m; \mathfrak{M}; \Delta)$, $\quad s \neq 0$,

$$Z^{\mathbb{S}}(s) \leqslant m + K - 1. \tag{4.125}$$

Proof. For $m=1$ the theorem is concerned with piecewise constants, and $k=K$. In this case the only kind of isolated zero possible is a jump through zero at a knot. If s vanishes on an interval (x_p, x_q), then the interval can count at most $q-p+1$. We conclude that any such s can have at most k zeros, and the theorem is proved in this case.

We now proceed by induction on m. Suppose the theorem has been established for order $m-1$ splines. Let $s \in \mathbb{S}(\mathcal{P}_m; \mathfrak{M}; \Delta)$ be such that $Z^{\mathbb{S}}(s) \geqslant m+K$. Suppose for a moment that s has no m-tuple knots. Then $s \in C(\mathbf{R})$, and by Rolle's Theorem 4.50 for splines, $Z^{D\mathbb{S}}(D_+ s) \geqslant m+K-1$. By the inductive hypothesis, we conclude that $Ds=0$. It follows that s must be a piecewise constant, but since s is continuous, it must actually be a constant. Now since s vanishes at least once, we conclude that $s \equiv 0$. We have proved the theorem in this case.

It remains to deal with the case where s may have m-tuple zeros. Given s with $Z^{\mathbb{S}}(s) \geqslant m+K$, by Lemma 4.52 there exists a continuous spline s_δ which also has K knots (none of which are m-tuple), and which also has $m+K$ zeros. The above argument shows that $s_\delta \equiv 0$. But since $s(x) = s_\delta(x)$, except in small neighborhoods of the m-tuple knots (in particular, in the middle third of each interval), it follows that $s \equiv 0$ also. ■

The bound given in Theorem 4.53 is the best possible, as the following theorem (to be proved in the next section) shows:

THEOREM 4.54

There exists a spline $s \in \mathbb{S}(\mathcal{P}_m; \mathfrak{M}; \Delta)$ with

$$Z^{\mathbb{S}}(s) = Z^1(s) = m+K-1, \tag{4.126}$$

where Z^1 counts only simple, distinct zeros and jump zeros.

The bound on the number of zeros of a spline in the space $\mathbb{S}(\mathcal{P}_m; \mathfrak{M}; \Delta)$ is precisely one less than the dimension of this space. This is the same relationship that we observed for polynomials (and, in general, for Tchebycheff systems). On the other hand, since $\mathbb{S}$ always contains splines that vanish on intervals (e.g., the plus functions), it does not form a T-space. The following theorem shows that $\mathbb{S}$ is a Weak Tchebycheff-space:

THEOREM 4.55

The space $\mathbb{S}(\mathcal{P}_m; \mathfrak{M}; \Delta)$ is a WT-space.

Proof. For any nontrivial $s \in \mathbb{S}(\mathcal{P}_m; \mathfrak{M}; \Delta)$, by Theorem 4.53

$$S^-(s) \leqslant Z^{\mathbb{S}}(s) \leqslant m+K-1.$$

Since $\mathbb{S}$ is of dimension $m+K$, Theorem 2.39 implies $\mathbb{S}$ is a WT-space. ■

In the following section we shall examine more closely determinants formed from the B-splines. There we shall see that the basis of B-splines $\{B_i\}_1^{m+K}$ for $\mathbb{S}$ forms an OCWT-system. An important tool for our analysis of B-spline determinants is the following simple corollary of Theorem 4.53:

THEOREM 4.56

Let $y_1 \leqslant y_2 \leqslant \cdots \leqslant y_{n+m}$ be given with $y_i < y_{i+m}$, $i=1,2,\ldots,n$. Suppose $\{N_i^m\}_1^n$ are the corresponding B-splines. Then for every $s=\sum_{i=1}^n c_i N_i^m$ that does not vanish on any subinterval of (y_1,y_{n+m}),

$$Z_{(y_1,y_{n+m})}(s) \leqslant n-1, \tag{4.127}$$

where Z counts isolated multiple zeros as in (4.115).

Proof. Let $\mathbb{S}$ be the space of splines with knots at $y_1,\ldots,y_{n+m}$. Then $s\in\mathbb{S}$, and by Theorem 4.53, $Z^{\mathbb{S}}(s) \leqslant n+2m-1$. But s vanishes on $(-\infty,y_1)$ and (y_{n+m},∞), and thus these are both m-tuple zeros. It follows that in (y_1,y_{n+m}) the spline s can have at most $n-1$ zeros. ■

If we apply Theorem 4.56 with $n=1$, we have another proof of the fact that the B-spline associated with $y_i,\ldots,y_{i+m}$ is nonzero throughout (y_i,y_{i+m}). Theorem 4.56 can also be used to discuss the derivatives of a B-spline.

THEOREM 4.57

Let N_i^m be the B-spline associated with the knots $y_i,\ldots,y_{i+m}$. Then

$$Z_{(y_i,y_{i+m})}(D_+^j N_i^m) \leqslant j, \qquad j=0,1,\ldots,m-1, \tag{4.128}$$

where Z counts multiplicities as in (4.115). Moreover, if $D_+^{j-1}N_i^m$ is continuous on the closed interval $[y_i,y_{i+m}]$, then $D_+^j N_i^m$ has exactly j zeros in (y_i,y_{i+m}).

Proof. By (4.23), $D_+N_i^m$ is a linear combination of N_i^{m-1} and N_{i+1}^{m-1}. Theorem 4.56 thus implies that $Z_{(y_i,y_{i+m})}(D_+s) \leqslant 1$. On the other hand, by Rolle's Theorem 4.50 for splines, if N_i^m is continuous on $[y_i,y_{i+m}]$, then $D_+N_i^m$ must have a zero between the zeros y_i and y_{i+m}. The argument can be repeated for higher derivatives. ■

Theorem 4.53 gives a bound on the number of zeros of a spline on $(-\infty,\infty)$. In general, we can give even better bounds on a finite interval

$[a,b]$, provided we have some information on the behavior of the derivatives of the spline at the points a and b. The following theorem is the analog of the Budan-Fourier Theorem 3.9 for polynomials:

THEOREM 4.58. Budan-Fourier Theorem For Polynomial Splines

Suppose $a=x_0<x_1<\cdots<x_k<x_{k+1}=b$, and let $\Delta=\{x_i\}_1^k$. Let $\mathfrak{M}=(m_1,\ldots,m_k)$ be a corresponding multiplicity vector with $1\leqslant m_i\leqslant m$, $i=1,2,\ldots,k$. Given $s\in\mathbb{S}(\mathcal{P}_m;\mathfrak{M};\Delta)$, let p_i denote the polynomial piece of s on (x_i,x_{i+1}), $i=0,1,\ldots,k$. Suppose that at least one of these polynomials is of exact order m, and that p_0 and p_k are of exact order d_0 and d_k, respectively. Then

$$Z_{(a,b)}^{\mathbb{S}}(s)\leqslant m+K-1-S^+\left[s(b-),D_-s(b),\ldots,D_-^{d_k-1}s(b)\right]$$
$$-S^+\left[s(a),-D_+s(a),\ldots,(-1)^{d_0-1}D_+^{d_0-1}s(a)\right], \tag{4.129}$$

where S^+ counts weak sign changes as in (2.46).

Proof. For $m=1$ the statement reduces to Theorem 4.53. We now proceed by induction on m, assuming the result has been proved for splines of order $m-1$. Suppose for the moment that s is continuous. Let

$$A_j=S^+\left[(-1)^jD_+^js(a),\ldots,(-1)^{d_0-1}D_+^{d_0-1}s(a)\right]$$
$$B_j=S^+\left[D_-^js(b),\ldots,D_-^{d_k-1}s(b)\right],$$

for $j=0,1,\ldots$ and let $\alpha=A_0-A_1$ and $\beta=B_0-B_1$. The numbers α and β can only be 0 or 1. As shown in the proof of the Budan-Fourier Theorem 3.9 for polynomials, $\alpha=1$ can happen only if a is a left Rolle's point for $s=p_0$, whereas $\beta=1$ can happen only if b is a right Rolle's point for $s=p_k$.

The essential ingredient in our proof is the inequality

$$Z_{(a,b)}^{\mathbb{S}}(s)\leqslant Z_{(a,b)}^{\mathfrak{D}\mathbb{S}}(D_+s)+1-\alpha-\beta, \tag{4.130}$$

which we now prove. There are two cases:

CASE 1. $Z^{\mathbb{S}}(s)=0$. Then (4.130) clearly holds if not both α and β are 1. But if they are both 1, then by the extended Rolle's Theorem 2.19, D_+s has a zero between the Rolle's points a and b.

CASE 2. $Z^{\mathbb{S}}(s)>0$. By Rolle's Theorem 4.50 for splines, D_+s has at least $Z^{\mathbb{S}}(s)-1$ zeros in (a,b). If both α and β are 0, this establishes (4.130). On the other hand, if $\alpha=1$, D_+s must have another zero between a and the first zero of s in (a,b). (Note: p_0 cannot be identically zero on an interval by assumption.) Similarly, if $\beta=1$, then D_+s has another zero between the last zero of s in (a,b), and the point b.

We are now ready to prove (4.129) for continuous s. Suppose (4.129) does not hold for such an s. Then using (4.130) we obtain

$$Z^{\mathcal{DS}}_{(a,b)}(D_+s) \geqslant Z^{\mathcal{S}}_{(a,b)}(s) - 1 + \alpha + \beta$$

$$\geqslant m + K - A_0 - B_0 + \alpha + \beta = m + K - 1 - A_1 - B_1.$$

This contradicts the inductive hypothesis, and the theorem is proved in this case.

Suppose now that (4.129) does not hold for a general s. Then we may replace s by a continuous spline s_δ with the same number of zeros and the same number of knots. (We may accomplish this with the help of Lemma 4.52, and in such a way that s and s_δ agree outside of small intervals around the multiple knots of s, and, in particular, near the points a and b.) In this case (4.129) also fails for s_δ, and the above argument leads to a contradiction as before. ■

The following example shows how improved bounds on the number of zeros a spline can have in an interval of the form (a,b) can be obtained with the use of the Budan-Fourier Theorem 4.58.

EXAMPLE 4.59

Let $[a,b]=[0,3]$, $\Delta=\{1,2\}$, and $\mathcal{M}=\{2,1\}$. Let s be the spline in $\mathcal{S}(\mathcal{P}_3;\mathcal{M};\Delta)$ shown in Figure 16.

Discussion. Here $m=3$. Theorem 4.53 asserts that $Z^{\mathcal{S}}(s) \leqslant m+K-1=5$. On the other hand, since $S^+[s(a), -D_+s(a), D_+^2 s(a)]=2$ while $S^+[s(b), D_-s(b), D_-^2 s(b)]=1$, the Budan-Fourier Theorem yields the bound $Z^{\mathcal{S}}_{(0,3)}(s) \leqslant 5-2-1=2$. ■

The following example illustrates why it is necessary to introduce the orders d_0 and d_k of the polynomials p_0 and p_k in the statement of the Budan-Fourier Theorem 4.58:

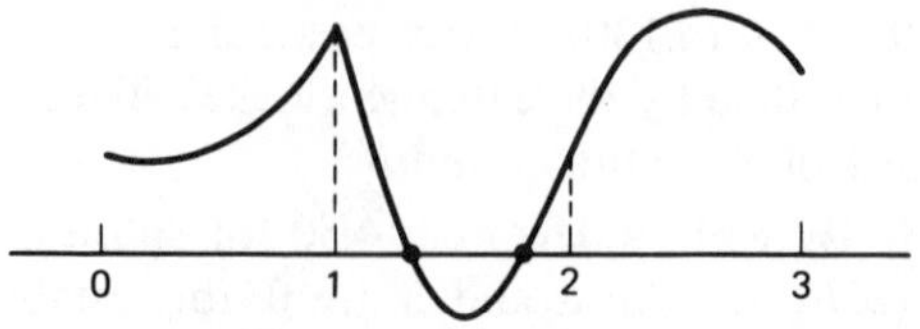

Figure 16. The spline in Example 4.59.

EXAMPLE 4.60

Let $[a,b]=[0,2]$ and $\Delta=\{1\}$, and $\mathfrak{M}=\{1\}$. Let s be the spline in $\mathcal{S}_3(\Delta)=\mathcal{S}(\mathcal{P}_3;\mathfrak{M};\Delta)$ shown in Figure 17.

Discussion. Here $m=3$, $d_0=3$, and $d_1=1$. We note that

$$S^+\left[s(a),-D_+s(a),D_+^2s(a)\right]=2$$

$$S^+\left[s(b),D_-s(b),D_-^2s(b)\right]=2.$$

Thus if we tried to subtract both of these, we would get the bound $Z_{(0,2)}^{\mathcal{S}}(s)\leqslant 3-2-2=-1$, which is incorrect since s actually has one zero in this interval. ∎

§ 4.8. MATRICES AND DETERMINANTS

Suppose $y_1<y_2<\cdots<y_{n+m}$ is a sequence of points with $y_i<y_{i+m}$, all i, and suppose $N_1^m,\ldots,N_n^m$ are the associated normalized B-splines. In this section we examine various matrices formed from these B-splines. We begin with the matrix that arises in Lagrange interpolation with $N_1^m,\ldots,N_n^m$ (cf. Problem 2.6).

THEOREM 4.61

Let $t_1<\cdots<t_n$. Then the matrix

$$M\begin{pmatrix} t_1,\ldots,t_n \\ N_1^m,\ldots,N_n^m \end{pmatrix}=\left(N_j^m(t_i)\right)_{i,j=1}^n \tag{4.131}$$

is nonsingular if and only if

$$t_i\in\sigma_i=\{x:\ N_i^m(x)\neq 0\},\qquad i=1,2,\ldots,n. \tag{4.132}$$

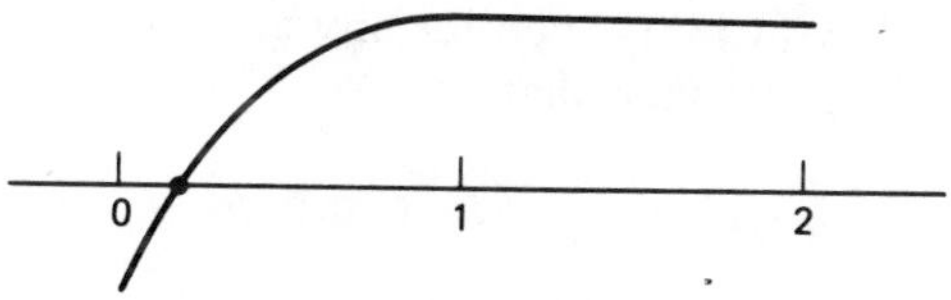

Figure 17. The spline in Example 4.60.

Proof. For convenience, we note (cf. Theorem 4.17) that

$$\sigma_i = \begin{cases} (y_i, y_{i+m}) & \text{if} \quad y_i < y_{i+m-1} \\ [y_i, y_{i+m}) & \text{otherwise,} \end{cases} \tag{4.133}$$

$i = 1, 2, \ldots, n$. The statement of the theorem is easily checked for $m = 1$. Suppose now that $m > 1$. We first show that if (4.132) fails, then the determinant D of the matrix M is zero. There are two cases.

CASE 1. Suppose t_j is too far left to lie in σ_j. Then $N_\mu^m(t_\nu) = 0$ for all $1 \leqslant \nu \leqslant j \leqslant \mu \leqslant n$. In this case the first j rows of D are clearly dependent, and so $D = 0$.

CASE 2. Suppose t_j is too far right to be in σ_j. In this case a similar argument shows that the elements in columns $1, 2, \ldots, j$ and rows $j, \ldots, n$ are all 0. Again, $D = 0$ follows.

Suppose now that (4.132) holds, but that M is nevertheless singular. Then there exist $c_1, \ldots, c_n$, not all zero, such that

$$s(t_i) = \sum_{j=1}^{n} c_j N_j^m(t_i) = 0, \qquad i = 1, 2, \ldots, n.$$

Let l be such that c_l is the first nonzero coefficient, and let $r = \min\{j \geqslant l: s(x) = 0 \text{ on an interval with left endpoint } y_{j+m}\}$. The fact that $s(x) = 0$ on (y_{r+m}, y_{r+m+1}) implies by the linear independence of the B-splines that $c_{r+1} = \cdots = c_{r+m-1} = 0$. Again, there are two cases:

CASE 1. $t_l > y_l$. Then $\tilde{s} = \sum_{i=l}^{r} c_i N_i^m$ has zeros at $t_l, \ldots, t_r$ in (y_l, y_{r+m}). This contradicts Theorem 4.56.

CASE 2. $t_l = y_l$. This can only happen if $y_l = \cdots = y_{l+m-1}$. But then we get the contradiction $0 = s(t_l) = c_l N_l^m(t_l) \neq 0$.

We have shown that M cannot be singular when (4.132) holds, and the theorem is established. ■

Applying this result to Lagrange interpolation, we obtain the following corollary:

COROLLARY 4.62

Suppose $t_1, \ldots, t_n$ satisfy (4.132). Then for any given $v_1, \ldots, v_n$ there exists a unique spline $s = \sum_{i=1}^{n} c_i N_i^m$ such that

$$s(t_i) = v_i, \qquad i = 1, 2, \ldots, n. \tag{4.134}$$

The coefficients of s can be determined by solving the system $M\mathbf{c} = v$, where M is defined in (4.131), and where $\mathbf{v} = (v_1, \ldots, v_n)^T$, $\mathbf{c} = (c_1, \ldots, c_n)^T$.

This matrix is $2m-1$ *banded* (cf. Remark 4.4). The entries of M can be computed stably and accurately using Algorithm 5.5.

Proof. The bandedness of the matrix M under the conditions (4.132) follows from the support properties of the B-splines. It has been shown (see Remark 4.6) that the matrix M is numerically well conditioned. ■

While Theorem 4.61 and its corollary are stated in terms of a general set of B-splines, it is clear that they can be applied immediately to the B-spline basis $\{B_i\}_1^n$, $n=m+K$, for the spline space $\mathbb{S}(\mathcal{P}_m;\mathfrak{M};\Delta)$.

THEOREM 4.63

Let $a \leqslant t_1 < t_2 < \cdots < t_n \leqslant b$. Then

$$M\begin{pmatrix} t_1,\ldots,t_n \\ B_1,\ldots,B_n \end{pmatrix} = (B_j(t_i))_{i,j=1}^n$$

is nonsingular if and only if

$$t_i \in \tilde{\sigma}_i = \{x\colon B_i(x) \neq 0\}, \qquad i=1,2,\ldots,n. \tag{4.135}$$

Proof. The only difference between the B's and the N's is that $B_n(b)>0$, whereas $N_n^m(b)$ can be zero if the extended partition defining the B's is taken with $b=y_{n+1}=\cdots=y_{n+m}$ (cf. Corollary 4.10). The proof that M is singular when (4.135) fails proceeds exactly as in the proof of Theorem 4.61. The converse also proceeds unchanged when $t_n<b$. On the other hand, if $t_n=b$, then we must have $b=y_{n+1}=\cdots=y_{n+m}$. Then the spline $s=\sum_1^n c_i B_i$ satisfies $0=s(t_n)=c_n B_n(t_n)$, and so $c_n=0$. But then $\hat{s}=\sum_{i=1}^{n-1} c_i B_i$ can be used to arrive at a contradiction, just as in the proof of Theorem 4.61. ■

We can now use Theorem 4.61 to establish Theorem 4.54 on the existence of splines with a maximal number of simple zeros.

Proof of Theorem 4.54. Given $\{y_i\}_1^{n+m}$, it is clear that we can always choose some $a \leqslant t_1 < t_2 < \cdots < t_n \leqslant b$ so that the condition (4.135) is satisfied. Then by the nonsingularity of M, we can solve the interpolation problem

$$s(t_i) = \sum_{j=1}^n c_j B_j(t_i) = (-1)^i, \qquad i=1,2,\ldots,n.$$

Then it is clear that s must have at least one zero between each pair of t's for a total of at least $n-1=m+K-1$. However, since s is nontrivial, it can have at most $m+K-1$ zeros, and we conclude that each of these zeros must be a simple, distinct zero; that is, $\hat{Z}(s)=Z^1(s)=m+K-1$. If there are no m-tuple knots, each of these zeros is an ordinary zero of the function. If there are m-tuple knots, then a zero at such a knot may be a jump zero. ■

In Theorem 4.55 we showed that the set of B-splines $\{N_i^m\}_1^n$ span a WT-space on $\mathbb{R}$. The following theorem provides an alternate proof of this fact, and moreover, also shows that $\{N_i^m\}_1^n$ is, indeed, a WT-system.

THEOREM 4.64

For any $t_1<t_2<\cdots<t_n$,

$$D\begin{pmatrix} t_1,\ldots,t_n \\ N_1^m,\ldots,N_n^m \end{pmatrix}=\det\left(N_j^m(t_i)\right)_{i,j=1}^n \geqslant 0, \tag{4.136}$$

and D is positive precisely when the conditions (4.132) hold.

Proof. Since we have already shown in Theorem 4.61 that $D=0$ whenever conditions (4.132) fail, it remains only to show that $D>0$ for all $\mathbf{t}=(t_1,\ldots,t_n)$ in the set $T^*=\{\mathbf{t}\colon a\leqslant t_1<t_2<\cdots<t_n\leqslant b$, and (4.132) is satisfied$\}$. We already know from Theorem 4.61 that D is never zero as $\mathbf{t}$ runs over T^*. We now show that D has one sign on T^*.

First, we claim that for all $1\leqslant i\leqslant n$ and $1\leqslant j\leqslant n$, the B-spline $B_j(t_i)$ as a function of t_i is a continuous function of t_i as it runs over the set σ_i defined in (4.133). Indeed, it is clear that the interval (y_i,y_{i+m}) cannot contain any m-tuple knots, so all B-splines are continuous for t_i in this set. On the other hand, since the B-splines are right continuous, they are all continuous as $t_i\downarrow y_i$ also. It follows that D is a continuous function of $\mathbf{t}$ as $\mathbf{t}$ runs over T^*. Since it never vanishes, it must have one sign throughout.

It remains to compute the sign of D. Let $y_1^\nu<y_2^\nu<\cdots<y_{n+m}^\nu$ be such that $y_i^\nu\to y_i$ as $\nu\to\infty$. Let $y_i^\nu<t_i<y_{i+1}^\nu$, $i=1,2,\ldots,n$. Then it is clear that $D_\nu>0$ since in this case the corresponding matrix defining D_ν has positive diagonal elements and all zeros above the diagonal. Now as $\nu\to\infty$ it also follows (cf. Theorem 4.26) that $D_\nu\to D$. ■

It is clear that the analog of Theorem 4.63 holds for the B-splines $B_1,\ldots,B_n$ on the interval $[a,b]$. The determinant associated with any $a\leqslant t_1<t_2<\cdots<t_n\leqslant b$ will be positive precisely when conditions (4.135) hold. This means that the B-spline basis functions $\{B_i\}_1^n$ form a WT-system on $[a,b]$.

The following theorem shows that the set of B-splines $\{N_i^m\}_1^n$ form an Order Complete Weak Tchebycheff system on $\mathbb{R}$:

THEOREM 4.65

For any integers $1 \leqslant \nu_1 < \cdots < \nu_p \leqslant n$ and any points $t_1 < t_2 < \cdots < t_p$,

$$D\begin{pmatrix} t_1,\ldots,t_p \\ N_{\nu_1},\ldots,N_{\nu_p} \end{pmatrix} \geqslant 0$$

and strict positivity holds if and only if

$$t_i \in \sigma_{\nu_i}, \qquad i=1,2,\ldots,p, \tag{4.137}$$

where $\{\sigma_i\}_1^n$ are the sets defined in (4.133).

Proof. The fact that $D=0$ when (4.137) fails is established exactly as in the proof of Theorem 4.61. We now assume that (4.137) holds, and proceed to show that $D>0$. Let g be the number of gaps in the sequence $\nu_1,\ldots,\nu_p$; that is, g = cardinality of the set $\{\nu_1,\nu_1+1,\ldots,\nu_p\}\backslash\{\nu_1,\ldots,\nu_p\}$. We proceed by induction on g and p.

To begin the induction, we note that the result holds for $g=0$ and all $1 \leqslant p \leqslant n$ by Theorem 4.64. Suppose now that the assertion has been established for determinants of size $p-1$, and for determinants of size p with at most $g-1$ gaps. We now prove the result for the determinant D with g gaps. There are three cases.

CASE 1. Suppose $t_j \notin \sigma_{\nu_{j+1}}$ for some $1 \leqslant j \leqslant p-1$. Then

$$D = D\begin{pmatrix} t_1,\ldots,t_j \\ N_{\nu_1},\ldots,N_{\nu_j} \end{pmatrix} D\begin{pmatrix} t_{j+1},\ldots,t_p \\ N_{\nu_{j+1}},\ldots,N_{\nu_p} \end{pmatrix}.$$

Each of these is positive by the induction hypothesis.

CASE 2. Suppose $t_j \notin \sigma_{\nu_{j-1}}$ for some $2 \leqslant j \leqslant p$. Then

$$D = D\begin{pmatrix} t_1,\ldots,t_{j-1} \\ N_{\nu_1},\ldots,N_{\nu_{j-1}} \end{pmatrix} D\begin{pmatrix} t_j,\ldots,t_p \\ N_{\nu_j},\ldots,N_{\nu_p} \end{pmatrix},$$

and again both of these are positive by the induction hypothesis.

CASE 3. Suppose

$$\begin{aligned} t_j &\in \sigma_{\nu_{j+1}}, \qquad j=1,2,\ldots,p-1 \\ t_j &\in \sigma_{\nu_{j-1}}, \qquad j=2,3,\ldots,p. \end{aligned} \tag{4.138}$$

Let i be one of the missing indices in the sequence $\nu_1,\ldots,\nu_p$, say $\nu_1 < \cdots < \nu_l < i < \nu_{l+1} < \cdots \nu_p$. Then by a basic determinantal identity (see Remark 3.2),

$$D\begin{pmatrix} t_1,\ldots,t_{l-1},t_l,t_{l+1},\ldots,t_{p-1} \\ N_{\nu_2},\ldots,N_{\nu_l},N_i,N_{\nu_{l+1}},\ldots,N_{\nu_{p-1}} \end{pmatrix} D\begin{pmatrix} t_1,\ldots,t_p \\ N_{\nu_1},\ldots,N_{\nu_p} \end{pmatrix}$$

$$= D\begin{pmatrix} t_1,\ldots,t_{p-1} \\ N_{\nu_2},\ldots,N_{\nu_p} \end{pmatrix} D\begin{pmatrix} t_1,\ldots,t_l,t_{l+1},t_{l+2},\ldots,t_p \\ N_{\nu_1},\ldots,N_{\nu_l},N_i,N_{\nu_{l+1}},\ldots,N_{\nu_{p-1}} \end{pmatrix}$$

$$+ D\begin{pmatrix} t_1,\ldots,t_{p-1} \\ N_{\nu_1},\ldots,N_{\nu_{p-1}} \end{pmatrix} D\begin{pmatrix} t_1,\ldots,t_{l-1},t_l,t_{l+1},\ldots,t_p \\ N_{\nu_2},\ldots,N_{\nu_l},N_i,N_{\nu_{l+1}},\ldots,N_{\nu_p} \end{pmatrix}.$$

The desired determinant is the second on the left-hand side. All the other determinants are either of order $p-1$ or have only $g-1$ gaps. Moreover, in view of (4.138), all of these other determinants are positive by the induction hypothesis. We conclude that $D>0$ also, and the theorem is proved. ■

The analog of Theorem 4.65 also holds for the B-splines $\{B_i\}_1^n$ of Corollary 4.10 if we require that $a \leqslant t_1 < \cdots < t_p \leqslant b$ and replace the σ's by $\tilde{\sigma}$'s. This establishes that $\{B_i\}_1^n$ is an OCWT-system on the interval $[a,b]$.

So far we have been examining the matrix that arises from Lagrange interpolation problems. It is also of interest to study the analogous matrix associated with Hermite interpolation problems. Since polynomial splines do not have arbitrarily many derivatives everywhere (in particular, higher derivatives may not exist at knots), it is convenient to introduce a slightly modified form of the Hermite Interpolation Problem 2.7.

PROBLEM 4.66. Modified Hermite Interpolation

Let $t_1 \leqslant t_2 \leqslant \cdots \leqslant t_n$ and real numbers $v_1,\ldots,v_n$ be given. Define

$$d_i = \max\{j: t_i = \cdots = t_{i-j}\}, \qquad i=1,2,\ldots,n. \tag{4.139}$$

Given sufficiently smooth functions$\{\varphi_i\}_1^n$, find $s=\sum_{i=1}^n c_i\varphi_i$ such that

$$D_+^{d_i} s(t_i) = v_i, \qquad i=1,2,\ldots,n. \tag{4.140}$$

Discussion. The only difference between this problem and the usual Hermite interpolation problem is that we have used right derivatives rather than ordinary derivatives. ■

The following theorem gives conditions under which the modified Hermite interpolation Problem 4.66 can be solved using a set of B-splines:

THEOREM 4.67

Let $N_1^m,\ldots,N_n^m$ be a set of B-splines of order m associated with the knots $y_1 \leqslant \cdots \leqslant y_{n+m}$. Let $t_1 \leqslant \cdots \leqslant t_n$ with $t_i < t_{i+m}$, all i. Then

$$D\begin{pmatrix} t_1,\ldots,t_n \\ N_1,\ldots,N_n \end{pmatrix} = \left(D_+^{d_i} N_j(t_i)\right)_{i,j=1}^n \geqslant 0, \tag{4.141}$$

and strict positivity holds if and only if

$$t_i \in \sigma_i = (y_i, y_{i+m}) \cup \left\{x : D_+^{d_i} N_i(x) \neq 0\right\}, \qquad i = 1,\ldots,n, \tag{4.142}$$

where $d_1,\ldots,d_n$ are defined in (4.139).

Proof. Theorem 4.17 implies that

$$\sigma_i = \begin{cases} [y_i, y_{i+m}), & \text{if} \quad d_i \geqslant m - \alpha_i \\ (y_i, y_{i+m}), & \text{otherwise,} \end{cases} \tag{4.143}$$

where

$$\alpha_i = \max\left\{j : y_i = \cdots = y_{i+j-1}\right\}, \tag{4.144}$$

$i = 1,2,\ldots,n$.

The fact that the determinant D is zero when condition (4.142) fails is established by the same kind of argument used in the proof of Theorem 4.61. Our next task is to show that when (4.142) holds, then D is nonzero. Suppose the contrary. Then there must be a nontrivial set $c_1,\ldots,c_n$ such that the spline $s = \sum_{i=1}^n c_i N_i^m$ satisfies

$$D_+^{d_i} s(t_i) = 0, \qquad i = 1,2,\ldots,n. \tag{4.145}$$

We now show that this leads to a contradiction. Let l be such that c_l is the first nonzero coefficient. Let $r = \min\{j \geqslant l : s(x) = 0$ on an interval with left endpoint $y_{j+m}\}$. The linear independence of the B-splines implies that $c_{r+1} = \cdots = c_{r+m-1} = 0$. As in the proof of Theorem 4.61 there are two cases.

CASE 1. $t_l > y_l$. Then $\tilde{s} = \sum_{i=l}^n c_i N_i^m$ has $r - l + 1$ zeros (counting multiplicites as in Definition 4.45) at $t_l,\ldots,t_r$ in (y_l, y_{r+m}). This contradicts Theorem 4.56.

CASE 2. $t_l = y_l$. Since (4.145) is satisfied, we must have $d_l \geqslant m - \alpha_l$. But then we get the contradiction $0 = D_+^{d_l} s(t_l) = c_l D_+^{d_l} N_l^m(t_l) \neq 0$.

We have proved that D is nonzero whenever (4.142) holds.

We now establish our assertion about positivity. The case where $t_1 < t_2 < \cdots < t_n$ follows from Theorem 4.61. But by Lemma 2.9 the sign of D does not change as the t's are allowed to coalesce. (Note: this is an application of Lemma 2.9 where the underlying functions are only right continuous.) ■

It is clear that the analog of Theorem 4.63 for the B-splines $\{B_i\}_1^n$ spanning the space $\mathcal{S}(\mathcal{P}_m; \mathfrak{M}; \Delta)$ holds provided we require $a \leqslant t_1 \leqslant t_2 \leqslant \cdots \leqslant t_n \leqslant b$, and substitute

$$\tilde{\sigma}_i = (y_i, y_{i+m}) \cup \{x: D^{d_i} B_i(x) \neq 0\}, \qquad i = 1, 2, \ldots, n \tag{4.146}$$

in (4.142). In this case we must take *left* derivatives at b.

We give one example to illustrate Hermite interpolation with B-splines.

EXAMPLE 4.68

Let $[a,b] = [0,3]$ and $\Delta = \{1,2\}$. Given $v_1, \ldots, v_4$, find $s \in \mathcal{S}_2(\Delta)$ such that $s(0) = v_1$, $D_+ s(0) = v_2$, $s(3) = v_3$, and $D_- s(3) = v_4$.

Discussion. We take the B-splines $\{B_i\}_1^4$ spanning $\mathcal{S}_2(\Delta)$ corresponding to the extended partition $\{0,0,1,2,3,3\}$. Then the required spline s is given by $s = \sum_{i=1}^4 c_i B_i$, where $c = (c_1, \ldots, c_4)^T$ is the solution of the system $M\mathbf{c} = \mathbf{v}$ with $\mathbf{v} = (v_1, \ldots, v_4)^T$ and

$$M = M\begin{pmatrix} 0,0,3,3 \\ B_1, \ldots, B_4 \end{pmatrix} = \begin{bmatrix} 1 & 0 & 0 & 0 \\ -1 & 1 & 0 & 0 \\ 0 & 0 & 0 & 1 \\ 0 & 0 & -1 & 1 \end{bmatrix}.$$

This matrix is nonsingular since its determinant has the value 1. ■

Since polynomial splines may have discontinuous derivatives at their knots, it is possible to consider a still more general kind of Hermite interpolation problem for splines in which *both left and right derivatives* of appropriate orders are specified at the *same* point. In particular, if a spline has a knot of multiplicity ρ at the point t, then s and $Ds, \ldots, D^{m-\rho-1}s$ are all continuous across t. On the other hand, $D_-^i s(t)$ and $D_+^i s(t)$ may have different values for $i = m - \rho, \ldots, m-1$.

PROBLEM 4.69 Extended Hermite Interpolation By Splines

Let $y_1 \leqslant y_2 \leqslant \cdots \leqslant y_{n+m}$ be given, and suppose $\{N_i^m\}_1^n$ are the associated normalized B-splines. Let $t_1 \leqslant t_2 \leqslant \cdots \leqslant t_n$, and define

$$\rho_i = \text{number of } y\text{'s equal to } t_i, \qquad i=1,2,\ldots,n. \tag{4.147}$$

Let $\theta_1,\ldots,\theta_n$ be a sequence of signs, and define

$$d_i = \begin{cases} \max\{j:\ t_i = \cdots = t_{i-j} \text{ with } \theta_i = \cdots = \theta_{i-j}\}, & \text{if } \theta_i = + \\ m-\rho_i+\max\{j:\ t_i = \cdots = t_{i+j} \text{ with } \theta_i = \cdots = \theta_{i+j}\}, & \text{if } \theta_i = - \end{cases}, \tag{4.148}$$

$i=1,\ldots,n$. Then given real numbers $v_1,\ldots,v_n$, we seek $s=\sum_{j=1}^n c_j N_j^m$ such that

$$D_{\theta_i}^{d_i} s(t_i) = v_i, \qquad i=1,2,\ldots,n. \tag{4.149}$$

Discussion. We make the following basic assumptions about the relationship between the t's and θ's in this problem:

$$0 \leqslant d_i \leqslant m-1; \tag{4.150}$$

$$\text{if } \theta_i = + \text{ and } \theta_{i+1} = -, \text{ then } t_i < t_{i+1}; \tag{4.151}$$

$$\text{if } \theta_i = - \text{ and } \theta_{i+1} = +, \text{ then } t_i = t_{i+m-\rho_i} \text{ and } \theta_{i+1} = \cdots = \theta_{i+m-\rho_i} = t. \tag{4.152}$$

In (4.150) we have required that the problem does not involve derivatives higher than order $m-1$. Condition (4.151) assures that the t's with associated signs are in a natural order. Finally, condition (4.152) is introduced so that a complete set of $m-\rho_i-1$ *right* derivatives are specified at a knot t_i of multiplicity ρ_i before any left derivatives are specified.

It is easy to check that if we choose plus signs for all points $t_i < b$ and minus signs for all points equal to b, the extended Hermite interpolation problem reduces to the modified Hermite Interpolation Problem 4.66. The solvability of the extended problem for arbitrary data depends on the nonsingularity of the matrix

$$M\begin{pmatrix} t_1,\ldots,t_n \\ \theta_1,\ldots,\theta_n \\ N_1,\ldots,N_n \end{pmatrix} = \left[D_{\theta_i}^{d_i} N_j^m(t_i) \right]_{i,j=1}^n. \tag{4.153}$$ ■

The analogous problem can be posed for interpolation with the basis splines $\{B_i\}_1^n$ for a spline space $\mathcal{S}(\mathcal{P}_m; \mathfrak{M}; \Delta)$ defined on an interval $[a,b]$. Then we would restrict the t's to lie in $[a,b]$, and we would also require that

$$\text{if } t_i = a, \text{ then } a = t_1 = \cdots = t_i \text{ and } \theta_1 = \cdots = \theta_i = +; \tag{4.154}$$

$$\text{if } t_i = b, \text{ then } t_i = \cdots = t_n = b \text{ and } \theta_i = \cdots = \theta_n = -. \tag{4.155}$$

These conditions assure that only right derivatives are specified at a, while only left derivatives are specified at b. Before discussing the matrix M in (4.153), we give an example of an extended Hermite interpolation problem.

EXAMPLE 4.70

Let $[a,b]=[-1,1]$ and $m=2$. Let B_1, B_2, B_3 be the basis of B-splines for the space $\mathcal{S}_2(\Delta)$ with $\Delta = \{0\}$. Consider the extended Hermite interpolation problem with $t_1 = t_2 = t_3 = 0, \theta_1 = -, \theta_2 = \theta_3 = +$.

Discussion. Here the interpolation problem calls for specifying the values of $D_- s(0)$, $s(0)$, and $D_+ s(0)$. This problem has a solution for any given data v_1, v_2, v_3 since the determinant of the associated matrix M in this case is

$$D\begin{pmatrix} 0, & 0, & 0 \\ -, & +, & + \\ B_1, & B_2, & B_3 \end{pmatrix} = \begin{vmatrix} D_- B_1(0) & D_- B_2(0) & D_- B_3(0) \\ B_1(0) & B_2(0) & B_3(0) \\ D_+ B_1(0) & D_+ B_2(0) & D_+ B_3(0) \end{vmatrix}$$

$$= \begin{vmatrix} -1 & 1 & 0 \\ 0 & 1 & 0 \\ 0 & -1 & 1 \end{vmatrix} = -1. \qquad \blacksquare$$

We now give precise conditions on when the matrix M which was defined in (4.153) in connection with extended Hermite interpolation problems is nonsingular.

THEOREM 4.71

The matrix M defined in (4.153) is nonsingular if and only if

$$t_i \in \sigma_i, \qquad i = 1, 2, \ldots, n, \tag{4.156}$$

where

$$\sigma_i = (y_i, y_{i+m}) \cup \{x: D_{\theta_i}^{d_i} N_i^m(x) \neq 0\}, \qquad i = 1, 2, \ldots, n.$$

Proof. First we note that by Theorem 4.17, the sets σ_i are given by

$$\sigma_i = \begin{cases} [y_i, y_{i+m}), & \text{if } \theta_i = + \text{and } d_i \geqslant m - \alpha_i \\ (y_i, y_{i+m}], & \text{if } \theta_i = - \text{and } d_i \geqslant m - \beta_{i+m} \\ (y_i, y_{i+m}), & \text{otherwise,} \end{cases} \tag{4.157}$$

where

$$\alpha_i = \max\{j: y_i = \cdots = y_{i+j-1}\}$$

$$\beta_{i+m} = \max\{j: y_{i+m} = \cdots = y_{i+m-j+1}\}.$$

$i = 1, 2, \ldots, n$.

The proof that the determinant D of the matrix M is zero when (4.156) fails proceeds in the same way as in our earlier theorems. The proof that D is not zero when (4.156) holds is also similar. In particular, if $D = 0$, then there exists a spline $s = \sum_{i=1}^{n} c_i N_i$ with coefficients, not all zero, such that

$$D_{\theta_i}^{d_i} s(t_i) = 0, \qquad i = 1, 2, \ldots, n. \tag{4.158}$$

We shall now show that s or a related spline $\tilde{s}$ has too many zeros.

Let l be such that c_l is the first nonzero coefficient of s. Define $r = \max\{i:\ l \leqslant i \leqslant \mu,\ c_i \neq 0\}$, where $\mu = \min\{j \geqslant l:\ s(x) = 0$ on an interval with left endpoint $y_{j+m}\}$. By definition, $c_{r+1} = \cdots = c_\mu = 0$. By the linear independence of the B-splines, $c_{\mu+1} = \cdots = c_{r+m-1} = 0$. It follows that $\tilde{s} = \sum_{i=l}^{r} c_i N_i^m$ agrees with s on the interval (y_l, y_{r+m}). Now there are three cases.

CASE 1. $t_l = y_l$. Then the assumption that (4.156) holds implies that $d_l \geqslant m - \alpha_l$. This yields the contradiction $0 = D_{\theta_l}^{d_l} s(t_l) = c_l D_{\theta_l}^{d_l} N_l^m(t_l) \neq 0$.

CASE 2. $t_r = y_{r+m}$. In this case $d_r \geqslant m - \beta_{r+m}$, and we have the contradiction $0 = D_{\theta_r}^{d_r} s(t_r) = c_r D_{\theta_r}^{d_r} N_r^m(t_r) \neq 0$.

CASE 3. $y_l < t_l$ and $t_r < y_{r+m}$. We now show that $\tilde{s}$ has too many zeros on the interval (y_l, y_{r+m}). Suppose that the point set $t_l, \ldots, t_r$ consists of the points $\tau_1 < \cdots < \tau_p$, where each τ_i is repeated l_i times with a minus sign, and r_i times with a plus sign. Then $r - l + 1 = \sum_{i=1}^{p} (l_i + r_i)$. We examine what happens at τ_i. Suppose ρ_i knots fall at τ_i (if no knots fall at τ_i, we take $\rho_i = 0$) . Depending on the relative sizes of l_i, ρ_i, and r_i, the homogeneous interpolation conditions (4.158) may force $\tilde{s}$ to have additional continuity at τ_i, which implies that it can be regarded as a spline with fewer than ρ_i knots at τ_i. Table 4 gives the number of continuity conditions e_i satisfied by $\tilde{s}$ at τ_i, the number of knots δ_i that can be removed, and a lower bound on the multiplicity z_i of the zero at τ_i, counting as in Definition 4.45.

Table 4 Continuity of $\tilde{s}$ in the Proof of Theorem 4.71.

Case	e_i	δ_i	z_i
$r_i < m-\rho_i,\ l_i = 0$	$m-\rho_i$	0	r_i
$m-\rho_i \leqslant r_i \leqslant m-\rho_i+l_i$	r_i	$-m+r_i+\rho_i$	$m-\rho_i+l_i$
$m-\rho_i+l_i < r_i$	$m-\rho_i+l_i$	l_i	r_i

We observe that in all cases $z_i = l_i + r_i - \delta_i$. Thus the number of zeros (counting multiplicities) of $\tilde{s}$ on (y_l, y_{r+m}) is $z = \sum_{i=1}^{p} z_i = r - l + 1 - \sum_1^p \delta_i$. On the other hand, since $\tilde{s}$ can be regarded as a spline with only $r+m-l-1-\sum_1^p \delta_i$ knots in this interval, it can be written as a combination of only z B-splines. This contradicts Theorem 4.56, and the converse is proved. ∎

Theorem 4.71 gives precise conditions on when the determinant D of the matrix M in (4.153) is nonzero. In some applications it is also important to know the sign of such determinants. The exact sign depends on the relationship between the θ's and the d's. To take account of this relationship, we define

$$D\begin{pmatrix} t_1,\ldots,t_n \\ \theta_1,\ldots,\theta_n \\ N_1,\ldots,N_n \end{pmatrix} = \det\left[\theta_i^{d_i} D_{\theta_i}^{d_i} N_j^m(t_i)\right]_{i,j=1}^{n}. \tag{4.159}$$

Except for the sign, this is the determinant of M.

THEOREM 4.72

The determinant D in (4.159) is always nonnegative, and it is positive precisely when condition (4.156) is satisfied.

Proof. By Theorem 4.71 we know that D is nonzero precisely when condition (4.156) is satisfied. Now if $a \leqslant t_1 < t_2 < \cdots < t_n \leqslant b$, then the fact that D is positive follows from Theorem 4.64. On the other hand, if we now let the t's coalesce, then by Lemma 2.9 the signs remain the same in the limit. Lemma 2.9 must be applied with care at the knots. In particular, the necessary Taylor expansions will have to be made to the left in case the t's have negative θ's associated with them. The factors $\theta_i^{d_i}$ introduced in the definition of D are designed to take account of the powers of -1 which arise in the Taylor expansion when working leftward. ∎

We can now prove an even stronger property about determinants formed from minors of the matrix M.

THEOREM 4.73

For any integers $1 \leqslant \nu_1 < \cdots < \nu_p \leqslant n$ and any points $a \leqslant t_1 \leqslant \cdots \leqslant t_p \leqslant b$,

$$D\begin{bmatrix} t_1,\ldots,t_p \\ \theta_1,\ldots,\theta_p \\ N_1,\ldots,N_p \end{bmatrix} \geqslant 0,$$

and strict positivity holds if and only if

$$t_i \in \sigma\left(D_{\theta_i}^{d_i} N_{\nu_i}^m\right) = (y_{\nu_i}, y_{\nu_i+m}) \cup \left\{x:\ D_{\theta_i}^{d_i} N_{\nu_i}^m(x) \neq 0\right\},$$

$$i = 1,2,\ldots,p. \tag{4.160}$$

Proof. The proof follows along the same lines as the proofs of Theorems 4.64 and 4.65 for simple t's. ■

The role of the signs $\theta_i^{d_i}$ in Theorem 4.72 and 4.73 can be better appreciated by looking at some simple examples; see Example 4.70. There are obvious analogs of Theorems 4.72 and 4.73 for the matrices and determinants formed from the set of basis splines $B_1,\ldots,B_n$ for a spline space $\mathcal{S}(\mathcal{P}_m; \mathfrak{M}; \Delta)$ defined on an interval $[a,b]$. We do not bother to state them.

§ 4.9. VARIATION-DIMINISHING PROPERTIES

In this section we show that there is a close connection between the shape of a polynomial spline s and the behavior of the coefficients of its B-spline expansion

$$s(x) = \sum_{i=1}^{n} c_i N_i^m(x). \tag{4.161}$$

We begin with two simple examples illustrating this connection.

EXAMPLE 4.74

If $c_i \geqslant 0$, $i = 1,2,\ldots,n$, then $s(x) \geqslant 0$ for all x.

Discussion. This assertion follows immediately from the fact that the B-splines take on only nonnegative values. ■

EXAMPLE 4.75

Suppose the sequence $c_1,\ldots,c_n$ is monotone increasing; that is, $c_{i+1} > c_i$, $i = 1,2,\ldots,n-1$. Then $s(x)$ is also a monotone-increasing function.

Discussion. Using (4.23), we can show (cf. Theorem 5.9) that

$$D_+s(x)=(m-1)\sum_{i=1}^{n+1}\frac{(c_i-c_{i-1})}{(y_{i+m-1}-y_i)}N_i^{m-1}(x),$$

where we set $c_0=c_{n+1}=0$ and ignore terms in the sum with a 0 denominator. Now if $c_1,\ldots,c_n$ is monotone increasing, then the coefficients of the B-spline expansion of D_+s are all nonnegative, and by Example 4.74, $D_+s(x)\geqslant 0$. ■

To state a general variation-diminishing result for B-spline expansions, we use the concept of sign changes introduced in § 2.4.

THEOREM 4.76

For any nontrivial vector $\mathbf{c}=(c_1,\ldots,c_n)$,

$$S_{\mathbf{R}}^-\left(\sum_{i=1}^{n}c_iN_i^m\right)\leqslant S^-(\mathbf{c}). \tag{4.162}$$

Proof. In Theorem 4.65 we have shown that the B-splines form an OCWT-system on $\mathbf{R}$. But then the assertion (4.162) follows immediately from Theorem 2.42. ■

If we are working on the interval $[a,b]$ with the B-spline basis $\{B_i\}_1^n$ for $\mathcal{S}(\mathcal{P}_m;\mathfrak{M};\Delta)$ in Corollary 4.10 [remember: B_n may differ slightly from N_n^m since $B_n(b)=\lim_{x\uparrow b}N_n^m(x)$], then the variation-diminishing property reads

$$S_{[a,b]}^-\left(\sum_{i=1}^{n}c_iB_i\right)\leqslant S^-(c_1,\ldots,c_n). \tag{4.163}$$

The result again follows from Theorem 2.42 since (cf. the discussion following Theorem 4.65) $\{B_i\}_1^n$ form an OCWT-system on $[a,b]$.

Theorem 4.76 yields both Example 4.74 and 4.75 as special cases. For example, if $c_1,\ldots,c_m$ are monotone increasing, then for any constant d, so is the sequence $c_1+d,\ldots,c_n+d$, and it follows that

$$\sum_{i=1}^{n}(c_i+d)N_i^m=\sum_{i=1}^{n}c_iN_i^m+d$$

can only have one sign change. Since this is true for arbitrary d, s must be monotone increasing.

Our proof of Theorem 4.76 is based on general results about OCWT-systems. If we are willing to do a little more work, then we can exploit the local support properties of the B-splines to establish a stronger version of this variation-diminishing property.

THEOREM 4.77

Suppose $s(x)=\sum_{i=1}^{n} c_i N_i^m(x)$ is such that

$$(-1)^j s(t_j)>0, \qquad j=1,2,\ldots,q \tag{4.164}$$

for some $t_1<t_2<\cdots<t_q$. Then there exist $1\leqslant i_1<i_2<\cdots<i_q\leqslant n$ such that

$$(-1)^j c_{i_j} N_{i_j}^m(t_i)>0, \qquad j=1,2,\ldots,q. \tag{4.165}$$

Proof. We say that s alternates on $t_1,\ldots,t_q$ when property (4.164) holds. We assume that s does not alternate on a larger set $t_0<t_1<\cdots<t_q$, since if it did, we could work with $t_0,\ldots,t_q$ instead. This assumption is equivalent to $s(t)\leqslant 0$ for $t<t_1$, which means that the first nonzero coefficient of s must be negative.

Let $S^-(\mathbf{c})=d-1$. By Theorem 4.76, $q\leqslant d$. Our proof proceeds by induction on d. If $d=1$, it is trivial. Suppose now that it has been proved for $d-2$ sign changes in $\mathbf{c}$. We now prove it for $d-1$ sign changes. To this end, suppose $c_1,\ldots,c_n$ have been divided into d groups as in (2.69). Since the B-splines form an OCWT-system, the set of functions $V=\{v_i\}_1^d$ defined in (2.70) form a CWT-system (cf. the proof of Theorem 2.42.). In view of the form of the v's, to prove (4.165) it will suffice to show that there exist $j_1<j_2<\cdots<j_q$ with $v_{j_i}(t_i)>0$, $i=1,\ldots,q$.

Suppose $v_i(t_i)>0$, $i=1,2,\ldots,r$. We may choose $j_i=i$, $i=1,2,\ldots,r$. If $r=q$, we are done. If not, we must show how to choose the remaining j_i's. There are two cases.

CASE 1. $v_{r+1}(t)=0$, all $t\geqslant t_{r+1}$. In this case the spline $\tilde{s}=\sum_{i=r+2}^{d}(-1)^i v_i$ alternates on the points $t_{r+1},\ldots,t_q$, and the inductive hypothesis may be used to choose $j_{r+1},\ldots,j_q$.

CASE 2. $v_{r+1}(t)>0$ for some $t>t_{r+1}$. Then by the local support structure of the v's, it follows that $v_{r+2}(t_i)=\cdots=v_d(t_i)$, $i=1,2,\ldots,r+1$. If $r=0$ this would mean that $s(t_1)=0$, contradicting our assumption about s. If $r>0$, then this would mean that $\hat{s}=\sum_{i=1}^{r}(-1)^i v_i$ alternates on $t_1,\ldots,t_{r+1}$. This would mean that it has r sign changes (but only $r-1$ sign changes in its coefficient vector), contradicting Theorem 2.39 for the WT-system $\{v_i\}_1^r$. ■

Theorem 4.77 subsumes Theorem 4.76, since if (4.164) holds (i.e., s has $q-1$ sign changes), then its coefficient vector must have at least $q-1$ sign changes. The assertion (4.165) is more precise, however. It says that c_{i_j}

must be the coefficient of a B-spline which is positive at t_j (i.e., c_{i_j} influences the behavior of s only near t_j).

§ 4.10. SIGN PROPERTIES OF THE GREEN'S FUNCTION

The Green's function

$$g_m(t;x) = \frac{(t-x)_+^{m-1}}{(m-1)!} \tag{4.166}$$

has played an important role in our development of spline bases. Since it is a Green's function (cf. Theorem 2.3), it is also important in the theory of differential equations. Hence it is of interest to examine the signs of determinants formed from g_m.

We recall from (2.21) to (2.24) that

$$D_t^j g_m(t;x) = g_{m-j}(t;x), \qquad j=0,1,\ldots,m-1$$

$$(-1)^j D_x^j g_m(t;x) = g_{m-j}(t;x), \qquad j=0,1,\ldots,m-1,$$

where D_x stands for the *right* derivative with respect to x, and D_t stands for the *left* derivative with respect to t. We may also take mixed derivatives $D_t^i D_x^j g_m(t;x)$, and these will be zero if $i+j \geqslant m$.

THEOREM 4.78

Let p be any positive integer, and suppose

$$\begin{aligned} &t_1 \leqslant t_2 \cdots \leqslant t_p \\ &y_1 \leqslant y_2 \cdots \leqslant y_p \end{aligned} \tag{4.167}$$

are given with $t_i < t_{i+m}$ and $y_i < y_{i+m}$, all i. Associated with these sets define

$$\begin{aligned} d_i &= \max\{j:\ t_{i-j} = \cdots = t_i\} \\ e_i &= \max\{j:\ y_{i-j} = \cdots = y_i\}, \end{aligned} \tag{4.168}$$

$i=1,2,\ldots,p$. Then the determinant

$$g_m\begin{pmatrix} t_1,\ldots,t_p \\ y_1,\ldots,y_p \end{pmatrix} = \det\left[D_t^{d_i} D_y^{e_j} g_m(t_i;y_j) \right]_{i,j=1}^{p} \tag{4.169}$$

is always nonnegative, and it is positive precisely when

$$y_i < t_i < y_{i+m}, \quad i = 1, 2, \ldots, p, \tag{4.170}$$

where equality is allowed on the left if $y_i = \cdots = y_{i+m-d_i-1}$, and the right side is ignored if $i+m>p$.

Proof. The proof is very much like the proofs of our earlier theorems dealing with determinants of B-splines, so we can be brief. The fact that the determinant is zero when (4.170) fails is established by examining subdeterminants, just as in the proofs of Theorems 4.61, 4.67, and so on. On the other hand, if (4.170) holds but the determinant is zero, then we can construct a spline with too many zeros.

Now we claim that the determinant D in (4.169) maintains one sign over all sets of t's and y's satisfying conditions (4.170). If the y's and t's are all distinct, this is clear from the continuity of the Green's function (and this the continuity of D), it remains to see what happens if we allow the y's or t's to coalesce. For given distinct y's, we may use Lemma 2.9 to see what happens as the t's come together. Once we have the fact that the sign does not change for coalesced t's, we may allow the y's to come together, again using Lemma 2.9.

Finally, to see what the sign of D is, we need only compute the sign for a particular set of t's and y's satisfying (4.170). Consider the case when $y_i < t_i < y_{i+1}$, $i = 1, 2, \ldots, p$. Then clearly D is the determinant of a matrix with positive diagonal elements, and zeros below the diagonal, and thus is itself positive. ■

§ 4.11. HISTORICAL NOTES

Section 4.1

The early history of piecewise polynomials and splines has been discussed in Section 1.6. The origin of the terminology "spline function" was discussed in Section 1.4. The space $\mathcal{S}(\mathcal{P}_m; \mathfrak{M}; \Delta)$ has been treated in numerous papers—there does not seem to be any standard notation for it. The idea of using a multiplicity vector such as $\mathfrak{M}$ to describe the smoothness at the knots is attributed to Curry and Schoenberg [1947]. We have elected to work with splines that are right continuous. This tradition began with Schoenberg [1946a, b].

Section 4.2

If we are willing to accept the definition of B-splines via divided differences as a *fait accompli*, then this section can be skipped (except for

Lemma 4.7 which is used later). Our purpose here has been to show that the B-splines arise naturally out of an attempt to construct splines with small support sets.

Section 4.3

It is hard to say when B-splines were first used. Schoenberg [1946a, p. 68] suggests that they were known to Laplace in connection with their role as probability density functions. Favard [1940] used them (without calling them splines). In the articles by Schoenberg [1946a, b] they were referred to as "basic spline curves." In [1967] he shortened the name to B-spline.

Schoenberg [1946a] dealt first with B-splines on equispaced knots, and, in fact, defined them via the Fourier transform formula (4.61). He also observed that they could be written as divided differences of the Green's function. The definition of B-splines for arbitrary knot sequences was suggested by Curry [1947], and carried out in the articles by Curry and Schoenberg [1947, 1966]. Most of the basic properties, including their linear independence, were already known to these authors. Schoenberg continued to champion their use as a theoretical tool throughout the 1960s. B-splines were used for computing interpolating splines by Greville [1964a].

The importance of B-splines (particularly for numerical applications) was greatly enhanced by the discovery of the basic recursion relation (4.22). It was discovered by at least three different authors simultaneously. Cox [1972] established it for simple knots via some direct calculations. The result for general knots (and the method of proof used here) is credited to deBoor [1972]. In his paper he mentions that Lois Mansfield had also discovered the recursion. The derivative formula (4.23) can also be found in the article by deBoor [1972]. The strong form of linear independence given in Theorem 4.18 was established by deBoor [1973a] using a certain dual set of linear functionals. We have elected to give a direct proof.

The fact that the normalized B-splines form a partition of unity was observed in the article by Marsden and Schoenberg [1966]. The related identity (4.33) is credited to Marsden [1970]. Estimates on the size of the normalized B-splines were given in the article by deBoor and Fix [1973] and in the work of Lyche and Schumaker [1975].

In many early papers B-splines were introduced via formula (4.42) rather than (4.16). The connection between these two definitions was shown in the text by Greville [1969b]. The Peano representation (4.38) for the divided difference was given in the articles by Schoenberg [1946a, b] and Curry and Schoenberg [1947, 1966] for equally spaced knots and for general knots, respectively. The orthogonality of the jth derivative of the

B-spline to polynomials of order j stated in (4.39) is credited to H. G. Burchard—see deBoor [1976b]. The double divided difference formula (4.43) for inner products of B-splines was used in the paper by Greville [1964a].

The continuity of the B-splines as a function of the knot locations is a part of the folklore about B-splines. A formal proof of this fact, using properties of divided differences, can be found in the article by deBoor [1976b]. We give a different proof based on the recursion relation.

Section 4.4

For more on the theory of spline functions defined on equally spaced knots, see the monograph by Schoenberg [1973] and references therein. Many of the properties listed in this section were obtained already in Schoenberg [1946a]. The inner products given in Table 2 were hand calculated.

Section 4.5

Perfect splines first arose as solutions of optimal interpolation problems, where we minimize $\|D^m f\|_{L_\infty[a,b]}$ over some set U of the form $U = \{f \in L_\infty^m[a,b]: \lambda_i f = y_i,\ i = 1,2,\ldots,N\}$ with $\{\lambda_i\}_1^N$, a set of linear functionals. In this connection we mention the paper by Favard [1940]. The minimization problem (4.76) is of this type. It was solved by Louboutin [1967]. The name seems to be attributed to Glaeser [1967, 1973]. The development presented here follows the article by Schoenberg [1971].

Section 4.6

The leading exponent of the use of dual linear functionals to study B-spline expansions has been deBoor [1966, 1968c, 1973a, 1975, 1976b, c]. The construction of the dual basis in Theorem 4.41 follows the article by deBoor [1976c] (see also deBoor [1966] and Jerome and Schumaker [1969] for some related constructions). The constants $D(m,\Delta)$, and so on, and their connection with the conditioning of the B-spline basis, were studied in deBoor's papers. The lower bounds given in Theorem 4.42 for these constants were obtained by Lyche [1978]. It is conjectured that $d_m = D(m)$ for all m. This has been shown for $m = 1,2,\ldots,10$ by deBoor [1975, 1976c], and it seems to be born out by numerical experience for higher values of m. The analysis of D_m^* for the case of equally spaced knots is credited to deBoor [1973a]. For more on Euler splines (and splines on equally spaced knots, in general), see the monograph by Schoenberg [1973] and references therein.

Section 4.7

The first paper giving bounds on the number of zeros of a spline seems to be by Johnson [1960]. He considered splines with simple knots and counted zeros (with multiplicities) according to a procedure introduced by Schoenberg [1958] for monosplines. In this count, intervals are considered to be shrunk to a point. The case of multiple knots (but counting only simple zeros) was discussed in the dissertation by Schumaker [1966] (see also Schumaker [1968b]). Braess [1971] showed that the same bound holds for splines with multiple knots, even if we count double zeros twice.

Stronger counting procedures for splines were introduced by Schumaker [1976b]. I was motivated to consider such counts by similar ones which had been used by Schoenberg [1958] and by Micchelli [1972] for monosplines (cf. Section 8.4 for results on zeros of monosplines). In Schumaker [1976b], a version of Theorem 4.53 was established in which interval zeros were counted as being of multiplicity m or $m+1$. The idea of also counting the number of knots in an interval zero when evaluating its multiplicity is credited to Pence [1976]. The knot-splitting Lemma 4.51 is taken from the article by Schumaker [1969]. The bound on the number of zeros of derivatives of B-splines given in Theorem 4.57 was established first (with a different proof) by Curry and Schoenberg [1966].

Budan-Fourier type theorems for splines were developed independently by Melkman [1974a], deBoor and Schoenberg [1976], and Schumaker [1976c]. The statements vary somewhat due to different counting procedures for multiple zeros and interval zeros. The result presented here is an improvement of the version in my 1976 paper, but with a much simpler proof.

Section 4.8

Results on determinants associated with splines were first obtained for the matrix associated with the Green's kernel. We discuss this development in the notes for Section 4.10. The first result on matrices formed from B-splines was obtained by Karlin [1968], page 503. There he established Theorem 4.67 by using results on the Green's function. Burchard also obtained Theorem 4.67 in his dissertation [1968]. A simple direct proof relying on Rolle's theorem can be found in the paper by deBoor [1976a]. The proof given here follows that given in the article by Schumaker [1976b], and it provides a nice application of our results on zeros of splines.

Theorem 4.65 asserts that the matrix (4.131) of B-spline values is *totally positive* (cf. Karlin [1968] for the definition and significance of this property). This theorem (without the exact conditions on when the determinants are positive) was also established by Karlin [1968, p. 527]. The stronger version presented here (and the proof of it) follows the work of deBoor [1976a].

The problem of extended Hermite interpolation (Problem 4.69) was studied independently by Melkman [1974b] and Lyche and Schumaker [1976]. We have followed the latter paper. Melkman's proofs are based on his version of the Budan-Fourier theorem for splines, and his results also cover certain interpolation problems with prescribed boundary conditions. For more on interpolation with side conditions, see Karlin [1971], Karlin and Pinkus [1976a], and Melkman [1977].

Section 4.9

The idea of variation-diminishing transformations is an old one. For a host of results and references, see Karlin [1968]. Theorem 4.76 for splines was first proved by Karlin [1968,] using the total positivity of the matrix of B-spline values and general results on variation-diminishing properties of totally positive matrices. The strengthened version of Theorem 4.76 is credited to deBoor [1976a].

Section 4.10

Green's functions are very important in the theory of differential equations. An early paper dealing with determinants formed from the Green's function was that by Krein and Finkelstein [1939], where Theorem 4.78 was proved for t's and y's without repetitions. The study of the determinant (4.169) in connection with splines was first carried out by Schoenberg and Whitney [1949, 1953]. They rediscovered the result for distinct t's and y's using the methods of Fourier analysis as a tool. A version of Theorem 4.78 allowing multiple y's was established in Schumaker's dissertation [1966]. The complete result allowing multiple t's as well was established first by Karlin and Ziegler [1966], using a complicated triple induction. The proof given here, based on zero properties of splines, follows Schumaker [1976b].

§ 4.12. REMARKS

Remark 4.1

The classical *symmetric functions* $\text{symm}_j(t_1,\ldots,t_p)$ are defined by the relation

$$Q(t)=(t-t_1)(t-t_2)\cdots(t-t_p)=\sum_{j=0}^{p} t^{p-j}\,\text{symm}_j(t_1,\ldots,t_p). \quad (4.171)$$

Thus $\text{symm}_j(t_1,\ldots,t_p) = D^{p-j}Q(0)/(p-j)!$. For example, we have

$$\text{symm}_0(t_1,\ldots,t_p) = 1;$$

$$\text{symm}_1(t_1,\ldots,t_p) = t_1 + \cdots + t_p;$$

$$\text{symm}_p(t_1,\ldots,t_p) = t_1 t_2 \cdots t_p.$$

An elementary calculation shows that, in general,

$$\text{symm}_j(t_1,\ldots,t_p) = \sum_{1 \leqslant i_1 < i_2 < \cdots < i_j \leqslant p} t_{i_1} t_{i_2} \cdots t_{i_j} \tag{4.172}$$

This is a sum of $\binom{p}{j}$ terms.

Remark 4.2

Tchebycheff polynomials play an important role in numerical analysis and approximation theory. For a complete treatment, see Rivlin [1974]. The Tchebycheff polynomial of the first kind is defined by $T_m(x) = \cos[m \cdot \arccos(x)]$. It is a polynomial of degree m; in particular, $T_m(x) = 2^{m-1}x^m + \cdots$ for $m \geqslant 1$. The first few Tchebycheff polynomials are given by $T_0(x) = 1$, $T_1(x) = x$, $T_2(x) = 2x^2 - 1$, and $T_3(x) = 4x^3 - 3x$. T_m is symmetric about the origin, and on $(-1, 1)$ it takes extreme values of ± 1 at the zeros of T_m', which are $\cos(i\pi/m)$, $i = 1, 2, \ldots, m-1$. We also have $T_m(-1) = (-1)^m$, while $T_m(1) = 1$.

The Tchebycheff polynomials satisfy (as do all sequences of orthogonal polynomials) an appropriate three-term recurrence relation. They also satisfy the differential equation

$$(1 - x^2)D^2T_m(x) - xDT_m(x) + m^2T_m(x) = 0. \tag{4.173}$$

Remark 4.3

The following observation is useful (cf. Schoenberg [1971]):

LEMMA

If $x_i = \cos[(m-i)\pi/m]$, $i = 0, 1, \ldots, m$, then

$$(-1)^m[x_0,\ldots,x_m]f = \frac{2^{m-2}}{m}\big(f(x_0) - 2f(x_1) + \cdots + (-1)^{m-1}2f(x_{m-1}) + (-1)^m f(x_m)\big). \tag{4.174}$$

Proof. We use the expansion (2.87). Here $\omega(t)=[(t^2-1)DT_m(t)]/m2^{m-1}$. Using the differential equation (4.173), we calculate

$$D\omega(x_i)=\frac{(-1)^{m-i}m}{2^{m-1}}, \qquad i=1,2,\ldots,m-1.$$

Now (4.174) follows since $D\omega(-1)=(-1)^m m/2^{m-2}$, while $D\omega(1)=m/2^{m-2}$. ■

Remark 4.4

A matrix $M=(M_{ij})_{i,j=1}^n$ is called $(2b-1)$ *banded* provided

$$M_{ij}=0 \qquad \text{whenever } |i-j|\geqslant b. \tag{4.175}$$

In this case, M has zeros everywhere except on the diagonal and on b super- and b subdiagonals. Banded matrices can be stored and manipulated efficiently.

Remark 4.5

Let $M=(M_{ij})_{i=1,j=1}^{m,n}$ be an m by n matrix with $m\leqslant n$. We say that M is *totally positive* provided all minor determinants (of all orders up to m) are nonnegative. More precisely, using the notation of Remark 3.2, we require that for all $1\leqslant p\leqslant m$,

$$M\begin{pmatrix} i_1,\ldots,i_p \\ j_1,\ldots,j_p \end{pmatrix}\geqslant 0 \qquad \text{for all} \quad \begin{matrix} 1\leqslant i_1<\cdots<i_p\leqslant m \\ 1\leqslant j_1<\cdots<j_p\leqslant n. \end{matrix}$$

For a comprehensive theory of totally positive matrices, see Karlin [1968]. If a matrix M is totally positive, then as a transformation it is variation diminishing; that is,

$$S^-(M\mathbf{c})\leqslant S^-(\mathbf{c}).$$

See Karlin [1968], Chapter 5. It has recently been shown by deBoor and Pinkus [1977] via a backward error analysis that for totally positive matrices M, the linear system $M\mathbf{c}=\mathbf{r}$ can be solved by Gauss elimination *without* partial pivoting. This can mean a substantial saving in computational effort without any loss in accuracy.

Remark 4.6

Theorem 4.65 shows that the matrix M formed from any set of n B-splines $N_1^m,\ldots,N_n^m$ at any set of points $t_1 \leqslant t_2 \leqslant \cdots \leqslant t_n$ is totally positive. Since M is $2m-1$ banded whenever it is nonsingular (cf. Corollary 4.62), it follows from Remarks 4.4 and 4.5 that the system $M\mathbf{c}=\mathbf{r}$ can be handled in band form and by Gauss elimination without partial pivoting. Such systems arise in performing Lagrange or Hermite interpolation with splines (cf. Corollary 4.62 and Problem 4.66).

Remark 4.7

If $N_0^{m+1},\ldots,N_m^{m+1}$ are the normalized B-splines associated with the extended partition

$$y_0=y_1=\cdots=y_m=a, \qquad b=y_{m+1}=\cdots=y_{2m+1},$$

then

$$N_i^{m+1}(x)=\binom{m}{i}\left(\frac{x-a}{b-a}\right)^i\left(\frac{b-x}{b-a}\right)^{m-i}, \qquad i=0,1,\ldots,m.$$

These are the polynomials appearing in the definition of the Bernstein polynomial of degree m (see Remark 2.5).

5

COMPUTATIONAL METHODS

One of the main reasons for the importance of polynomial spline functions is the fact that they are easy to deal with on a digital computer. In this chapter we document this assertion by examining some algorithms for storing, evaluating, and manipulating splines on a computer.

§ 5.1. STORAGE AND EVALUATION

Throughout this chapter we work with the $m+K$ dimensional linear space $\mathcal{S}(\mathcal{P}_m;\mathfrak{M};\Delta)$ defined in Definition 4.1. As shown in Theorem 4.9, if $\{N_i^m\}_1^{m+K}$ are normalized B-splines associated with an extended partition $\{y_i\}_1^{2m+K}$, then every $s\in\mathcal{S}(\mathcal{P}_m:\mathfrak{M};\Delta)$ has a *unique expansion* of the form

$$s(x)=\sum_{i=1}^{n} c_i N_i^m(x), \qquad \text{all } y_m \leqslant x < y_{m+K+1}, \tag{5.1}$$

where for convenience we write $n=m+K$. We call (5.1) the *B-spline expansion* of s.

Because of the unique connection between a spline s and its B-spline expansion coefficients $c_1,\dots,c_n$, to *store* a spline s on the computer, it suffices to store the coefficient vector $\mathbf{c}=(c_1,\dots,c_n)$. There are, of course, other ways of uniquely representing polynomial splines (e.g., via the one-sided basis discussed in Theorem 4.5 or via the piecewise polynomial expansion discussed in Section 5.3). But as we shall see in this chapter, for most applications the B-spline expansion is preferred.

We turn now to the question of *evaluating* a given B-spline expansion at a given point. Our first result shows that because of the local support properties of the B-splines, to compute $s(x)$ we need only compute a sum involving m of the B-splines.

THEOREM 5.1

Let s be as in (5.1), and suppose for some $m \leqslant l < n$ that $y_l \leqslant x < y_{l+1}$. Then

$$s(x) = \sum_{i=l+1-m}^{l} c_i N_i^m(x). \tag{5.2}$$

The values of the B-splines needed to compute $s(x)$ can be found by generating the triangular array

$$\begin{array}{cccc} & & & Q_l^1(x) \\ & & & Q_{l-1}^2(x) Q_l^2(x) \\ & \cdots & & \\ & Q_{l+2-m}^{m-1}(x) & & \\ N_{l+1-m}^m(x) & N_{l+2-m}^m(x) & \cdots & N_{l-1}^m(x) N_l^m(x). \end{array} \tag{5.3}$$

This array can be generated stably and efficiently.

Proof. By the support properties of the B-splines, for $y_l \leqslant x < y_{l+1}$ only the B-splines $N_{l+1-m}^m, \ldots, N_l^m$ have value at x, hence $s(x)$ is as in (5.2). The triangular array (5.3) can be generated recursively, starting with the fact that $Q_l^1(x) = 1/(y_{l+1} - y_l)$. Each of the succeeding rows 2 through $m-1$ can be computed using the recursion relation (4.22). This is a numerically stable process since only convex combinations of nonnegative quantities are involved. The last row of the array can be computed from the next-to-last row using the recursion

$$N_i^m(x) = (x - y_i) Q_i^{m-1}(x) + (y_{i+m} - x) Q_{i+1}^{m-1}(x), \tag{5.4}$$

which follows from (4.22) after multiplication by $(y_{i+m} - y_i)$. ∎

At times it is required to compute the value

$$s(y_{l+1}-) = \lim_{x \uparrow y_{l+1}} s(x). \tag{5.5}$$

If y_{l+1} has multiplicity m, this will generally be different from the value of $s(y_{l+1}+)$. In practice, this situation arises, for example, if we choose the extended partition with

$$y_1 = \cdots = y_m = a < b = y_{n+1} = \cdots = y_{n+m} \tag{5.6}$$

and desire the value of the spline at $x = b$ (cf. Corollary 4.10 and Example 4.11). The following corollary of Theorem 5.1 shows how to compute (5.5):

COROLLARY 5.2

Let s be as in (5.1), and suppose $y_l < y_{l+1}$. Then

$$s(y_{l+1}-) = \sum_{i=l+1-m}^{l} c_i N_i^m(y_{l+1}-), \tag{5.7}$$

and the values of the required B-splines can be found by generating the array (5.3) with $x = y_{l+1}$.

Proof. We observe that $Q_l^1(y_{l+1}-) = 1/(y_{l+1} - y_l)$; cf. (4.17). Taking the limit as $x \uparrow y_{l+1}$, we observe that both of the recursions (4.22) and (5.4) are valid at $y_{l+1}-$. ■

It is clear that to make use of Theorem 5.1 for the evaluation of $s(x)$, we first have to locate the interval $[y_l, y_{l+1})$ in which x lies. This is a standard search problem. One possibility is to use simple bisection.

ALGORITHM 5.3. Bisection—Given that x Lies in $[y_p, y_q)$ to Find l Such that $y_l \leq x < y_{l+1}$

1. $l \leftarrow p$, $u \leftarrow q$;
2. If $u - l \leq 1$, quit;
3. mid$\leftarrow \lfloor (l+u)/2 \rfloor$;
4. If $x < y$ (mid), then $u \leftarrow$mid and go to step (2);
5. $l \leftarrow$mid and go to step (2).

Discussion. The number of operations required to perform bisection depends on where x is located. It can never exceed ν, where ν is such that $2^{\nu-1} < q - p \leq 2^{\nu}$. ■

Often it is required to evaluate a B-spline expansion at a fairly large number of x's (e.g., in graphing it). In this case the particular x we are interested in very likely will lie in the same interval as the last x that was considered, or in an adjacent interval. Algorithm 5.3 can be modified to take advantage of this information.

ALGORITHM 5.4. Given x in $[y_p, y_q)$ and a Guess for l to Find l Such that $y_l \leq x < y_{l+1}$

1. If $x \geq y(l)$, then
   (a) if $x < y(l+1)$, quit;
   (b) if $x < y(l+2)$, then $l \leftarrow l+1$ and quit;
   (c) perform bisection on $[y_{l+2}, y_q)$;
2. If $x \geq y(l-1)$, then $l \leftarrow l-1$ and quit;
3. Perform bisection on $[y_p, y_{l-1})$.

Discussion. On input, l must have some definite value. If it is a good guess for the correct l, then the search will be considerably faster than in Algorithm 5.3. ■

Having found the correct l to use with x in (5.2), we now need to find the values of the corresponding B-splines. In view of Theorem 5.1, a stable and efficient way for accomplishing this is to generate the triangular array (5.3). The following algorithm may be used:

ALGORITHM 5.5. Given x in $[y_l, y_{l+1})$ to Generate $N_{l+1-m}^m(x),\ldots,N_l^m(x)$

1. For $j \leftarrow 1$ step 1 until $m-1$
   $Q(j) \leftarrow 0$;
2. $Q(m) \leftarrow 1/[y(l+1)-y(l)]$, $Q(m+1) \leftarrow 0$;
3. For $j \leftarrow 2$ step 1 until $m-1$
   for $i \leftarrow m-j+1$ step 1 until m
   (a) denom$\leftarrow y(i+l-m+j)-y(i+l-m)$;
   (b) $a1 \leftarrow [x-y(i+l-m)]/\text{denom}$;
   (c) $a2 \leftarrow 1-a1$;
   (d) $Q(i) \leftarrow a1 * Q(i) + a2 * Q(i+1)$;
4. For $i \leftarrow 1$ step 1 until m
   $Q(i) \leftarrow [x-y(i+l-m)] * Q(i) + [y(i+l)-x] * Q(i+1)$.

Discussion. The denominators computed in step 3(a) can never be zero; indeed, denom $\geqslant y(l+1)-y(l)$. Since we are interested only in the last row of the array (5.3), we have been able to use a one-dimensional array Q, overlaying the values as we proceed. The operation count (counting only multiplications and divisions) is $(3m^2+m-4)/2$. As observed in Corollary 5.2, if this algorithm is carried out with $x = y_{l+1}$, then it produces the values $N_{l+1-m}^m(y_{l+1}-),\ldots,N_l^m(y_{l+1}-)$. ■

We can now give a complete algorithm for evaluating the B-spline expansion of a spline $s \in \mathbb{S}(\mathcal{P}_m;\mathfrak{M};\Delta)$ at a given point x in $[a,b]$.

ALGORITHM 5.6. Evaluation of $s(x)$ for Given $a \leqslant x \leqslant b$

1. Compute l using Algorithm 5.4;
2. Use Algorithm 5.5 to compute $N_{l+1-m}^m(x),\ldots,N_l^m(x)$;
3. Compute $s =$ the sum in (5.2).

Discussion. If $x = b$, we should set $l = n$ in step (1). The summation in step (3) requires m operations, and hence the total operation count for steps (2) and (3) is $(3m^2+3m-4)/2$. ■

There is an alternate algorithm for computing $s(x)$ that does not make use of the array (5.3) of B-spline values, and which is more efficient than Algorithm 5.6. It is based on the following theorem:

THEOREM 5.7

Let s be as in (5.1). Then for any $1 \leqslant j \leqslant m$,

$$s(x) = \sum_{i=1}^{n+j-1} c_i^{[j]}(x) N_i^{m-j+1}(x), \tag{5.8}$$

where

$$c_i^{[1]} = c_i, \qquad i = 1, 2, \ldots, n, \tag{5.9}$$

and the $c_i^{[j]}(x)$ can be computed recursively by setting $c_{n+j}^{[j]} = c_0^{[j]} = 0$, all j, and using

$$c_i^{[j+1]}(x) = \begin{cases} 0, & \text{if } y_{i+m-j} - y_i = 0 \\ \dfrac{(x - y_i)c_i^{[j]}(x) + (y_{i+m-j} - x)c_{i-1}^{[j]}(x)}{(y_{i+m-j} - y_i)}, & \text{otherwise} \end{cases} \tag{5.10}$$

for $i = 1, 2, \ldots, n+j$ and $j = 1, 2, \ldots, m-1$. In particular, if $y_l \leqslant x < y_{l+1}$, then

$$s(x) = c_l^{[m]}(x), \tag{5.11}$$

and this value can be computed by generating the array

$$\begin{array}{ccc} c_{l+1-m}^{[1]}(x) & \cdots & c_l^{[1]}(x) \\ & c_{l+2-m}^{[2]}(x) \cdots & c_l^{[2]}(x) \\ & \cdots & c_l^{[m]}(x). \end{array} \tag{5.12}$$

If this array is generated using the value $x = y_{l+1}$, then we obtain

$$s(y_{l+1}-) = \lim_{x \uparrow y_{l+1}} s(x) = c_l^{[m]}(x). \tag{5.13}$$

Proof. Using the recursion (4.22), we obtain

$$s(x) = \sum_{i=1}^{n} c_i^{[1]} \left[\frac{(x - y_i)N_i^{m-1}(x)}{(y_{i+m-1} - y_i)} + \frac{(y_{i+m} - x)N_{i+1}^{m-1}(x)}{(y_{i+m} - y_{i+1})} \right],$$

where terms in the sum with zero denominator are to be interpreted as zero. Collecting the coefficients of $N_i^{m-1}(x)$, we see that (5.8) holds for $j=2$, with $c_i^{[2]}$ given by (5.9). The process can be continued to find similar expansions in terms of the lower-order B-splines. The formula (5.11) for $s(x)$ follows from (5.8) with $j=m$ along with the fact that if $y_l \leqslant x < y_{l+1}$, then the only first-order B-spline with value at x is $N_l^1(x)$ (and its value is 1).

To see what happens when we take $x=y_{l+1}$, we need only take the limit in (5.8) as $x \uparrow y_{l+1}$, and we conclude that (5.10) also holds at $x=y_{l+1}-$. Since $N_l^1(y_{l+1}-)=1$, (5.13) follows. ■

Theorem 5.7 leads to the following alternate method for evaluating a spline:

ALGORITHM 5.8. To Evaluate the B-Spline Expansion (5.1) of s at Given x in $[a,b]$

1. Find the correct l using Algorithm 5.4;
2. For $j \leftarrow 1$ step 1 until m
   $cx(j) \leftarrow c(j+l-m)$;
3. For $j \leftarrow 2$ step 1 until m
   for $i \leftarrow m$ step -1 until j
   (a) denom$\leftarrow y(i+l-j+1)-y(i+l-m)$;
   (b) $a1 \leftarrow [x-y(i+l-m)]/$denom,
   $a2 \leftarrow 1-a1$,
   $cx(i) \leftarrow a1 * cx(i) + a2 * cx(i-1)$.
4. $s \leftarrow cx(m)$.

Discussion. The denominator in step 3(b) is never zero since $y_l < y_{l+1}$. As we work only with convex combinations of nonnegative quantities, this algorithm is also extremely stable. Ignoring the work required to find l, the operation count is $(3m^2-3m)/2$. This is considerably cheaper than Algorithm 5.6—see Table 5 for comparison. If $x=b$, we should set $l=n$ in step (1). ■

Table 5. A Comparison of the Operation Counts of Algorithms 5.6 and 5.8

m	Algorithm 5.6 $\frac{3m^2+3m-4}{2}$	Algorithm 5.8 $\frac{3m^2-3m}{2}$
2	7	3
3	16	9
4	28	18
5	43	30

§ 5.2. DERIVATIVES

In this section we discuss the computation of derivatives of a B-spline expansion. Theorem 4.49 shows that the right derivative of a polynomial spline of order m is a polynomial spline of order $m-1$. The following theorem shows how to find its B-spline expansion:

THEOREM 5.9

Let $s=\sum_{i=1}^{n} c_i N_i^m$, and suppose $1 \leqslant d \leqslant m$. Then for all $y_m \leqslant x < y_n$,

$$D_+^{d-1}s(x)=\sum_{i=d}^{n} c_i^{(d)} N_i^{m-d+1}(x), \tag{5.14}$$

where $c_i^{(1)}=c_i$, $i=1,2,\ldots,n$, and

$$c_i^{(j)}=\begin{cases}(m-j+1)\dfrac{(c_i^{(j-1)}-c_{i-1}^{(j-1)})}{(y_{i+m-j+1}-y_i)}, & \text{if } (y_{i+m-j+1}-y_i)>0 \\ 0, & \text{otherwise}\end{cases} \tag{5.15}$$

for $i=j,\ldots,n$ and $j=2,3,\ldots,d$.

Proof. For $d=1$, there is nothing to prove. Now, using (4.23),

$$D_+s(x)=(m-1)\sum_{i=1}^{n} c_i\left[Q_i^{m-1}(x)-Q_{i+1}^{m-1}(x)\right]$$

$$=(m-1)\sum_{i=2}^{n}(c_i-c_{i-1})Q_i^{m-1}(x)=(m-1)\sum_{i=2}^{n} c_i^{(2)} N_i^{m-1}(x).$$

For x in the interval $[y_m,y_n)$, the B-splines $Q_1^{m-1}(x)$ and $Q_{n+1}^{m-1}(x)$ have no value, hence we were able to leave these terms out in rearranging the first sum. This same argument can be repeated to compute the higher derivatives. ■

It is a relatively easy task to convert Theorem 5.9 to a numerical algorithm for computing the coefficients of the B-spline expansions of all derivatives up to order $m-1$ of a given spline. Derivatives of order m and higher are of no interest as they are zero wherever they exist. Let

$$D_+^{d-1}s(x)=\sum_{j=d}^{n} cd(d,j)N_j^{m-d+1}(x), \qquad a \leqslant x < b. \tag{5.16}$$

ALGORITHM 5.10. Computation of the matrix $[cd(j,i)]_{j=1,i=j}^{m}{}^{n}$

1. For $i \leftarrow 1$ step 1 until n
   $cd(1,i) \leftarrow c(i)$;
2. For $j \leftarrow 2$ step 1 until m
   $mj \leftarrow m-j+1$
   for $i \leftarrow n$ step -1 until j
   (a) demon$\leftarrow y(i+mj)-y(i)$;
   (b) $cd(j,i) \leftarrow \begin{cases} 0, & \text{if denom}=0 \\ mj * \dfrac{[cd(j-1,i)-cd(j-1,i-1)]}{\text{denom}}, & \text{otherwise.} \end{cases}$

Discussion. This algorithm requires $(m-1)(2n-m)$ operations. ■

We can now give an algorithm for computing the values of derivatives of a B-spline expansion at a given point.

ALGORITHM 5.11. Computation of $D^{d-1}s(x)$ for Given $a \leqslant x \leqslant b$

1. Compute l by Algorithm 5.4.
2. Use either Algorithm 5.6 or Algorithm 5.8 with $\tilde{m}=m-d+1$ and c set equal to the dth row of the matrix cd.

Discussion. If we use Algorithm 5.6, then the operation count to compute $D^{d-1}s(x)$ is $(3\tilde{m}^2-3\tilde{m}-4)/2$. If Algorithm 5.8 is used, the count is $(3\tilde{m}^2-3\tilde{m})/2$. ■

Our next algorithm deals with the problem of computing the entire set of derivatives $s(x), Ds(x),\ldots,D^{m-1}s(x)$.

ALGORITHM 5.12. Computation of $s(x), Ds(x),\ldots,D^{m-1}s(x)$ at Given $a \leqslant x \leqslant b$

1. Compute l by Algorithm 5.4;
2. $D^{m-1}s(x) \leftarrow cd(m,l)$;
3. For $d \leftarrow 1$ step 1 until $m-1$ compute $D^{d-1}s(x)$ using Algorithm 5.11.

Discussion. Not counting the work to find l and assuming that the array cd has been precomputed, this algorithm requires $\sum_{i=2}^{m}(3i^2-3i)/2=(m+1)m(m-1)/2$ operations. ■

There is a second approach to computing the set $s(x), Ds(x),\ldots,D^{m-1}s(x)$ which (as we shall see) is more efficient for $m>6$. The idea here is to use the fact that for $y_l \leqslant x < y_{l+1}$ and any $1 \leqslant j \leqslant m$,

$$D_+^{m-j}s(x) = \sum_{i=l+1-j}^{l} cd(m-j+1,i)(y_{i+j}-y_i)Q_i^j(x). \tag{5.17}$$

The values of the Q's needed to compute these derivatives are all contained in the array (5.3). Thus following the ideas developed in Algorithm 5.5, we have the following alternative to Algorithm 5.11:

ALGORITHM 5.13. Computation of $s(x), Ds(x), \ldots, D^{m-1}s(x)$ at given $a \leqslant x \leqslant b$

1. Compute l by Algorithm 5.4;
2. $D^{m-1}s(x) \leftarrow cd(m,l)$;
3. For $j \leftarrow 1$ step 1 until $m-1$
   $Q(j) \leftarrow 0$;
4. $Q(m) \leftarrow 1/[y(l+1)-y(l)]$; $Q(m+1) \leftarrow 0$;
5. For $j \leftarrow 2$ step 1 until $m-1$
   (a) for $i \leftarrow m-j+1$ step 1 until m
      (i) denom$\leftarrow y(i+l-m+j)-y(i+l-m)$;
      (ii) $a1 \leftarrow (x-y(i+l-m))/$denom;
      (iii) $a2 \leftarrow 1-a1$;
      (iv) $Q(i) \leftarrow a1 * Q(i) + a2 * Q(i+1)$;
   (b) $D^{m-j}s(x) \leftarrow \Sigma_{v=l+1-j}^{l} cd(m-j+1,v)*[y(v+j)-y(v)]*$
   $Q(m-l+v)$;
6. For $i \leftarrow 1$ step 1 until m
   $Q(i) \leftarrow [x-y(i+l-m)]*Q(i)+[y(i+l)-x]*Q(i+1)$;
7. $s(x) \leftarrow \Sigma_{v=l+1-m}^{l} c(v)*Q(m-l+v)$.

Discussion. Ignoring the work to compute l and the matrix cd, the operation count for this algorithm is $(5m^2+m-8)/2$. We compare Algorithms 5.12 and 5.13 in Table 6 where we see that Algorithm 5.13 is more efficient for $m \geqslant 6$. In practice, however, this is not very significant since we usually work with $m<6$. ∎

Table 6. Comparison of the Operation Counts of Algorithms 5.12 and 5.13

m	Algorithm 5.12 $\frac{(m+1)m(m-1)}{2}$	Algorithm 5.13 $\frac{5m^2+m-8}{2}$
2	3	7
3	12	20
4	30	38
5	60	61
6	105	89

§ 5.3. THE PIECEWISE POLYNOMIAL REPRESENTATION

While the B-spline expansion is the preferred way to deal with splines on a computer in most cases, there are some situations where it may be advantageous to generate the coefficients of each of the polynomial pieces (e.g., when we have to evaluate the spline at a large number of points, as might be the case if we want to graph it).

DEFINITION 5.14. Piecewise Polynomial Representation

Let $\Delta=\{x_1<x_2<\cdots<x_k\}$ be a partition of the interval $[a,b]$ into pieces $I_0=[a,x_1), I_1=[x_1,x_2),\dots,I_k=[x_k,b]$. Suppose $s_0,s_1,\dots,s_k$ are the polynomials representing a spline $s\in\mathcal{S}(\mathcal{P}_m;\mathfrak{M};\Delta)$ on each subinterval $I_0,\dots,I_k$, respectively. In particular, suppose

$$s_0(x)=\sum_{j=1}^{m} cw_{0j}(x-x_1)^{j-1} \tag{5.18}$$

and

$$s_i(x)=\sum_{i=1}^{m} cw_{ij}(x-x_i)^{j-1}, \qquad i=1,2,\dots,k. \tag{5.19}$$

We call this the *piecewise polynomial representation* of s.

It is evident from Definition 5.14 that the piecewise polynomial representation of a spline s is completely determined by the matrix

$$CW=(cw_{ij})_{i=0,j=1}^{k\ \ m}. \tag{5.20}$$

Thus this representation provides an alternate way of storing a polynomial spline. In general, this representation is bulkier than the B-spline expansion in that $m(k+1)$ values must be stored here as compared with $m+K$ for the B-spline expansion.

Given the B-spline expansion of a spline, it is a relatively easy matter to compute the matrix CW of its piecewise representation. From (5.18) and (5.19) we find that

$$cw_{ij}=\frac{D_+^{j-1}s(x_i)}{(j-1)!}, \qquad \begin{array}{l} j=1,2,\dots,m \\ i=1,2,\dots,k, \end{array} \tag{5.21}$$

and

$$cw_{0j}=\frac{D_-^{j-1}s(x_1)}{(j-1)!}, \qquad j=1,2,\dots,m. \tag{5.22}$$

This leads to the following numerical method:

ALGORITHM 5.15. Conversion From the B-Spline Expansion to the Piecewise Polynomial Representation

1. Compute the matrix cd by Algorithm 5.10;
2. For $j \leftarrow 1$ step 1 until m
   compute $\text{fac}(j) = (j-1)!$;
3. For $i \leftarrow 1$ step 1 until k
   (a) compute $s(x_i), \ldots, D^{m-1}s(x_i)$ by Algorithm 5.12;
   (b) for $j \leftarrow 1$ step 1 until m
   $CW(i,j) \leftarrow D^{j-1}s(x_i)/\text{fac}(j)$;
4. Compute $s(x_1-), D_-s(x_1), \ldots, D_-^{m-1}s(x_1)$ by Algorithm 5.12;
5. For $j \leftarrow 1$ step 1 until m
   $CW(0,j) \leftarrow D_-^{j-1}s(x_1)/\text{fac}(j)$.

Discussion. Not counting the calculation of fac or cd, the operation count for this algorithm is $(k+1)m(m^2+1)/2$. ■

Once we have the piecewise polynomial representation of a spline, we can use Horner's scheme to evaluate it at any given point. This requires (cf. Algorithm 3.2) only $m-1$ operations as compared to the $3m(m-1)/2$ operations required by the most efficient of the methods discussed in §5.2 using the B-spline expansion. This difference can become quite significant if it is required to compute s at a large number of points. Derivatives can also be computed from the piecewise polynomial representation (using Horner's scheme or synthetic division—see Remark 3.1) with significantly fewer operations than required by the methods of §5.2.

The decision as to whether it pays to carry out the conversion to the piecewise polynomial representation can only be made once the user has decided how often the spline and/or its derivatives will be evaluated. For $m = 2,3,4$ the breakpoint is about two evaluations per interval.

§ 5.4. INTEGRALS

In this section we deal with definite and indefinite integrals and with inner products of splines. We begin by showing that the indefinite integral of a spline is again a spline.

THEOREM 5.16

Let $s \in \mathcal{S}(\mathcal{P}_m; \mathfrak{M}; \Delta)$. Then

$$D_a^{-1}s(x) = \int_a^x s(t)\,dt \in \mathcal{S}(\mathcal{P}_{m+1}; \mathfrak{M}; \Delta). \tag{5.23}$$

Proof. It is clear that $D_a^{-1}s$ is a piecewise polynomial of order $m+1$. But if $s,\ldots,D^j s$ are continuous at a knot x_i, then $D_a^{-1}s,\ldots,D^{j+1}D_a^{-1}s$ will also be continuous at x_i, and we conclude that $D_a^{-1}s \in \mathcal{S}(\mathcal{P}_{m+1};\mathfrak{M};\Delta)$. ■

In view of Theorem 5.16 the antiderivative of a spline can be expanded in terms of B-splines of order $m+1$. The following theorem shows how the coefficients of this expansion can be computed from the coefficients of the B-spline expansion of s:

THEOREM 5.17

Suppose $s=\sum_{i=1}^{n} c_i N_i^m$. Then for all $x<y_{n+1}$,

$$D_{y_1}^{-1}s(x)=\int_{y_1}^{x} s(t)\,dt=\sum_{i=1}^{n} c_i^{(-1)}N_i^{m+1}(x), \tag{5.24}$$

where

$$c_i^{(-1)}=\sum_{j=1}^{i} c_j\frac{(y_{m+j}-y_j)}{m}, \qquad i=1,2,\ldots,n. \tag{5.25}$$

Proof. Choose $y_0<y_1$ and $y_{n+m+1}>y_{n+m}$. Then on $[y_1,y_{n+1}]$ we know that $D_{y_1}^{-1}s(x)$ can be expanded as a linear combination of the B-splines $N_0^{m+1},\ldots,N_n^{m+1}$; suppose

$$D_{y_1}^{-1}s(x)=\sum_{i=0}^{n} c_i^{(-1)}N_i^{m+1}(x). \tag{5.26}$$

We note that $c_0^{(-1)}=D_{y_1}^{-1}s(y_1)/N_0^{m+1}(y_1)=0$. To find the other coefficients, we observe that $s=D_+D^{-1}s$, and so by applying Theorem 5.9 to the B-spline expansion (5.26),

$$c_i=\begin{cases} m\dfrac{(c_i^{(-1)}-c_{i-1}^{(-1)})}{(y_{i+m}-y_i)}, & \text{if } y_{i+m}-y_i>0\\ 0, & \text{otherwise,}\end{cases}$$

and (5.25) follows. ■

We note that the following corollary of Theorem 5.17 gives the B-spline expansion of the antiderivative of a single B-spline:

COROLLARY 5.18

Suppose $y_k < y_{k+m}$. Then

$$D_{y_k}^{-1} N_k^m(x) = \begin{cases} 0, & x < y_k \\ \dfrac{(y_{m+k} - y_m)}{m} \displaystyle\sum_{i=k}^{k+m-1} N_i^{m-1}(x), & y_k \leqslant x < y_{k+m} \\ \dfrac{(y_{m+k} - y_m)}{m}, & y_{m+k} \leqslant x. \end{cases} \quad (5.27)$$

Proof. Let $c_k = 1$ and $c_i = 0$, all $i \neq k$ in Theorem 5.17. Then

$$c_i^{(-1)} = \begin{cases} 0 & \text{if } i < k \\ \dfrac{(y_{m+k} - y_m)}{m} & \text{if } i \geqslant k. \end{cases}$$

■

Theorem 5.17 is easily converted to a numerical algorithm. Let

$$mD_{y_1}^{-1} s(x) = m \int_{y_1}^{x} s(t)\, dt = \sum_{i=1}^{n} CI(i) N_i^{m+1}(x). \quad (5.28)$$

ALGORITHM 5.19. Computation of the Coefficients of $mD_{y_1}^{-1}s$

1. Sum←0;
2. For i←1 step 1 until n
   (a) sum←sum$+[y(i+m) - y(i)]*c(i)$;
   (b) $CI(i)$←sum.

Discussion. This is simply formula (5.25), where to save work we have not divided by m at every step. The number of operations is clearly n. ■

Now we can describe an algorithm for computing definite integrals.

ALGORITHM 5.20. To Compute $I = \int_c^d s(t)\,dt$ for Given $a \leqslant c < d \leqslant b$

1. Use Algorithm 5.8 to compute $Id = mD_{y_1}^{-1}s(d)$;
2. Similarly compute $Ic = mD_{y_1}^{-1}s(c)$;
3. $I = (Id - Ic)/m$.

Discussion. It is assumed that the coefficients $\{CI(j)\}_1^n$ of the antiderivative of s have already been computed by Algorithm 5.19. The operation count (ignoring the computation of the matrix CI and the location of the intervals in which c and d lie) is $3m^2 - 3m + 1$. ■

We now discuss the question of computing *inner products* of B-splines with each other, and with other functions. In particular, suppose we want to compute

$$\int_{y_i}^{y_{i+m}} N_i^m(x) f(x)\, dx \tag{5.29}$$

for a function f defined on (y_i, y_{i+m}). If f is a polynomial, then one approach to computing (5.29) is to use the explicit formulae for the moments of the B-spline given in (4.39). Since this involves computing the functions $\rho_\nu(y_i, \ldots, y_{i+m})$, this will only be practical for polynomials of very low order.

In general, we recommend the use of Gauss quadrature. To this end, we recall the following result from numerical analysis:

THEOREM 5.21. Gauss Quadrature

There exist points

$$-1 < z_1 < z_2 < \cdots < z_m < 1$$

(symmetric about zero) and positive weights $w_1, \ldots, w_m$ so that the quadrature formula

$$\int_{-1}^{1} f \approx Qf = \sum_{j=1}^{m} w_j f(z_j)$$

for integrating functions f on $[-1,1]$ is exact for polynomials of order $2m$; that is,

$$\int_{-1}^{1} f(x)\, dx = Qf \qquad \text{for all } f \in \mathcal{P}_{2m}.$$

Discussion. This result is proved in most numerical analysis texts. The weights are also symmetric. For low values of m we give the sample points and weights in Tables 7 and 8, respectively. If f is defined on an interval $[a,b]$, then we may make a change of variables to obtain

$$\int_a^b f(x)\, dx \approx Qf = \frac{(b-a)}{2} \sum_{j=1}^{m} w_j f\left[\frac{z_j(b-a)}{2} + \frac{(b+a)}{2}\right], \tag{5.30}$$

and this formula will also be exact for all $f \in \mathcal{P}_{2m}$. ■

Table 7. Gauss Quadrature Sample points

m	z_2	z_3	z_4	z_5
2	0.5773502692			
3	0	0.7745966692		
4		0.3399810436	0.8611363116	
5		0	0.5384693101	0.9061798459

Table 8. Gauss Quadrature Weights

m	w_2	w_3	w_4	w_5
2	1			
3	0.8888888889	0.5555555556		
4		0.6521451549	0.3478548451	
5		0.5688888889	0.4786286705	0.2369268851

While for arbitrary functions f the Gauss quadrature rule in (5.30) only produces an approximation to the inner product (5.29), it can be used to compute inner products of B-splines *exactly*. In particular, if we write

$$G_{ij}=\int_{y_i}^{y_{i+m}} N_i^m(x)N_j^m(x)\,dx=\sum_{\nu=i}^{i+m-1}\int_{y_\nu}^{y_{\nu+1}} N_i^m(x)N_j^m(x)\,dx, \quad (5.31)$$

then the value of G_{ij} can be computed exactly by applying Gauss quadrature to each of the m integrals in this sum. By the support properties of the B-splines, it is clear that G_{ij} is zero except for $i+1-m \leqslant j \leqslant i+m-1$.

The matrix

$$G=(G_{ij})_{i,j=1}^{n}, \quad (5.32)$$

with G_{ij} given by (5.31), is called the *Gram matrix* associated with the functions $\{N_i\}_1^n$. It is positive definite, symmetric, and is $2m-1$ banded. We recommend the following algorithm for computing the Gram matrix:

ALGORITHM 5.22. Computation of the Gram Matrix (5.32)

1. For $v \leftarrow m$ step 1 until n
   if $y_{v+1} > y_v$, then for $u \leftarrow 1$ step 1 until m
   (a) use Algorithm 5.5 to compute the B-splines with value at the point $t_{uv}=z_u(y_{v+1}-y_v)/2+(y_{v+1}+y_v)/2$;
   (b) for all i and j such that both N_i^m and N_j^m have value at t_{uv}, add $w_u N_i^m(t_{uv})N_j^m(t_{uv})$ to G_{ij}.

Discussion. Since G is a banded matrix, it should be stored in banded form. This algorithm requires order nm^3 operations, and since it is based on B-splines, it is extremely stable. The Gram matrix can also be computed by taking divided differences (in both variables) of the Green's function $(x-y)_+^{2m-1}$ (see Theorem 4.25). If the computation is properly arranged, this can be done in order nm^2 operations. However, it is well known that the use of divided differences is a relatively unstable process. For a numerical comparison of the two approaches, see Remark 5.1. The use of divided differences can only be recommended for small m and for knots with relatively uniform spacing. For equally spaced knots the inner products can be given explicitly—see the following section. ∎

§ 5.5. EQUALLY SPACED KNOTS

As we saw in Section 4.4, there are some substantial simplifications in various formulae involving B-splines in the case of equally spaced knots. In this section we show how these simplifications can be exploited to design more efficient algorithms. We begin with a definition.

DEFINITION 5.23. Uniform Partition

Given an interval $[a,b]$ and an integer $k \geqslant 1$, let

$$x_i = a + ih, \qquad i = 0, 1, \ldots, k+1, \text{ where } h = \frac{(b-a)}{(k+1)}. \tag{5.33}$$

Then we say that $\Delta = \{x_i\}_1^k$ defines a *uniform partition of* $[a,b]$ *with mesh size* h.

In this section we deal with the space $\mathcal{S}_m(\Delta)$ of splines with simple knots (cf. Definition 1.2) at the points of a uniform partition Δ. As a basis for this space we shall take the normalized B-splines $\{N_i^m\}_1^n$, $n = m + k$, corresponding to the extended partition

$$y_i = a + (i-m)\cdot h, \qquad i = 1, 2, \ldots, k+2m. \tag{5.34}$$

The advantage of this choice of extended partition is that all B-splines are translates of the basic B-spline N^m defined in §4.4 (cf. Theorem 4.32). Thus every spline $s \in \mathcal{S}_m(\Delta)$ has a unique expansion of the form

$$s(x) = \sum_{i=1}^{n} c_i N^m\left(\frac{x - y_i}{h}\right). \tag{5.35}$$

It follows from (5.35) that to *store* s on a computer we may store its coefficients $c_1, \ldots, c_n$. To *evaluate* $s(x)$ we may use the following simplified version of Algorithm 5.8:

ALGORITHM 5.24. Evaluation of $s(x)$ for Given $a \leqslant x \leqslant b$

1. Find l using Algorithm 5.4;
2. $xdh \leftarrow (x-a)/h - l + m$;
3. For $j \leftarrow 1$ step 1 until m
   $cx(j) \leftarrow c(j+l-m)$;
4. For $j \leftarrow 2$ step 1 until m
   for $i \leftarrow m$ step -1 until j
   $cx(i) \leftarrow (xdh+m-i)*cx(i)+(i-j+1-xdh)*cx(i-1)$;
5. $s \leftarrow cx(m)/(m-1)!$.

Discussion. This algorithm takes a total of $m(m-1)+2$ operations if we assume that the value of $(m-1)!$ has been precomputed. This is a slight savings over Algorithm 5.8 for $m>2$. In particular, for $m=4$ (cubic splines) this algorithm requires 14 operations as compared with 18 for Algorithm 5.8. ∎

It is also possible to evaluate $s(x)$ using Algorithm 5.6. In this regard it is useful to have a streamlined version of Algorithm 5.5 for computing B-splines. The following algorithm makes use of the recursion relation (4.52):

ALGORITHM 5.25. Generation of $N_{l+1-m}^m(x),\ldots,N_l^m(x)$ for $y_l \leqslant x < y_{l+1}$

1. $xdh \leftarrow (x-a)/h - l + 2m$;
2. For $j \leftarrow 1$ step 1 until $m-1$
   $Q(j) \leftarrow 0$;
3. $Q(m) \leftarrow 1$, $Q(m+1) \leftarrow 0$;
4. For $j \leftarrow 2$ step 1 until m
   for $i \leftarrow m-j+1$ step 1 until m
   $Q(i) \leftarrow (xdh-i)*Q(i)+(i+j-xdh)*Q(i+1)$;
5. For $i \leftarrow 1$ step 1 until m
   $Q(i) \leftarrow Q(i)/(m-1)!$.

Discussion. Assuming that the value of $(m-1)!$ has been precomputed, this algorithm requires m^2+2m-1 operations. This is somewhat cheaper than Algorithm 5.5 if $m>3$. For example, for $m=4$ it uses 23 operations as compared to 24 for Algorithm 5.5. ∎

If Algorithm 5.25 is going to be used for evaluation of splines, then it may be reasonable to delete step (5), since we can always divide out the factor $(m-1)!$ after summing up the B-splines times the coefficients.

Dealing with the *derivatives* of B-spline expansions for uniform partitions is also somewhat simpler than in the general case. First, we note

$$D_+ N^m(x) = N^{m-1}(x) - N^{m-1}(x-1). \tag{5.36}$$

Now if s has the B-spline expansion (5.35), then for all $a \leqslant x \leqslant b$,

$$D_+^d s(x) = h^{-d} \sum_{i=d+1}^{m+k} \nabla^d c_i N^{m-d}\left(\frac{x-y_i}{h}\right), \qquad d=0,1,\ldots,m-1, \quad (5.37)$$

where ∇^d is the backward difference operator defined by

$$\nabla^d f(x) = \sum_{i=0}^{d} (-1)^i \binom{d}{i} f(x-i). \qquad (5.38)$$

Let

$$cd(d,i) = \nabla^{d-1} c_i, \qquad \begin{array}{l} d=1,2,\ldots,m \\ i=d,\ldots,n, \; n=m+k. \end{array} \qquad (5.39)$$

These are the desired coefficients without the h^{-d} factor. The array cd is produced by the following algorithm:

ALGORITHM 5.26. To Compute the Array *cd* Defined in (5.39)

1. For $i \leftarrow 1$ step 1 until n
   $cd(1,i) \leftarrow c(i)$;
2. For $j \leftarrow 2$ step 1 until m
   for $i \leftarrow j$ step 1 until n
   $cd(j,i) \leftarrow cd(j-1,i) - cd(j-1,i-1)$.

Discussion. This algorithm does not require any multiplications or divisions. ■

Algorithm 5.24 may now be used to compute derivatives (cf. Algorithm 5.12). It is also possible to design a streamlined version of Algorithm 5.15 for converting the B-spline expansion of s to a piecewise polynomial representation.

The computation of *indefinite* and *definite integrals* of B-spline expansions with equidistant knots proceeds exactly as in the general case. For example, the indefinite integral is given by

$$D_{y_1}^{-1} s(x) = \int_{y_1}^{x} s(t)\,dt = h \sum_{i=1}^{n} \sum_{j=1}^{i} c_j N^{m+1}\left(\frac{x-y_i}{h}\right). \qquad (5.40)$$

Inner products of B-splines on uniform partitions can be computed explicitly as shown in Section 4.4.

§ 5.6. HISTORICAL NOTES

Section 5.1

It is difficult to say when the first numerical computations with splines were made. It seems likely that Eagle [1928], Quade and Collatz [1938], and Schoenberg [1946a, b] made at least some hand calculations. The use of splines on large-scale digital computers had to wait, of course, for such machines to become generally available. Serious computer calculations with splines seem to have begun with researchers trying to model the mechanical spline on the computer—see, for example, MacLaren [1958], Theilheimer and Starkweather [1961], Birkhoff and Garabedian [1960], Asker [1962], Fowler and Wilson [1963], Berger and Webster [1963], and Berger, Webster, Tapia, and Atkins [1966].

Most early methods for computing with splines used some form of piecewise polynomial representation. The possibility of using one-sided basis elements was also tried, but was quickly discarded as being too numerically unstable. At first there was not much use of the B-splines. In one early application, Greville [1964a] used them to compute interpolating natural splines (of order $2m$) by representing the mth derivative of the desired spline as a B-spline expansion. Despite all of the nice features of B-splines, the use of piecewise polynomial representations remained popular (cf. Carraso and Laurent [1968], Cox [1971], and Reinsch [1967]). Perhaps users were somewhat wary about the fact that B-splines had to be computed by divided differences.

The importance of B-splines for numerical computations was immediately and enormously enhanced by the discovery of the recursion (4.22) by Cox [1972] and deBoor [1972]. Both authors observed Theorem 5.1 and described the associated algorithm for stably computing B-splines. Cox [1972] concentrated on the case of distinct knots, and gave a detailed backward error analysis to back up the assertion that the method is stable. deBoor [1972] allowed multiple knots, and later published a package (see deBoor [1977]) of FORTRAN programs based on his paper. Most of the ideas of this section come from deBoor's papers. We have elected to describe the algorithms in an informal language so that the user can write his own programs—we give operation counts to help in selecting which algorithm to program.

Section 5.2

Theorem 5.9 is established in the article by deBoor [1972]. For equally spaced knots it appears already in the paper by Schoenberg [1946a] [cf. (5.37)]. Cox [1971] has given some recursion relations for the derivatives of

B-splines, which together with (4.23) lead to a variety of methods for computing them. Butterfield [1976] has carried out a careful comparison of these methods based on backward error analysis.

Section 5.3

In defining the piecewise polynomial representation of a polynomial spline, we have decided to work with expansions of the polynomials in terms of powers of $(x-x_i)$. While it is well known that the power basis for polynomials is not well suited for numerical computations if m is moderately large or if the interval is moderately long, we expect that, in practice, there should be no difficulty with this choice since we will, in fact, be working with small m and on small intervals. An alternative would be to work with Tchebycheff polynomial expansions—see Cox [1971] or Rivlin [1974]. Here we have discussed only the problem of converting a B-spline expansion of a spline to a piecewise polynomial expansion. deBoor [1977, 1978] also discusses the reverse process, although it seems much less likely to be required.

Section 5.4

Theorem 5.17, giving formulae for the B-spline coefficients of the antiderivative of a spline, was found by deBoor, Lyche, and Schumaker [1976]. The formula (5.27) for the antiderivative of a B-spline had been found earlier by Gaffney [1974]. Our discussion of the computation of inner products of B-splines is based on the article by deBoor, Lyche, and Schumaker [1976], where an explicit ALGOL program can be found. For some quadrature formulae for computing integrals against B-splines, see Phillips and Hanson [1974].

§ 5.7. REMARKS

Remark 5.1

To illustrate the difficulties that can arise in computing inner products of B-splines using the double divided difference formula given in Theorem 4.25, deBoor, Lyche, and Schumaker [1976] considered the problem of computing the value

$$\int_5^9 N_i^4(x)N_i^4(x)\,dx \qquad \text{corresponding to the B-spline } N_i^4(x)$$

of order 4 over the knots $y_i,\ldots,y_{i+4}=5,6,6+10^{-r},8,9$ for various values of r. This integral was computed first by using Theorem 4.25, and then by using Gauss quadrature as described in Algorithm 5.22.

The results of this experiment are shown in Table 9. The values found using Algorithm 5.22 are all correct to nine decimal digits, while the values produced by the divided difference formula are increasingly inaccurate as r becomes larger (and the spacing of the knots becomes more nonuniform). For ease of comparison we have underlined the first incorrect digit in each value produced by the divided difference method.

Table 9. A Comparison of Two Methods For Computing the Inner Product of B-Splines

r	Gauss Quadrature	Divided Differences
0	4.194444445	4.194444445
1	4.066497736	4.066497716
2	4.040109644	4.040109621
3	4.037345542	4.037344000
4	4.037067900	4.037048708
5	4.037040124	4.036818046
6	4.037037346	4.037239721
7	4.037037068	4.020766567
8	4.037037040	4.041030623
9	4.037037037	2.016235838

6
APPROXIMATION POWER OF SPLINES

In this chapter we examine the relationship between the smoothness of a function, and how well it can be approximated by polynomial splines. We include direct theorems, lower bounds, inverse theorems, and saturation results. In addition, we characterize some classical spaces of smooth functions in terms of their order of approximation by polynomial splines.

§ 6.1. INTRODUCTION

Suppose $\mathcal{F}$ is a class of smooth functions defined on the interval $[a,b]$, and that $\mathcal{S}$ is a space of polynomial splines defined on the same interval. In this chapter we are interested in relating the smoothness of f to how well it can be approximated by splines in $\mathcal{S}$. In order to provide a measure of how well $f \in \mathcal{F}$ can be approximated, we define

$$d(f,\mathcal{S})_X = \inf_{s \in \mathcal{S}} \|f-s\|_X, \tag{6.1}$$

where X is some normed linear space containing both $\mathcal{F}$ and $\mathcal{S}$. In practice, we shall be interested primarily in the spaces $X = C[a,b]$ or $X = L_p[a,b]$, with $1 \leqslant p < \infty$.

The most important part of this chapter is Section 6.4, where we obtain upper bounds on the size of $d(f,\mathcal{S})$. Our results will be similar to the direct theorems established in Chapter 3 for polynomial approximation. There we established bounds of the form

$$d(f,\mathcal{P}_m)_X \leqslant C\omega_f\left(\frac{1}{m}\right),$$

where C is a constant independent of m and f, and where $\omega_f(t)$ is some function describing the smoothness of f. We were especially interested in the behavior of the bound as $m \to \infty$.

For spline approximation, however, we are not interested in large values of m. To get accurate approximations using splines we would prefer to keep m fixed at a rather low value, and increase the number of knots. Hence our direct theorems for spline approximation will have the form

$$d(f,\mathcal{S})_X \leqslant C_1\omega_f(\overline{\Delta}), \qquad \text{all } f\in\mathcal{F}, \tag{6.2}$$

where $\overline{\Delta}$ is the *mesh size* of the partition $\Delta=\{x_i\}_0^{k+1}$ associated with the spline space $\mathcal{S}$, defined by

$$\overline{\Delta}=\max_{0\leqslant i\leqslant k}(x_{i+1}-x_i). \tag{6.3}$$

Our main *direct theorems* for spline approximation are contained in Section 6.4. We shall concentrate on giving bounds for the space of splines $\mathcal{S}_m(\Delta)$ with simple knots. Indeed, since

$$\mathcal{S}_m(\Delta)\subseteq\mathcal{S}(\mathcal{P}_m;\mathfrak{M};\Delta), \qquad \text{all } \mathfrak{M},$$

it follows that

$$d[f,\mathcal{S}(\mathcal{P}_m;\mathfrak{M};\Delta)]\leqslant d[f,\mathcal{S}_m(\Delta)],$$

and our upper bounds will automatically be valid for all spline spaces $\mathcal{S}(\mathcal{P}_m;\mathfrak{M};\Delta)$.

As companions for the upper bounds on $d(f,\mathcal{S})_X$, we shall also give corresponding *lower bounds* of the form

$$d[F,\mathcal{P}\mathcal{P}_m(\Delta)]_X \geqslant C_2\omega_f(\overline{\Delta}), \qquad \text{some } F\in\mathcal{F}, \tag{6.4}$$

where $\mathcal{P}\mathcal{P}_m(\Delta)$ is the space of piecewise polynomials of order m associated with the partition Δ. Since

$$\mathcal{S}(\mathcal{P}_m;\mathfrak{M};\Delta)\subseteq\mathcal{P}\mathcal{P}_m(\Delta), \qquad \text{all } \mathfrak{M},$$

the lower bounds for $\mathcal{P}\mathcal{P}_m(\Delta)$ will automatically produce lower bounds for all spline spaces $\mathcal{S}(\mathcal{P}_m;\mathfrak{M};\Delta)$.

In those cases where we can establish both (6.2) and (6.4), we will be assured that our bounds have the correct order. Moreover, by comparing the sizes of C_1 and C_2, we can get some information on how good the constants are. In this case we are also able to conclude that all of the spline spaces between $\mathcal{P}\mathcal{P}_m(\Delta)$ and $\mathcal{S}_m(\Delta)$ have the same approximation power, independent of how the multiplicity vector $\mathfrak{M}$ is selected.

It is natural to expect that the order of approximation attainable with polynomial splines will increase with the smoothness of the class of functions $\mathcal{F}$ being approximated. Up to a limit this is true. We will show in Sections 6.8 to 6.9, however, that if $\mathcal{F} \cap \mathcal{P}_m \neq \phi$, then the maximal order of convergence possible for the class $\mathcal{F}$ is $\overline{\Delta}^m$, no matter how much smoothness $\mathcal{F}$ is assumed to have. This is a *saturation result*. Results of this kind will follow from various *inverse theorems* in which we will estimate the modulus of smoothness of a function in terms of how well it can be approximated by piecewise polynomials or by splines.

In Section 2.10 we observed that the sequence of polynomial spaces $\mathcal{P}_1, \mathcal{P}_2, \ldots$ is an asymptotically optimal sequence of approximating spaces for several classical spaces of smooth functions (in the sense of n-widths). In this chapter we shall show that the spline spaces $\mathcal{S}_m(\Delta_1), \mathcal{S}_m(\Delta_2), \ldots$ for appropriate sequences of finer and finer partitions $\Delta_1, \Delta_2, \ldots$ are also *asymptotically optimal approximating spaces*.

§ 6.2. PIECEWISE CONSTANTS

In order to illustrate the relationship between the smoothness of functions and their order of approximability by splines, in this section and the following one we consider two simple special cases. Here we examine approximation by piecewise constants; in Section 6.3 we consider approximation by piecewise linear splines.

Let

$$\Delta = \{a = x_0 < x_1 < \cdots < x_k < x_{k+1} = b\}$$

be a partition of the interval $[a,b]$, and let $\mathcal{S}_1(\Delta)$ be the corresponding space of piecewise constant functions. Corresponding to Δ, we define the mesh spacing by

$$\overline{\Delta} = \max_{0 \leqslant i \leqslant k} (x_{i+1} - x_i). \tag{6.5}$$

Then we can establish the following direct theorem, giving bounds on the approximation order for several typical spaces of smooth functions:

THEOREM 6.1

For any Δ,

$$d[f, \mathcal{S}_1(\Delta)]_\infty \leqslant \tfrac{1}{2}\omega(f; \overline{\Delta})_\infty, \qquad \text{all } f \in B[a,b]; \tag{6.6}$$

$$d[f, \mathcal{S}_1(\Delta)]_q \leqslant \tfrac{1}{2}\overline{\Delta}^{1-1/p+1/q}\|Df\|_p, \quad \text{all } f \in L_p^1[a,b], \quad 1 \leqslant p \leqslant q \leqslant \infty; \tag{6.7}$$

$$d[f, \mathcal{S}_1(\Delta)]_\infty \leqslant \tfrac{1}{2}\overline{\Delta}\|Df\|_\infty, \qquad \text{all } f \in C^1[a,b]. \tag{6.8}$$

Proof. See Section 2.1 for the definition of these various spaces. Suppose now that $f \in B[a,b]$. Let

$$s(x) = \frac{(m_i + M_i)}{2}, \qquad x \in I_i, \qquad i = 0, 1, \ldots, k, \tag{6.9}$$

where

$$m_i = \inf_{x \in I_i} f(x), \qquad M_i = \sup_{x \in I_i} f(x),$$

where $I_0, \ldots, I_k$ are the subintervals of $[a,b]$ defined by the partition Δ. Clearly $s \in \mathcal{S}_1(\Delta)$. Moreover, for all $x \in I_i$,

$$|s(x) - f(x)| \leqslant \frac{(M_i - m_i)}{2} \leqslant \frac{1}{2}\omega(f; \overline{\Delta}).$$

This proves (6.6). (We could also obtain this directly from Whitney's theorem—see Theorem 3.16.)

Suppose now that $f \in L^1_\infty[a,b]$. Then there exists $\xi_i \in I_i$ such that

$$s(x) = f(\xi_i) = \frac{(m_i + M_i)}{2}, \qquad \text{all } x \in I_i,$$

hence

$$|f(x) - s(x)| \leqslant |f(x) - f(\xi_i)| \leqslant \int_{x_i}^{x_{i+1}} |Df(t)|\,dt \leqslant \overline{\Delta}\|Df\|_\infty \tag{6.10}$$

for $x \in I_i$. Since this holds for all $i = 0, 1, \ldots, k$, we have proved (6.7) for $p = q = \infty$. The estimate (6.8) follows since $C^1[a,b] \subseteq L^1_\infty[a,b]$.

To prove (6.7) for general $1 \leqslant p \leqslant q \leqslant \infty$, we apply Hölder's inequality to (6.10) to obtain

$$|f(x) - s(x)| \leqslant \overline{\Delta}^{1-1/p}\left(\int_{x_i}^{x_{i+1}} |Df(t)|^p\,dt\right)^{1/p}, \qquad x \in I_i.$$

This implies

$$\int_{x_i}^{x_{i+1}} |f(x) - s(x)|^q\,dx \leqslant \overline{\Delta}^{q-q/p+1}\left(\int_{x_i}^{x_{i+1}} |Df(t)|^p\,dt\right)^{q/p},$$

and summing over $i = 0, 1, \ldots, k$ yields

$$\|f - s\|_q \leqslant \overline{\Delta}^{1-1/p+1/q}\left[\sum_{i=0}^{k}\left(\int_{x_i}^{x_{i+1}} |Df(t)|^p\,dt\right)^{q/p}\right]^{1/q}.$$

Finally, applying Jensen's inequality (see Remark 6.2) to the last sum, we find that it is bounded by

$$\left[\sum_{i=0}^{k}\int_{x_i}^{x_{i+1}}|Df(t)|^p\,dt\right]^{1/p}=\|Df\|_p.$$

We have proved (6.7). ■

We can also give error bounds for piecewise approximation of L_p functions. Since we shall give L_p results in §6.4 for splines of arbitrary order, we do not bother with them here.

To illustrate how lower bounds corresponding to the upper bounds given in Theorem 6.1 can be obtained, we establish lower bounds for functions in $C[a,b]$ and $C^1[a,b]$. Lower bounds for functions in the Sobolev spaces can be established similarly.

THEOREM 6.2

Given any Δ, there exists a function $F_1 \in C[a,b]$ such that

$$d\left[F_1,\mathcal{S}_1(\Delta)\right]_\infty \geqslant \tfrac{1}{2}\omega\left(F_1;\overline{\Delta}\right)_\infty. \tag{6.11}$$

Moreover, for all $\varepsilon>0$, there exists a function $F_\varepsilon \in C^1[a,b]$ such that

$$d\left[F_\varepsilon,\mathcal{S}_1(\Delta)\right]_\infty \geqslant \frac{(1-\varepsilon)\overline{\Delta}}{2}\|DF_\varepsilon\|_\infty. \tag{6.12}$$

Proof. Let $[x_i,x_{i+1}]$ be a subinterval of $[a,b]$ corresponding to Δ with $x_{i+1}-x_i=\overline{\Delta}$. We define

$$F_1(x)=\begin{cases}0, & x\leqslant x_i\\ (x-x_i)/(x_{i+1}-x_i), & x_i\leqslant x\leqslant x_{i+1}\\ 1, & x_{i+1}\leqslant x\end{cases}$$

Then since $d[F_1,\mathcal{S}_1(\Delta)]_\infty=1/2$ while $\omega(F_1;\overline{\Delta})_\infty=1$, we have proved (6.11).

Given $\varepsilon>0$, we define $g(x)$ to be zero for $x\notin[x_i,x_{i+1}]$, to be 1 on $[x_i+\varepsilon\overline{\Delta},x_{i+1}-\varepsilon\overline{\Delta}]$, and to be linear in between. Then the function $F_\varepsilon(x)=\int_{x_i}^{x}g(t)\,dt$ is zero for $x\leqslant x_i$, is $(1-\varepsilon)\overline{\Delta}$ for $x\geqslant x_{i+1}$, and is monotone increasing in between. It follows that $d[F_\varepsilon,\mathcal{S}_1(\Delta)]_\infty=\overline{\Delta}(1-\varepsilon)/2$. On the other hand, $\|DF_\varepsilon\|_\infty=\|g\|_\infty=1$, and (6.12) follows. ■

The lower bounds in Theorem 6.2 show that the upper bounds (6.6) and (6.8) for functions in $C[a,b]$ and in $C^1[a,b]$ are the best possible, including the constants.

We turn now to the inverse question of estimating the smoothness of a function in terms of how well it can be approximated by piecewise constants. As a first step in this direction, we establish the following theorem, where we use

$$\underline{\Delta} = \min_{0 \leqslant i \leqslant k} (x_{i+1} - x_i): \tag{6.13}$$

THEOREM 6.3

For any $f \in C[a,b]$,

$$\omega(f;\overline{\Delta})_\infty \leqslant 4\left\lceil \frac{\overline{\Delta}}{\underline{\Delta}} \right\rceil d[f;\mathbb{S}_1(\Delta)]_\infty, \tag{6.14}$$

where $\lceil \overline{\Delta}/\underline{\Delta} \rceil = \min\{j: \overline{\Delta}/\underline{\Delta} \leqslant j\}$.

Proof. Suppose $y - x \leqslant \underline{\Delta}$. then the interval (x,y) can contain at most one knot. If it does not contain any knots, then

$$|f(y) - f(x)| \leqslant |f(y) - s(y)| + |s(x) - f(x)| \leqslant 2d[f, \mathbb{S}_1(\Delta)].$$

On the other hand, if (x,y) contains the knot x_i, then

$$|f(y) - f(x)| \leqslant |f(y) - f(x_i)| + |f(x_i) - f(x)| \leqslant 4d[f, \mathbb{S}_1(\Delta)].$$

In either case we conclude that $\omega(f;\underline{\Delta}) \leqslant 4d[f, \mathbb{S}_1(\Delta)]$. Now, combining this with the fact that

$$\omega(f;\overline{\Delta}) \leqslant \omega\left(f; \frac{\overline{\Delta}\cdot\underline{\Delta}}{\underline{\Delta}}\right) \leqslant \left\lceil \frac{\overline{\Delta}}{\underline{\Delta}} \right\rceil \omega(f;\underline{\Delta}),$$

we obtain (6.14). ■

Theorem 6.3 by itself does not actually say very much about the smoothness of the function f. In order to have information on the smoothness of f, we need to know the behavior of $\omega(f;t)$ for at least some sequence of t's converging to zero (rather than just at a single point as in Theorem 6.3). To get such information, we must know something about the size of $d[f, \mathbb{S}_1(\Delta_\nu)]$ for a *sequence* of partitions Δ_ν. In view of the dependence of the bound in (6.14) on the ratio $\overline{\Delta}_\nu/\underline{\Delta}_\nu$, it is clear that not every sequence of partitions will do, and we will need some control on the mesh ratios. We are led to make the following definition:

DEFINITION 6.4. Quasi-Uniform Partitions

Let $\Delta_1, \Delta_2, \ldots$ be a sequence of partitions of $[a,b]$. If $\sigma > 0$ is a constant such that

$$\frac{\overline{\Delta}_\nu}{\underline{\Delta}_\nu} \leqslant \sigma \qquad \text{all } \nu, \tag{6.15}$$

then we say that Δ_ν is a *σ-quasi-uniform sequence of partitions*.

In addition to controlling the mesh ratios, in order to get information on $\omega(f;t)$ for small t, we have to assume that $\overline{\Delta}_\nu \to 0$. On the other hand, we do not want the ratio $\overline{\Delta}_\nu / \overline{\Delta}_{\nu+1}$ to be too big. Hence we make the following definition:

DEFINITION 6.5. Steadiness of a Sequence of Partitions

We say that the sequence Δ_ν of partitions *goes steadily to zero* provided there exist constants $1 \leqslant \alpha < \infty$ and $1 < \beta < \infty$ such that for all ν

$$\alpha \overline{\Delta}_{\nu+1} \leqslant \overline{\Delta}_\nu \leqslant \beta \overline{\Delta}_{\nu+1}. \tag{6.16}$$

Before proceeding to our first inverse theorem for piecewise constant approximation, we give two important examples of sequences of partitions that go steadily to zero.

EXAMPLE 6.6. Uniform Partitions

For $\nu = 1, 2, \ldots$, let

$$\Delta_\nu = \left\{ a + i\overline{\Delta}_\nu \right\}_{i=0}^{\nu}, \qquad \overline{\Delta}_\nu = \underline{\Delta}_\nu = \frac{(b-a)}{\nu}.$$

Discussion. For each ν, Δ_ν divides $[a,b]$ into ν subintervals of equal length. A uniform partition is 1-quasi-uniform. This sequence of partitions goes steadily to zero with constants $\alpha = 1$ and $\beta = 2$. ■

EXAMPLE 6.7. Nested Uniform Partitions

For $\nu = 1, 2, \ldots$, let

$$\Delta_\nu = \left\{ a + i\overline{\Delta}_\nu \right\}_{i=0}^{2^\nu}, \qquad \overline{\Delta}_\nu = \frac{(b-a)}{2^\nu}.$$

Discussion. We call this sequence of partitions *nested* since $\Delta_\nu \subseteq \Delta_{\nu+1}$. It goes to zero steadily with constants $\alpha = \beta = 2$. ∎

The following theorem is an *inverse theorem* for approximation by $\mathcal{S}_1(\Delta_\nu)$ on a sequence of partitions Δ_ν.

THEOREM 6.8

Suppose Δ_ν is a sequence of σ-quasi-uniform partitions of $[a,b]$ with mesh size going steadily to zero. Also suppose $f \in C[a,b]$ is such that

$$d[f, \mathcal{S}_1(\Delta_\nu)]_\infty \leqslant \phi(\overline{\Delta}_\nu), \qquad \text{all } \nu = 1,2,\ldots, \tag{6.17}$$

where ϕ is a monotone-increasing function on $(0, \overline{\Delta}_1)$. Then

$$\omega(f;t) \leqslant 4\lceil \beta \rceil \lceil \sigma \rceil \phi(t) \qquad \text{all } 0 \leqslant t \leqslant \overline{\Delta}_1, \tag{6.18}$$

where β is the constant in Definition 6.5 of steadiness and $\lceil \beta \rceil = \min\{j: \beta \leqslant j\}$.

Proof. Let ν be such that $\overline{\Delta}_{\nu+1} \leqslant t < \overline{\Delta}_\nu$. By the monotonicity of ω and ϕ, (6.17) coupled with Theorem 6.3 implies

$$\begin{aligned}\omega(f;t) &\leqslant \omega(f;\overline{\Delta}_\nu) \leqslant \omega(f;\beta\overline{\Delta}_{\nu+1}) \leqslant \lceil \beta \rceil \omega(f;\overline{\Delta}_{\nu+1}) \\ &\leqslant 4\lceil \beta \rceil \lceil \sigma \rceil d[f, \mathcal{S}_1(\Delta_{\nu+1})] \leqslant 4\lceil \beta \rceil \lceil \sigma \rceil \phi(\overline{\Delta}_{\nu+1}) \leqslant 4\lceil \beta \rceil \lceil \sigma \rceil \phi(t).\end{aligned}$$ ∎

The following simple corollary of Theorem 6.8 shows that there is a limit to how well functions can be approximated by piecewise constants, no matter how smooth the function may be. This kind of result is called a *saturation theorem*.

THEOREM 6.9

Let Δ_ν be a sequence of partitions as in Theorem 6.8. Suppose $f \in C[a,b]$ is such that for all ν

$$d[f, \mathcal{S}_1(\Delta_\nu)]_\infty \leqslant C\overline{\Delta}_\nu \psi(\overline{\Delta}_\nu), \tag{6.19}$$

where C is a constant and $\psi(t)$ is a function with $t\psi(t)$ monotone and $\psi(t) \to 0$ as $t \to 0$. Then f must be a constant; that is, $f \in \mathcal{P}_1$.

Proof. Theorem 6.8 together with (6.19) implies that $\omega(f;t)/t \to 0$ as $t \to 0$. By properties of modulus of smoothness [cf. (2.122)], this implies f is a constant. ∎

Theorem 6.9 asserts that the maximal order of convergence attainable with piecewise constants defined on a sequence of quasi-uniform partitions Δ_ν going steadily to zero is one. This order of convergence is attained as soon as f belongs to $L^1_\infty[a,b]$, as shown in Theorem 6.1. Higher order of convergence occurs only for constants (which are approximated exactly). This does not exclude the possibility, however, that for a *given function f* some higher order of approximation might be possible if an appropriate (non-quasi-uniform) sequence of meshes is used. See Chapter 7 for results in the case where the knots are free parameters.

We may now combine our direct and inverse theorems to obtain a characterization of the Lipschitz space Lip^α defined in (2.128).

THEOREM 6.10

Let Δ_ν be a sequence of partitions as in Theorem 6.8. Then a function $f \in C[a,b]$ belongs to Lip^α if and only if for all ν

$$d[f, \mathcal{S}_1(\Delta_\nu)]_\infty \leqslant C(\overline{\Delta}_\nu)^\alpha \tag{6.20}$$

for some constant C.

Proof. If $f \in \text{Lip}^\alpha$, then $\omega(f;t) \leqslant Ct^\alpha$, and (6.20) follows from Theorem 6.1. On the other hand, Theorem 6.8 shows that if f satisfies (6.20), then $\omega(f;t) = \mathcal{O}(t^\alpha)$, and thus $f \in \text{Lip}^\alpha$. ■

The assumption that $f \in C[a,b]$ cannot be removed in Theorems 6.8 through 6.10 without some further assumption on the sequence of meshes. Consider the following example:

EXAMPLE 6.11

Approximate $f(x) = (x-(a+b)/2)^0_+$ by piecewise constants.

Discussion. Clearly $\omega(f;t)=1$ for all $0<t<b-a$. On the other hand, if Δ is a partition that contains the point $(a+b)/2$, then $d[f,\mathcal{S}_1(\Delta)]=0$. Thus (6.14) cannot hold. Similarly, if Δ_ν is a sequence of partitions each of which contains $(a+b)/2$ (which might easily be the case with nested partitions, for example), then Theorem 6.10 cannot hold. Indeed, (6.20) holds trivially with any α, but since f is not continuous, it certainly does not belong to any Lipschitz class. ■

It is possible to establish analogs of Theorems 6.8 to 6.10 for $f \in B[a,b]$ and appropriate Δ_ν. First we prove a sharper version of Theorem 6.3.

THEOREM 6.12

For any $f \in B[a,b]$,

$$\omega(f;\overline{\Delta}) \leqslant \left\lceil \frac{\overline{\Delta}}{\underline{\Delta}} \right\rceil \left(4d[f;\mathbb{S}_1(\Delta)]_\infty + \max_{1 \leqslant j \leqslant k} \operatorname{jump}[f]_{x_j}\right). \tag{6.21}$$

where

$$\operatorname{jump}[f]_{x_j} = f(x_j+) - f(x_j-), \qquad j = 1,2,\ldots,k.$$

Proof. The proof is very much like the proof of Theorem 6.3. Let $y - x \leqslant \underline{\Delta}$. Then (x,y) contains at most one knot. If there are no knots in (x,y), then $|f(y)-f(x)| \leqslant 2d[f,\mathbb{S}_1(\Delta)]$ as before. If x_i is a knot in (x,y), then

$$|f(y)-f(x)| \leqslant |f(y)-f(x_i+)| + |f(x_i+)-f(x_i-)| + |f(x_i-)-f(x)|.$$

This implies that $\omega(f;\underline{\Delta}) \leqslant 4d[f,\mathbb{S}_1(\Delta)] + \max_{1 \leqslant j \leqslant k} \operatorname{jump}[f]_{x_j}$. Now (6.21) follows by the estimate $\omega(f;\overline{\Delta}) \leqslant \lceil \overline{\Delta}/\underline{\Delta} \rceil \omega(f;\underline{\Delta})$. ∎

We can now prove an improved version of Theorem 6.8 (with a stronger hypothesis on the partitions).

THEOREM 6.13

Let $\Delta_\nu = \{x_i^\nu\}_0^{k_\nu+1}$ be a sequence of σ-quasi-uniform partitions going steadily to zero, and suppose that for all $1 \leqslant i \leqslant k_\nu$ there exists $n_{i,\nu} > \nu$ so that $x_i^\nu \notin \Delta_{n_{i,\nu}}$. Then if $f \in B[a,b]$ is such that

$$d[f,\mathbb{S}_1(\Delta)]_\infty \leqslant \phi(\overline{\Delta}_\nu), \tag{6.22}$$

with ϕ monotone increasing on $(0,\overline{\Delta}_1)$, it follows that

$$\omega(f;t) \leqslant 6\lceil \sigma \rceil \lceil \beta \rceil \phi(t), \qquad \text{all } 0 < t < \overline{\Delta}_1. \tag{6.23}$$

Proof. Let ν be such that $\overline{\Delta}_{\nu+1} \leqslant t < \overline{\Delta}_\nu$. For each i we know $x_i^{\nu+1} \notin \Delta_{n_{i,\nu+1}}$, and thus

$$|f(x_i^{\nu+1}+) - f(x_i^{\nu+1}-)| \leqslant 2d\left[f,\mathbb{S}_1(\Delta_{n_{i,\nu+1}})\right] \leqslant 2\phi(t).$$

It follows that

$$\omega\left(f;\bar{\Delta}_{\nu+1}\right) \leqslant 6\lceil \sigma \rceil \phi(t),$$

and arguing as in the proof of Theorem 6.8 we obtain (6.23). ■

COROLLARY 6.14

Let Δ_ν be a sequence of partitions of $[a,b]$ as in Theorem 6.13. Then a function $f \in B[a,b]$ belongs to Lip^α if and only if (6.20) holds.

Proof. The proof proceeds exactly as in Theorem 6.10, using Theorem 6.13 in place of Theorem 6.8. ■

We observe that both Theorem 6.13 and its corollary can be applied in the case of a sequence of uniform partitions as in Example 6.6 (since it is easily seen that given any x_i^ν in Δ_ν, there is a later partition that does not contain x_i^ν). On the other hand, the results do not apply for the sequence of nested partitions given in Example 6.7.

In this section we have concentrated on inverse and saturation results for the ∞-modulus of continuity. Similar results can be established for the p-modulus of continuity in terms of $d[f, \mathbb{S}_1(\Delta_\nu)]_p$—see Section 6.9.

§ 6.3. PIECEWISE LINEAR FUNCTIONS

In this section we further illustrate the connection between the smoothness of a function and the order of its approximation by splines. Here we deal with linear splines (the case $m=2$) and with the uniform norm only. Our first theorem contains upper bounds for several spaces of smooth functions.

THEOREM 6.15

Let Δ be a partition of $[a,b]$ with mesh spacing $\bar{\Delta}$. Then

$$d\left[f, \mathbb{S}_2(\Delta)\right]_\infty \leqslant \omega\left(f;\bar{\Delta}\right), \qquad \text{all } f \in C\left[a,b\right]; \tag{6.24}$$

$$d\left[f, \mathbb{S}_2(\Delta)\right]_\infty \leqslant \omega_2\left(f;\bar{\Delta}/2\right), \qquad \text{all } f \in C\left[a,b\right]; \tag{6.25}$$

$$d\left[f, \mathbb{S}_2(\Delta)\right]_\infty \leqslant \frac{\bar{\Delta}}{2}\omega\left(Df;\bar{\Delta}\right), \qquad \text{all } f \in L^1_\infty\left[a,b\right]; \tag{6.26}$$

$$d\left[f, \mathbb{S}_2(\Delta)\right]_\infty \leqslant \bar{\Delta}\|Df\|_\infty, \qquad \text{all } f \in L^1_\infty\left[a,b\right]; \tag{6.27}$$

$$d\left[f, \mathbb{S}_2(\Delta)\right]_\infty \leqslant \frac{\bar{\Delta}^2}{8}\|D^2 f\|_\infty, \qquad \text{all } f \in L^2_\infty\left[a,b\right]. \tag{6.28}$$

Proof. Given $f \in C[a,b]$, let

$$s(x) = \begin{cases} f_i + \dfrac{(f_{i+1} - f_i)(x - x_i)}{(x_{i+1} - x_i)}, & x_i \leqslant x \leqslant x_{i+1}, \\ i = 0, 1, \dots, k, \end{cases} \tag{6.29}$$

where, in general, we write $f_i = f(x_i)$, $i = 0, 1, \dots, k+1$. This is the piecewise linear polynomial that interpolates f at the points $a = x_0 < x_1 < \cdots < x_k < x_{k+1} = b$; that is,

$$s(x_i) = f(x_i), \qquad i = 0, 1, \dots, k+1.$$

Now, for any $0 \leqslant i \leqslant k$ and $x_i \leqslant x \leqslant x_{i+1}$, by the continuity of f there must exist ξ_x in $I_i = [x_i, x_{i+1}]$ such that $s(x) = f(\xi_x)$. But then

$$|f(x) - s(x)| = |f(x) - f(\xi_x)| \leqslant \omega\big(f; \overline{\Delta}\big) \qquad \text{for } x \in I_i.$$

We have established (6.24).

Suppose now that

$$M = \|f - s\|_\infty = \max_{a \leqslant x \leqslant b} \delta(x), \qquad \delta(x) = |f(x) - s(x)|.$$

Since δ is continuous, there exists η (which we suppose lies in the interval $I_i = [x_i, x_{i+1}]$) such that $M = \delta(\eta)$. Then if $x_i \leqslant \eta \leqslant (x_i + x_{i+1})/2$, it follows that

$$|f(\eta - h) - 2f(\eta) + f(\eta + h)| = |\delta(\eta - h) - 2\delta(\eta) + \delta(\eta + h)| \geqslant M,$$

where $h = \eta - x_i$. This implies that $M \leqslant \omega_2(f; h) \leqslant \omega_2(f; \overline{\Delta}/2)$. If x is in the second half of I_i, a similar argument with $h = x_{i+1} - \eta$ yields the same estimate, and we have proved (6.25), (cf. Theorem 3.17).

Assume now that $f \in L^1_\infty[a,b]$. Then for any $0 \leqslant i \leqslant k$ and $x_i \leqslant x \leqslant (x_i + x_{i+1})/2$,

$$|f(x) - s(x)| \leqslant \int_{x_i}^{x} |Df(t) - Ds(t)|\, dt \leqslant \frac{1}{2}\overline{\Delta} \sup_{x_i \leqslant t \leqslant x_{i+1}} |Df(t) - Ds(t)|.$$

But $Ds(t)$ is a constant in I_i, and since s and f agree in value at x_i and x_{i+1}, it follows that $\inf_{t \in I_i} Df(t) \leqslant Ds(t) \leqslant \sup_{t \in I_i} Df(t)$, hence

$$\sup_{t \in I_i} |Df(t) - Ds(t)| \leqslant \omega\big(Df; \overline{\Delta}\big).$$

Substituting this in the above, we obtain $|f(x)-s(x)| \leqslant \frac{1}{2}\overline{\Delta}\omega(Df;\overline{\Delta})$. If x is in the second half of the interval I_i, then we use

$$|f(x)-s(x)| \leqslant \int_x^{x_{i+1}} |Df(t)-Ds(t)|\,dt,$$

and proceed as before. Since i is arbitrary, we have proved (6.26). The estimate (6.27) follows immediately.

Finally, suppose $f \in L_\infty^2[a,b]$. Then by the explicit remainder formula given in (3.5) for polynomial interpolation, we have

$$f(x)-s(x)=(x-x_i)(x-x_{i+1})[x_i,x_{i+1},x]f.$$

By (2.93), $[x_i,x_{i+1},x]f \leqslant \frac{1}{2}\|D^2f\|_\infty$. On the other hand, it is clear that for $x_i \leqslant x \leqslant x_{i+1}$, $|(x-x_i)(x-x_{i+1})| \leqslant \overline{\Delta}^2/4$, and (6.28) follows. ■

We now give some lower bounds that are companions to the upper bounds of Theorem 6.15. We do not bother to compute the best possible constants.

THEOREM 6.16

Given any Δ, there exists a function $F_1 \in L_\infty^1[a,b]$ such that

$$d[F_1, \mathcal{P}\mathcal{P}_2(\Delta)]_\infty \geqslant \frac{1}{2}\omega(F_1;\overline{\Delta}), \tag{6.30}$$

$$d[F_1, \mathcal{P}\mathcal{P}_2(\Delta)]_\infty \geqslant \frac{1}{4}\omega_2(F_1;\overline{\Delta}), \tag{6.31}$$

$$d[F_1, \mathcal{P}\mathcal{P}_2(\Delta)]_\infty \geqslant \frac{\overline{\Delta}}{8}\omega(DF_1;\overline{\Delta}), \tag{6.32}$$

and

$$d[F_1, \mathcal{P}\mathcal{P}_2(\Delta)]_\infty \geqslant \frac{\overline{\Delta}}{4}\|DF_1\|_\infty. \tag{6.33}$$

Moreover, there exists a function $F_2 \in L_\infty^2[a,b]$ with

$$d[F_2, \mathcal{P}\mathcal{P}_2(\Delta)]_\infty \geqslant \frac{1}{32}\overline{\Delta}^2\|D^2F_2\|_\infty. \tag{6.34}$$

Proof. Given Δ, let $I_i=[x_i,x_{i+1}]$ be a subinterval of length $\bar{\Delta}$. Let

$$F_1(x)=\begin{cases}2(x-x_i)/(x_{i+1}-x_i), & x_i\leqslant x\leqslant \bar{x}_i=(x_i+x_{i+1})/2\\ 2(x_{i+1}-x)/(x_{i+1}-x_i), & \bar{x}_i\leqslant x\leqslant x_{i+1}\\ 0, & \text{otherwise.}\end{cases}$$

We note that $d[F_1,\mathcal{P}\mathcal{P}_2(\Delta)]=1/2$, $\omega(F_1;\bar{\Delta})=1$, $\omega_2(F_1;\bar{\Delta})=2$, $\omega(DF_1;\bar{\Delta})=4/\bar{\Delta}$, and $\|DF_1\|_\infty=2/\bar{\Delta}$. The inequalities (6.30) to (6.33) follow. This function is the normalized B-spline with knots x_i, $\bar{x}_i$, x_{i+1}.

We now define F_2. Let $F_2(x)=B_3^*([2x-(x_i+x_{i+1})]/(x_{i+1}-x_i))$, where B_3^* is the perfect B-spline of order 3—see Example 4.35. Then since $F_2((x_i+x_{i+1})/2)=1$, $d[F_2,\mathcal{P}\mathcal{P}_2(\Delta)]=1/2$. On the other hand, $\|D^2F_2\|_\infty=(2/\bar{\Delta})^2\|D^2B_3^*\|_\infty$ while $\|D^2B_3^*\|_\infty=4$. Putting this information together, we obtain (6.34). ∎

In Sections 6.8 to 6.9 we give inverse theorems, saturation theorems, and characterization theorems for approximation by splines of arbitrary order and for approximation in any p-norm, $1\leqslant p\leqslant\infty$.

§ 6.4. DIRECT THEOREMS

In this section we give bounds on how well functions in various smooth spaces can be approximated by splines in the space $\mathcal{S}_m(\Delta)$. To establish our error bounds, we construct a *linear operator* Q mapping $B[a,b]$ into $\mathcal{S}_m(\Delta)$. We need the following lemma:

LEMMA 6.17

Let $\Delta=\{a=x_0<x_1<\cdots<x_k<x_{k+1}=b\}$ be a partition of the interval $[a,b]$. Then there exists an associated partition $\Delta^*=\{a=x_0^*<x_1^*<\cdots<x_l^*<x_{l+1}^*=b\}$ with $\Delta^*\subseteq\Delta$ such that

$$\frac{\bar{\Delta}}{2}\leqslant\underline{\Delta}^*\leqslant\bar{\Delta}^*\leqslant\frac{3\bar{\Delta}}{2}. \tag{6.35}$$

The partition Δ^* is 3-quasi-uniform (cf. Definition 6.4).

Proof. Let $x_0^*=a$, and define $x_1^*,\ldots,x_l^*$ recursively by

$$x_j^*=\min\left\{x_i\colon x_{j-1}^*+\frac{\bar{\Delta}}{2}\leqslant x_i\leqslant x_{j-1}^*+\bar{\Delta}\text{ and }x_i<b-\frac{\bar{\Delta}}{2}\right\}.$$

This process does not stop until $b-3\bar{\Delta}/2\leqslant x_l^*$. Now let $x_{l+1}^*=b$. The property (6.35) follows by construction. ∎

Lemma 6.17 shows that any partition can be thinned out to get a quasi-uniform partition. We return now to the construction of Q. Corresponding to the partition Δ^*, define the extended partition

$$y_1=\cdots=y_m=a,\quad y_{m+1}=x_1^*,\dots,y_{m+l}=x_l^*,\quad b=y_{m+l+1}=\cdots=y_{2m+l}. \tag{6.36}$$

For convenience write $n=m+l$. Associated with this extended partition, let $B_1,\dots,B_n$ be the set of normalized B-splines forming a basis for $\mathcal{S}_m(\Delta^*)$.

For each $i=1,2,\dots,n$, let

$$\tau_{ij}=y_i+(y_{i+m}-y_i)\frac{(j-1)}{(m-1)},\qquad j=1,2,\dots,m, \tag{6.37}$$

and

$$\alpha_{ij}=\sum_{\nu=1}^{j}\frac{\xi_i^{(\nu)}D^{\nu-1}\psi_{i,j}(0),}{(\nu-1)!}\qquad j=1,2,\dots,m, \tag{6.38}$$

where

$$\xi_i^{(\nu)}=\frac{(-1)^{\nu-1}(\nu-1)!}{(m-1)!}\varphi_{i,m}^{(m-\nu)}(0),$$

and

$$\varphi_{i,m}(t)=\prod_{r=1}^{m-1}(t-y_{i+r}),\qquad \psi_{i,j}(t)=\prod_{r=1}^{j-1}(t-\tau_{ir}),\qquad \psi_{i,1}(t)\equiv 1.$$

Let

$$\lambda_i f=\sum_{j=1}^{m}\alpha_{ij}[\tau_{i1},\dots,\tau_{ij}]f,\qquad i=1,\dots,n. \tag{6.39}$$

THEOREM 6.18

For any $f\in B[a,b]$ define

$$Qf(x)=\sum_{i=1}^{n}(\lambda_i f)B_i(x). \tag{6.40}$$

Then Q is a linear operator mapping $B[a,b]$ into $\mathcal{S}_m(\Delta^*)\subseteq\mathcal{S}_m(\Delta)$. Moreover,

$$Qp=p\qquad\text{for all } p\in\mathcal{P}_m. \tag{6.41}$$

Proof. By construction, Qf is a spline in $\mathcal{S}_m(\Delta^*)$. This spline space is a subspace of $\mathcal{S}_m(\Delta)$ since $\Delta^* \subseteq \Delta$. We now prove (6.41). First we claim that for all $i=1,2,\ldots,n$,

$$\text{if } p(x)=\sum_{r=1}^{m} c_r x^{r-1}, \qquad \text{then } \lambda_i p=\sum_{r=1}^{m} c_r \xi_i^{(r)}. \tag{6.42}$$

To prove this, we show that it holds for each of the polynomials $\psi_{i,1},\ldots,\psi_{i,m}$ (which clearly span $\mathcal{P}_m$). For each $j=1,2,\ldots,m$ we have

$$\psi_{i,j}(t)=\sum_{r=1}^{j} \frac{t^{r-1}}{(r-1)!}\psi_{i,j}^{(r-1)}(0),$$

while

$$\lambda_i\psi_{i,j}=\sum_{r=1}^{m} \alpha_{ir}[\tau_{i1},\ldots,\tau_{ir}]\psi_{i,j}=\alpha_{ij}=\sum_{r=1}^{m} \frac{\xi_i^{(r)}\psi_{i,j}^{(r-1)}(0)}{(r-1)!}$$

and (6.42) follows.

Now suppose $p(x)=\sum_{r=1}^{m} c_r x^{r-1}$. Then using Marsden's identity (4.34), we have

$$Qp=\sum_{i=1}^{n}(\lambda_i p)B_i=\sum_{i=1}^{n}\sum_{r=1}^{m} c_r\xi_i^{(r)}B_i=\sum_{r=1}^{m} c_r\sum_{i=1}^{n}\xi_i^{(r)}B_i=p. \qquad \blacksquare$$

The following lemma gives some information on the size of the coefficients α_{ij} used in the construction of the linear functionals λ_i:

LEMMA 6.19

Let $\{\alpha_{ij}\}_{i=1,j=1}^{n,m}$ be defined as in (6.38). Then

$$|\alpha_{ij}| \leqslant (y_{i+m}-y_i)^{j-1} \leqslant (m\bar{\Delta})^{j-1}, \qquad \begin{matrix} j=1,2,\ldots,m \\ i=1,\ldots,n \end{matrix}. \tag{6.43}$$

Proof. For any i we have $\alpha_{i1}=1$, so (6.43) holds for $j=1$. For $j>1$ we use an identity on polynomials (see Remark 6.1) to write

$$\alpha_{ij}=(-1)^{j-1}\frac{(m-j)!}{(m-1)!}\sum_{\nu_1,\ldots,\nu_{j-1}}(y_{\nu_1}-\tau_{i1})\cdots(y_{\nu_{j-1}}-\tau_{i,j-1}),$$

where the sum is taken over all choices of distinct $\nu_1,\ldots,\nu_{j-1}$ from $\{i+1,\ldots,i+m-1\}$. This is a sum of exactly $(m-1)!/(m-j)!$ terms. The largest any one term can be is $(y_{i+m}-y_i)$, and (6.43) follows. $\blacksquare$

We are now ready to give error bounds for how well Qf approximates f. Our first theorem is *local* in nature; that is, the error bounds in an interval I_l will depend only on the behavior of f in a somewhat larger interval including I_l.

THEOREM 6.20

Let $m \leqslant l \leqslant n$, $I_l=[y_l,y_{l+1}]$, and $\tilde{I}_l=[y_{l+1-m},y_{l+m}]$. Then for any $1 \leqslant \sigma \leqslant m$,

$$\left.\begin{array}{l} \|D^r(f-Qf)\|_{L_q[I_l]} \\ \quad r=0,\ldots,\sigma-1, \\ \|D^rQf\|_{L_q[I_l]} \\ \quad r=\sigma,\ldots,m-1 \end{array}\right\} \leqslant C_1(\overline{\Delta})^{\sigma-r-1+1/q}\omega_{m-\sigma+1}(D^{\sigma-1}f;\overline{\Delta})_{L_\infty[\tilde{I}_l]} \tag{6.44}$$

for all $f \in C^{\sigma-1}[\tilde{I}_l]$. The constant C_1 depends only on m.

Proof. For convenience we write

$$E_r(t)=\begin{cases} D^rf(t)-D^rQf(t), & r=0,1,\ldots,\sigma-1 \\ D^rQf(t), & r=\sigma,\ldots,m-1. \end{cases}$$

Fix I_l, and let $t \in I_l$. To estimate $E_r(t)$, we are going to first approximate f by a polynomial on I_l. By Theorem 3.19 there exists $p_f \in \mathcal{P}_m$ such that

$$|D^j(f-p_f)(t)| \leqslant C_2(\overline{\Delta}^*)^{\sigma-j-1}\omega_{m-\sigma+1}(D^{\sigma-1}f;\overline{\Delta}^*)_{C[\tilde{I}_l]}, \tag{6.45}$$

$j=0,1,\ldots,\sigma-1$, all $t \in \tilde{I}_l$. We write $R(t)=f(t)-p_f(t)$. Since $Qp_f=p_f$, it follows easily that

$$E_r(t)=\begin{cases} D^rR(t)-D^rQR(t), & r=0,1,\ldots,\sigma-1 \\ D^rQf(t), & r=\sigma,\ldots,m-1. \end{cases}$$

It remains to estimate $D^rQR(t)$.

By the definition of Q,

$$|D^rQR(t)| \leqslant \sum_{i=l+1-m}^{l} \sum_{j=1}^{m} |\alpha_{ij}||\lambda_{ij}R||D^rB_i(t)|.$$

We have estimates of the $|\alpha_{ij}|$ in Lemma 6.19. By Theorem 4.22,

$$|D^rB_i(t)| \leqslant \frac{\Gamma_{m,r}}{(\underline{\Delta}^*)^r},$$

where $\Gamma_{m,r}$ is the constant given in (4.37). We now examine $|\lambda_{ij}R|$.

For $j=1,2,\ldots,\sigma$ we have

$$|\lambda_{ij}R|=|[\tau_{i1},\ldots,\tau_{ij}]R|=\frac{|D^{j-1}R(\eta_{ij})|}{(j-1)!}\leqslant\frac{\|D^{j-1}R\|_{C[\tilde{I}_l]}}{(j-1)!}.$$

For $\sigma<j\leqslant m$ Theorem 2.56 gives

$$|\lambda_{ij}R|\leqq\sum_{\nu=0}^{j-\sigma}\frac{\binom{j-\sigma}{\nu}|[\tau_{i,\nu+1},\ldots,\tau_{i,\nu+\sigma}]R|}{\gamma_\sigma\cdots\gamma_{j-1}},$$

where $\gamma_\mu=\min_{1\leqslant\nu\leqslant j-\mu}|\tau_{i,\nu+\mu}-\tau_{i,\nu}|$. As each γ_μ is bounded below by $\underline{\Delta}^*/m$ [cf. the definition of the τ's in (6.37)], we obtain

$$|\lambda_{ij}R|\leqslant\left(\frac{2m}{\underline{\Delta}^*}\right)^{j-\sigma}\frac{\|D^{\sigma-1}R\|_{C[\tilde{I}_l]}}{(\sigma-1)!},\qquad j=\sigma+1,\ldots,m.$$

Combining our estimates on $|\alpha_{ij}|$, $|\lambda_{ij}R|$, $|D^rB_i|$, and $|D^jR|$, we obtain

$$|D^rQR(t)|\leqslant C_2m\omega_{m-\sigma+1}(D^{\sigma-1}f;\bar{\Delta}^*)_{C[\tilde{I}_l]}\cdot\Gamma_{m,r}(\underline{\Delta}^*)^{-r}$$

$$\cdot\left[\sum_{j=1}^{\sigma}\frac{(m\bar{\Delta}^*)^{j-1}(\bar{\Delta}^*)^{\sigma-j}}{(j-1)!}+\sum_{j=\sigma+1}^{m}\left(\frac{2m}{\underline{\Delta}^*}\right)^{j-\sigma}\frac{(m\bar{\Delta}^*)^{j-1}}{(\sigma-1)!}\right].$$

We now use the fact $\bar{\Delta}^*/\underline{\Delta}^*\leqslant 3$ while $\bar{\Delta}^*\leqslant\bar{\Delta}$ to obtain

$$|E_r(t)|\leqslant C_1\bar{\Delta}^{\sigma-r-1}\omega_{m-\sigma+1}(D^{\sigma-1}f;\bar{\Delta})_{C[\tilde{I}_l]},$$

$r=0,1,\ldots,m-1$. For $q=\infty$ we are done. The result for $q<\infty$ follows by integration. ∎

We also have the following global version of Theorem 6.20:

COROLLARY 6.21

Let $1\leqslant\sigma\leqslant m$. Then for all $f\in C^{\sigma-1}[a,b]$,

$$\left.\begin{array}{l}\|D^r(f-Qf)\|_{C[a,b]}\\ \quad r=0,\ldots,\sigma-1\\ \|D^rQf\|_{C[a,b]}\\ \quad r=\sigma,\ldots,m-1\end{array}\right\}\leqslant C_1(\bar{\Delta})^{\sigma-r-1}\omega_{m-\sigma+1}(D^{\sigma-1}f;\bar{\Delta})_{C[a,b]}.\qquad(6.46)$$

The constant C_1 depends only on m.

Proof. We simply apply Theorem 6.20 to each subinterval of $[a,b]$ and then take the maximum. ■

We have made no attempt in Theorem 6.20 and Corollary 6.21 to obtain the best possible constants. On the other hand, it will be shown in the following section that there are lower bounds corresponding to the upper bounds given here, and thus the orders of approximation given are the best possible. In fact, it will be shown in §6.7 that the orders of approximation obtained here agree with the asymptotic orders obtainable by arbitrary finite dimensional spaces, and thus the splines are asymptotically optimal spaces in the sense of n-widths (cf. Section 2.10).

The bounds given in (6.44) and (6.46) show that Q is a remarkable operator: (1) it delivers approximations that admit *local error bounds*; (2) the derivatives of the spline approximate the derivatives of f *simultaneously*; (3) the operator Q is *linear*; and (as the next theorem shows) (4) it is *bounded* from $C[a,b]$ into $C[a,b]$.

THEOREM 6.22

For every $f \in C[a,b]$,

$$\|Qf\|_{C[a,b]} \leqslant (2m)^m \|f\|_{C[a,b]}. \tag{6.47}$$

Proof. For all $i=1,2,\ldots,n$ and all $j=1,2,\ldots,m$, by Theorem 2.56,

$$|\lambda_{ij} f| = |[\tau_{i1},\ldots,\tau_{ij}]f| \leqslant \left(\frac{2(m-1)}{y_{i+m}-y_i} \right)^{j-1} \|f\|_{C[a,b]}.$$

Coupling this with the estimate (6.43) for α_{ij}, we obtain

$$|\lambda_i f| \leqslant \sum_{j=1}^{m} |\alpha_{ij}||\lambda_{ij} f| \leqslant \sum_{j=1}^{m} (2m)^{j-1} \|f\|_{C[a,b]}$$
$$\leqslant (2m)^m \|f\|_{C[a,b]}.$$

Now since $\sum_{i=1}^{n} B_i(x) \equiv 1$, (6.47) follows. ■

While the operator Q does reproduce $\mathcal{P}_m$, usually $Qs \neq s$ for all $s \in \mathcal{S}_m(\Delta)$, and thus, in general, it is not a *projector* (cf. the following example). It is possible to construct projectors if desired—see Remark 6.3.

EXAMPLE 6.23

Let $m=2$. Then for $i=1,2,\ldots,n$,

$$\lambda_i f = f(y_i) + \frac{(y_{i+1}-y_i)[f(y_{i+2}-f(y_i)]}{(y_{i+2}-y_i)}.$$

Discussion. In this case the B-splines have the property that $B_i(y_{j+1})=\delta_{ij}$, $i,j=1,2,\ldots,n$. It follows that

$$Qf(y_{i+1})=\lambda_i f, \qquad i=1,2,\ldots,n.$$

Thus Qf is the piecewise linear spline that *interpolates* the values $\lambda_i f$ at y_{i+1}, $i=1,2,\ldots,n$. Moreover, $\lambda_i f = P_i(y_{i+1})$, where P_i is the linear polynomial interpolating the values of f at the points y_i and y_{i+2}. See Figure 18. ■

We turn now to some estimates of how well Qf approximates functions f in various Sobolev spaces. First we prove a local result.

THEOREM 6.24

Let $1 \leqslant p,\ q \leqslant \infty$, and $1 \leqslant \sigma \leqslant m$. Let $m \leqslant l \leqslant n$, $I_l=[y_l,y_{l+1}]$, and $\tilde{I}_l=[y_{l+1-m},y_{l+m}]$. Then for any function $f \in L_p^\sigma[\tilde{I}_l]$,

$$\left.\begin{array}{c} \|D^r(f-Qf)\|_{L_q[I_l]} \\ r=0,\ldots,\sigma-1 \\ \|D^rQf\|_{L_q[I_l]} \\ r=\sigma,\ldots,m-1 \end{array}\right\} \leqslant C_1(\bar{\Delta})^{\sigma-r+1/q-1/p}\omega_{m-\sigma}\left(D^\sigma f;\bar{\Delta}\right)_{L_p[\tilde{I}_l]}. \tag{6.48}$$

The constant C_1 depends only on m and p.

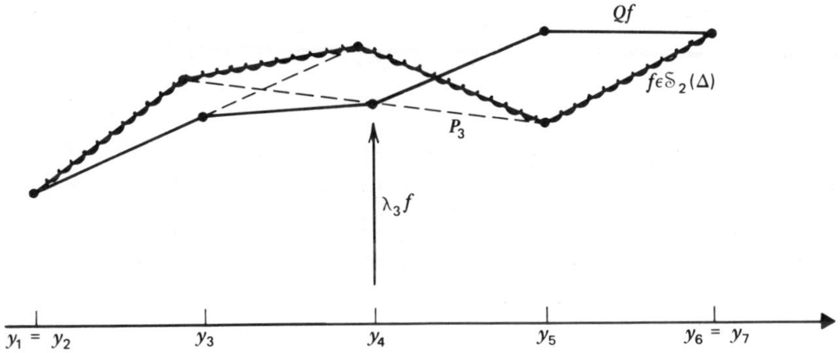

Figure 18. The operator Q does not reproduce $\mathcal{S}_m(\Delta)$.

Proof. By the Whitney Theorem 3.20 there exists $p_f \in \mathcal{P}_m$ such that for all $t \in I_l$,

$$|D^j(f-p_f)(t)| \leqslant C_2 \bar{\Delta}^{\sigma-j-1/p} \omega_{m-\sigma}(D^\sigma f; \bar{\Delta})_{L_p[\tilde{I}_l]},$$

$j=0,1,\ldots,\sigma-1$. Now if we use this inequality in place of (6.45), then the proof of Theorem 6.20 carries over directly to establish

$$|E_r(t)| \leqslant C_1 \bar{\Delta}^{\sigma-r-1/p} \omega_{m-\sigma}(D^\sigma f; \bar{\Delta})_{L_p[\tilde{I}_l]}.$$

Now integrating this over I_l, we obtain (6.48). ■

Next we give a global version of Theorem 6.24.

THEOREM 6.25

Let $1 \leqslant p \leqslant q \leqslant \infty$ and $1 \leqslant \sigma \leqslant m$. Then for every $f \in L_p^\sigma[a,b]$,

$$\left.\begin{array}{c} \|D^r(f-Qf)\|_{L_q[a,b]} \\ r=0,\ldots,\sigma-1 \\ \|D^r Qf\|_{L_q[a,b]} \\ r=\sigma,\ldots,m-1 \end{array}\right\} \leqslant C_1 \bar{\Delta}^{\sigma-r+1/q-1/p} \omega_{m-\sigma}(D^\sigma f; \bar{\Delta})_{L_p[a,b]} \tag{6.49}$$

The constant C_1 depends only on m and p.

Proof. By Theorem 6.24, for each $m \leqslant l \leqslant n$,

$$\int_{y_l}^{y_{l+1}} |E_r(t)|^q \, dt \leqslant C^q \bar{\Delta}^{q(\sigma-r+1/q-1/p)} \left(\int_{y_{l+1-m}}^{y_{l+m}} |D^\sigma f(t)|^p \, dt \right)^{q/p}.$$

If we sum these inequalities over $l=m,\ldots,n$ and take the qth root, we obtain

$$\|E_r\|_{L_q[a,b]} \leqslant K(\bar{\Delta})^{\sigma-r+1/q-1/p} \left(\sum_{l=m}^{n} \|D^\sigma f\|_{L_p[\tilde{I}_l]}^q \right)^{1/q}.$$

For $p \leqslant q \leqslant \infty$ Jensen's inequality (see Remark 6.2) implies

$$\left(\sum_{l=m}^{n} \|D^\sigma f\|_{L_p[\tilde{I}_l]}^q \right)^{1/q} \leqslant \left(\sum_{l=m}^{n} \|D^\sigma f\|_{L_p[\tilde{I}_l]}^p \right)^{1/p} \leqslant (2m-1)\|D^\sigma f\|_{L_p[a,b]}.$$

Note: $\tilde{I}_l \subset [y_{l+1-m}, y_{l+m}]$, so each piece of the interval $[a,b]$ is added into the sum at most $(2m-1)$ times. Substituting this in the above yields

$$\left.\begin{array}{l} \|D^r(f-Qf)\|_{L_q[a,b]} \\ \quad r=0,\dots,\sigma-1 \\ \|D^rQf\|_{L_q[a,b]} \\ \quad r=\sigma,\dots,m-1 \end{array}\right\} \leqslant C_2(\bar{\Delta})^{\sigma-r+1/q-1/p}\|D^\sigma f\|_{L_p[a,b]}. \tag{6.50}$$

To replace $\|D^\sigma f\|$ by $\omega_{m-\sigma}(D^\sigma f;\bar{\Delta})$, we use the K-functional of §2.9. First suppose $0 \leqslant r \leqslant \sigma-1$. Let $g \in L_q^m[a,b]$. Then

$$\|D^r(f-Qf)\|_q \leqslant \|D^r(f-g)\|_q + \|D^r(g-Qg)\|_q + \|D^rQ(f-g)\|_q.$$

Since $f-g \in L_q^r[a,b]$, (6.50) implies

$$\|D^rQ(f-g)\|_q \leqslant C_2\|D^r(f-g)\|_q.$$

It also asserts that

$$\|D^r(g-Qg)\|_q \leqslant C_2\bar{\Delta}^{m-r}\|D^m g\|_q.$$

Combining these estimates, we obtain

$$\|D^r(f-Qf)\|_q \leqslant (1+C_2)\left[\|D^r(f-g)\|_q + \bar{\Delta}^{m-r}\|D^{m-r}D^r g\|_q\right].$$

Now, as we vary g over $L_q^m[a,b]$, $D^r g$ varies over $L_q^{m-r}[a,b]$. Since the infimum of the expression in brackets as $D^r g$ varies over $L_q^{m-r}[a,b]$ is the K-functional applied to $D^r f$, we find that

$$\|D^r(f-Qf)\|_q \leqslant 2(1+C_2)K_{m-r,q}(\bar{\Delta})D^r f \leqslant C_3\omega_{m-r}(D^r f;\bar{\Delta})_q.$$

But by properties of the modulus of smoothness,

$$\omega_{m-r}(D^r f;\bar{\Delta})_q \leqslant C_4\bar{\Delta}^{\sigma-r}\omega_{m-\sigma}(D^\sigma f;\bar{\Delta})_q \leqslant C_5\bar{\Delta}^{\sigma-r+1/q-1/p}\omega_{m-\sigma}(D^\sigma f;\bar{\Delta})_p,$$

and substituting this in the above leads to the inequality (6.49) for $0 \leqslant r \leqslant \sigma-1$.

Suppose now that $\sigma \leqslant r \leqslant m$. Then for any $g \in L_q^m[a,b]$,

$$\begin{aligned}\|D^rQf\|_q &\leqslant \|D^rQ(f-g)\|_q + \|D^rQg\|_q \\ &\leqslant C_2\left[\bar{\Delta}^{\sigma-r+1/q-1/p}\|D^\sigma(f-g)\|_p + \bar{\Delta}^{m-r+1/q-1/p}\|D^{m-\sigma}D^\sigma g\|_p\right],\end{aligned}$$

where we have used (6.50) for $f-g$ and for g. Since g is arbitrary, we conclude that

$$\|D^rQf\|_q \leqslant 2C_2\bar{\Delta}^{\sigma-r+1/q-1/p}K_{m-\sigma,p}(\bar{\Delta})(D^\sigma f)$$
$$\leqslant 2C_2\bar{\Delta}^{\sigma-r+1/q-1/p}\omega_{m-\sigma}(D^\sigma f;\bar{\Delta})_p.$$

This is (6.49) for $\sigma \leqslant r \leqslant m-1$, and the theorem is proved. ■

Since Theorem 6.25 gives estimates for how well f *and* its derivatives are approximated, we may restate it in terms of the Sobolev norm:

$$\|f\|_{L_q^r[a,b]} = \sum_{i=0}^{r} \|D^i f\|_{L_q[a,b]}. \tag{6.51}$$

COROLLARY 6.26

Let Q be as in Theorem 6.18 and suppose $1 \leqslant p \leqslant q \leqslant \infty$. Then for any $0 \leqslant r \leqslant \sigma - 1 \leqslant m-1$ there exists a constant C independent of Δ such that for all $f \in L_p^\sigma[a,b]$,

$$\|f-Qf\|_{L_q^r[a,b]} \leqslant C(\bar{\Delta})^{\sigma-r+1/q-1/p}\|D^\sigma f\|_{L_p[a,b]}$$
$$\leqslant C(\bar{\Delta})^{\sigma-r+1/q-1/p}\|f\|_{L_p^\sigma[a,b]}. \tag{6.52}$$

We conclude this section with an estimate for how well functions in the space $L_p[a,b]$ can be approximated by polynomial splines.

THEOREM 6.27

Let $1 \leqslant p \leqslant \infty$. Then there exists a constant C_1 (depending only on m and p) such that for $f \in L_p[a,b]$,

$$d[f, \mathcal{S}_m(\Delta)]_p \leqslant C_1\omega_m(f;\bar{\Delta})_p. \tag{6.53}$$

Proof. By (6.50) we have

$$d[f, \mathcal{S}_m(\Delta)]_p \leqslant C_2\bar{\Delta}^m\|D^m f\|_{L_p[a,b]}$$

for all $f \in L_p^m[a,b]$. Applying Theorem 2.68 we obtain (6.53). ■

Theorem 6.27 gives a result on the L_p-distance of a function f to the space $\mathcal{S}_m(\Delta)$. We cannot show that $f - Qf$ satisfies (6.53)—indeed, Qf is not even defined for arbitrary functions in $L_p[a,b]$. On the other hand, the same order can be achieved with an appropriate bounded linear operator $\tilde{Q}$ mapping $L_p[a,b]$ into $\mathcal{S}_m(\Delta)$; see Remark 6.4.

§ 6.5. DIRECT THEOREMS IN INTERMEDIATE SPACES

In the previous section we have given an assortment of theorems detailing how well polynomial splines approximate functions in the classical spaces $C[a,b],\ldots,C^m[a,b]$ as well as in the Sobolev spaces $L_p[a,b],\ldots,L_p^m[a,b]$. Using the theory of intermediate spaces, these results can be further refined to produce direct theorems for various smooth spaces of functions lying between these spaces. To illustrate how this can be done, we need to review the main features of the theory of intermediate spaces.

Suppose X_0 and X_1 are two Banach spaces with norms $\|\cdot\|_0$ and $\|\cdot\|_1$, respectively. Suppose both are contained in a common linear Hausdorff space $\mathcal{X}$ such that the identity mapping from X_i into $\mathcal{X}$ is continuous for $i=0,1$. (We say X_i are *continuously imbedded in* $\mathcal{X}$.) Now we may consider the linear space

$$X_0+X_1=\{f=f_0+f_1, f_i\in X_i,\ i=0,1\}$$

with norm

$$\|f\|_{X_0+X_1}=\inf_{\substack{f_i\in X_i\\ f=f_0+f_1}}(\|f_0\|_0+\|f_1\|_1).$$

If $t>0$, we may define a functional on the Banach space X_0+X_1 by

$$\mathcal{K}(t)f=\inf_{\substack{f_0\in X_0, f_1\in X_1\\ f=f_0+f_1}}(\|f_0\|_0+t\|f_1\|_1)$$

THEOREM 6.28

Fix $1\leqslant p'\leqslant\infty$ and $0<\theta<1$. Then the set

$$(X_0,X_1)_{\theta,p'}=\{f\in X_0+X_1\colon \|f\|_{(X_0,X_1)_{\theta,p'}}<\infty\}$$

with the norm

$$\|f\|_{(X_0,X_1)_{\theta,p'}}=\begin{cases}\left(\int_0^1\left[t^{-\theta}\mathcal{K}(t)f\right]^{p'}t^{-1}\,dt\right)^{1/p'}, & 1\leqslant p'<\infty\\ \operatorname*{ess\,sup}_{0<t<1} t^{-\theta}\mathcal{K}(t)f, & p'=\infty\end{cases}$$

is a *Banach space*.

Discussion. For the basic theory of intermediate spaces, see Peetre [1963], Butzer and Berens [1967], or Löfstrom [1970]. ■

We illustrate this process with an important example.

EXAMPLE 6.29

Fix $1 \leqslant p', p \leqslant \infty$, and $0<\theta<1$, and define

$$B_p^{\sigma,p'}[a,b]=(L_p[a,b],L_p^m[a,b])_{\theta,p'} \quad , \tag{6.54}$$

where $\sigma=\theta m$. We call $B_p^{\sigma,p'}$ a *Besov space.*

Discussion. To describe the norm on the Besov spaces in more detail, define

$$|f|_{\sigma,p,p',m}=\begin{cases}\left(\int_0^1[t^{-\sigma}\omega_m(f,t)_p]^{p'}\dfrac{dt}{t}\right)^{1/p'}, & 1\leqslant p'<\infty\\ \operatorname*{ess\,sup}_{t>0} t^{-\sigma}\omega_m(f,t)_p \quad , & p'=\infty.\end{cases} \tag{6.55}$$

Then $B_p^{\sigma,p'}[a,b]$ forms a Banach space relative to the norm

$$\|f\|_{B_p^{\sigma,p'}[a,b]}=\|f\|_{L_p[a,b]}+|f|_{\sigma,p,p',m} \tag{6.56}$$

Equivalent norms can be defined on $B_p^{\sigma,p'}$ that do not involve m directly. For example, it is known that there exist constants c_1 and c_2 such that for all $f\in B_p^{\sigma,p'}[a,b]$,

$$c_1\|f\|_{B_p^{\sigma,p'}}\leqslant\|f\|_{L_p^{\lfloor\sigma\rfloor}}+|D^{\lfloor\sigma\rfloor}f|_{\sigma-\lfloor\sigma\rfloor,p,p',1}\leqslant c_2\|f\|_{B_p^{\sigma,p'}} \tag{6.57}$$

if σ is not an integer, and

$$c_1\|f\|_{B_p^{\sigma,p'}}\leqslant\|f\|_{L_p^{\sigma-1}}+|D^{\sigma-1}f|_{1,p,p',2}\leqslant c_2\|f\|_{B_p^{\sigma,p'}} \tag{6.58}$$

if σ is an integer. For $1\leqslant p\leqslant\infty$, the Besov space $B_p^{\sigma,p'}$ satisfies

$$L_p^{\sigma-1}[a,b]\subseteq B_p^{\sigma,p'}[a,b]\subseteq L_p^{\sigma}[a,b].$$

If $p'=p=\infty$, then the Besov spaces reduce to spaces of functions whose derivatives satisfy Lipschitz or Zygmund conditions. In particular,

$$B_\infty^{m+\alpha,\infty}[a,b]=\mathrm{Lip}^{m,\alpha}[a,b], \qquad 0<\alpha<1,$$

and

$$B_{\infty}^{m+1,\infty}[a,b]=\mathfrak{L}^m[a,b]=\{f\colon D^m f\in\mathfrak{L}[a,b]\}$$

[see (2.128) and (2.129)]. ∎

The key to obtaining bounds on how well polynomial splines approximate functions in intermediate spaces (and, in particular, in the Besov spaces) is the following general theorem from interpolation space theory:

THEOREM 6.30

Suppose $0<\theta<1$ and $1\leqslant p'\leqslant\infty$. If T is a bounded linear operator mapping X_i into Y_i with norm M_i, then it is also a bounded linear operator mapping $(X_0,X_1)_{\theta,p'}$ into $(Y_0,Y_1)_{\theta,p'}$ with norm $M\leqslant M_0^{\theta}M_1^{1-\theta}$.

We can now give bounds on $f-Qf$ for f in a Besov space, where Q is the spline operator defined in §6.4, Theorem 6.18.

THEOREM 6.31

Let $1\leqslant p\leqslant q\leqslant\infty$ and $1\leqslant p',q'\leqslant\infty$. Suppose $1\leqslant\sigma\leqslant m$ and that $0<\tau<\lfloor\sigma-1\rfloor$. Then there exists a constant C_1 such that for all functions $f\in B_p^{\sigma,p'}[a,b]$,

$$\|f-Qf\|_{B_q^{\tau,q'}[a,b]}\leqslant C_1(\bar{\Delta})^{\sigma-\tau-1/p+1/q}\|f\|_{B_p^{\sigma,p'}[a,b]}. \tag{6.59}$$

Proof. We proceed in two steps using Theorem 6.30. First, we prove that for any $1\leqslant\sigma\leqslant m$ and $0\leqslant r\leqslant\sigma-1$ (where r is an integer),

$$\|f-Qf\|_{L_q^r[a,b]}\leqslant C_2(\bar{\Delta})^{\sigma-r+1/q-1/p}\|f\|_{B_p^{\sigma,p'}[a,b]}. \tag{6.60}$$

For this we apply Theorem 6.30 with $X_0=L_p^{r+1}[a,b]$, $X_1=L_p^m[a,b]$, $Y_0=Y_1=L_q^r[a,b]$, and $T=I-Q$. Now by (6.52),

$$\|Tf\|_{Y_0}\leqslant C_3(\bar{\Delta})^{1+\delta}\|f\|_{X_0}$$

$$\|Tf\|_{Y_1}\leqslant C_4(\bar{\Delta})^{m-r+\delta}\|f\|_{X_1},$$

where $\delta=1/q-1/p$. Let $0<\theta<1$ be such that $\sigma=\theta(r+1)+(1-\theta)m$. Then $\theta(1+\delta)+(1-\theta)(m-r+\delta)=\sigma-r+\delta$. Theorem 6.30 yields (6.60).

For the second step, we choose $X_0 = X_1 = B_p^{\sigma,p'}[a,b]$, $Y_0 = L_q[a,b]$, $Y_1 = L_q^{\lfloor \sigma-1 \rfloor}\lfloor a,b \rfloor$, and T as above. Then by (6.60),

$$\|Tf\|_{Y_0} \leqslant C_5(\bar{\Delta})^{\sigma+\delta}\|f\|_{X_0}$$

$$\|Tf\|_{Y_1} \leqslant C_6(\bar{\Delta})^{\sigma-\lfloor \sigma-1 \rfloor+\delta}\|f\|_{X_1}.$$

Now for each $0<\theta<1$, let $\tau=\theta\lfloor \sigma-1 \rfloor$. Since $\theta(\sigma-\lfloor \sigma-1 \rfloor+\delta)+(1-\theta)(\sigma+\delta)=\sigma-\tau+\delta$, (6.59) follows. ■

§ 6.6. LOWER BOUNDS

In this section we give explicit lower bounds for how well certain smooth functions can be approximated by piecewise polynomials. These bounds will show that the results given in Section 6.4 are of optimal order. Our first result provides lower bounds that are companions to the upper bounds established in Corollary 6.21. Define

$$d_j[f,\mathcal{P}\mathcal{P}_m(\Delta)]_p = \inf_{s\in\mathcal{P}\mathcal{P}_m(\Delta)} \|D^j(f-s)\|_{L_p[a,b]}. \tag{6.61}$$

THEOREM 6.32

There is a constant $C_1>0$ (depending only on m) such that for any partition Δ of $[a,b]$ and any $1\leqslant\sigma\leqslant m$ there exists a function $F\in C^{\sigma-1}[a,b]$ with

$$C_1\bar{\Delta}^{\sigma-j-1}\omega_{m-\sigma+1}(D^{\sigma-1}F;\bar{\Delta}) \leqslant d_j[F,\mathcal{P}\mathcal{P}_m(\Delta)]_\infty, \quad j=0,1,\ldots,\sigma-1. \tag{6.62}$$

Proof. We make use of the perfect B-spline $B_{\sigma+1}^*$ defined in (4.67). Given a partition Δ, suppose that ν is chosen so that $x_{\nu+1}-x_\nu=\bar{\Delta}$. Now define

$$F(x)=\begin{cases} \left(\dfrac{\bar{\Delta}}{2}\right)^\sigma B_{\sigma+1}^*\left[\dfrac{2(x-x_\nu)-(x_{\nu+1}-x_\nu)}{(x_{\nu+1}-x_\nu)}\right], & x_\nu\leqslant x\leqslant x_{\nu+1}. \\ 0, & \text{otherwise} \end{cases} \tag{6.63}$$

By construction, $F\in L_\infty^\sigma[a,b]$ and

$$\|D^\sigma F\|_{L_\infty[a,b]}=2^{\sigma-1}\sigma!. \tag{6.64}$$

By property (2.119) of moduli of smoothness, this implies

$$\omega_{m-\sigma+1}(D^{\sigma-1}F;\bar{\Delta})_\infty \leqslant \bar{\Delta}\|D^\sigma F\|_\infty \leqslant C_2\bar{\Delta}. \tag{6.65}$$

Now fix $0 \leqslant j \leqslant \sigma - 1$ and let

$$d_j = \inf_{g \in \mathcal{P}_m} \|D^j(B^*_{\sigma+1} - g)\|_{L_\infty[-1,1]}$$

Since $B^*_{\sigma+1}$ is not a polynomial of order m, $C_4 = \min_{0 \leqslant j \leqslant \sigma-1} d_j > 0$. On the other hand, by change of variables,

$$C_4 \leqslant d_j \leqslant \max_{-1 \leqslant x \leqslant 1} |D^j[B^*_{\sigma+1}(x) - g(x)]| = \left(\frac{\bar{\Delta}}{2}\right)^{-\sigma+j} \max_{x_\nu \leqslant t \leqslant x_{\nu+1}} |D^j(F(t) - \tilde{g}(t))|$$

where

$$\tilde{g}(t) = \left(\frac{\bar{\Delta}}{2}\right)^{\sigma-j} g\left[\frac{2(t-x_\nu)-(x_{\nu+1}-x_\nu)}{x_{\nu+1}-x_\nu}\right] \tag{6.66}$$

is also a polynomial of order m. Since this holds for all $g \in \mathcal{P}_m$, we conclude that

$$C_4 \leqslant \bar{\Delta}^{-\sigma+j} d_j[F, \mathcal{P}\mathcal{P}_m(\Delta)]_\infty.$$

Combining this with (6.65), we obtain (6.62). ■

A similar argument can be used to obtain lower bounds in the L_q-norm.

THEOREM 6.33

Let $0 \leqslant \sigma \leqslant m$ and $1 \leqslant p, q \leqslant \infty$. Then there exists a constant $C_1 > 0$ (depending only on m, σ, p, and q) such that for all partitions Δ of $[a,b]$ there is a corresponding function $F \in L_p^\sigma[a,b]$ with

$$C_1 \bar{\Delta}^{\sigma-j+1/q-1/p} \omega_{m-\sigma}(D^\sigma F; \bar{\Delta})_{L_p[a,b]} \leqslant d_j[F, \mathcal{P}\mathcal{P}_m(\Delta)]_q, \quad j = 0, 1, \ldots, \sigma - 1. \tag{6.67}$$

Proof. Let F be the function constructed in (6.63). Then in view of (6.64) and the fact that F is zero outside of an interval of length $\bar{\Delta}$, we have

$$\|D^\sigma F\|_{L_p[a,b]} \leqslant \bar{\Delta}^{1/p} \|D^\sigma F\|_{L_\infty[a,b]} \leqslant C_2 \bar{\Delta}^{1/p}. \tag{6.68}$$

Now fix $1 \leqslant q < \infty$, and define

$$C_3 = \min_{0 \leqslant j \leqslant \sigma-1} \inf_{g \in \mathcal{P}_m} \|D^j(B^*_{\sigma+1} - g)\|_{L_q[-1,1]}.$$

But then, for any $0 \leqslant j \leqslant \sigma-1$ and any $g \in \mathcal{P}_m$,

$$C_3 \leqslant \left(\int_{-1}^{1} |D^j[B_{\sigma+1}^*(x) - g(x)]|^q \, dx \right)^{1/q} = \left(2 \int_{x_\nu}^{x_{\nu+1}} |D^j[F(t) - \tilde{g}(t)]|^q \frac{dt}{\overline{\Delta}} \right)^{1/q} \cdot$$

$$\cdot \left(\frac{\overline{\Delta}}{2} \right)^{-\sigma+j} \leqslant \left(\frac{\overline{\Delta}}{2} \right)^{-\sigma+j-1/q} \| D^j(F - \tilde{g}) \|_{L_q[a,b]},$$

where $\tilde{g}$ is defined in (6.66). We conclude that

$$C_3 \leqslant \overline{\Delta}^{-\sigma+j-1/q} d_j[F, \mathcal{P}\mathcal{P}_m(\Delta)]_{L_q[a,b]}.$$

Combining this with (6.68), we obtain

$$C_4 \overline{\Delta}^{\sigma-j+1/q-1/p} \| D^\sigma F \|_{L_p[a,b]} \leqslant d_j[F, \mathcal{P}\mathcal{P}_m(\Delta)]_{L_q[a,b]}. \tag{6.69}$$

Finally, to obtain (6.67), we observe that

$$\omega_{m-\sigma}(D^\sigma F; \overline{\Delta})_p \leqslant 2^{m-\sigma} \| D^\sigma F \|_p. \qquad \blacksquare$$

As a corollary, we have the following companion to Corollary 6.26:

COROLLARY 6.34

Let $1 \leqslant p, q \leqslant \infty$ and $0 \leqslant r \leqslant \sigma \leqslant m$. Then there is a constant $C_1 > 0$ (depending only on p, q, r, σ, m) such that for all partitions Δ of $[a,b]$ there exists a corresponding function $F \in L_p^\sigma[a,b]$ with

$$C_1 \overline{\Delta}^{\sigma-r+1/q-1/p} \| F \|_{L_p^\sigma[a,b]} \leqslant d[F, \mathcal{P}\mathcal{P}_m(\Delta)]_{L_q^r[a,b]}. \tag{6.70}$$

Proof. Let F be defined as in (6.63). Then summing the inequalities (6.69) for $j = 0, 1, \ldots, r$, we obtain

$$C_5 \overline{\Delta}^{\sigma-r+1/q-1/p} \| D^\sigma F \|_{L_p[a,b]} \leqslant \sum_{j=0}^{r} d_j[F, \mathcal{P}\mathcal{P}_m(\Delta)]_{L_q[a,b]}$$

$$\leqslant d[F, \mathcal{P}\mathcal{P}_m(\Delta)]_{L_q^r[a,b]}.$$

But since F and its derivatives up to order $\sigma - 1$ all vanish at the point x_ν (cf. the definition of F), it follows by applying Hölder's inequality to the Taylor expansion that

$$\| D^j F \|_{L_p[a,b]} \leqslant \frac{(b-a)^{\sigma-j}}{(\sigma-j-1)!} \| D^\sigma F \|_{L_p[a,b]}, \qquad j = 0, 1, \ldots, \sigma.$$

This implies that

$$\|F\|_{L_p^\sigma[a,b]} \leqslant \left(\sum_{j=0}^{\sigma} \frac{(b-a)^{\sigma-j}}{(\sigma-j-1)!} \right) \|D^\sigma F\|_{L_p[a,b]}.$$

Combining these estimates, we obtain (6.70). ■

§ 6.7. *n*-WIDTHS

In this section we show that polynomial splines provide asymptotically optimal sequences of approximating spaces (in the sense of *n*-widths) for several classes of smooth functions. Throughout this section we shall deal with the sequence of spaces

$$X_n = \mathcal{S}_m(\Delta_{n-m}), \quad n = m+1, m+2, \ldots$$

where Δ_{n-m} is the uniform partition of the interval $[a,b]$ with $n-m$ knots (cf. Example 6.6). For each $n > m$, X_n is an n-dimensional linear space.

Our first result concerns the unit ball $UL_p^\sigma[a,b]$ of the Sobolev space $L_p^\sigma[a,b]$ (cf. Theorem 2.77).

THEOREM 6.35

Fix $1 \leqslant \sigma \leqslant m$ and $1 \leqslant q \leqslant p \leqslant \infty$. Then X_n is an asymptotically optimal sequence for approximating $UL_p^\sigma[a,b]$ in $L_q[a,b]$. In particular,

$$d(UL_p^\sigma[a,b], X_n)_{L_q[a,b]} \approx \left(\frac{1}{n}\right)^\sigma. \tag{6.71}$$

Proof. For $n \geqslant 2m$ we have $\overline{\Delta}_{n-m} = (b-a)/(n-m) \leqslant 2(b-a)/n$. Thus by Theorem 6.27, for every $f \in UL_p^\sigma[a,b]$,

$$d(f, X_n)_q \leqslant d(f, X_n)_p \leqslant C_1 (\overline{\Delta}_{n-m})^\sigma \leqslant C_1 [2(b-a)]^\sigma \left(\frac{1}{n}\right)^\sigma.$$

Theorem 2.77 asserts $d_n(UL_p^\sigma[a,b], L_q[a,b]) \approx C_2 (1/n)^\sigma$; and (6.71) follows. ■

Theorem 6.35 also holds for $1 \leqslant p < q \leqslant 2$. On the other hand, for $2 \leqslant p < q \leqslant \infty$ Theorem 2.77 asserts that

$$d_n(UL_p^\sigma[a,b], L_q[a,b])_\infty \approx \left(\frac{1}{n}\right)^\sigma,$$

whereas by (6.50) and (6.69), the best possible order for X_n is

$$d(UL_p^\sigma[a,b],X_n)_q \approx \left(\frac{1}{n}\right)^{\sigma+1/q-1/p}.$$

Thus for these values of p and q the spline spaces X_n are not asymptotically optimal for approximation of UL_p^σ in L_q. The same is true for $1 \leqslant p \leqslant 2 < q \leqslant \infty$. In this case, Theorem 2.77 asserts that

$$d_n(UL_p^\sigma[a,b], L_q[a,b]) \approx \left(\frac{1}{n}\right)^{\sigma+1/2-1/p},$$

whereas (6.50) and (6.69) show that the best possible order for X_n is

$$d(UL_p^\sigma[a,b], X_n)_q \approx \left(\frac{1}{n}\right)^{\sigma+1/q-1/p}.$$

Our next result shows that the spaces of splines X_n in Theorem 6.35 also provide asymptotically optimal approximation for the set $\Lambda_\omega^\sigma[a,b]$ in $C[a,b]$.

THEOREM 6.36

Let $1 \leqslant \sigma \leqslant m$, and let $\Lambda_\omega^\sigma[a,b]$ be the set of smooth functions defined in (3.20). Then

$$d(\Lambda_\omega^\sigma[a,b], X_n)_\infty \approx d_n(\Lambda_\omega^\sigma[a,b], C[a,b]) \approx \left(\frac{1}{n}\right)^\sigma \omega\left(\frac{1}{n}\right).$$

Proof. The n-width of Λ_ω^σ is computed in Example 3.15. On the other hand, Corollary 6.21 shows that for every $f \in \Lambda_\omega^\sigma[a,b]$,

$$d(f, X_n)_\infty \leqslant C_1(\overline{\Delta}_{n-m})^\sigma \omega(\overline{\Delta}_{n-m}) \leqslant C_2\left(\frac{1}{n}\right)^\sigma \omega\left(\frac{1}{n}\right). \qquad \blacksquare$$

§ 6.8. INVERSE THEORY FOR $p=\infty$

In this section we obtain inverse, saturation, and characterization theorems for approximation by polynomial splines in the L_∞-norm. Similar results for the L_p-norms are given in the following section.

The idea of inverse theorems is to estimate the modulus of smoothness of a function in terms of how well it can be approximated by a sequence of spline spaces. As we shall see, the analysis depends to some extent on how much smoothness we require of the spline spaces. We begin by discussing approximation with a sequence of piecewise polynomial spaces $\mathcal{P}\mathcal{P}_m(\Delta_\nu)$.

As Example 6.11 showed, it is impossible to get estimates on the smoothness of a function in terms of $d[f,\mathcal{P}\mathcal{P}_m(\Delta_\nu)]_\infty$ without some assumption on the behavior of the sequence of partitions Δ_ν. We make the following definition:

DEFINITION 6.37. Mixing Condition

For $\nu=1,2,\cdots$ let

$$\Delta_\nu=\left\{a=x_0^\nu<x_1^\nu<\cdots<x_{k_\nu}^\nu<x_{k_\nu+1}^\nu=b\right\} \tag{6.72}$$

be a sequence of partitions of the interval $[a,b]$. We say that Δ_ν satisfies the *mixing condition* provided there exists a constant $0<\rho<1$ such that for all ν and all $1\leqslant i\leqslant k_\nu$,

$$\sup_{n>\nu} d(x_i^\nu,\Delta_n)\geqslant\rho\bar{\Delta}_\nu, \tag{6.73}$$

where

$$d(x,\Delta_n)=\min_{0\leqslant j\leqslant k_n+1}|x-x_j^n|.$$

Before proceeding to our first inverse theorem, we show that the sequence of uniform partitions (see Example 6.6) satisfies the mixing condition.

THEOREM 6.38

The sequence of uniform partitions

$$\Delta_\nu=\left\{a+i\bar{\Delta}_\nu\right\}_{i=0}^{\nu},\qquad \bar{\Delta}_\nu=\underline{\Delta}_\nu=\frac{(b-a)}{\nu} \tag{6.74}$$

satisfies the mixing condition with $\rho=1/6$.

Proof. We may assume $[a,b]=[0,1]$. Let $1\leqslant\alpha\leqslant\nu$ and $0\leqslant\beta\leqslant\alpha-1$ be integers, and suppose that $(6\beta+2)/6\alpha\leqslant i/\nu\leqslant(6\beta+4)/6\alpha$. Then we easily check that

$$\frac{i+\beta}{\nu+\alpha}<\frac{i}{\nu}<\frac{i+\beta+1}{\nu+\alpha}.$$

In fact,

$$\varepsilon_1 = \frac{i}{\nu} - \frac{i+\beta}{\nu+\alpha} = \frac{i\alpha - \nu\beta}{\nu(\nu+\alpha)} \geqslant \frac{1}{6\nu} = \frac{\overline{\Delta}_\nu}{6};$$

$$\varepsilon_2 = \frac{i+\beta+1}{\nu+\alpha} - \frac{i}{\nu} = \frac{\nu+\nu\beta - i\alpha}{\nu(\nu+\alpha)} \geqslant \frac{1}{6\nu} = \frac{\overline{\Delta}_\nu}{6}.$$

This shows that $x_i^\nu = i/\nu$ satisfies (6.73) with $n = \nu + \alpha$ and with $\rho = 1/6$.

It remains to check that all of the points $x_i = i/\nu, i = 1, \ldots \nu - 1$ are contained in an interval $I_{\alpha\beta} = [(6\beta+2)/6\alpha, (6\beta+4)/6\alpha]$ for some choice of α and β. To this end, we show that $I_r = \cup_{\alpha=1}^{r} \cup_{\beta=1}^{\alpha-1} I_{\alpha\beta} = (2/6r, (6r-2)/6r)$. For $r = 1$ this is trivial. Now we proceed by induction on r. It is easily checked that the intervals

$$I_{r+1,0} = \left(\frac{2}{6(r+1)}, \frac{4}{6(r+1)} \right)$$

and

$$I_{r+1,r} = \left(\frac{6r+2}{6(r+1)}, \frac{6r+4}{6(r+1)} \right)$$

both overlap the interval I_r. This shows that I_{r+1} also has the stated form. Now $I_\nu = (1/3\nu, 1 - 1/3\nu)$, and this contains all the points $x_i^\nu, 1 \leqslant i \leqslant k_\nu = \nu - 1$. ■

For sequences that are mixed we can now estimate $\omega_m(f; \overline{\Delta}_\nu)$.

THEOREM 6.39

Let Δ_ν be a sequence of partitions of $[a,b]$ satisfying the mixing condition. Then for any $f \in B[a,b]$,

$$\omega_m\big(f; \overline{\Delta}_\nu\big) \leqslant 2^{m+1} \left\lceil \frac{m}{\rho} \right\rceil^m \sup_{n > \nu} d\big[f, \mathcal{P}\mathcal{P}_m(\Delta_n)\big]_\infty. \tag{6.75}$$

Proof. For each n let $s_n \in \mathcal{P}\mathcal{P}_m(\Delta_n)$ be such that $\|f - s_n\| \leqslant 2d[f, \mathcal{P}\mathcal{P}_m(\Delta_n)]$. Now fix ν, and let $h \leqslant \rho\overline{\Delta}_\nu/m$ and $a \leqslant x \leqslant b - mh$. Let i be such that $x_i^\nu \leqslant x < x_{i+1}^\nu$. If $x < x_{i+1}^\nu - \rho\overline{\Delta}_\nu$, then

$$|\Delta_h^m f(x)| = |\Delta_h^m (f - s_\nu)(x)| \leqslant 2^m \|f - s_\nu\| \leqslant 2^{m+1} d\big[f, \mathcal{P}\mathcal{P}_m(\Delta_\nu)\big]_\infty.$$

On the other hand, if $x_{i+1}^{\nu}-\rho\bar{\Delta}_{\nu}\leqslant x$, then by the mixing condition there exists n and j so that $x_j^n\leqslant x<x+mh<x_{j+1}^n$, and

$$|\Delta_h^m f(x)|=|\Delta_h^m(f-s_n)(x)|\leqslant 2^{m+1}d[f,\mathcal{P}\mathcal{P}_m(\Delta_n)]_\infty.$$

We conclude that

$$\omega_m\left(f;\frac{\rho\bar{\Delta}_\nu}{m}\right)\leqslant 2^{m+1}\sup_{n\geqslant\nu} d[f,\mathcal{P}\mathcal{P}_m(\Delta_n)]_\infty.$$

Combining this with

$$\omega_m(f;\bar{\Delta}_\nu)=\omega_m\left(f;\frac{m}{\rho}\frac{\rho\bar{\Delta}_\nu}{m}\right)\leqslant\left[\frac{m}{\rho}\right]^m\omega_m\left(f;\frac{\rho\bar{\Delta}_\nu}{m}\right),$$

we obtain (6.75). ∎

The following theorem shows that the estimate (6.75) cannot hold unless the mixing condition does.

THEOREM 6.40

Suppose Δ_ν is a sequence of partitions such that for all $f\in B[a,b]$ and for all ν in some infinite set V,

$$\omega_2(f;\bar{\Delta}_\nu)\leqslant C\sup_{n\geqslant\nu} d[f,\mathcal{P}\mathcal{P}(\Delta_n)]_\infty. \tag{6.76}$$

Then Δ_ν must satisfy the mixing condition.

Proof. Fix ν and i with $1\leqslant i\leqslant k_\nu$. Let $f(x)=(x-x_i^\nu)_+$, and let n be such that

$$\frac{1}{2C}\omega_2(f;\bar{\Delta}_\nu)\leqslant d[f,\mathcal{P}\mathcal{P}(\Delta_n)]_\infty.$$

Let j be such that $x_j^n\leqslant x_i^\nu<x_{j+1}^n$. Let $h=x_i^\nu-x_j^n$ and $\tilde{h}=x_{j+1}^n-x_i^\nu$. Now there are two cases.

CASE 1. If $h<\tilde{h}$, then with $p(x)=x-x_j^n$, we see that

$$d[f,\mathcal{P}\mathcal{P}(\Delta_n)]\leqslant\|f-p\|=h.$$

CASE 2. If $\tilde{h}\leqslant h$, then with $p(x)=0$, we have

$$d[f,\mathcal{P}\mathcal{P}(\Delta_n)]\leqslant\|f-p\|=\tilde{h}.$$

We conclude that $\omega_2(f;\bar{\Delta}_\nu)/2C \leqslant \min(h,\tilde{h}) = d(x_i^\nu, \Delta_n)$. But we may easily check that $\omega_2(f;\bar{\Delta}_\nu) = \bar{\Delta}_\nu$, and it follows that Δ_ν satisfies the mixing condition with constant $\rho = 1/2C$. ∎

We now give a complete inverse theorem in which $\omega_m(f;t)$ is estimated for all small t.

THEOREM 6.41

Let Δ_ν be a sequence of partitions going steadily to zero as in Definition 6.5 and satisfying the mixing condition. Suppose in addition that $f \in B[a,b]$ is such that

$$d[f, \mathcal{P}\mathcal{P}_m(\Delta_\nu)]_\infty \leqslant \phi(\bar{\Delta}_\nu), \tag{6.77}$$

where ϕ is a monotone increasing function on $(0,\bar{\Delta}_1)$. Then

$$\omega_m(f;t) \leqslant C_1\phi(t) \qquad \text{for all } 0<t<\bar{\Delta}_1. \tag{6.78}$$

Proof. Let i be such that $\bar{\Delta}_{i+1} \leqslant t < \bar{\Delta}_i$. Then

$$\omega_m(f;t) \leqslant \omega_m(f;\bar{\Delta}_i) \leqslant \omega_m(f;\beta\bar{\Delta}_{i+1}) \leqslant \lceil \beta \rceil^m \omega_m(f;\bar{\Delta}_{i+1}).$$

On the other hand, by the monotonicity of ϕ and Theorem 6.39,

$$\begin{aligned}\omega_m(f;\bar{\Delta}_{i+1}) &\leqslant 2^{m+1}\left\lceil \frac{m}{\rho} \right\rceil^m \sup_{n \geqslant i+1} d[f, \mathcal{P}\mathcal{P}_m(\Delta_n)]_\infty \\ &\leqslant 2^{m+1}\left\lceil \frac{m}{\rho} \right\rceil^m \sup_{n \geqslant i+1} \phi(\bar{\Delta}_n) \leqslant 2^{m+1}\left\lceil \frac{m}{\rho} \right\rceil^m \phi(\bar{\Delta}_{i+1}).\end{aligned}$$

Since $\phi(\bar{\Delta}_{i+1}) \leqslant \phi(t)$, (6.78) follows with $C_1 = 2^{m+1}\lceil \beta \rceil^m \lceil m/\rho \rceil^m$. ∎

The following saturation result shows that (except for polynomials that are approximated exactly) no matter how smooth a function f may be, it cannot be approximated to order better than $\bar{\Delta}^m$ by a sequence $\mathcal{P}\mathcal{P}_m(\Delta_\nu)$ of piecewise linear polynomials with Δ_ν satisfying the conditions of Theorem 6.41.

THEOREM 6.42

Let Δ_ν be a sequence of partitions of $[a,b]$ as in Theorem 6.41. Suppose $f \in B[a,b]$ is such that

$$d[f, \mathcal{P}\mathcal{P}_m(\Delta_\nu)]_\infty \leqslant \bar{\Delta}_\nu^m \psi(\bar{\Delta}_\nu) \tag{6.79}$$

for some function $\psi(t)$ with $t^m\psi(t)$ monotone and $\psi(t)\to 0$ as $t\to 0$. Then $f\in\mathcal{P}_m$; that is f is a polynomial of order m.

Proof. Theorem 6.41 implies that if (6.79) holds, then $\omega_m(f;t)/t^m \leq C\psi(t) \to 0$ as $t\to 0$. By properties of moduli of smoothness, cf. (2.122), this implies that $f\in\mathcal{P}_m$. ■

We emphasize again that the limit on the order of convergence attainable using a sequence of piecewise polynomial spaces given in Theorem 6.42 holds only for sequences of partitions satisfying the hypotheses of Theorem 6.41. Higher-order convergence can occur for a specific function with a proper choice of knot locations (cf. Chapter 7).

Putting the direct theorems of §6.4 together with the inverse theorems established here, we can now characterize some of the classical smooth spaces in terms of how well they can be approximated by piecewise polynomials.

THEOREM 6.43

Suppose Δ_ν is a sequence of partitions of $[a,b]$ as in Theorem 6.41. Then

$$d[f,\mathcal{P}\mathcal{P}_m(\Delta_\nu)]_\infty \leq C(\overline{\Delta}_\nu^{k+\alpha}) \tag{6.80}$$

if and only if

$$\begin{aligned}
&f\in \mathrm{Lip}^{k,\alpha}[a,b], && \text{when } 0\leq k\leq m-1 \text{ and } 0<\alpha<1; && (6.81)\\
&f\in \mathcal{Z}^{k-1}[a,b], && \text{when } 1\leq k\leq m-1 \text{ and } \alpha=0; && (6.82)\\
&f\in \mathrm{Lip}^{m-1,1}[a,b], && \text{when } k=m,\alpha=0; && (6.83)\\
&f\in \mathcal{P}_m, && \text{when } k+\alpha>m. && (6.84)
\end{aligned}$$

For the definition of these spaces, see (2.128) and (2.129).

Proof. The assertions that (6.80) follows from (6.81) to (6.84) are direct theorems, and they were proved in Section 6.4. Conversely, if (6.80) holds, then by Theorem 6.41, $\omega_m(f;t)=\mathcal{O}(t^{k+\alpha})$. If $k+\alpha>m$, this implies $f\in\mathcal{P}_m$ by the saturation Theorem 6.42. In all other cases it implies that f belongs to the spaces indicated in (6.81) to (6.83). ■

So far we have been working with the space $\mathcal{P}\mathcal{P}_m(\Delta)$ of piecewise polynomials of order m. We have seen that the mixing condition plays an important role in the inverse theory for such spaces. In the remainder of this section we shall examine inverse theorems for approximating spaces $\mathcal{S}_\nu$ contained in $\mathcal{P}\mathcal{P}_m(\Delta)\cap C^l[a,b]$ with $0\leq l\leq m-2$. In this case we will be able to establish inverse theorems without a mixing condition. The results, however, are weaker than those obtained above, and they do not lead to complete characterization theorems in all cases.

In preparation for our first inverse result, we now give an estimate for the modulus of smoothness of a piecewise polynomial.

THEOREM 6.44

Let Δ be a partition of $[a,b]$, and let $0<\varepsilon\leqslant\underline{\Delta}$. Then for any $s\in\mathcal{P}\mathcal{P}_m(\Delta)$,

$$\omega_m(s;\varepsilon)\leqslant(2m)^m\sum_{j=0}^{m-1}\varepsilon^j J(D^j s), \tag{6.85}$$

where

$$J(D^j s)=\max_{1\leqslant i\leqslant k}|D^j s(x_i+)-D^j s(x_i-)|.$$

If $s\in\mathcal{P}\mathcal{P}_m(\Delta)\cap C^l[a,b]$, then we have the estimate

$$\omega_m(s,\varepsilon)\leqslant(2m)^m\sum_{j=l+1}^{m-1}\varepsilon^j J(D^j s). \tag{6.86}$$

Proof. We observe that $\omega_m(f;\varepsilon)\leqslant m^m\omega_m(f;\varepsilon/m)$. Now let $h\leqslant\varepsilon/m$. Then for any x, the interval $(x,x+mh)$ contains at most one knot of s. If it does not contain any knots, then $\Delta_h^m s(x)=0$. Suppose it contains one knot, say x_i. Then for all t in $(x,x+mh)$, s can be written in the form

$$s(t)=p(t)+\sum_{j=1}^{m}\frac{c_j(t-x_i)_+^{m-j}}{(m-j)!},$$

where $p\in\mathcal{P}_m$ and $c_j=\text{jump}[D^{m-j}s]_{x_i}$, $j=0,1,\ldots,m$. Thus

$$\Delta_h^m s(x)=\sum_{r=0}^{m}\binom{m}{r}(-1)^{m-r}\sum_{j=1}^{m}\frac{c_j(x+rh-x_i)_+^{m-j}}{(m-j)!}.$$

Since $|x+rh-x_i|\leqslant\varepsilon$, (6.85) follows. If $s\in C^l[a,b]$, then s and its derivatives up to the lth order have no jumps, and we obtain (6.86). ∎

Our next task is to estimate the size of the jumps for a sequence of splines with $\overline{\Delta}_\nu\to0$.

THEOREM 6.45

Let Δ_ν be a sequence of partitions of $[a,b]$ such that $\overline{\Delta}_\nu\downarrow0$, $\Delta_0=\{a,b\}$. For each ν let $\mathcal{S}_\nu\in\mathcal{P}\mathcal{P}_m(\Delta_\nu)\cap C^l[a,b]$ be a linear space of splines. Given

$f\in B[a,b]$, let $s_\nu\in\mathbb{S}_\nu$ be such that $\|f-s_\nu\|\leqslant 2\varepsilon_\nu$. Then for $j=l+1,\ldots,m-1$,

$$J(D^j s_\nu)\leqslant C_1\sum_{r=1}^{\nu}\frac{(\varepsilon_r+\varepsilon_{r-1})}{\underline{\Delta}_r^j}, \tag{6.87}$$

where C_1 depends only on m.

Proof. We proceed by induction. For $\nu=0$ the result is trivial since $\Delta_0=[a,b]$, and neither s_0 nor its derivatives have any jumps. Now suppose the result has been established for $\nu-1$. We then obtain (6.87) for $j=m-1$ and all ν if we can establish

$$J(D^{m-1}s_\nu)\leqslant C_2\underline{\Delta}_\nu^{1-m}(\varepsilon_\nu+\varepsilon_{\nu-1})+J(D^{m-1}s_{\nu-1}). \tag{6.88}$$

To prove this, let x_i^ν be one of the knots of Δ_ν. Let j be such that $x_j^{\nu-1}\leqslant x_i^\nu<x_{j+1}^{\nu-1}$. There are three cases.

CASE 1. $x_{j+1}^{\nu-1}-x_i^\nu\geqslant\underline{\Delta}_\nu/4$ and $x_i^\nu-x_j^{\nu-1}\geqslant\underline{\Delta}_\nu/4$. Let $I=[x_i^\nu-\underline{\Delta}_\nu/4,x_i^\nu)$ and $J=[x_i^\nu,x_i^\nu+\underline{\Delta}_\nu/4)$. Then $g=s_\nu-s_{\nu-1}$ is a polynomial on both I and J. Moreover,

$$[D^{m-1}s_\nu]_i=[D^{m-1}g]_i=D^{m-1}g(x_i^\nu+)-D^{m-1}g(x_i^\nu-)$$

since $s_{\nu-1}$ has no knot at x_i^ν. (Here we have written $[\;]_i$ for the jump at x_i^ν). By the Markov inequality (cf. Theorem 3.3),

$$\|D^{m-1}g\|_{L_\infty[I]}\leqslant C_3\left(\frac{\underline{\Delta}_\nu}{4}\right)^{1-m}\|g\|_{L_\infty[I]},$$

and a similar estimate holds for the interval J. Since

$$\|g\|\leqslant\|s_\nu-f\|+\|s_{\nu-1}-f\|\leqslant\varepsilon_\nu+\varepsilon_{\nu-1},$$

(6.88) follows in this case.

CASE 2. $x_i^\nu-x_j^{\nu-1}\leqslant\underline{\Delta}_\nu/4$. In this case $x_{j+1}^{\nu-1}-x_i^\nu\geqslant\underline{\Delta}_\nu/4$ since $\underline{\Delta}_{\nu-1}\geqslant\underline{\Delta}_\nu$. Now let $I=[x_j^{\nu-1}-\underline{\Delta}_\nu/4,x_j^\nu)$ and $J=[x_i^\nu,x_i^\nu+\underline{\Delta}_\nu/4)$. Again, g is a polynomial on each of these intervals. Now we estimate

$$[D^{m-1}s_\nu]_i=D^{m-1}g(x_i^\nu+)-D^{m-1}g(x_i^\nu-)+[D^{m-1}s_{\nu-1}]_i.$$

If $x_j^{\nu-1}=x_i^\nu$, we can estimate the values of $D^{m-1}g$ just as in Case 1, and (6.88) follows. If $x_j^{\nu-1}<x_i^\nu$, then using the fact that $D^{m-1}g$ is constant on

$[x_j^{\nu-1}, x_i^\nu]$, we note that

$$[D^{m-1}s_\nu]_i = D^{m-1}g(x_i^\nu +) - D^{m-1}g(x_j^{\nu-1} -) + [D^{m-1}s_{\nu-1}]_j.$$

Using Markov's inequality, we again obtain (6.88).

CASE 3. $x_{j+1}^{\nu-1} - x_i^\nu \leqslant \underline{\Delta}_\nu / 4$. This case is the mirror image of Case 2.

To prove (6.88) for general j, suppose it is established now for $m-1, \ldots, j+1$. Again, it suffices to show that

$$J(D^j s_\nu) \leqslant C \underline{\Delta}_\nu^{-j}(\varepsilon_\nu + \varepsilon_{\nu-1}) + J(D^j s_{\nu-1}). \tag{6.89}$$

The proof of this breaks into the same three cases as above. Case 1 is virtually identical except that the Markov inequality is used to estimate $D^j g$. Suppose now that we are in Case 2. If $x_j^{\nu-1} = x_i^\nu$, then the analysis is as in Case 1. Suppose now that $x_j^{\nu-1} < x_i^\nu$. Then we write

$$\begin{aligned}[D^j s_\nu]_i &= D^j g(x_i^\nu +) - D^j g(x_j^{\nu-1} -) + [D^j s_{\nu-1}]_j \\ &\quad + D^j g(x_j^{\nu-1} +) - D^j g(x_i^\nu -).\end{aligned}$$

We estimate the value of $D^j g$ at $x_i^\nu +$ and $x_j^{\nu-1} -$ using the Markov inequality on the intervals I and J as before. It remains to deal with the last two terms. By Taylor's expansion, we have

$$D^j g(x_i^\nu -) - D^j g(x_j^{\nu-1} +) = \sum_{q=1}^{m-j-1} \frac{D^{j+q} g(x_j^{\nu-1} +)}{q!} (x_i^\nu - x_j^{\nu-1})^q.$$

But

$$\begin{aligned}|D^{j+q} g(x_j^{\nu-1} +)| &\leqslant |D^{j+q} g(x_j^{\nu-1} -)| + [D^{j+q} s_{\nu-1}]_j \\ &\leqslant C \underline{\Delta}_\nu^{-j+q}(\varepsilon_\nu + \varepsilon_{\nu-1}) + J(D^{j+q} s_{\nu-1})\end{aligned}$$

(using the Markov inequality). Since $x_i^\nu - x_j^{\nu-1} \leqslant \underline{\Delta}_\nu$, we obtain

$$|D^j g(x_i^\nu -) - D^j g(x_j^{\nu-1} +)| \leqslant C \sum_{r=1}^{\nu} \frac{(\varepsilon_r + \varepsilon_{r-1})}{\underline{\Delta}_r^j},$$

where we have used (6.87) for $\nu - 1$ and for $m-1, \ldots, j+1$. Combining these results yields (6.89), and the proof is complete. ∎

We can now combine Theorems 6.44 and 6.45 to give an inverse theorem for rather general sequences of partitions.

THEOREM 6.46

Let Δ_ν be a sequence of partitions of $[a,b]$ with $\underline{\Delta}_\nu \downarrow 0$ and $\Delta_0=\{a,b\}$. For each ν let $\mathbb{S}_\nu$ be a linear space of splines contained in $\mathcal{P}\mathcal{P}_m(\Delta_\nu)\cap C^l[a,b]$. Given $f\in B[a,b]$, let $\varepsilon_\nu = d(f,\mathbb{S}_\nu)$. Then

$$\omega_m(f;\underline{\Delta}_\nu) \leqslant C_1 \underline{\Delta}_\nu^{l+1} \sum_{r=1}^{\nu} \frac{\varepsilon_r+\varepsilon_{r-1}}{\underline{\Delta}_r^{l+1}}, \tag{6.90}$$

where C_1 is a constant depending only on m.

Proof. Let $s_\nu \in \mathbb{S}_\nu$ be such that $\|f-s_\nu\| \leqslant 2\varepsilon_\nu$. Then

$$\omega_m(f,\underline{\Delta}_\nu) \leqslant \omega_m(s_\nu,\underline{\Delta}_\nu)+\omega_m(f-s_\nu,\underline{\Delta}_\nu).$$

Clearly $\omega_m(f-s_\nu,\underline{\Delta}_\nu)\leqslant 2^{m+1}\varepsilon_\nu$. On the other hand, by Theorems 6.44 and 6.45,

$$\omega_m(s_\nu,\underline{\Delta}_\nu) \leqslant C_2 \sum_{j=l+1}^{m-1} \underline{\Delta}_\nu^j \sum_{r=1}^{\nu} \frac{(\varepsilon_r+\varepsilon_{r-1})}{\underline{\Delta}_r^j}.$$

Now for all $r=1,\ldots,\nu$, $\underline{\Delta}_\nu/\underline{\Delta}_r \leqslant 1$, hence the sum over j can be estimated by the term with $j=l+1$. ■

We now characterize some classical smooth spaces in terms of approximation by splines on a sequence of partitions not necessarily satisfying the mixing condition.

THEOREM 6.47

Let Δ_ν be a sequence of σ-quasi-uniform partitions with $\Delta_0=\{a,b\}$. Suppose Δ_ν satisfies the steadiness condition

$$\alpha\bar{\Delta}_\nu \leqslant \underline{\Delta}_{\nu-1} \leqslant \bar{\Delta}_{\nu-1} \leqslant \beta\bar{\Delta}_\nu \qquad \text{with } 1<\alpha<\infty \text{ and } 1<\beta<\infty. \tag{6.91}$$

Let $0\leqslant l\leqslant m-2$, and let $\mathbb{S}_\nu$ be linear spaces of splines in $\mathcal{P}\mathcal{P}_m(\Delta_\nu)\cap C^l[a,b]$ but not in C^{l+1}. Finally, suppose $f\in B[a,b]$ is such that

$$d(f,\mathbb{S}_\nu)=\mathcal{O}\left(\bar{\Delta}_\nu^{k+\alpha}\right). \tag{6.92}$$

Then

$$f\in \mathrm{Lip}^{k,\alpha}[a,b] \qquad \text{if } 0\leqslant k\leqslant l \text{ and } 0<\alpha<1; \tag{6.93}$$

$$f\in \mathfrak{Z}^{k-1}[a,b] \qquad \text{if } 1\leqslant k\leqslant l \text{ and } \alpha=0; \tag{6.94}$$

$$f\in \mathfrak{Z}^{l}[a,b] \qquad \text{if } k+\alpha>l+1. \tag{6.95}$$

When $k+\alpha=l+1$,

$$\omega_m(f;t)=\mathcal{O}\left[t^{l+1}|\log(t)|\right]. \tag{6.96}$$

Proof. By (6.92) we have $\varepsilon_r=\mathcal{O}(\overline{\Delta}_r^{k+\alpha})$. By the steadiness assumption, $\overline{\Delta}_{r-1}\leqslant\beta\overline{\Delta}_r$, and so $\varepsilon_{r-1}=\mathcal{O}(\overline{\Delta}_r^{k+\alpha})$ also. Theorem 6.46 then implies

$$\omega_m(f;\underline{\Delta}_\nu)\leqslant C_2\underline{\Delta}_\nu^{l+1}\sum_{r=1}^{\nu}\frac{\overline{\Delta}_r^{k+\alpha}}{\underline{\Delta}_r^{l+1}}. \tag{6.97}$$

Now for $k+\alpha<l+1$, we have

$$\omega_m(f;\underline{\Delta}_\nu)\leqslant C_2\underline{\Delta}_\nu^{k+\alpha}\sum_{r=1}^{\nu}\left(\frac{\overline{\Delta}_r}{\underline{\Delta}_r}\right)^{k+\alpha}\left(\frac{\underline{\Delta}_\nu}{\underline{\Delta}_r}\right)^{l+1-k-\alpha}.$$

But by (6.91), $\underline{\Delta}_\nu/\underline{\Delta}_r\leqslant(1/\alpha)^{\nu-r}$. Then, since the geometric series $\sum_{j=0}^{\infty}(1/\alpha)^{(l+1-k-\alpha)j}$ converges to a finite number, we conclude that $\omega_m(f;\underline{\Delta}_\nu)=\mathcal{O}(\underline{\Delta}_\nu^{k+\alpha})$. By the same kind of argument used in the proof of Theorem 6.8, we can convert this to $\omega_m(f;t)=\mathcal{O}(t^{k+\alpha})$, and (6.93) and (6.94) follow.

If $k+\alpha>l+1$, then

$$\omega_m(f;\underline{\Delta}_\nu)\leqslant C_2\underline{\Delta}_\nu^{l+1}\sum_{r=1}^{\nu}\left(\frac{\overline{\Delta}_r}{\underline{\Delta}_r}\right)^{l+1}\overline{\Delta}_r^{k+\alpha-l-1}$$

$$\leqslant C_3\sigma^{l+1}\underline{\Delta}_\nu^{l+1}\sum_{r=1}^{\nu}\left(\frac{1}{\alpha}\right)^{r(k+\alpha-l-1)}$$

$$\leqslant C_4\underline{\Delta}_\nu^{l+1}.$$

In this case, we obtain $\omega_m(f;t)=\mathcal{O}(t^{l+1})$, and (6.95) is proved.

Finally, if $k+\alpha=l+1$, then by the same argument, we obtain

$$\omega_m(f;\underline{\Delta}_\nu)\leqslant C_2\sigma^{l+1}\underline{\Delta}_\nu^{l+1}\left(\sum_{r=1}^{\nu}1\right).$$

Now by (6.91), $\underline{\Delta}_\nu\leqslant(b-a)/\alpha^\nu$, hence $\nu\log(\alpha)\leqslant\log((b-a)/\underline{\Delta}_\nu)$. Combining this with the above, we have $\omega_m(f;\underline{\Delta}_\nu)\leqslant C_4\underline{\Delta}_\nu^{l+1}|\log(\underline{\Delta}_\nu)|$. This converts directly to (6.96). ■

The inverse assertions in Theorem 6.47 are the correct companions for the direct theorems proved earlier as long as $k+\alpha<l+1$. For these cases,

the direct and inverse results combine to give characterizations of $\text{Lip}^{k,\alpha}[a,b]$ and $\mathfrak{Z}^{k-1}[a,b]$ in terms of approximation by linear spaces of splines in $C^l[a,b]$ defined on sequences of partitions not necessarily satisfying the mixing condition. For $k+\alpha \geqslant l+1$, however, the inverse results do not match the direct theorems. This is the sacrifice we must make for dropping the mixing condition.

The hypothesis (6.91) on Δ_ν is satisfied for nested uniform partitions with $\alpha=\beta=2$. Given a sequence of σ-quasi-uniform partitions that go steadily to zero, we can always find a subsequence satisfying (6.91).

It can be shown by example that the assertion (6.96) of Theorem 6.47 cannot be improved; that is, there exist functions that can be approximated to order $\overline{\Delta}_\nu^{l+1}$ for which $\omega_2(f;t)=\mathcal{O}[t^{l+1}|\log(t)|]$. The assertion (6.95) is also sharp. Indeed, if (6.95) could be strengthened to $\omega_m(f;t)=\mathcal{O}(t^{l+1+\alpha})$ with $\alpha>0$, then for a nested sequence of knots it would follow that

$$\mathcal{S}_\nu \subseteq \text{Lip}^{l+1,\alpha}[a,b] \subseteq C^{l+1}[a,b],$$

which contradicts the definition of l.

Without the assumption of mixing, it is not possible to establish a saturation result for arbitrary functions. But we can show that for sufficiently smooth functions, the maximal order of approximation is $\overline{\Delta}_\nu^m$ unless the function is a polynomial of order m.

THEOREM 6.48

Let Δ_ν be a sequence of partitions with $\overline{\Delta}_\nu \downarrow 0$. Suppose $f \in C^m[a,b]$ is a function such that

$$d[f, \mathcal{P}\mathcal{P}_m(\Delta_\nu)]_\infty \leqslant C_1 \overline{\Delta}_\nu^m \psi(\overline{\Delta}_\nu), \tag{6.98}$$

where ψ is a monotone-increasing function with $\psi(t)\to 0$ as $t \to 0$. Then $f \in \mathcal{P}_m$.

Proof. Let s_ν be such that $\|f-s_\nu\| \leqslant 2\varepsilon_\nu := 2d[f, \mathcal{P}\mathcal{P}_m(\Delta_\nu)]$. Now, $\omega_m(f;\underline{\Delta}_\nu) \leqslant \omega_m(f-s_\nu;\underline{\Delta}_\nu)+\omega_m(s_\nu;\underline{\Delta}_\nu)$, while $\omega_m(f-s_\nu;\underline{\Delta}_\nu) \leqslant 2^{m+1}\varepsilon_\nu$. By Theorem 6.44,

$$\omega_m(s_\nu;\underline{\Delta}_\nu) \leqslant C_2 \sum_{j=0}^{m-1} \underline{\Delta}_\nu^j J(D^j s_\nu).$$

We must estimate $J(D^j s_\nu)$ for $j=0,1,\ldots,m-1$. For each $1 \leqslant i \leqslant k_\nu$, let $[s_\nu]_i^j = \text{jump}[D^j s_\nu]_{x_i^\nu}$. Fix i, and let $\varepsilon = \theta \underline{\Delta}_\nu$ with $\theta < 1$. By Whitney's Theorem 3.18 there exists $g_i \in \mathcal{P}_m$ such that

$$\|f-g_i\|_{L_\infty[I_i]} \leqslant C_3 \varepsilon^m \|D^m f\|_{L_\infty[I_i]},$$

where $I_i=(x_i^\nu-\varepsilon, x_i^\nu+\varepsilon)$. The function $s_\nu - g_i$ is a polynomial on each of the intervals $I_i^- = (x_i^\nu - \varepsilon, x_i^\nu)$ and $I_i^+ = (x_i^\nu, x_i^\nu+\varepsilon)$. Using the Markov inequality, we obtain

$$\varepsilon^j[s_\nu]_i^j = \varepsilon^j[(s_\nu - g_i)]_i^j \leqslant C_4(\|s_\nu - g_i\|_{L_\infty[I_i^-]} + \|s_\nu - g_i\|_{L_\infty[I_i^+]}).$$

But $\|s_\nu - g_i\| \leqslant \|s_\nu - f\| + \|f-g_i\|$ and $\|s_\nu - f\| \leqslant 2\varepsilon_\nu$. Combining these facts, we obtain

$$\omega_m(s_\nu; \underline{\Delta}_\nu) \leqslant C_5 \sum_{j=0}^{m-1} \theta^{-j}\big(\varepsilon_\nu + \varepsilon^m \|D^m f\|_{L_\infty[a,b]}\big).$$

Taking $\theta = [\psi(\overline{\Delta}_\nu)]^{1/m}$, this implies

$$\omega_m(f; \underline{\Delta}_\nu) \leqslant C_6 \sum_{j=0}^{m-1} \big[\psi(\overline{\Delta}_\nu)\big]^{\frac{m-j}{m}} = o\big(\overline{\Delta}_\nu^m\big).$$

Since $\overline{\Delta}_\nu \downarrow 0$ with $\nu \to \infty$, it follows that $f \in \mathcal{P}_m$. ■

§ 6.9. INVERSE THEORY FOR $1 \leqslant p < \infty$

In this section we carry out the same program as in Section 6.8, but now we use the p-norm instead of the ∞-norm. The basic ideas are the same, but the details are a bit more complicated. We begin with an important estimate for the smoothness of a piecewise polynomial in terms of the jumps in the various derivatives at the knots.

THEOREM 6.49

Let Δ be any partition of the interval $[a,b]$, and let $0 < \varepsilon < \underline{\Delta}$. Then for any spline $s \in \mathcal{P}\mathcal{P}_m(\Delta)$,

$$\omega_m(s;\varepsilon)_p \leqslant C_1 \varepsilon^{1/p} \sum_{j=0}^{m-1} \varepsilon^j J_p(D^j s), \tag{6.99}$$

where

$$J_p(D^j s) = \left(\sum_{i=1}^{k} |\text{jump}[D^j s]_{x_i}|^p \right)^{1/p}. \tag{6.100}$$

Proof. We first show that for any $0 \leqslant j \leqslant m-1$,

$$\omega_{m-j}(D^j s;\varepsilon)_p \leqslant C_2\left[\varepsilon^{1/p} J_p(D^j s) + \omega_{m-j-1}(D^{j+1}s;\varepsilon)_p\right]. \tag{6.101}$$

Fix j. Let $g(x) = \sum_{i=1}^{k}[s]_i^j (x-x_i)_+^0$, where we write $[s]_i^j = \text{jump}[D^j s]_{x_i}$. Then $[g]_i = [s]_i^j$, $i=1,2,\ldots,k$. Moreover,

$$\omega_{m-j}(D^j s;\varepsilon)_p \leqslant \omega_{m-j}(g;\varepsilon)_p + \omega_{m-j}(D^j s - g;\varepsilon)_p.$$

For the first term we note that

$$\omega_{m-j}(g;\varepsilon) \leqslant 2^{m-j-1}\omega_1(g;\varepsilon).$$

Since $D(D^j s - g) = D^{j+1}s$, almost everywhere, for the second term we have

$$\omega_{m-j}(D^j s - g;\varepsilon)_p \leqslant \omega_{m-j-1}(D^{j+1}s;\varepsilon)_p.$$

Now as x runs over the interval $[a, b-h]$ (with $h<\varepsilon$), the function $|g(x+h)-g(x)|$ is a piecewise constant. It takes on the value $|[g]_i| = |[s]_i^j|$ on the interval $I_i = (x_i - h, x_i)$, $i=1,2,\ldots,k$, and is otherwise zero. Thus

$$\omega_1(g;\varepsilon)_p = \sup_{h \leqslant \varepsilon}\left(\int_a^{b-h} |g(x+h)-g(x)|^p\,dx\right)^{1/p} \leqslant (2\varepsilon)^{1/p} J_p(D^j s).$$

Combining these estimates, we obtain (6.101). Now (6.99) follows by stringing the inequalities (6.101) together. ■

In order to apply Theorem 6.49, we need estimates on the size of the jumps in the various derivatives of a sequence of splines s_ν. As in the uniform norm case, such estimates cannot be obtained without assuming *either* that the splines have some kind of global smoothness, *or* that the partitions satisfy some sort of mixing condition. We consider the case of mixed partitions first.

DEFINITION 6.50. *p*-Mixing Condition

Let $\Delta_\nu = \{x_i\}_{i=0}^{k_\nu+1}$ be a sequence of partitions of $[a,b]$, and let $1 \leqslant p < \infty$. We say that Δ_ν satisfies the *p-mixing condition* provided there exists $C_1 > 0$ such that for every ν there is a sequence $\alpha^\nu = (\alpha_{\nu+1}^\nu, \alpha_{\nu+2}^\nu, \ldots)$ with $\sum_{n=\nu+1}^{\infty} \alpha_n^\nu \leqslant 1$ so that for all $1 \leqslant i \leqslant k_\nu$,

$$\sum_{n=\nu+1}^{\infty} \alpha_n^\nu d(x_i^\nu, \Delta_n)^{mp+1} \geqslant C_1 \overline{\Delta}_\nu^{mp+1}. \tag{6.102}$$

There are many examples of sequences satisfying the p-mixing condition.

The following theorem shows that the sequence of uniform partitions is one such example:

THEOREM 6.51

The sequence of uniform partitions $\Delta_\nu = \{a+ih\}_{i=0}^{\nu}$ with $h=(b-a)/\nu$ satisfies the p-mixing condition for any $1 \leqslant p < \infty$.

Proof. We choose

$$\alpha_n^\nu = \begin{cases} 1/2, & n=\nu+1 \\ 1/2\nu, & n=\nu+2,\dots,2\nu \\ 0, & n=2\nu+1,\dots \end{cases}.$$

Fix i, and let $d_{i,n} = d(x_i^\nu, \Delta_n)$, $n=\nu+1,\dots$. We note that $\sum_{n=\nu+1}^{\infty}\alpha_n^\nu \leqslant 1$. Now the proof of (6.102) divides into three cases. Assume $[a,b]=[0,1]$.

CASE 1. $1/3 \leqslant x_i^\nu = i/\nu \leqslant 2/3$. Then x_i^ν is in the interval $I_{1,0}$ defined in the proof of Theorem 6.38, and by the estimates there, $d_{i,\nu+1} \geqslant \overline{\Delta}_\nu/6$. Thus

$$\sum_{n=\nu+1}^{\infty} \alpha_n^\nu (d_{i,n})^{mp+1} \geqslant \frac{1}{2}\left(\frac{1}{6}\right)^{mp+1} \overline{\Delta}_\nu^{mp+1}.$$

CASE 2. $1/\nu \leqslant x_i^\nu \leqslant 1/3$. First we compute $d_{i,\nu+\alpha}$ for each $1 \leqslant \alpha \leqslant \nu$. Given such an α, there must exist an integer $0 \leqslant \beta \leqslant i$ such that α is in one of the intervals

$$J_\beta^+ = \left[\frac{\beta\nu}{i}, \frac{(\beta+1/2)\nu}{i}\right) \quad \text{or} \quad J_\beta^- = \left[\frac{(\beta-1/2)\nu}{i}, \frac{\beta\nu}{i}\right).$$

But then

$$\frac{i+\beta}{\nu+\alpha} \leqslant \frac{i}{\nu} < \frac{i+\beta+1}{\nu+\alpha} \quad \text{or} \quad \frac{i+\beta-1}{\nu+\alpha} < \frac{i}{\nu} < \frac{i+\beta}{\nu+\alpha},$$

and thus

$$d_{i,\nu+\alpha} \geqslant \begin{cases} \dfrac{i(\alpha-\beta\nu/i)}{2\nu^2}, & \alpha \in J_\beta^+ \\[2ex] \dfrac{i(\beta\nu/i-\alpha)}{2\nu^2}, & \alpha \in J_\beta^-. \end{cases}$$

This implies

$$\sum_{n=\nu+1}^{2\nu} (d_{i,n})^{mp+1} = \sum_{\beta=0}^{i} \left[\sum_{\alpha \in J_\beta^-} (d_{i,n})^{mp+1} + \sum_{\alpha \in J_\beta^+} (d_{i,n})^{mp+1} \right]$$

$$\geqslant \sum_{\beta=0}^{i} \sum_{s=1}^{\lfloor \nu/2i \rfloor} \left(\frac{is}{2\nu^2} \right)^{mp+1} = \left(\frac{i}{2\nu^2} \right)^{mp+1} (i+1) \sum_{s=1}^{\lfloor \nu/2i \rfloor} s^{mp+1}.$$

Now

$$\sum_{s=1}^{\lfloor \nu/2i \rfloor} s^{mp+1} \geqslant \int_0^{\lfloor \nu/2i \rfloor} s^{mp+1} ds \geqslant \frac{\left[(\nu/2i) - 1 \right]^{mp+2}}{mp+2}.$$

Since $i/\nu \leqslant 1/3$, $(\nu/2i - 1) \geqslant \nu/6i$. Combining these facts, we obtain

$$\sum_{n=\nu+1}^{2\nu} \alpha_n^\nu (d_{i,n})^{mp+1} \geqslant \frac{1}{2\nu} \sum_{n=\nu+1}^{2\nu} (d_{i,n})^{mp+1} \geqslant \frac{1}{\nu} C_2 \left(\frac{i}{2\nu^2} \right)^{mp+1} \left(\frac{\nu}{6i} \right)^{mp+2}$$

$$\geqslant C_1 \left(\frac{1}{\nu} \right)^{mp+1},$$

which is (6.102).

CASE 3. $2/3 \leqslant i/\nu \leqslant (\nu-1)/\nu$. This case is almost identical to Case 2. ■

With the p-mixing condition, we can now establish the analog of Theorem 6.39.

THEOREM 6.52

Let Δ_ν be a sequence of quasi-uniform partitions of $[a,b]$ satisfying the p-mixing condition. Then for every $f \in L_p[a,b]$,

$$\omega_m\left(f; \overline{\Delta}_\nu\right)_p \leqslant C_1 \sup_{n \geqslant \nu} d\left[f, \mathcal{P}\mathcal{P}_m(\Delta_n) \right]_p. \tag{6.103}$$

Proof. Let $s_\nu \in \mathcal{P}\mathcal{P}_m(\Delta_\nu)$ be such that $\|f - s_\nu\|_p \leqslant 2\varepsilon_\nu$, where $\varepsilon_\nu = d[f, \mathcal{P}\mathcal{P}_m(\Delta_\nu)]_p$. Then

$$\omega_m\left(f; \overline{\Delta}_\nu\right)_p \leqslant \omega_m\left(f - s_\nu; \overline{\Delta}_\nu\right)_p + \omega_m\left(s_\nu; \overline{\Delta}_\nu\right)_p.$$

For the first term we have

$$\omega_m\left(f - s_\nu; \overline{\Delta}_\nu\right)_p \leqslant 2^m \|f - s_\nu\|_p \leqslant 2^{m+1} \varepsilon_\nu.$$

To estimate the second term, we use Theorem 6.49 to get

$$\omega_m(f;\bar{\Delta}_\nu)_p \leqslant \lceil \sigma \rceil^m \omega_m(f;\underline{\Delta}_\nu)_p \leqslant C_2 \bar{\Delta}_\nu^{1/p} \sum_{j=0}^{m-1} \bar{\Delta}_\nu^j J_p(D^j s).$$

We now estimate $J_p(D^j s_\nu)$. Fix i, and suppose $n \geqslant \nu$ is such that $d_{i,n} = d(x_i^\nu, \Delta_n) > 0$. Then

$$[s_\nu]_i^j := \text{jump}[D^j s_\nu]_{x_i} = [s_\nu - s_n]_i^j,$$

and $s_\nu - s_n$ is a polynomial of order m on each of the intervals $I_{i,n}^- = (x_i^\nu - d_{i,n}, x_i^\nu)$ and $I_{i,n}^+ = (x_i^\nu, x_i^\nu + d_{i,n})$. Applying the Markov inequality (see Theorem 3.3) to each of these intervals, we obtain

$$|[s_\nu]_i^j| \leqslant C_4 \frac{1}{(d_{i,n})^{j+1/p}} \left(\|s_\nu - s_n\|_{L_p[I_{i,n}^-]} + \|s_\nu - s_n\|_{L_p[I_{i,n}^+]} \right).$$

We conclude that

$$\bar{\Delta}_\nu^{j+1/p} J_p(D^j s_\nu) \leqslant C_3 \left[\sum_{i=1}^{k_\nu} \frac{\bar{\Delta}_\nu^{jp+1} \|s_\nu - s_n\|_{L_p[I_{i,n}]}^p}{(d_{i,n})^{jp+1}} \right]^{1/p}.$$

Since $\bar{\Delta}_\nu / d_{i,n} \geqslant 1$, the maximum of this expression for $1 \leqslant j \leqslant m$ occurs for $j = m$. Thus using (6.102) to estimate $\bar{\Delta}_\nu^{mp+1}$, we obtain

$$\bar{\Delta}_\nu^{j+1/p} J_p(D^j s_\nu) \leqslant C_2 \left\{ \sum_{i=1}^{k_\nu} \frac{\left[\sum_{n=\nu+1}^{\infty} \alpha_n^\nu (d_{i,n})^{mp+1} \right]}{(d_{i,n})^{mp+1}} \|s_\nu - s_n\|_{L_p[I_{i,n}]}^p \right\}^{1/p}$$

$$\leqslant C_2 \sum_{n=\nu+1}^{\infty} \alpha_n^\nu \|s_\nu - s_n\|_{L_p[a,b]}^p \leqslant C_1 \sup_{n > \nu} \varepsilon_n. \quad \blacksquare$$

We can now prove the following analog of Theorem 6.41:

THEOREM 6.53

Let Δ_ν be a sequence of quasi-uniform partitions going steadily to zero and satisfying the p-mixing condition. Suppose $f \in L_p[a,b]$ is such that

$$d[f, \mathcal{P}\mathcal{P}_m(\Delta_\nu)]_p \leqslant \phi(\bar{\Delta}_\nu), \tag{6.104}$$

where ϕ is a monotone-increasing function with $\phi(t) \to 0$ as $t \to 0$. Then

$$\omega_m(f;t)_p \leqslant C_1\phi(t). \tag{6.105}$$

Proof. The estimate (6.105) follows directly from (6.104) and Theorem 6.52, exactly as in the proof of Theorem 6.41. ∎

The inverse Theorem 6.53 can now be used to establish a saturation result and to characterize various Lipschitz and Zygmund classes.

THEOREM 6.54

Let Δ_ν be a sequence of partitions as in Theorem 6.53. Then $f \in L_p[a,b]$ satisfies

$$d[f, \mathcal{P}\mathcal{P}_m(\Delta_\nu)]_p = \mathcal{O}\left(\bar{\Delta}_\nu^{k+\alpha}\right) \tag{6.106}$$

if and only if

$$f \in \mathrm{Lip}_p^{k,\alpha}[a,b] \qquad \text{when } 0 \leqslant k \leqslant m-1 \text{ and } 0 < \alpha < 1; \tag{6.107}$$

$$f \in \mathfrak{Z}_p^{k-1}[a,b] \qquad \text{when } 1 \leqslant k \leqslant m-1 \text{ and } \alpha = 0; \tag{6.108}$$

$$f \in \mathrm{Lip}_p^{m-1,1}[a,b] \qquad \text{when } k = m, \alpha = 0; \tag{6.109}$$

$$f \in \mathcal{P}_m \qquad \text{when } k + \alpha > m. \tag{6.110}$$

Proof. The direct assertions follow from theorems proved in Section 6.4. Conversely, if (6.106) holds, then by Theorem 6.53, $\omega_m(f;t)_p = \mathcal{O}(t^{k+\alpha})$. This implies that f is in the various spaces listed in (6.107) to (6.109), depending on the value of k and α. ∎

We note that the sequence of uniform partitions satisfies all of the hypotheses of Theorem 6.54. As in the case of $p = \infty$, certain functions can be approximated to higher order if the sequence of partitions does not satisfy these hypotheses (cf. Chapter 7).

In the remainder of this section we examine the extent to which an inverse theory can be developed without assuming that the sequence of partitions satisfies the p-mixing condition. We have the following inverse theorem (cf. Theorem 6.46 for the case $p = \infty$):

THEOREM 6.55

Let Δ_ν be a sequence of partitions of $[a,b]$ with $\underline{\Delta}_\nu \downarrow 0$ and $\Delta_0 = \{a,b\}$. For each ν let $\mathcal{S}_\nu$ be a linear space of spline functions contained in $\mathcal{P}\mathcal{P}_m(\Delta_\nu) \cap$

$C^l[a,b], l \geqslant 0$. Given $f \in L_p[a,b]$, let $\varepsilon_\nu = d(f, \mathbb{S}_\nu)_p$. Then

$$\omega_m(f; \underline{\Delta}_\nu)_p \leqslant C_1 \underline{\Delta}_\nu^{l+1+1/p} \sum_{r=1}^{\nu} \frac{\varepsilon_r + \varepsilon_{r-1}}{\underline{\Delta}_r^{l+1+1/p}}, \tag{6.111}$$

where C_1 is a constant that depends only on m and p.

Proof. The proof follows that of Theorem 6.46. Let $s_\nu \in \mathbb{S}_\nu$ be such that $\|f - s_\nu\|_p \leqslant 2\varepsilon_\nu$. Then

$$\omega_m(f; \underline{\Delta}_\nu)_p \leqslant \omega_m(s_\nu; \underline{\Delta}_\nu)_p + \omega_m(f - s_\nu; \underline{\Delta}_\nu)_p.$$

Now, $\omega_m(f - s_\nu; \underline{\Delta}_\nu) \leqslant 2^{m+1}\varepsilon_\nu$. By Theorem 6.49 and the fact that the derivatives of s_ν up to order l are all continuous, we have

$$\omega_m(f; \underline{\Delta}_\nu)_p \leqslant C \underline{\Delta}_\nu^{1/p} \sum_{j=l+1}^{m-1} \underline{\Delta}_\nu^j J_p(D^j s_\nu).$$

It remains to estimate $J_p(D^j s_\nu)$ for $j = l+1, \ldots, m-1$. Here, the proof of Theorem 6.45 can be carried over with only minor changes. First consider $j = m-1$. Then by induction it suffices to show that

$$J_p(D^{m-1} s_\nu) \leqslant C \frac{\varepsilon_\nu + \varepsilon_{\nu-1}}{\underline{\Delta}_\nu^{m-1+1/p}} + J_p(D^{m-1} s_{\nu-1}). \tag{6.112}$$

To show this, let $1 \leqslant i \leqslant k_\nu$. We must examine three cases, as in the proof of Theorem 6.45. In all cases the Markov inequality (cf. Theorem 3.3) gives

$$|[D^{m-1} s_\nu]_i| \leqslant |[D^{m-1} s_{\nu-1}]_i| + C \underline{\Delta}_\nu^{m-1} \|g\|_{L_p[I_i]},$$

where $g = s_\nu - s_{\nu-1}$ and I_i is a subinterval of $[x_i^\nu - \underline{\Delta}_\nu/2, x_i^\nu + \underline{\Delta}_\nu/2]$. Now, using Hölder's inequality, we have

$$\left(\sum_{i=1}^{k_\nu} |[D^{m-1} s_{\nu-1}]_i|^p \right)^{1/p} \leqslant \left(\sum_{i=1}^{k_\nu} |[D^{m-1} s_{\nu-1}]|_i^p \right) + C \underline{\Delta}_\nu^{m-1} \left(\sum_{i=1}^{k_\nu} \|g\|_{L_p[I_i]}^p \right)^{1/p}.$$

Since the I_i are disjoint,

$$\begin{aligned} \left(\sum_{i=1}^{k_\nu} \|g\|_{L_p[I_i]}^p \right)^{1/p} &\leqslant \|g\|_{L_p[a,b]} \leqslant \|s_\nu - f\|_{L_p[a,b]} + \|s_{\nu-1} - f\|_{L_p[a,b]} \\ &\leqslant \varepsilon_\nu + \varepsilon_{\nu-1}, \end{aligned}$$

and by induction, we obtain

$$J_p(D^j s_\nu) \leq C \underline{\Delta}_\nu^{1/p} \sum_{r=l+1}^{\nu} \frac{\varepsilon_r + \varepsilon_{r-1}}{\underline{\Delta}_r^{j+1/p}} \tag{6.113}$$

for $j = m-1$. Arguments similar to those used in the proof of Theorem 6.45 can now be used to establish (6.113) for $1 \leq j \leq m-2$. Then

$$\omega_m(f; s_\nu)_p \leq C \sum_{j=l+1}^{m-1} \sum_{r=1}^{\nu} \left(\frac{\underline{\Delta}_\nu}{\underline{\Delta}_r} \right)^{j+1/p} (\varepsilon_r + \varepsilon_{r-1}).$$

Since $\underline{\Delta}_\nu / \underline{\Delta}_r \leq 1$, the largest term occurs for $j = l+1$, and (6.111) follows. ∎

We can now translate the inverse assertion of Theorem 6.55 into an inverse theorem involving classical smooth spaces.

THEOREM 6.56

Let Δ_ν be a sequence of quasi-uniform partitions with $\Delta_0 = \{a,b\}$. In addition, suppose

$$\alpha \overline{\Delta}_\nu \leq \underline{\Delta}_{\nu-1} \leq \overline{\Delta}_{\nu-1} \leq \beta \overline{\Delta}_\nu \qquad \text{with } 1 < \alpha < \infty \text{ and } 1 < \beta < \infty. \tag{6.114}$$

Let $0 \leq l \leq m-2$, and let $\mathcal{S}_\nu$ be linear spaces of splines in $\mathcal{P}\mathcal{P}_m(\Delta_\nu) \cap C^l[a,b]$, but not in $C^{l+1}[a,b]$. Finally, suppose $f \in L_p[a,b]$ is such that

$$d(f, \mathcal{S}_\nu)_p = \mathcal{O}\left(\overline{\Delta}_\nu^{k+\alpha}\right). \tag{6.115}$$

Then

$$f \in \text{Lip}_p^{k,\alpha}[a,b] \qquad \text{if } 0 \leq k \leq l+1 \text{ and } 0 < \alpha < 1; \tag{6.116}$$

$$f \in \mathfrak{Z}_p^{k-1}[a,b] \qquad \text{if } 1 \leq k \leq l+1 \text{ and } \alpha = 0; \tag{6.117}$$

$$f \in \mathfrak{Z}_p^{l}[a,b] \qquad \text{if } k+\alpha > l+1+1/p. \tag{6.118}$$

If $k+\alpha = l+1+1/p$, then

$$\omega_m(f;t)_p = \mathcal{O}\left[t^{l+1} |\log(t)| \right]. \tag{6.119}$$

Proof. By the same arguments used in the proof of Theorem 6.47, if $k+\alpha < l+1+1/p$, the assumption (6.115) together with the inverse Theorem 6.55 implies $\omega_m(f;t)_p = \mathcal{O}(t^{k+\alpha})$. The assertions (6.116) and (6.117) follow. When $k+\alpha>l+1+1/p$, we get $\omega_m(f;t)_p=\mathcal{O}(t^{l+1})$, which gives (6.118). The case $k+\alpha=l+1+1/p$ also follows, as in the proof of Theorem 6.47. ■

As in the case of $p=\infty$, treated in Section 6.8, the inverse assertions of Theorem 6.56 are the correct companions for the direct theorems proved earlier as long as $k+\alpha<l+1+1/p$. For these cases the direct and inverse theorems combine to give complete characterizations of the spaces $\text{Lip}_p^{k,\alpha}[a,b]$ and $\mathfrak{Z}_p^{k-1}[a,b]$, respectively. For $k+\alpha\geqslant l+1+1/p$, however, the inverse and direct theorems do not match. As for $p=\infty$, these inverse theorems cannot be improved.

We close this section with a saturation theorem for approximation by a sequence of piecewise polynomial spaces $\mathcal{P}\mathcal{P}_m(\Delta_\nu)$. We do not assume that the partitions Δ_ν satisfy a mixing condition. The result shows that (except for polynomials) smooth functions cannot be approximated to an order higher than $\bar{\Delta}_\nu^m$.

THEOREM 6.57

Let Δ_ν be a sequence of partitions with $\bar{\Delta}_\nu\downarrow 0$. Suppose $f\in L_p^m[a,b]$ is such that

$$d[f,\mathcal{P}\mathcal{P}_m(\Delta_\nu)]_p\leqslant C_1\bar{\Delta}_\nu\psi(\bar{\Delta}_\nu), \tag{6.120}$$

where ψ is a monotone-increasing function such that $\psi(t)\to 0$ as $t\to 0$. Then $f\in\mathcal{P}_m$.

Proof. The proof is similar to the proof of Theorem 6.48. In particular, if s_ν is such that $\|f-s_\nu\|_p\leqslant 2\varepsilon_\nu:=2d[f,\mathcal{P}\mathcal{P}_m(\Delta_\nu)]_p$, then

$$\omega_m(f;\underline{\Delta}_\nu)_p\leqslant 2^{m+1}\varepsilon_\nu+\omega_m(s_\nu;\underline{\Delta}_\nu)_p.$$

Using Theorem 6.49, we have

$$\omega_m(s_\nu;\underline{\Delta}_\nu)_p\leqslant C\sum_{j=0}^{m-1}\underline{\Delta}_\nu^j J_p(D^js_\nu).$$

To estimate $J_p(D^js_\nu)$, we note that by Whitney's Theorem 3.18 there exists a polynomial $g_i\in\mathcal{P}_m$ such that $\|f-g_i\|_{L_p[I_i]}\leqslant\varepsilon^m\|D^mf\|_{L_p[I_i]}$, $i=1,\ldots,k$, where $I_i=[x_i^\nu-\varepsilon,x_i^\nu+\varepsilon]$. Now, using Markov's inequalities (3.2) and (3.3), we have

$$\varepsilon^j[s_\nu]_i^j = \varepsilon^j[s_\nu - g_i]_i^j \leqslant C\big(\|s_\nu - g_i\|_{L_\infty[I_i^-]} + \|s_\nu - g_i\|_{L_\infty[I_i^+]}\big)$$

$$\leqslant C\varepsilon^{-1/p}\|s_\nu - g_i\|_{L_p[I_i]} \leqslant C\varepsilon^{-1/p}\big(\|f - s_\nu\|_{L_p[I_i]} + \varepsilon^m\|D^m f\|_{L_p[I_i]}\big).$$

Applying the discrete Hölder inequality (see Remark 2.1) to the pth power of this inequality and then summing over $i = 1, 2, \ldots, k_\nu$, we obtain

$$\varepsilon^j J_p(D^j s_\nu) \leqslant C\varepsilon^{-1/p}(\|f - s\|_p + \varepsilon^m\|D^m f\|_p).$$

It follows that with $\varepsilon = \theta\underline{\Delta}\nu$,

$$\omega_m(f; \underline{\Delta}_\nu)_p \leqslant C\sum_{j=0}^{m-1}\theta^{-j}(\varepsilon_\nu + \varepsilon^m\|D^m f\|_p),$$

and the rest of the proof proceeds exactly as in Theorem 6.48. ∎

§ 6.10. HISTORICAL NOTES

Section 6.2

The approximation power of piecewise constant functions was studied in the lecture notes of Kahane [1961]. His notes include direct theorems, the inverse theorem (for equally spaced partitions only), and the characterization of Lip^α (for $f \in C[a, b]$). An inverse theory for $B[a, b]$ was developed later by Nitsche [1969b] and DeVore and Richards [1973b]. The characterization of Lip^α in the case of $f \in B[a, b]$ was given in the latter paper. A different proof was given later by Shisha [1974b].

Section 6.3

The approximation power of piecewise linear polynomials was also discussed by Kahane [1961], and independently by Brudnyi and Gopengauz [1963]. Both papers include inverse results (although only for equispaced partitions) as well as direct theorems. See also Malozemov [1966, 1967].

Section 6.4

Prior to the mid-1960s there were only a few isolated papers dealing with the question of how well classes of smooth functions can be approximated by piecewise polynomials or splines. We have already mentioned the papers by Kahane [1961] and Brudnyi and Gopengauz [1963] in connection with approximation by piecewise constants and piecewise linear polynomials. Other early papers include those by Peterson [1962] and Smoluk [1964].

The intensive development of the theory of interpolating splines that began in the early 1960s (cf. the discussion in the historical notes for Section 1.4), led to a concommitant development of error bounds for them. As we are not discussing interpolation by splines here, we will not give a complete historical account of these developments. Some of the early contributors include Ahlberg and Nilson [1963], Birkhoff and deBoor [1964], Ahlberg, Nilson, and Walsh [1965 a, b], Atkinson [1968], Sharma and Meir [1966], Cheney and Schurer [1968, 1970], Birkhoff, Schultz, and Varga [1968], Swartz [1968], Hall [1968], and Varga [1969]. The refinement of these bounds to the point where they yield the direct theorems given in this section (where we have mixed norms, p-moduli of smoothness, and local results) remained a rather lengthy process involving considerable machinery. To get an idea of how this progressed, see the papers by Swartz [1970], Hedstrom and Varga [1971], Swartz and Varga [1972], Demko and Varga [1974], and Scherer [1974b].

A number of authors have constructed approximation schemes of the form (6.40). For example, Birkhoff [1976] and Birkhoff and deBoor [1968b] developed methods of this type based on local moments. deBoor [1968c] studied a broad class of such operators abstractly, and also discussed an explicit formula of this type based on point evaluations of f. Another explicit method (based on point evaluations of f and its derivatives) was introduced by deBoor and Fix [1973]. They called such methods *quasi-interpolation* methods. Still other examples of quasi-interpolation methods can be found in the article by Lyche and Schumaker [1975] along with a detailed analysis of their approximation power.

We have elected to develop direct theorems for spline approximation by using the explicit quasi-interpolation operator Q defined in Theorem 6.18. This approach permits, in my opinion, the shortest and cleanest development of the complete range of direct theorems including the mixed norms, the p-moduli of continuity, and the local results. The proofs in this section are based on the methods of Lyche and Schumaker [1975]. The idea of thinning out a partition to get a (sub-) partition that is quasi-uniform is credited to Sharma and Meir [1966].

Section 6.5

Error bounds for spline approximation in Besov spaces were obtained first by Hedstrom and Varga [1971]. Their approach was to use the interpolation method described in this section on certain spline interpolation operators. For further results in intermediate spaces, see Demko and Varga [1974] and Scherer [1974b].

Section 6.6

Lower bounds for the order of approximation of smooth functions by piecewise polynomials have been discussed in a number of papers. The basic idea is contained already in the article by Birkhoff, Schultz, and Varga [1968].

Section 6.8

Inverse, saturation, and characterization theorems for approximation by piecewise constants were obtained by Kahane [1961]. Similar results for piecewise polynomial approximation were obtained by Kahane [1961], Brudnyi and Gopengauz [1963], and Nitsche [1969a, b].

The first inverse results for splines of order m were saturation theorems obtained by Ahlberg, Nilson, and Walsh [1967b, p. 174], Golomb [1968], and Gaier [1970]. Gaier's result was established for arbitrary integrable functions—the others worked with smoother classes of functions. Nitsche [1969b] also obtained saturation results, as well as inverse theorems, but instead of working on sequences of partitions he considered the order of approximation on all possible partitions. Characterization theorems for some classical smooth spaces in terms of approximation by splines on equally spaced partitions were obtained by Scherer [1970b] and by Richards [1972]. Scherer's proofs are based on a general approximation theoretic theorem of Butzer and Scherer [1969, 1970].

The first paper to deal with nonequally spaced partitions was that of DeVore and Richards [1973a, b]. They observed that such results could not be proved without some kind of mixing condition (cf. Example 6.11) and proceeded to introduce the condition given in Definition 6.37. The fact that equally spaced partitions satisfy this mixing condition has often been asserted, but there appears to be no published proof. (The arguments used by Richards [1972] serve to show that an appropriate *subsequence* of the equally spaced partitions satisfies the mixing condition). The estimate (6.75) for $\omega_m(f, \cdot)$ can be simplified to one involving only two terms on the right if we consider a sequence of equally spaced partitions (see Gaier [1970]. In Theorem 6.43 we have given only a few of the possible characterization results. For example, Scherer [1974a] has given similar characterizations for generalized Lipschitz spaces.

The first inverse theorems for nonmixed sequences of partitions were obtained by Johnen and Scherer [1976] for nested partitions. A general treatment of nonmixed partitions, which we have followed here, was given by DeVore and Scherer [1976]. See also Scherer [1976, 1977]. The saturation result for nonmixed partitions comes from this last paper.

Section 6.9

Except for some results on piecewise constants and on piecewise linear polynomials, most of the early results for splines dealt with the uniform norm. The first results for the p-norms for splines of order m were obtained by Scherer [1970a], again using the abstract theory of Butzer and Scherer [1969, 1970]. Direct proofs of inverse, saturation, and characterization results were first given by Butler and Richards [1972]. Both of these papers dealt with sequences of equispaced partitions. Theorem 6.52 can be strengthened in the equally spaced case to an estimate involving only two terms on the right—see Scherer [1974b]. The early results for nonequally spaced partitions involved a so-called "strong-mixing condition," see DeVore and Scherer [1976], Johnen and Scherer [1976], and Scherer [1976]. The idea of replacing this complicated condition (which apparently was already too strong to encompass uniform partitions) by the p-mixing condition of Definition 6.50 is credited to Scherer [1977]. The idea for the condition (as well as for the proof that uniform partitions satisfy it) comes from some calculations performed by Butler and Richards [1972].

For nonmixed partitions, the first result is again credited to Johnen and Scherer [1976] where nested partitions were considered. Our development here follows the work of DeVore and Scherer [1976]—see also Scherer [1976, 1977]. The saturation results for nonmixed partitions follows the work of Scherer [1977].

§ 6.11. REMARKS

Remark 6.1

The following identity for multivariable polynomials was needed in the proof of Lemma 6.19.

LEMMA 6.58

Let $2 \leqslant j \leqslant m$, and suppose $y_1 < y_2 < \cdots < y_{m-1}$ and $\tau_1 < \tau_2 < \cdots < \tau_{j-1}$ are given. Let

$$G(\tau_1, \ldots, \tau_{j-1}) = \frac{(m-j)!}{(m-1)!} \sum (y_{i_1} - \tau_1) \cdots (y_{i_{j-1}} - \tau_{j-1}),$$

where the sum is taken over all choices of distinct $i_1, \ldots, i_{j-1}$ in the set

$\{1,2,\ldots,m-1\}$. Then

$$G=\sum_{\nu=1}^{j}(-1)^{j-\nu}\frac{(m-\nu)!(\nu-1)!}{(m-1)!}s_{\nu-1}(y)s_{j-\nu}(\tau),$$

where $s_{\nu-1}(y)=\text{symm}_{\nu-1}(y_1,\ldots,y_{m-1})$ and $s_{j-\nu}(\tau)=\text{symm}_{j-\nu}(\tau_1,\ldots,\tau_{j-1})$. For properties of these symmetric functions, see Remark 4.1.

Proof. For any choice of $\nu_1,\ldots,\nu_r$ we may easily compute

$$\frac{\partial G(0)}{\partial\tau_{\nu_1}\cdots\partial\tau_{\nu_r}}=\begin{cases}0, & \text{if any two } \nu\text{'s are equal}\\ \dfrac{(-1)^r(j-1-r)!(m-j+r)!}{(m-1)!}s_{j-1-r}(y), & \\ & \text{otherwise.}\end{cases}$$

We note that these partial derivatives do not depend on the τ's. Now by the multidimensional Taylor expansion,

$$G=G(0)+\sum\frac{\partial G}{\partial\tau_i}\tau_i+\sum\sum\frac{\partial G}{\partial\tau_i\partial\tau_j}\tau_i\tau_j+\cdots$$

$$=\sum_{r=0}^{j-1}\frac{(-1)^r(j-1-r)!(m-j+r)!}{(m-1)!}s_{j-1-r}(y)s_r(\tau),$$

which after a change of summation index is the desired expansion if G. ■

Remark 6.2

If $1\leqslant p\leqslant q$, then for any real numbers $w_i>0$, $i=0,1,\ldots,N$,

$$\left(\sum_{i=0}^{N}w_i^p\right)^{1/p}\geqslant\left(\sum_{i=0}^{N}w_i^q\right)^{1/q}.$$

This inequality is called Jensen's inequality. For a proof, see Beckenbach and Bellman [1961, p. 18].

Remark 6.3

If Y is a linear subspace of the linear space X, we say that an operator P mapping X into Y is a *projector* provided $Pg=g$ for all $g\in Y$. As we saw in Example 6.23, the spline approximation operator Q defined in (6.40) maps $C[a,b]$ in $\mathbb{S}=\text{span}\{B_i\}_1^n$, but is generally *not* a projector. The following lemma gives conditions on the $\{\lambda_i\}$ which guarantee that Q will be a projector onto $\mathbb{S}$:

LEMMA 6.59

Q defined in (6.40) is a projector onto $\mathbb{S}$ if and only if $\{\lambda_i\}_1^n$ form a dual basis to $\{B_i\}_1^n$.

Proof. Q is a projector if and only if $QB_j = \sum_{i=1}^n (\lambda_i B_j) B_i = B_j$ for all $j = 1,2,\ldots,n$. This is equivalent to the requirement that $\lambda_i B_j = \delta_{ij}$, $i, j = 1,2,\ldots,n$. ∎

In the following remark we use Lemma 6.59 to construct a projector of $L_p[a,b]$ onto the splines which delivers optimal error bounds. For some other explicit constructions, see Lyche and Schumaker [1975].

Remark 6.4

Combining Lemma 6.59 with the dual linear functionals of Theorem 4.41, we can construct a projector of $L_p[a,b]$ functions onto splines producing optimal order approximations.

THEOREM 6.60

Let $\{y_i\}_1^{n+m}$ be an extended partition of $[a,b]$, and let $\{N_i^m\}_1^n$ be the associated normalized B-splines. Let $\{\lambda_i\}_1^n$ be the linear functionals defined in Theorem 4.41 forming a dual basis for $\mathbb{S} = \text{span}\,\{N_i^m\}_1^n$. Then for any $1 \leqslant p < \infty$,

$$Qf = \sum_{i=1}^{n} (\lambda_i f) N_i^m(x)$$

defines a bounded linear projector of $L_p[a,b]$ onto $\mathbb{S}$. Moreover, for all $f \in L_p[a,b]$,

$$\|f - Qf\|_p \leqslant C d(f, \mathbb{S})_p.$$

Proof. Q is defined on $L_p[a,b]$ since $\lambda_i f$ is defined for all $f \in L_p[a,b]$. In fact,

$$|\lambda_i f| \leqslant \|f\|_{L_p[I_i]} \|D^m \psi_i\|_{L_q[I_i]},$$

where $I_i = (y_i, y_{i+m})$. Applying the estimates of $\|D^m \psi_i\|_\infty$ obtained in the proof of Theorem 4.41, we have

$$|\lambda_i f| \leqslant C \|f\|_{L_p[I_i]} h_i^{-1/p},$$

where
$$h_i = y_{i+m} - y_i.$$

But then

$$|Qf(x)| \leqslant \sum |\lambda_i f| N_i^m(x) \leqslant C \| f \|_{L_p[I_i]} h_i^{-1/p}$$

and integrating the pth power over I_i, we obtain

$$\| Qf \|_{L_p[I_i]}^p \leqslant C \| f \|_{L_p[I_i]}.$$

Summing over $i = m, m+1, \ldots, n$, we obtain

$$\| Qf \|_{L_p[a,b]} \leqslant C \| f \|_{L_p[a,b]},$$

which shows that Q is bounded. The fact that Q is a projector follows from Lemma 6.59. Finally, the fact that Qf approximates f to order $d(f, \mathbb{S})_p$ follows from the elementary inequality

$$\| f - Qf \|_p \leqslant \| f - s \|_p + \| Q(f - s) \|_p \leqslant (1 + \| Q \|) \| f - s \|_p,$$

which holds for all $s \in \mathbb{S}$; that is,

$$\| f - Qf \|_p \leqslant (1 + \| Q \|) d(f, \mathbb{S})_p. \qquad \blacksquare$$

Remark 6.5

It is easy to construct examples to show that the requirement that a sequence of partitions go to zero steadily (as in Definition 6.5) is a necessary condition in order to obtain saturation results.

7

APPROXIMATION POWER OF SPLINES (FREE KNOTS)

In this chapter we study the question of how well smooth functions can be approximated by polynomial splines where the knots of the spline are regarded as free parameters that can be adjusted to the particular function being approximated.

§ 7.1. INTRODUCTION

The aim of this chapter is to study how well smooth functions f on an interval $[a,b]$ can be approximated by polynomial splines of order m with k knots. To express this more precisely, let $\mathcal{S}_{m,k}$ be the space of *polynomial splines of order m with k knots in* (a,b) defined by

$$\mathcal{S}_{m,k} = \left\{ \begin{array}{l} s\text{: } s \text{ is a polynomial spline of order } m \text{ with } k \\ \text{knots in } (a,b), \text{ counting multiplicities} \end{array} \right\}, \tag{7.1}$$

and let

$$d(f, \mathcal{S}_{m,k})_p = \inf_{s \in \mathcal{S}_{m,k}} \|f - s\|_{L_p[a,b]}. \tag{7.2}$$

Our goal here is to relate the smoothness of f to the behavior of $d(f, \mathcal{S}_{m,k})_p$ as $k \to \infty$.

Since we are primarily interested in asymptotic results, it will be convenient to introduce two additional spaces closely related to $\mathcal{S}_{m,k}$. Let

$$\mathcal{S}^1_{m,k} = \left\{ \begin{array}{l} s\text{: } s \text{ is a polynomial spline of order } m \text{ with } k \\ \text{simple knots in } (a,b) \end{array} \right\}, \tag{7.3}$$

and

$$\mathcal{P}\mathcal{P}_{m,k} = \left\{ \begin{array}{l} s\colon s \text{ is a piecewise polynomial of order } m \text{ with} \\ k \text{ knots in } (a,b) \end{array} \right\}. \tag{7.4}$$

The following theorem shows that the distances $d(f, \mathcal{P}\mathcal{P}_{m,k})_p$, $d(f, \mathbb{S}_{m,k})_p$, and $d(f, \mathbb{S}^1_{m,k})_p$ all have the same asymptotic behavior, and thus we will be able to restrict our attention to the somewhat more tractable space $\mathcal{P}\mathcal{P}_{m,k}$:

THEOREM 7.1

For any function $f \in L_p[a,b]$,

$$d(f, \mathcal{P}\mathcal{P}_{m,k})_p \approx d(f, \mathbb{S}_{m,k})_p \approx d(f, \mathbb{S}^1_{m,k})_p \tag{7.5}$$

as $k \to \infty$.

Proof. Clearly, $\mathbb{S}^1_{m,k} \subseteq \mathbb{S}_{m,k} \subseteq \mathcal{P}\mathcal{P}_{m,k}$, and thus

$$d(f, \mathcal{P}\mathcal{P}_{m,k})_p \leqslant d(f, \mathbb{S}_{m,k})_p \leqslant d(f, \mathbb{S}^1_{m,k})_p.$$

On the other hand, if $s \in \mathcal{P}\mathcal{P}_{m,k}$, then by Lemma 4.51 each of the m-tuple knots of s can be pulled apart into m simple knots to obtain a spline $\tilde{s} \in \mathbb{S}^1_{m,mk}$ which differs from s only in small intervals around each knot of s. This can be done in such a way that

$$\|f - \tilde{s}\|_\infty = \|f - s\|_\infty.$$

It follows that

$$d(f, \mathcal{P}\mathcal{P}_{m,k})_p \approx d(f, \mathbb{S}^1_{m,mk})_p,$$

which in turn implies (7.5). ∎

In view of Theorem 7.1, in the remainder of this chapter we shall restrict our attention to $\mathcal{P}\mathcal{P}_{m,k}$. It is clear from the definition that $\mathcal{P}\mathcal{P}_{m,k}$ is a rather large space; indeed,

$$\mathcal{P}\mathcal{P}_{m,k} = \bigcup_{\Delta} \mathcal{P}\mathcal{P}_m(\Delta),$$

where the union is taken over all partitions of $[a,b]$ with k break points. It is also clear that $\mathcal{P}\mathcal{P}_{m,k}$ is *not* a linear space. On the other hand, if s_1 and s_2 are two elements of $\mathcal{P}\mathcal{P}_{m,k}$, then their sum $s_1 + s_2$ is an element of $\mathcal{P}\mathcal{P}_{m,2k}$. Furthermore, even though $\mathcal{P}\mathcal{P}_{m,k}$ is not linear, it is a computer-compatible space in the sense discussed in §1.5—indeed, each element

$s \in \mathcal{P}\mathcal{P}_{m,k}$ is uniquely determined by its set of knots and an associated coefficient vector. These same remarks apply to the spaces $\mathcal{S}_{m,k}$ and $\mathcal{S}^1_{m,k}$.

In view of the fact that $\mathcal{P}\mathcal{P}_{m,k}$ contains all linear spaces of piecewise polynomials with k knots (and thus the space of piecewise polynomials with equally spaced knots, for example), it follows from the results of Chapter 6 that wide classes of smooth functions can be approximated to high order by $\mathcal{P}\mathcal{P}_{m,k}$. Thus, for example, we know that

$$d(f, \mathcal{P}\mathcal{P}_{m,k})_p = \mathcal{O}\left(\frac{1}{k+1}\right)^{\sigma}, \qquad \text{all } f \in L_p^{\sigma}[a,b], \qquad \text{for } 1 \leqslant \sigma \leqslant m.$$

On the other hand, since $\mathcal{P}\mathcal{P}_{m,k}$ is so large, it is reasonable to expect that such orders of approximation should be obtainable for much wider classes of smooth functions (by taking advantage of the freedom of the knots). We shall see that this is indeed the case.

The main results of this chapter are the direct and inverse theorems of §7.4 relating the smoothness of f (as expressed in terms of an appropriate modulus of smoothness defined in §7.3) to the asymptotic behavior of $d(f, \mathcal{P}\mathcal{P}_{m,k})_p$. These theorems give a complete characterization of the function classes corresponding to various orders of convergence. In §7.5 we show that (as in the linear case) order m convergence is the maximum that can be obtained (for smooth functions). Because of the unusual form of the moduli of smoothness, we cannot generally describe the classes where a particular order of convergence takes place in terms of classical notions of smoothness. In §7.6 we examine some classes of functions where mth order convergence holds.

§ 7.2. PIECEWISE CONSTANTS

In this section we consider approximation using the space $\mathcal{P}\mathcal{P}_{1,k} = \mathcal{S}_{1,k}$. We begin with an example that shows that substantial gains in approximation order can be achieved when using free knots.

EXAMPLE 7.2

Let $0 < \alpha < 1$, and define $f_\alpha(x) = x^\alpha$. Approximate f_α on the interval [0, 1] by piecewise constants.

Discussion. It is clear (cf. Figure 19) that if $\Delta_k = \{i/(k+1)\}_{i=0}^{k+1}$ is the uniform partition of [0, 1] with k knots, then

$$d[f_\alpha, \mathcal{P}\mathcal{P}_1(\Delta_k)]_\infty = \left(\frac{1}{k+1}\right)^{\alpha}.$$

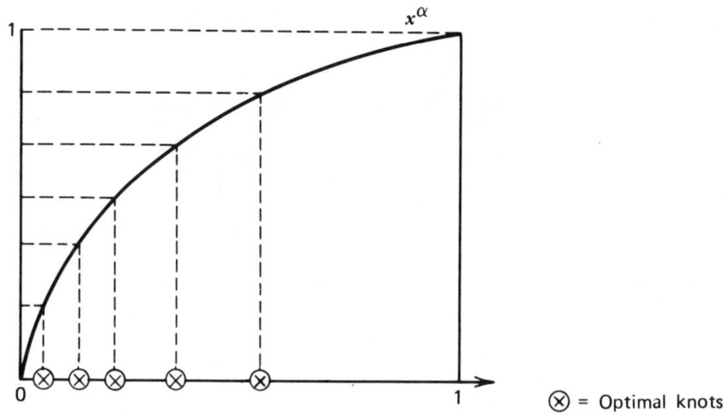

Figure 19. Approximation of x^α by piecewise constants.

On the other hand, if we take

$$\Delta_k^* = \left\{ \left(\frac{i}{k+1} \right)^{1/\alpha} \right\}_{i=0}^{k+1},$$

then

$$d(f_\alpha, \mathcal{P}\mathcal{P}_{1;k})_\infty = d[f, \mathcal{P}\mathcal{P}_1(\Delta_k^*)]_\infty = \frac{1}{k+1}.$$

Thus by allowing free knots we obtain order 1 convergence rather than order α. ■

Example 7.2 suggests (cf. Figure 19) that if f is any continuous monotone-increasing function on $[0, 1]$, then with the partition

$$\Delta_k^* = \{x_i^* = f^{-1}(i/(k+1))\}_{i=0}^{k+1}$$

we have

$$d(f, \mathcal{P}\mathcal{P}_{1,k})_\infty = d[f, \mathcal{P}\mathcal{P}_1(\Delta_k^*)]_\infty = \frac{f(b) - f(a)}{k+1}.$$

This idea can be applied to an even larger class of functions, which we now introduce.

Given a function defined on the interval $[a,b]$, let

$$V_a^b(f) = \sup\left\{ \sum_{i=0}^{n-1} |f(t_{i+1}) - f(t_i)| : n > 1 \text{ and } a \leqslant t_1 < \cdots < t_n \leqslant b \right\}. \tag{7.6}$$

We call $V_a^b(f)$ the *variation of f*. It is easily verified that

$$V_a^b(f) \leqslant V_a^c(f) + V_c^b(f), \qquad \text{all } a \leqslant c \leqslant b; \tag{7.7}$$

$$V_a^b(f+g) \leqslant V_a^b(f) + V_a^b(g); \tag{7.8}$$

$$V_a^b(\alpha f) \leqslant |\alpha|\, V_a^b(f); \tag{7.9}$$

$$V_a^b(f) = \int_a^b |Df(x)|\, dx, \qquad \text{all } f \in L_1^1[a,b]; \tag{7.10}$$

$$V_a^b(f) \leqslant \varliminf_{k\to\infty} V_a^b(s_k) \qquad \text{if } s_k(x) \to f(x), \text{ for all } x \in [a,b]. \tag{7.11}$$

We define the set of functions of *bounded variation on* $[a,b]$ by

$$BV[a,b] = \{ f \in B[a,b] : V_a^b(f) < \infty \}.$$

Clearly $BV[a,b]$ is a linear space that by (7.10) contains $L_1^1[a,b]$. On the other hand, $BV[a,b]$ is much larger than $L_1^1[a,b]$, and includes many functions that are not even continuous (e.g., all monotone-increasing functions). It is known that every $f \in BV[a,b]$ can be written as the difference of two monotone-increasing functions. It also follows from the definition that every $f \in BV[a,b]$ can have at most a finite number of jump discontinuities.

THEOREM 7.3

If $f \in C[a,b] \cap BV[a,b]$, then

$$d(f, \mathcal{S}_{1,k})_\infty \leqslant \frac{1}{2}\left(\frac{1}{k+1}\right) V_a^b(f). \tag{7.12}$$

Conversely, if $f \in C[a,b]$ is such that

$$d(f, \mathcal{S}_{1,k})_\infty \leqslant C\left(\frac{1}{k+1}\right), \qquad \text{all } k > 0, \tag{7.13}$$

then $f \in BV[a,b]$, and, in fact,

$$V_a^b(f) = \lim_{k\to\infty} 2(k+1)d(f,\mathcal{S}_{1,k})_\infty. \tag{7.14}$$

Finally, if $f \in C[a,b]$ and

$$(k+1)d(f,\mathcal{S}_{1,k})_\infty \to 0 \qquad \text{as } k\to\infty, \tag{7.15}$$

then f is a constant on $[a,b]$.

Proof. If $f \in C[a,b] \cap BV[a,b]$, then there exists a partition $a = x_0 < x_1 < \cdots < x_{k+1} = b$ of $[a,b]$ such that

$$V_{x_i}^{x_{i+1}}(f) = \frac{V_a^b(f)}{(k+1)}, \qquad i = 0,1,\ldots,k.$$

Now we define a piecewise constant approximation of f by

$$s(x) = \begin{cases} (M_i + m_i)/2 & \text{on } [x_i, x_{i+1}), \\ \quad i = 0,1,\ldots,k, \end{cases}$$

where $M_i = \max_{x_i \leqslant x \leqslant x_{i+1}} f(x)$ and $m_i = \min_{x_i \leqslant x \leqslant x_{i+1}} f(x)$. Clearly,

$$|f(x) - s(x)| \leqslant \tfrac{1}{2}(M_i - m_i) \leqslant \tfrac{1}{2}\left[V_{x_i}^{x_{i+1}}(f) \right] \leqslant \frac{V_a^b(f)}{2(k+1)},$$

all $x_i \leqslant x \leqslant x_{i+1}$, and all $i = 0,1,\ldots,k$. We have proved (7.12).

Suppose now that $f \in C[a,b]$ and that (7.13) holds. Then for each k we can find $s_k \in \mathcal{S}_{1,k}$ such that

$$\|f - s_k\|_\infty \leqslant d(f,\mathcal{S}_{1,k})_\infty + \frac{1}{(k+1)^2} \leqslant C\left(\frac{1}{k+1}\right) + \frac{1}{(k+1)^2}.$$

The sequence s_k of piecewise constants converge uniformly to f, and moreover,

$$V_a^b(s_k) = \sum_{i=0}^{k} |s(x_{i+1}) - s(x_i)| \leqslant 2(k+1)\|f - s_k\| \leqslant 2(C+1)$$

for all k. It follows from (7.11) that

$$V_a^b(f) \leqslant \varliminf_{k\to\infty} V_a^b(s_k) \leqslant 2(C+1),$$

so $f \in BV[a,b]$. In fact, using (7.12),

$$V_a^b(f) \leqslant \varliminf_{k\to\infty} V_a^b(s_k)$$
$$\leqslant \varlimsup_{k\to\infty} \left(2(k+1)d(f,\mathbb{S}_{1,k})_\infty + \frac{2}{(k+1)} \right) \leqslant V_a^b(f),$$

and (7.14) follows. If (7.15) holds, then by (7.14) it follows that $V_a^b(f)=0$, which can only happen if f is constant. ■

Since Theorem 7.3 contains both direct and inverse assertions, we obtain the following *characterization result*:

$$f \in C[a,b] \text{ belongs to } BV[a,b] \text{ if and only if } d(f,\mathbb{S}_{1,k}) = \mathcal{O}\left(\frac{1}{k+1}\right).$$

Statement (7.15) is a *saturation result*. It shows that the maximal order of convergence obtainable for continuous functions using piecewise constants is 1 (except for the trivial class of constants, which are approximated exactly). There are, of course, some *discontinuous* functions that are approximated exactly if k is sufficiently large; for example, all piecewise constants.

Because of the saturation phenomenon, for functions that are $C^1[a,b]$ or smoother, there is nothing to be gained by using splines with free knots. On the other hand, Theorem 7.3 shows that for less smooth functions, the order of convergence using free knots can be considerably better than with a fixed predetermined set of knots. We illustrated this for a specific function in Example 7.2. The following example illustrates it for a large class of relatively smooth functions:

EXAMPLE 7.4

Let $f \in L_p^1[a,b]$, $1 \leqslant p \leqslant \infty$.

Discussion. By (7.10) and Hölder's inequality, we have

$$V_a^b(f) \leqslant \int_a^b |Df(x)|\,dx \leqslant (b-a)^{1-1/p} \|Df\|_{L_p[a,b]}.$$

Coupling this with Theorem 7.3, we obtain

$$d(f,\mathcal{P}\mathcal{P}_{1,k})_\infty \leqslant \frac{(b-a)}{2(k+1)}^{1-1/p} \|Df\|_{L_p[a,b]},$$

which is order 1 convergence. On the other hand, by the results of Chapter 6, we know that using splines with equally spaced knots, for example, the maximal order of convergence is $1-1/p$. ■

So far we have concentrated on the case of the uniform norm. We turn now to some analogous results for the q-norms, $1 \leqslant q < \infty$.

THEOREM 7.5

Let $f \in BV[a,b]$, and let $1 \leqslant q < \infty$. Then

$$d(f, \mathcal{P}\mathcal{P}_{1,k})_q \leqslant \frac{CV_a^b(f)}{k+1}. \tag{7.16}$$

Proof. We recall that $f \in BV[a,b]$ implies it can have at most a finite number of jump discontinuities, say at $t_1 < t_2 < \cdots < t_d$. Now we can modify f in a neighborhood of each t_i so that the resulting function $\tilde{f}$ satisfies

$$\|f-\tilde{f}\|_{L_q[a,b]} \leqslant \frac{V_a^b(f)}{2(k+1)} \qquad \text{and } V_a^b(\tilde{f}) \leqslant V_a^b(f). \tag{7.17}$$

We can accomplish this, for example, by replacing f by a linear function in $(t_i - \varepsilon_i, t_i + \varepsilon_i)$ with ε_i sufficiently small. Then if $s_{\tilde{f}}$ is the spline constructed as in the proof of Theorem 7.3 associated with $\tilde{f}$, we have

$$\|f - s_{\tilde{f}}\|_{L_q[a,b]} \leqq \|f-\tilde{f}\|_{L_q[a,b]} + \|\tilde{f} - s_{\tilde{f}}\|_{L_q[a,b]}.$$

Using Theorem 7.3 for $\tilde{f}$, we obtain (7.16), where we may take $C = \frac{1}{2}[1 + (b-a)^{1/q}]$. ■

Theorem 7.3 asserts that $d(f, \mathcal{P}\mathcal{P}_{1,k})_q = \mathcal{O}(1/(k+1))$ for all $f \in BV[a,b]$. The following example shows that the converse is not true; that is, there exist some functions that are not in $BV[a,b]$ which can nevertheless be approximated to order 1 by piecewise constants.

EXAMPLE 7.6

Let $f \in BV[a,b]$ and g be a function that is a constant almost everywhere. Then

$$d(f+g, \mathcal{P}\mathcal{P}_{1,k})_q = \mathcal{O}\left(\frac{1}{k+1}\right). \tag{7.18}$$

Discussion. Clearly, if s is a spline that approximates f to order 1, and if c is the value that g takes almost everywhere, then $s+c$ provides an order 1 approximation of $f+g$. The function g can be quite nasty. For example, the function g defined to be 1 on the rationals and 0 on the irrationals is constant almost everywhere, but does not belong to $BV[a,b]$. ■

As in the uniform norm case, there is a limit to how well smooth functions can be approximated in the q-norms by piecewise constants—we prove a general saturation theorem in §7.5.

§ 7.3 VARIATIONAL MODULI OF SMOOTHNESS

In order to discuss the order of approximation of smooth functions by polynomial splines with free knots, it is convenient to introduce a new kind of modulus of smoothness. Given $m \geqslant 1$, we define

$$\nu_m(f;t)_\infty = \inf_{\varphi} \sup\{\omega_m(f;|I|)_{L_\infty[I]}: I=[c,d] \leqslant [a,b] \text{ and } \varphi(d)-\varphi(c)<t\}, \tag{7.19}$$

where the infimum is taken over all monotone-increasing functions on $[a,b]$ with $V_a^b(\varphi)=\varphi(b)-\varphi(a) \leqslant b-a$. We call $\nu_m(f;t)$ the *variational modulus of smoothness of f of order m*. The following theorem shows that $\nu_m(f;t)$ behaves very much like the classical modulus of smoothness $\omega_m(f;t)$:

THEOREM 7.7

For any $m \geqslant 1$,

$$\nu_m(f;t_1) \leqslant \nu_m(f;t_2), \qquad \text{all } 0<t_1<t_2 \leqslant b-a; \tag{7.20}$$

$$\nu_m(f;t) \leqslant \omega_m(f;t); \tag{7.21}$$

$$\nu_m(f;t) \leqslant 2^{m-j}\nu_j(f;t), \qquad 1 \leqslant j \leqslant m; \tag{7.22}$$

$$\nu_m(f;t) \leqslant Ct\nu_{m-1}(Df;2t) \qquad \text{if } Df \in B[a,b]; \tag{7.23}$$

$$\nu_1(f;t) \leqslant CtV_a^b(f) \text{ if } f \in BV[a,b]. \tag{7.24}$$

Proof. Property (7.20) follows directly from the definition, while (7.21) can be proved by taking $\varphi(x)=x$. Property (7.22) follows immediately from the analogous property of $\omega_m(f;\cdot)$. The proof of (7.23) is somewhat more delicate. Given any $\varphi \in BV[a,b]$, let $\tilde{\varphi}(x)=[\varphi(x)+x]/2$. Then

$$\begin{aligned}\nu_m(f;t) &\leq \sup\{\omega_m(f;|I|)_{L_\infty[I]}: I=[c,d] \text{ satisfies } \tilde{\varphi}(d)-\tilde{\varphi}(c)<t\} \\ &= \sup\{\omega_m(f;|I|)_{L_\infty[I]}: I=[c,d] \text{ satisfies } \varphi(d)-\varphi(c)+(d-c)<2t\} \\ &\leq \sup\{|I|\omega_{m-1}(Df;|I|)_{L_\infty[I]}: I=[c,d] \text{ satisfies } \varphi(d)-\varphi(c)+|I|<2t\} \\ &\leq 2t\sup\{\omega_{m-1}(Df;|I|)_{L_\infty[I]}: I=[c,d] \text{ satisfies } \varphi(d)-\varphi(c)<2t\}.\end{aligned}$$

Now taking the infimum over all monotone φ with $V_a^b(\varphi)\leq b-a$, we obtain (7.23).

To prove (7.24), let $\varphi(x)=(b-a)V_a^x(f)/V_a^b(f)$. Then

$$\nu_1(f;t)\leq \sup\left\{\omega_1(f;|I|)_{L_\infty[I]}: I=[c,d] \text{ satisfies } \frac{(b-a)}{V_a^b(f)}\left[V_a^d(f)-V_a^c(f)\right]<t\right\}.$$

Since $\omega_1(f;|I|)_{L_\infty[I]}\leq V_c^d(f)$, (7.24) follows. ■

It is also convenient to introduce a p-version of $\nu_m(f;t)$. We define

$$\nu_m(f;t)_p = \inf_\varphi \sup\{(\sum_{i=0}^{n}\omega_m(f;|I_i|)_{L_p[I_i]}^p)^{1/p}: \ I_i=[c_i d_i] \text{ is a disjoint set of subintervals of } [a,b] \text{ with } \varphi(d_i)-\varphi(c_i)<t,\ i=0,\dots,n\}, \tag{7.25}$$

where again the infimum is taken over all monotone-increasing functions φ on $[a,b]$ with $V_a^b(\varphi)\leq b-a$.

THEOREM 7.8

For any $m\geq 1$ and $1\leq p<\infty$,

$$\nu_m(f;t_1)_p\leq \nu_m(f;t_2)_p, \qquad \text{all } 0<t_1\leq t_2\leq b-a; \tag{7.26}$$

$$\nu_m(f;t)_p\leq 2^{m-j}\nu_j(f;t)_p, \qquad 1\leq j\leq m; \tag{7.27}$$

$$\nu_m(f;t)_p\leq Ct\nu_{m-1}(Df;2t)_p, \qquad \text{if } f\in L_p^1[a,b]; \tag{7.28}$$

$$\nu_1(f;t)_p\leq Ct\|Df\|_{L_1[a,b]}, \qquad \text{if } f\in L_1^1[a,b]. \tag{7.29}$$

Proof. Properties (7.26) to (7.28) follow just as in the $p=\infty$ case. To prove (7.29), let $\varphi(x)=(b-a)\int_a^x|Df|/\int_a^b|Df|$, and note that

$$\nu_1(f;t)_p \leqslant \sup\left\{\left(\sum_{i=0}^{n}\omega_1(f;|I_i|)_{L_p[I_i]}^p\right)^{1/p} : I_i=[c_i,d_i) \text{ satisfies } \frac{(b-a)}{\int_a^b|Df|}\int_{c_i}^{d_i}|Df|<t\right\}.$$

Now,

$$\omega_1(f;|I_i|)_{L_p[I_i]}^p \leqslant \sup_h \int_{c_i}^{d_i-h}\left(\int_x^{x+h}|Df|\right)^p \leqslant (d_i-c_i)\left(\int_{c_i}^{d_i}|Df|\right)^p$$

$$\leqslant (d_i-c_i)\left(\frac{t}{b-a}\int_a^b|Df|\right)^p.$$

Substituting this in the above and noting that $\sum_{i=0}^n(d_i-c_i)\leqslant b-a$, we obtain (7.29). ■

An important property of the classical moduli of smoothness is the fact that $\omega_m(f;t)_p$ can go to zero faster than t^m only if $f\in\mathcal{P}_m$. We now state an analogous result for $\nu_m(f;t)_p$. We delay its proof until Section 7.5.

THEOREM 7.9

If $f\in C^m[a,b]$ and $\nu_m(f;t)=o(t^m)$, then $f\in\mathcal{P}_m$. Similarly, if $f\in L_p^m[a,b]$ and $\nu_m(f;t)_p=o(t^m)$, then $f\in\mathcal{P}_m$.

§ 7.4. DIRECT AND INVERSE THEOREMS

We are ready to relate the order of approximation of a function f by piecewise polynomials of order m with k free knots to the smoothness of f as measured by $\nu_m(f;t)_p$.

THEOREM 7.10

Suppose $f\in L_p[a,b]$ if $1\leqslant p<\infty$ or that $f\in C[a,b]$ if $p=\infty$. Then

$$d(f,\mathcal{P}\mathcal{P}_{m,k})_p \approx \nu_m\left(f;\frac{b-a}{k+1}\right)_p \qquad \text{as } k\to\infty. \tag{7.30}$$

Proof. Consider first the case of $p=\infty$. Let $s\in\mathcal{P}\mathcal{P}_{m,k}$ be such that

$$\|f-s\|_\infty \leqslant d(f,\mathcal{P}\mathcal{P}_{m,k})_\infty+\varepsilon.$$

Suppose the knots of s are at $\{x_i\}_1^k$, where $a=x_0<x_1<\cdots<x_{k+1}=b$. Then $\varphi(x)=(b-a)\Sigma_{i=1}^{k+1}(x-x_i)_+^0/(k+1)$ is a monotone-increasing function with $\varphi(b)-\varphi(a)=b-a$, and

$$\nu_m\left(f;\frac{b-a}{k+1}\right)\leqslant \sup\left\{\omega_m(f;|I|)_{L_\infty[I]}\colon I=[c,d] \text{ and } \varphi(d)-\varphi(c)<\frac{b-a}{k+1}\right\}.$$

Now, by the definition of φ (it has jumps of $(b-a)/(k+1)$ at each x_i), the condition that $\varphi(d)-\varphi(c)<(b-a)/(k+1)$ implies that only intervals which do not contain any of the knots $\{x_i\}_1^k$ enter into the supremum. But for any such interval we have

$$\omega_m(f;|I|)_{L_\infty[I]}\leqslant 2^m\|f-s\|_{L_\infty[I]}\leqslant 2^m\|f-s\|_\infty.$$

We conclude that

$$\nu_m\left(f;\frac{b-a}{k+1}\right)\leqslant 2^m d(f,\mathcal{P}\mathcal{P}_{m,k})_\infty. \tag{7.31}$$

To complete the proof of (7.30) in the case of $p=\infty$, we now prove an inequality in the opposite direction. Suppose φ is a monotone-increasing function on $[a,b]$ with $\varphi(b)-\varphi(a)=b-a$ such that

$$\sup\left\{\omega_m(f;|I|)_{L_\infty[I]}\colon I=[c,d] \text{ and } \varphi(d)-\varphi(c)<\frac{b-a}{k+1}+\varepsilon\right\}$$
$$\leqslant 2\nu_m\left(f;\frac{b-a}{k+1}+\varepsilon\right).$$

Let $\Delta=\{x_i^*=\varphi^{-1}(i(b-a)/(k+1))\}_{i=0}^{k+1}$. Now on each interval $I_i=[x_i^*,x_{i+1}^*]$, the Whitney Theorem 3.18 asserts the existence of a polynomial $p_i\in\mathcal{P}_m$ with

$$\|f-p_i\|_{L_\infty[I_i]}\leqslant C\omega_m(f;|I_i|)_{L_\infty[I_i]}.$$

Let

$$s^*(x)=\{p_i(x),\qquad x_i^*\leqslant x<x_{i+1}^*,\qquad i=0,1,\ldots,k.$$

Since I_i is such that $\varphi(d_i)-\varphi(c_i)=(b-a)/(k+1)<(b-a)/(k+1)+\varepsilon$, it follows that

$$d(f,\mathcal{P}\mathcal{P}_{m,k})_\infty \leqslant \|f-s^*\| \leqslant \max_{0\leqslant i\leqslant k}\|f-p_i\|_{L_\infty[I_i]} \leqslant C\max_{0\leqslant i\leqslant k}\omega_m(f;|I_i|)_{L_\infty[I_i]}$$

$$\leqslant 2Cv_m\left(f;\frac{b-a}{k+1}+\varepsilon\right).$$

Since $\varepsilon>0$ is arbitrary, this implies

$$d(f,\mathcal{P}\mathcal{P}_{m,k})_\infty \leqslant 2Cv_m\left(f;\frac{b-a}{k+1}\right). \tag{7.32}$$

Combining this with (7.31), we have proved (7.30) in the case of $p=\infty$.

The proof of (7.30) for $1\leqslant p<\infty$ is quite similar. First, to prove the analog of (7.31), suppose $s\in\mathcal{P}\mathcal{P}_{m,k}$ has knots $\{x_i\}_1^k$ and is such that

$$\|f-s\|_p \leqslant d(f,\mathcal{P}\mathcal{P}_{m,k})_p+\varepsilon.$$

Then with $\varphi(x)=(b-a)\Sigma_{i=1}^{k+1}(x-x_i)_+^0/(k+1)$, we have

$$v_m\left(f;\frac{b-a}{k+1}\right)_p \leqslant \sup\left\{\begin{array}{l}\left(\Sigma\omega_m(f;|I_i|)_{L_p[I_i]}^p\right)^{1/p}: I_i=[c_i,d_i)\\ \text{and } \varphi(d_i)-\varphi(c_i)<\dfrac{b-a}{k+1}\end{array}\right\}.$$

Again the condition on the I_i implies that each of them contains no knots in its interior, hence

$$\omega_m(f;|I_i|)_{L_p[I_i]} \leqslant 2^m\|f-s\|_{L_p[I_i]}.$$

But then

$$v_m\left(f;\frac{b-a}{k+1}\right)_p \leqslant 2^m\sup\left\{\left(\Sigma\|f-s\|_{L_p[I_i]}^p\right)^{1/p}: I_i \text{ as above}\right\}$$

$$\leqslant 2^m\|f-s\|_p \leqslant 2^m d(f,\mathcal{P}\mathcal{P}_{m,k})_p+\varepsilon.$$

As $\varepsilon>0$ is arbitrary, we have proved

$$v_m\left(f;\frac{b-a}{k+1}\right)_p \leqslant 2^m d(f,\mathcal{P}\mathcal{P}_{m,k})_p. \tag{7.33}$$

It remains to prove the analog of (7.32) for $1 \leqslant p < \infty$. Let φ be a monotone-increasing function on $[a,b]$ with $\varphi(b)-\varphi(a)=b-a$ such that

$$\sup\left\{\left(\Sigma\omega_m(f;|I_i|)^p_{L_p[I_i]}\right)^{1/p}: I_i=[c_i,d_i) \text{ and } \varphi(d_i)-\varphi(c_i)<\frac{b-a}{k+1}+\varepsilon\right\}$$
$$\leqslant 2\nu_m\left(f;\frac{b-a}{k+1}+\varepsilon\right)_p.$$

Define $\{x_i^*\}_0^{k+1}$ as in the $p=\infty$ case, and let $p_i \in \mathcal{P}_m$ be the polynomials in the Whitney Theorem 3.18 such that

$$\|f-p_i\|_{L_p[I_i]} \leqslant C\omega_m(f;|I_i|)_{L_p[I_i]},$$

$i=0,1,\ldots,k$. Then with s^* defined to be p_i on I_i, we have

$$d(f,\mathcal{P}\mathcal{P}_{m,k})_p \leqslant \|f-s^*\|_p = \left(\sum_{i=0}^{k}\|f-p_i\|^p_{L_p[I_i]}\right)^{1/p}$$
$$\leqslant C\left(\sum_{i=0}^{k}\omega_m(f;|I_i|)^p_{L_p[I_i]}\right)^{1/p} \leqslant 2C\nu_m\left(f;\frac{b-a}{k+1}+\varepsilon\right).$$

As $\varepsilon>0$ is arbitrary, this implies

$$d(f,\mathcal{P}\mathcal{P}_{m,k})_p \leqslant 2C\nu_m\left(f;\frac{b-a}{k+1}\right)_p. \tag{7.34}$$

Combining (7.33) and (7.34), we have proved (7.30) for all $1 \leqslant p < \infty$. ∎

Theorem 7.10 allows us to describe the order of convergence of polynomial spline approximation to a given function directly in terms of its smoothness. In particular, we have the following characterization theorem:

THEOREM 7.11

Suppose $f \in C[a,b]$ if $p=\infty$ or $f \in L_p[a,b]$ if $1 \leqslant p < \infty$. Then for any $0<\theta\leqslant m$,

$$d(f,\mathcal{P}\mathcal{P}_{m,k})_p = \mathcal{O}\left(\frac{b-a}{k+1}\right)^\theta \tag{7.35}$$

if and only if

$$\nu_m(f;t)_p = \mathcal{O}(t^\theta). \tag{7.36}$$

Proof. If (7.36) holds, then (7.35) follows immediately from Theorem 7.10. To prove the converse, suppose $t>0$ is given, and let k be such that $(b-a)/k \leqslant t<(b-a)/(k+1)$. Then by the monotonicity properties (7.20) and (7.26),

$$\nu_m(f;t)_p \leqslant \nu_m\left(f;\frac{b-a}{k}\right)_p \leqslant Cd(f,\mathcal{P}\mathcal{P}_{m,k-1})_p$$

$$\leqslant C\left(\frac{b-a}{k}\right)^{\theta} \leqslant C_2(t^{\theta}).$$ ■

To illustrate how Theorem 7.11 can be used to obtain rates of convergence for approximation of smooth functions by piecewise polynomials (or splines) with free knots, we introduce still another space of smooth functions. Given any integer $\sigma \geqslant 1$, let

$$BV^{\sigma-1}[a,b]=\{f\in L_1^{\sigma-1}[a,b]: D^{\sigma-1}f\in BV[a,b]\}. \qquad (7.37)$$

In view of (7.10), we note that $L_1^{\sigma}[a,b]\subseteq BV^{\sigma-1}[a,b]$.

THEOREM 7.12

Suppose $f\in BV^{\sigma-1}[a,b]$, some $1\leqslant\sigma\leqslant m$. Then

$$d(f,\mathcal{P}\mathcal{P}_{m,k})_p=\mathcal{O}\left(\frac{b-a}{k+1}\right)^{\sigma},$$

all $1\leqslant p\leqslant\infty$.

Proof. It suffices to prove the result for $p=\infty$ as $d(f,\mathcal{P}\mathcal{P}_{m,k})_p \leqslant (b-a)^{1/p}d(f,\mathcal{P}\mathcal{P}_{m,k})_\infty$. Now, using properties (7.22) to (7.24) of $\nu_m(f;t)$, we have

$$\nu_m(f;t)\leqslant C_1\nu_\sigma(f;t)\leqslant C_2t^{\sigma-1}\nu_1(D^{\sigma-1}f;2^{\sigma-1}t)$$

$$\leqslant C_3t^{\sigma}V_a^b(D^{\sigma-1}f).$$

Theorem 7.11 now implies the desired result. ■

Theorem 7.12 shows that order m convergence holds for all functions in the rather large class $BV^{m-1}[a,b]$. In §7.6 we show that order m convergence obtains for an even larger class of functions. On the other hand, the results of the following section show that (for smooth functions) the maximum order of convergence is m.

§ 7.5. SATURATION

We begin with a lemma that gives an explicit expression for the L_p-error in approximating a function $f \in C^m[a,b]$ by polynomials.

LEMMA 7.13

Suppose $f \in C^m[a,b]$ and that $1 \leqslant p \leqslant \infty$. Then there exists $a \leqslant \xi \leqslant b$ such that

$$d(f, \mathcal{P}_m)_{L_p[a,b]} = c_{p,m}(b-a)^{m+1/p}|D^m f(\xi)|, \tag{7.38}$$

where

$$c_{p,m} = d\left(\frac{x^m}{m!}, \mathcal{P}_m\right)_{L_p[0,1]}.$$

Proof. Fix $1 \leqslant p \leqslant \infty$, and let $a \leqslant x_1^* \leqslant x_2^* \leqslant \cdots \leqslant x_m^* \leqslant b$ be such that

$$\|(x-x_1^*)\cdots(x-x_m^*)\|_{L_p[a,b]} = \inf_{\{x_i\}} \|(x-x_1)\cdots(x-x_m)\|_{L_p[a,b]}.$$

Let $Q(x) = (x-x_1^*)\cdots(x-x_m^*)$, and suppose that $q_f \in \mathcal{P}_m$ is the polynomial interpolating f at $\{x_i^*\}_1^m$. Then by (3.5) and (2.93), we have

$$f(x) - q_f(x) = Q(x)\frac{D^m f(\xi_x)}{m!}, \quad \text{some } a \leqslant \xi_x \leqslant b,$$

and it follows that

$$d(f, \mathcal{P}_m)_{L_p[a,b]} \leqslant \|f - q_f\|_p \leqslant \frac{\|Q\|_p}{m!} \max_{a \leqslant x \leqslant b} |D^m f(x)|.$$

Now suppose that $p_f \in \mathcal{P}_m$ is such that $\|f - p_f\|_p = d(f, \mathcal{P}_m)_p$. We shall show in a moment that p_f must interpolate f at some m points; that is, there exist $a \leqslant \tilde{x}_1 \leqslant \tilde{x}_2 \leqslant \cdots \leqslant \tilde{x}_m \leqslant b$ such that

$$f(\tilde{x}_i) - p_f(\tilde{x}_i) = 0, \qquad i = 1, 2, \ldots, m.$$

But then

$$d(f, \mathcal{P}_m)_{L_p[a,b]} = \|f - p_f\|_p \geqslant \frac{\|(x-\tilde{x}_1)\cdots(x-\tilde{x}_m)\|_p}{m!} \min_{a \leqslant x \leqslant b} |D^m f(x)|$$

$$\geqslant \frac{\|Q\|_p}{m!} \min_{a \leqslant x \leqslant b} |D^m f(x)|.$$

Combining this with our previous estimate of $d(f, \mathcal{P}_m)_p$ from above and using the continuity of $D^m f$, we conclude that

$$d(f, \mathcal{P}_m)_{L_p[a,b]} = \frac{\|Q\|_p}{m!} |D^m f(\xi)|, \qquad \text{some } a \leqslant \xi \leqslant b.$$

Now a change of variable argument shows that

$$\frac{\|Q\|_{L_p[a,b]}}{m!} = \frac{(b-a)}{m!}^{m+1/p} \min\|(t-t_1)\cdots(t-t_m)\|_{L_p[0,1]}$$

$$= (b-a)^{m+1/p} d\left(\frac{x^m}{m!}, \mathcal{P}_m\right)_{L_p[0,1]},$$

and (7.38) follows.

It remains to establish our assertion that p_f interpolates f at m points. For $p = \infty$ this follows from the fact that $f - p_f$ must alternate between $\pm\|f - p_f\|_\infty$ at least m times on $[a,b]$, (cf. Remark 7.5). Suppose now $1 \leqslant p < \infty$. Then (cf. Remark 7.6), it is known that p_f must satisfy the orthogonality condition

$$\int_a^b x^r |\delta(x)|^{p-1} \operatorname{sgn}[\delta(x)]\, dx = 0, \qquad r = 0, 1, \ldots, m-1, \tag{7.39}$$

where $\delta(x) = f(x) - p_f(x)$. This condition for $r = 0$ implies that $\delta(x)$ must have at least one zero on $[a,b]$. Suppose it has only the zeros $a \leqslant x_1 \leqslant \cdots \leqslant x_n \leqslant b$ with $n < m$. Then clearly $(x - x_1)\cdots(x - x_n)\delta(x)$ is nonnegative throughout $[a,b]$, and

$$\int_a^b (x - x_1)\cdots(x - x_n)|\delta(x)|^{p-1} \operatorname{sgn}[\delta(x)]\, dx > 0,$$

contradicting (7.39). This completes the proof that p_f interpolates f at m points, and the lemma is proved. ■

We can now prove our saturation result.

THEOREM 7.14

Suppose $f \in C^m[a,b]$ and that

$$d(f, \mathcal{P}\,\mathcal{P}_{m,k})_p = o\left(\frac{b-a}{k+1}\right)^m.$$

Then f is a polynomial of order m.

Proof. For convenience, set $\sigma=(m+1/p)^{-1}$. Given any partition $\Delta=\{a=x_0<x_1<\cdots<x_{k+1}=b\}$, we define

$$B_{p,m}(f,\Delta)=\sum_{i=0}^{k} d(f,\mathcal{P}_m)^{\sigma}_{L_p[I_i]}, \tag{7.40}$$

where $I_i=[x_i,x_{i+1})$, $i=0,1,\ldots,k$. By Lemma 7.13 we note that

$$B_{p,m}(f,\Delta)=c^{\sigma}_{p,m}\sum_{i=0}^{k}(x_{i+1}-x_i)|D^m f(\xi_i)|^{\sigma},$$

where $\xi_i\in I_i$, $i=0,1,\ldots,k$ and $c_{p,m}$ is defined in Lemma 7.13. Let $\Delta_k=\{i(b-a)/(k+1)\}_{i=0}^{k+1}$ for each k. Then $B_{p,m}(f,\Delta_k)$ is clearly a Riemann sum, and so

$$\lim_{k\to\infty} B_{p,m}(f,\Delta_k)^{1/\sigma}=c_{p,m}\left(\int_a^b |D^m f(x)|^{\sigma}\,dx\right)^{1/\sigma}=c_{p,m}\|D^m f\|_{L_\sigma[a,b]}.$$

On the other hand, applying the Hölder inequality with dual exponents $1/m\sigma$ and p/σ to (7.40), we have

$$\begin{aligned} B_{p,m}(f,\Delta) &\leqslant \left(\sum_{i=0}^{k} 1\right)^{m\sigma}\left(\sum_{i=0}^{k} d(f,\mathcal{P}_m)^{p}_{L_p[I_i]}\right)^{\sigma/p} \\ &\leqslant (k+1)^{m\sigma} d(f,\mathcal{P}\mathcal{P}_{m,k})^{\sigma}_{L_p[a,b]}. \end{aligned} \tag{7.41}$$

It follows that

$$c_{p,m}\|D^m f\|_{L_\sigma[a,b]}\leqslant \lim_{k\to\infty} B_{p,m}(\Delta_k)^{1/\sigma}\leqslant \lim_{k\to\infty}(k+1)^m d(f,\mathcal{P}\mathcal{P}_{m,k})_p.$$

Combining this fact with our assumption that $d(f,\mathcal{P}\mathcal{P}_{m,k})_p=o[(b-a)/(k+1)]^m$, it follows that $\|D^m f\|_{L_\sigma[a,b]}=0$. Since $f\in C^m[a,b]$, we conclude that $f\in\mathcal{P}_m$; that is, f is a polynomial of order m. ■

With some additional work (see Remark 7.2) it can be shown that for $m\geqslant 1$, Theorem 7.14 continues to hold assuming only that $f\in L_1^m[a,b]$. In any case, the theorem shows that for smooth functions, m is the maximum rate of convergence obtainable using polynomial splines of order m. This saturation result does not preclude the possibility that certain nonpolynomial functions which do not belong to $C^m[a,b]$ (or $L_1^m[a,b]$) can be approximated to order better than m.

EXAMPLE 7.15

Let f be a polynomial spline of order m with knots at $a<x_1<\cdots<x_K<b$.

Discussion. We do not say anything about the multiplicity of the knots. Thus f may lie in $C^{m-1}[a,b]$ or it may not even be continuous. In any case, it is clear that

$$d(f,\mathcal{P}\mathcal{P}_{m,k})_p=0 \qquad \text{for all } k\geqslant mK,$$

and thus that

$$d(f,\mathcal{P}\mathcal{P}_{m,k})_p=\mathrm{o}\left(\frac{b-a}{k+1}\right)^m \qquad \text{as } k\to\infty. \qquad \blacksquare$$

We can now prove that the only smooth functions with $\nu_m(f;t)_p=\mathrm{o}(t^m)$ are the polynomials.

Proof of Theorem 7.9. Suppose $f\in C^m[a,b]$ is such that $\nu_m(f;t)_p=\mathrm{o}(t^m)$. Then by Theorem 7.11 it follows that $d(f,\mathcal{P}\mathcal{P}_{m,k})_p=\mathrm{o}[(b-a)/(k+1)]^m$, which by Theorem 7.14 implies that $f\in\mathcal{P}_m$. $\blacksquare$

§ 7.6. SATURATION CLASSES

In view of the saturation Theorem 7.14, it is of interest to examine the class of functions for which approximation by piecewise polynomials of order m in the p-norm is of order m. By Theorem 7.11 this class is given by

$$\mathrm{Sat}(p,m)=\left\{f\colon d(f,\mathcal{P}\mathcal{P}_{m,k})_p=\mathcal{O}\left(\frac{b-a}{k+1}\right)^m\right\}=\{f\colon \nu_m(f;t)_p=\mathcal{O}(t^m)\}.$$

Theorem 7.3 shows that

$$\mathrm{Sat}(\infty,1)=C[a,b]\cap BV[a,b].$$

Unfortunately, for other values of m and p, no precise description of $\mathrm{Sat}(p,m)$ in terms of classical notions of smoothness is known. By the results of §7.4 we know that $\mathrm{Sat}(p,m)$ contains all the functions in $BV^{m-1}[a,b]$. Example 7.15 shows that it also contains all polynomial spline functions of order m with a finite number of knots, counting multiplicities. The following example shows that it contains other functions that are not in $BV^{m-1}[a,b]$:

EXAMPLE 7.16

Let $0<\alpha<1$ and $f_\alpha(x)=x^\alpha$. Then on the interval [0, 1], $f_\alpha \in \mathrm{Sat}(p,m)$ for all $m\geqslant 1$ and $1\leqslant p\leqslant\infty$.

Discussion. We show that

$$d(f_\alpha,\mathcal{P}\mathcal{P}_{m,k})_\infty=\mathcal{O}\left(\frac{b-a}{k+1}\right)^m.$$

Let $x_i=[i/(k+1)]^{m/\alpha}$, $i=0,1,\ldots,k+1$, and let $I_i=[x_i,x_{i+1}]$. For $1\leqslant i\leqslant k$, let $p_i\in\mathcal{P}_m$ be the polynomial approximating f_α on I_i as in the Whitney Theorem 3.18. Then

$$\|f_\alpha-p_i\|_{L_\infty[I_i]}\leqslant C_4\omega_m(f;|I_i|)_{L_\infty[I_i]}\leqslant C_3|I_i|^m\|D^m f_\alpha\|_{L_\infty[I_i]}.$$

Let $p_0=0$, and set

$$s(x)=p_i(x)\qquad \text{for } x_i\leqslant x<x_{i+1},\qquad i=0,1,\ldots,k.$$

It is clear that $\|f_\alpha-s\|_{L_\infty[I_0]}\leqslant[1/(k+1)]^m$. On the other hand, for $i=1,2,\ldots,k$,

$$\begin{aligned}|I_i|^m\|D^m f\|_{L_\infty[I_i]}&\leqslant(x_{i+1}-x_i)^m(\alpha)\cdots(\alpha-m+1)x_i^{\alpha-m}\\&\leqslant(m-1)!\left[(i+1)^{m/\alpha}-i^{m/\alpha}\right]^m\left(\frac{1}{k+1}\right)^m i^{m(\alpha-m)/\alpha}.\end{aligned}$$

Using the fact that $(i+1)^{m/\alpha}-i^{m/\alpha}\leqslant C_2(i+1)^{m/\alpha-1}$ for some constant C_2 (depending only on m and α), we obtain

$$\|f_\alpha-s\|_{L_\infty[0,1]}\leqslant C_1\left(\frac{1}{k+1}\right)^m\max_{1\leqslant i\leqslant k}\left(\frac{i+1}{i}\right)^{m(m/\alpha-1)}\leqslant C\left(\frac{1}{k+1}\right)^m. \qquad \blacksquare$$

Example 7.16 is quite impressive since it asserts that x^α can be approximated to order $[(b-a)/(k+1)]^m$ by splines of order m with free knots, while the best we can do with splines of order m with equally spaced knots is $[(b-a)/(k+1)]^\alpha$.

In order to help answer the question of what other functions lie in $\mathrm{Sat}(p,m)$, we now introduce a generalization of the space $BV^{m-1}[a,b]$.

Given any $1 \leqslant p \leqslant \infty$, let $\sigma = 1/(m+1/p)$ and

$$V_p^m[a,b] = \{f \in L_p[a,b]: |f|_{V_p^m[a,b]} < \infty\}, \qquad 1 \leqslant p < \infty$$

$$V_\infty^m[a,b] = \{f \in C[a,b]: |f|_{V_\infty^m[a,b]} < \infty\}, \qquad p = \infty. \tag{7.42}$$

Here

$$|f|_{V_p^m[a,b]} = \|f\|_{L_p^\sigma[a,b]} + N_{p,m}(f), \tag{7.43}$$

while

$$N_{p,m}(f) = \sup_{\Delta} B_{p,m}(f,\Delta), \tag{7.44}$$

where $B_{p,m}(f,\Delta)$ is defined in (7.40) and the supremum is taken over all partitions of $[a,b]$.

It is clear that $V_p^m[a,b]$ is a linear space, and it is easily seen that (7.43) defines a seminorm on it. Since it can be shown by standard arguments that $V_p^m[a,b]$ is complete, it follows that it is in fact a Frechét space (cf. Remark 7.1).

The following theorem shows that every function in $V_p^m[a,b]$ can be approximated to order m by $\mathcal{P}\mathcal{P}_{m,k}$, and thus that $V_p^m[a,b] \subseteq \text{Sat}(p,m)$:

THEOREM 7.17

For any $f \in V_p^m[a,b]$,

$$d(f,\mathcal{P}\mathcal{P}_{m,k})_p \leqslant \left(\frac{1}{k+1}\right)^m N_{p,m}(f)^{1/\sigma},$$

$\sigma = 1/(m+1/p)$.

Proof. We say that a partition $\Delta = \{a = x_0 < x_1 < \cdots < x_{k+1} = b\}$ is *balanced* provided that for some d_k,

$$d(f,\mathcal{P}_m)_{L_p[I_i]} = d_k, \qquad i = 0,1,\ldots,k,$$

where $I_i = [x_i, x_{i+1})$. We claim that for any given function f, there are balanced partitions for all $k = 0,1,\ldots$. To prove this, we proceed by induction. Clearly, there is a balanced partition for $k=0$. Now suppose $d_{k-1}(\alpha)$ corresponds to the balanced partition of $[a,\alpha]$ with $k-1$ knots. Let $\beta = \inf\{\alpha: d(f,\mathcal{P}_m)_{L_p[\alpha,b]} = d_{k-1}(\alpha)\}$. It follows that if $a = x_0 < x_1 < \cdots < x_k = \beta$ provides a balanced partition of $[a,\beta]$, then $a = x_0 < x_1 < \cdots < x_{k+1} = b$ provides a balanced partition of $[a,b]$.

To complete the proof, suppose Δ^* is a balanced partition of $[a,b]$ with k knots. Then

$$d(f,\mathcal{P}_m)_{L_p[I_i]}=\left(\frac{d[f,\mathcal{S}\mathcal{P}_m(\Delta^*)]^p_{L_p[a,b]}}{k+1}\right)^{1/p},$$

$i=0,1,\ldots,k$. But then

$$B_{p,m}(f,\Delta^*)=\sum_{i=0}^{k} d(f,\mathcal{P}_m)^\sigma_{L_p[I_i]}=(k+1)\left(\frac{d[f,\mathcal{S}\mathcal{P}_m(\Delta^*)]^p_p}{k+1}\right)^{\sigma/p}$$

$$=(k+1)^{m\sigma}d[f,\mathcal{S}\mathcal{P}_m(\Delta^*)]^\sigma_p.$$

We conclude that

$$d(f,\mathcal{S}\mathcal{P}_{m,k})_p\leqslant d[f,\mathcal{S}\mathcal{P}_m(\Delta^*)]_p\leqslant\frac{1}{(k+1)^m}B_{p,m}(f,\Delta^*)^{1/\sigma}$$

$$\leqslant\left(\frac{1}{k+1}\right)^m N_{p,m}(f)^{1/\sigma}. \qquad \blacksquare$$

We turn now to a discussion of the kinds of functions that lie in $V_p^m[a,b]$. In the case of $m=1$ and $p=\infty$, we can be very precise.

EXAMPLE 7.18

$V^1_\infty[a,b]=C[a,b]\cap BV[a,b]$.

Discussion. We have already seen that $V^1_\infty[a,b]\subseteq \mathrm{Sat}(\infty,1)=C[a,b]\cap BV[a,b]$. Suppose now that $f\in C[a,b]\cap BV[a,b]$. Then for any partition $\Delta=\{a=x_0<x_1<\cdots<x_{k+1}=b\}$ of $[a,b]$, we have

$$B_{\infty,1}(f,\Delta)=\sum_{i=0}^{k} d(f,\mathcal{P}_1)_{L_\infty[I_i]}\leqslant\sum_{i=0}^{k} V_{x_i}^{x_{i+1}}(f)\leqslant V_a^b(f).$$

It follows that $N_{\infty,1}(f)\leqslant V_a^b(f)$, and thus $f\in V^1_\infty[a,b]$. $\blacksquare$

We saw in §7.4 that the functions in $BV^{m-1}[a,b]$ can be approximated to order m by splines of order m. The following theorem provides another proof of this fact. It shows that $BV^{m-1}[a,b]\subseteq V_p^m[a,b]$.

THEOREM 7.19

Let $1 \leqslant p \leqslant \infty$. Then for any $f \in BV^{m-1}[a,b]$,

$$N_{p,m}(f)^{1/\sigma} \leqslant \frac{(b-a)^{1/\sigma - 1}}{(m-1)!} V_a^b(D^{m-1}f).$$

Proof. By a slight extension of the Taylor expansion in Theorem 2.1, there exists $p_f \in \mathcal{P}_m$ such that

$$f(x) = p_f(x) + \int_a^x \frac{(x-t)^{m-1} dD^{m-1}f(t)}{(m-1)!}.$$

It follows that

$$d(f, \mathcal{P}_m)_{L_p[a,b]} \leqslant \left(\int_a^b \left[\int_a^x \frac{|x-t|^{m-1}}{(m-1)!} |dD^{m-1}f(t)| \right]^p dx \right)^{1/p}$$

$$\leqslant \frac{(b-a)^{1/\sigma - 1}}{(m-1)!} V_a^b(D^{m-1}f).$$

A similar proof establishes the same inequality for $p = \infty$.

Now let Δ be any partition of $[a,b]$ into $k+1$ subintervals. Then

$$B_{p,m}(f,\Delta) = \sum_{i=0}^{k} d(f, \mathcal{P}_m)^{\sigma}_{L_p[I_i]}$$

$$\leqslant \left[\frac{1}{(m-1)!} \right]^{\sigma} \sum_{i=0}^{k} (x_{i+1} - x_i)^{1-\sigma} V_{x_i}^{x_{i+1}}(D^{m-1}f)^{\sigma}$$

$$\leqslant \left[\frac{1}{(m-1)!} \left[\sum_{i=0}^{k} (x_{i+1} - x_i) \right]^{1/\sigma - 1} \sum_{i=0}^{k} V_{x_i}^{x_{i+1}}(D^{m-1}f) \right]^{\sigma}$$

$$\leqslant \left[\frac{1}{(m-1)!} (b-a)^{1/\sigma - 1} V_a^b(D^{m-1}f) \right]^{\sigma}.$$

Taking the infimum over all Δ and then the σth root, we obtain the desired inequality. ∎

Theorem 7.19 shows that $V_p^m[a,b]$ contains a large class of smooth functions. Example 7.16 suggests that it may also contain functions (such as x^{α}) that are smooth except at a finite number of points. Our next theorem shows that this is indeed the case.

THEOREM 7.20

Suppose $1 \leqslant p \leqslant \infty$ and that $f \in C[a,b]$ if $p=\infty$ or $f \in L_p[a,b]$ otherwise. In addition, suppose $f \in L_1^m[a,b-\varepsilon]$ for all $\varepsilon > 0$, and that there exists a monotone-increasing function g such that $|D^m f(x)| \leqslant g(x)$ almost everywhere on (a,b). Then $f \in V_p^m[a,b]$, and, in fact,

$$N_{p,m}(f) \leqslant C_p \| g \|_{L_\sigma[a,b]}^{\sigma}, \qquad \sigma = \frac{1}{m+1/p}, \tag{7.45}$$

where $C_\infty = 1/m!$ and $C_p = (m!(mp+1)^{1/p})^{-1}$ for $1 \leqslant p < \infty$.

Proof. For any $a \leqslant c \leqslant x < b$ we have by the Taylor expansion that

$$|f(x) - p_f(x)| = \left| \int_c^x \frac{(x-t)^{m-1} D^m f(t)\, dt}{(m-1)!} \right|$$

$$\leqslant \varphi(x) := \int_c^x \frac{(x-t)^{m-1} g(t)}{(m-1)!}\, dt,$$

where $p_f(x) = \sum_{j=0}^{m-1} [D^j f(c)(x-c)^j]/j! \in \mathcal{P}_m$. It can be shown (cf. Lemma 7.21 below) that for any $c < d \leqslant b$, $\|\varphi\|_{L_p[c,d]} \leqslant C_p \| g \|_{L_\sigma[c,d]}$, and it follows that

$$d(f, \mathcal{P}_m)_{L_p[c,d]} \leqslant C_p \| g \|_{L_\sigma[c,d]}.$$

Now if $\Delta = \{a = x_0 < x_1 < \cdots < x_{k+1} = b\}$ is any partition of $[a,b]$, then

$$B_{p,m}(f,\Delta) = \sum_{i=0}^{k} d(f, \mathcal{P}_m)_{L_p[I_i]}^{\sigma} \leqslant (C_p)^{\sigma} \sum_{i=0}^{k} \| g \|_{L_\sigma[I_i]}^{\sigma} \leqslant (C_p)^{\sigma} \| g \|_{L_\sigma[a,b]}^{\sigma}.$$

Taking the supremum over all partitions, we have (7.45). ■

The way Theorem 7.20 is stated, it actually deals with functions that have a singularity at point b, but it is clear that a similar singularity at a could be handled in the same way. The function $f_\alpha(x) = x^\alpha$ on $[0,1]$ is then covered by the theorem since $\|D^m f\|_\sigma < \infty$ as long as $\alpha > -1/p$. The theorem also applies to functions with a finite number of singularities since in choosing the knots of the approximating piecewise polynomial we are free to put one at each singularity, thereby reducing the problem to separate problems of the type covered by the theorem.

The following lemma was needed in the proof of Theorem 7.20:

LEMMA 7.21

Let g be a positive monotone-increasing function on the interval $[c,d]$, and let

$$\varphi(x)=\int_c^x (x-t)^{m-1}g(t)\,dt. \tag{7.46}$$

Then for all $1 \leqslant p \leqslant \infty$,

$$\|\varphi\|_{L_p[c,d]} \leqslant d_p \|g\|_{L_\sigma[c,d]}, \qquad \sigma=\frac{1}{m+1/p} \tag{7.47}$$

where $d_\infty = 1/m$ and $d_p = 1/m(mp+1)^{1/p}$, $1 \leqslant p < \infty$.

Proof. We consider first the case of $p=\infty$ and $\sigma = 1/m$. Clearly (7.46) holds for $m=1$. Let

$$\psi(x)=\frac{1}{m}\left[\int_c^x |g(t)|^{1/m}\,dt\right]^m.$$

Both $\varphi(x)$ and $\psi(x)$ are monotone-increasing functions, and $\varphi(c)=\psi(c)$. We now compare their derivatives. We have

$$\frac{d\varphi(x)}{dx}=(m-1)\int_c^x (x-t)^{m-2}g(t)\,dt,$$

while

$$\frac{d\psi(x)}{dx}=\left[\int_c^x g(t)^{1/m}\,dt\right]^{m-1} g(x)^{1/m} \geqslant \left[\int_c^x g(t)^{1/(m-1)}\,dt\right]^{m-1}.$$

It follows from the inductive hypothesis that $d\varphi(x)/dx \leqslant d\psi(x)/dx$ for all x, and thus $\varphi(x) \leqslant \psi(x)$, all $c \leqslant x \leqslant d$. We have proved (7.47) in the case $p=\infty$.

Suppose now that $1 \leqslant p < \infty$, and let

$$\tilde{\varphi}(y)=\int_c^y \left|\int_c^x (x-t)^{m-1}g(t)\,dt\right|^p dx$$

$$\tilde{\psi}(y)=\frac{1}{m^p(mp+1)}\left[\int_c^y g(t)^\sigma\,dt\right]^{p/\sigma}.$$

Both of these functions are monotone-increasing functions of y, and $\tilde{\varphi}(c)=\tilde{\psi}(c)$. Now,

$$\frac{d\tilde{\varphi}(y)}{dy}=\left|\int_c^y (y-t)^{m-1}g(t)\,dt\right|^p$$

$$\frac{d\tilde{\psi}(y)}{dy}=\frac{p/\sigma}{m^p(mp+1)}\left[\int_c^y g(t)^\sigma\,dt\right]^{(p/\sigma)-1}g(y)^\sigma$$

$$\geqslant\frac{1}{m^p}\left[\int_c^y g(t)^{1/m}\,dt\right]^{mp}.$$

By the result for $p=\infty$, we conclude that $d\tilde{\varphi}(y)/dy\leqslant d\tilde{\psi}(y)/dy$, and thus $\tilde{\varphi}(y)\leqslant\tilde{\psi}(y)$ for all $c\leqslant y\leqslant d$. This statement for $y=d$ is precisely (7.47). ■

§ 7.7. HISTORICAL NOTES

Section 7.1

The study of approximation by piecewise polynomials with free knots began with the case of piecewise constants. Some early papers dealing with this problem include those by Kahane [1961], Ream [1961], and Stone [1961]. Piecewise linear polynomials and piecewise polynomials of order m were studied later by Brudnyi and Gopengauz [1963], Tihomirov [1965], Birman and Solomjak [1966, 1967], Sacks and Ylvisaker [1966, 1968, 1970], Phillips [1968], Rice [1969a], Freud and Popov [1969], and Subbotin and Chernyk [1970], as well as in a number of later papers that we mention below. The observation that it suffices to work with piecewise polynomials in order to get results for splines (cf. Theorem 7.1) seems to have been part of the folklore—see, for example, Rice [1969a], Burchard and Hale [1975], or Burchard [1977].

Section 7.2

The direct and inverse theorems given here for piecewise constants are credited to Kahane [1961] for $p=\infty$ and to Birman and Solomjak [1966] for $1\leqslant p<\infty$. The saturation theorem for $p<\infty$ follows from general results of Burchard [1977].

Section 7.3

The idea of introducing a new kind of modulus of smoothness in order to deal with spline approximation with free knots is credited to Popov

[1975a, b], where only the case of $p=\infty$ was considered. Our definition of $\nu_m(f;t)$ is a minor variant of his. The introduction of $\nu_m(f;t)_p$ for $p<\infty$ seems to be new.

Section 7.4

Our main direct and inverse Theorem 7.10 follows the ideas of Popov [1975a, b], where the case of $p=\infty$ is done. Direct theorems for functions in $C^m[a,b]$ or $L_p^m[a,b]$ were obtained in a number of early papers—see Brudnyi and Gopengauz [1963], Tihomirov [1965], Birman and Solomjak [1966, 1967], and Brudnyi [1971]. Direct theorems for $BV^{m-1}[a,b]$ were established in the papers by Subbotin and Chernyk [1970], Freud and Popov [1969], and Sendov and Popov [1970a, b].

Section 7.5

Lemma 7.13 is credited to Phillips [1970], while our main saturation result Theorem 7.14 comes from the work of Burchard and Hale [1975]. The result for $L_1^m[a,b]$ functions follows from certain asymptotic formulae derived by Burchard and Hale [1975]—see Remark 7.2.

Section 7.6

Example 7.16 is credited to Rice [1969a]. It motivated Dodson [1972] and deBoor [1973b] to establish order m convergence for the kind of functions found in Theorem 7.20. Sat(p,m) was studied in greater detail by Burchard [1974], Burchard and Hale [1975], and Burchard [1977]. The spaces $V_p^m[a,b]$ are introduced in the paper by Burchard and Hale [1975], and Theorem 7.17 is established there. These papers also contain exact asymptotic formulae (see Remark 7.2), and the 1977 paper by Burchard also includes certain inverse theorems (cf. Remark 7.4). Theorem 7.20 (as well as a refined version of it) can be found in the paper by Burchard and Hale [1975]. Lemma 7.21 is credited to Dodson [1972] and deBoor [1973b]. We also note that there have been several papers giving little oh saturation classes—see Remark 7.3. Spaces related to $V_p^m[a,b]$ have also been introduced by Brudnyi [1974] and by Bergh and Peetre [1974].

§ 7.8. REMARKS

Remark 7.1

A Frechét space is a complete linear space X with an invariant metric ρ; that is, a functional $\rho: X\to\mathbf{R}$ such that $\rho(f+g)\leqslant\rho(f)+\rho(g)$, $\rho(\alpha f)=|\alpha|\rho(f)$, and $\rho(f-g,0)=\rho(f)-\rho(g)$. A Frechét space is a Banach space if ρ is a norm. See, for example Dunford and Schwartz [1957].

Remark 7.2

A number of authors have obtained exact asymptotic expressions for $d(f,\mathcal{P}\mathcal{P}_{m,k})_p$ for certain classes of smooth functions. For example, Burchard and Hale [1975] have shown that

$$\lim_{k\to\infty}(k+1)^m d(f,\mathcal{P}\mathcal{P}_{m,k})_p = B_{p,m}(f)^{1/\sigma}, \qquad \sigma=\frac{1}{m+1/p} \tag{7.48}$$

$$B_{p,m}(f)=\overline{\lim_{\bar{\Delta}\to 0}}\, B_{p,m}(f,\Delta)$$

for all functions in the linear space

$$AC_p^{m-1}[a,b]=\text{closure } L_1^m[a,b] \text{ in } V_p^m[a,b]. \tag{7.49}$$

It is also shown there that $AC_p^{m-1}[a,b]$ is a Frechét space, and that it contains any $f\in L_1^m[a,b]$ (and, in fact, any f satisfying the hypotheses of Theorem 7.20). For such f they show that (7.48) holds with

$$B_{p,m}(f)=d\left(\frac{x^m}{m!},\mathcal{P}_m\right)_{L_p[0,1]}^{\sigma}\|D^m f\|_{L_\sigma[a,b]}^{\sigma}.$$

It follows from (7.48) in this case that if $d(f,\mathcal{P}\mathcal{P}_{m,k})_p = o[(b-a)/(k+1)]^m$, then $\|D^m f\|_\sigma = 0$, hence $f\in\mathcal{P}_m$. This is a strengthened version of the saturation Theorem 7.14. When $\sigma<1$, Burchard [1977] has shown that $BV^{m-1}[a,b]\subseteq AC_p^{m-1}[a,b]$, and that if $f\in BV^{m-1}[a,b]$ is such that $D^m f(x)=0$ almost everywhere, then $B_{p,m}(f)=0$, hence

$$\lim_{k\to\infty}(k+1)^m d(f,\mathcal{P}\mathcal{P}_{m,k})_p = 0.$$

For $m=1$ and $p=\infty$, the asymptotic formula (7.48) asserts that

$$\lim_{k\to\infty}(k+1)d(f,\mathcal{P}\mathcal{P}_{1,k})_\infty = \tfrac{1}{2}\left[V_a^b(f)\right], \qquad \text{all } f\in C[a,b],$$

a result established by Kahane [1961]. For $m=2$ and $p=2$, (7.48) was established by Ream [1961]. For $p=2$ and general m, see Sacks and Ylvisaker [1966, 1968, 1969, 1970]. For some other special cases, see Phillips [1970], McClure [1975], Dodson [1972], and deBoor [1973b].

Remark 7.3

Burchard [1977] has shown that

$$f\in AC_p^{m-2}[a,b] \qquad \text{implies } d(f,\mathcal{P}\mathcal{P}_{m,k})_p = o\left(\frac{b-a}{k+1}\right)^{m-1}. \tag{7.50}$$

This shows that little oh convergence of order $m-1$ occurs for a very large class of functions. He also shows that

$$AC_p^{m-1}[a,b] \subseteq V_p^m[a,b] \subseteq AC_p^{m-2}[a,b].$$

A weaker version of (7.50) was established by Freud and Popov [1970], who had in turn extended a theorem of Korneicuk for $f \in \mathrm{Lip}^1$ with $m=2$ and $p=\infty$. Since $AC_\infty^0[a,b] = AC[a,b]$, (7.50) improves Korneicuk's result.

Remark 7.4

Burchard [1977] also contains several inverse theorems. For example, it is shown that

$$d(f, \mathcal{P}\mathcal{S}_{m,k})_p = \mathcal{O}\left(\frac{b-a}{k+1}\right)^{m-1+\theta}, \qquad \theta > 0 \qquad \text{implies } f \in AC_p^{m-2}[a,b]. \tag{7.51}$$

For other results of this type as well as an analog of the K-functional, Stechkin-type results, connections with generalized derivatives, and results on asymptotically optimal knot distributions, see Burchard [1977]. For some different inverse theorems, see Brudnyi [1974] and Bergh and Peetre [1974].

Remark 7.5

The following alternation theorem of Tchebycheff can be found in any book on classical approximation theory: Suppose $f \in C[a,b]$ and that $\mathcal{U}_m$ is a T space. Then $u \in \mathcal{U}_m$ is the best approximation of f in the uniform norm on $[a,b]$ if and only if

$$(f-u)(t_i) = \|f-u\|_{L_\infty[a,b]}, \qquad i = 0, 1, \ldots, m$$

for some $a \leqslant t_0 < t_1 < \cdots < t_m \leqslant b$.

Remark 7.6

Best approximation in the p-norm, $1 \leqslant p < \infty$, can be characterized by a certain orthogonality condition. We have the following well-known result; see, for example, Holmes [1972], p. 77: Let $f \in L_p[a,b]$, and suppose $\mathcal{U}_m = \mathrm{span}\{u_i\}_1^m$. Then $u \in \mathcal{U}_m$ is the unique best approximation of f in the L_p-norm on $[a,b]$ if and only if

$$\int_a^b |(f-u)(t)|^{p-1} \mathrm{sgn}(f-u)(t) u_i(t) = 0, \qquad i = 1, 2, \ldots, m.$$

8

OTHER SPACES OF POLYNOMIAL SPLINES

In this chapter we discuss several spaces of piecewise polynomial functions that have proved useful in applications. These include periodic splines, natural splines, g-splines, monosplines, and discrete splines.

§ 8.1. PERIODIC SPLINES

In many problems in applied mathematics it is necessary to approximate a function that is known to be periodic. As it is usually desirable to work with periodic approximation functions in such cases, we devote this section to the study of a space of periodic splines.

Let $a<b$. By identifying b with a, we may regard the interval $[a,b)$ as a circle with circumference $L=b-a$. Now given $\Delta=\{a<x_1<x_2<\cdots<x_k<b\}$, we may think of Δ as partitioning the circle into k subintervals, $I_i=[x_i,x_{i+1})$, $i=1,2,\ldots,k-1$, and $I_k=[x_k,x_1)$. Given $m\geqslant 1$ and $\mathfrak{M}=(m_1,\ldots,m_k)$, a vector of integers with $1\leqslant m_i\leqslant m$, $i=1,2,\ldots,k$, we define

$$\mathring{\mathcal{S}}(\mathcal{P}_m;\mathfrak{M};\Delta)=\left\{\begin{array}{l}s\text{: there exist polynomials } s_1,\ldots,s_k \text{ of order } m\\ \text{so that } s(x)=s_i(x) \text{ on } I_i,\ i=1,2,\ldots,k \text{ and}\\ D^{j-1}s_{i-1}(x_i)=D^{j-1}s_i(x_i),\ j=1,\ldots,m-m_i,\ i=\\ 1,2,\ldots,k, \text{ where we take } s_0=s_k\end{array}\right\}. \tag{8.1}$$

We call $\mathring{\mathcal{S}}$ the space of *periodic polynomial splines of order m with knots at* $x_1,\ldots,x_k$ *of multiplicity* $m_1,\ldots,m_k$.

It is clear that $\mathring{\mathcal{S}}(\mathcal{P}_m;\mathfrak{M};\Delta)$ is a linear space of piecewise polynomials. In fact, it follows directly from the definition that it is a linear subspace of the space of polynomial splines defined in (4.1). We have

$$\mathring{\mathcal{S}}(\mathcal{P}_m;\mathfrak{M};\Delta)=\left\{\begin{array}{c}s\in\mathcal{S}(\mathcal{P}_m;\mathfrak{M};\Delta)\text{: } s^{(j)}(b)=s^{(j)}(a),\\ j=0,1,\ldots,m-1\end{array}\right\}, \tag{8.2}$$

This observation will be of value in dealing with periodic splines on a computer.

Since $\mathcal{S}(\mathcal{P}_m;\mathfrak{M};\Delta)$ is of dimension $m+K$ with $K=\sum_{i=1}^k m_i$, it is to be expected from (8.2) that the dimension of $\mathring{\mathcal{S}}(\mathcal{P}_m;\mathfrak{M};\Delta)$ is K. The following theorem verifies this:

THEOREM 8.1

Let $K=\sum_{i=1}^k m_i$. Then

$$\mathring{\mathcal{S}}(\mathcal{P}_m;\mathfrak{M};\Delta) \qquad \text{is a linear space of dimension } K. \tag{8.3}$$

Proof. The proof follows the same pattern as for Theorem 4.4, with

$$A\mathbf{c}=\begin{bmatrix} A_1 & & & & -A_1 \\ & A_2 & -A_2 & & \\ & & \cdots & & \\ & & & A_k & -A_k \end{bmatrix}\begin{bmatrix} c_1 \\ \vdots \\ c_k \end{bmatrix}=0.$$

Now the matrix A is a transformation of $\mathbf{R}^{mk}$ into $\mathbf{R}^{\Sigma_1^k(m-m_i)}$ with rank equal to the number of rows $\Sigma_1^k(m-m_i)$. It follows that the dimension of $\mathring{\mathcal{S}}$ is $mk-\sum_{i=1}^k(m-m_i)=K$. ■

Following the blueprint of Chapter 4, we turn to the construction of a basis for $\mathring{\mathcal{S}}(\mathcal{P}_m;\mathfrak{M};\Delta)$. While it is possible to deal with the case of $K\leqslant m$, we shall consider only the case of $K>m$. This will permit us to find a basis of B-splines. Let

$$y_{m+1},\dots,y_{m+K}=\overbrace{x_1,\dots,x_1}^{m_1},\dots,\overbrace{x_k,\dots,x_k}^{m_k}, \tag{8.4}$$

and

$$y_i=y_{i+K}-L, \qquad y_{K+m+i}=y_{m+i}+L, \qquad i=1,2,\dots,m, \tag{8.5}$$

where $L=b-a$. We define the *periodic B-splines* by

$$\mathring{N}_i^m(x)=N_i^m(x), \qquad i=m+1,\dots,K, \tag{8.6}$$

and

$$\mathring{N}_i^m(x)=\begin{cases} N_i^m(x), & a\leqslant x<y_{i+m}, \\ N_{i+K}^m(x), & y_{i+m}\leqslant x<b, \end{cases} \qquad i=1,2,\dots,m. \tag{8.7}$$

THEOREM 8.2

The periodic B-splines $\{\mathring{N}_i^m\}_1^K$ form a basis for $\mathring{\mathcal{S}}(\mathcal{P}_m;\mathfrak{M};\Delta)$.

Proof. Each of the B-splines $\mathring{N}_{m+1}^m,\ldots,\mathring{N}_K^m$ is automatically periodic since it vanishes outside $[x_1,x_k]$. By the choice of the y's in (8.5), we observe that $N_i^m(x)=N_{i+K}^m(x+L)$, $i=1,2,\ldots,m$. It follows that the $\mathring{N}_i^m$ are also periodic for $i=1,2,\ldots,m$. It remains to check that these K functions are linearly independent. But this follows from Theorem 4.18, and, in fact, on each subinterval I_i defined by the partition, the set of all B-splines with values in the interval are linearly independent. ■

Theorem 8.2 asserts that each periodic spline $s\in\mathring{\mathcal{S}}(\mathcal{P}_m;\mathfrak{M};\Delta)$ can be written uniquely in the form

$$s(x)=\sum_{i=1}^{K} c_i\mathring{N}_i^m(x). \tag{8.8}$$

Thus to store and work with a periodic spline on the computer, it would suffice to store the vector $\mathbf{c}=(c_1,\ldots,c_K)$. In practice, however, it is more convenient to work with a representation involving nonperiodic B-splines. The next theorem shows how:

THEOREM 8.3

Let $s=\sum_{i=1}^K c_i\mathring{N}_i^m$, and define $c_{i+K}=c_i$, $i=1,2,\ldots,m$. Then s can also be uniquely written in the form

$$s=\sum_{i=1}^{m+K} c_iN_i^m. \tag{8.9}$$

Proof. This follows directly from the definition of the $\mathring{N}_i$'s. ■

Theorem 8.3 shows that periodic splines can be manipulated on a computer using the algorithms developed in Chapter 5 for nonperiodic splines. In particular, we may use these routines to evaluate the spline and its derivatives, to convert to a piecewise polynomial representation, and to find antiderivatives and integrals.

We turn now to zero properties of periodic splines. Given $s\in\mathring{\mathcal{S}}(\mathcal{P}_m;\mathfrak{M};\Delta)$, let $\mathring{Z}(s)$ denote the number of zeros of s on the circle $[a,b]$, counting multiplicities as in Definition 4.46. The $\circ$ on this notation is to remind us that we are working on a circle so there are no end intervals; that is, if s vanishes on $[a,\bar{t})$ and on $[\underline{t},b)$, then this should be considered as just one interval $[\underline{t},\bar{t})$. We also observe that since this counting procedure is defined so that s changes sign at odd order zeros and does not change sign at even order ones, $\mathring{Z}(s)$ must always be an even number.

THEOREM 8.4

If $s \in \mathring{\mathcal{S}}(\mathcal{P}_m; \mathfrak{M}; \Delta)$ and is not identically zero, then

$$\mathring{Z}(s) \leqslant \begin{cases} K-1 & \text{if } K \text{ is odd} \\ K & \text{if } K \text{ is even.} \end{cases}$$

Proof. For $m=1$ we are dealing with periodic piecewise constant functions whose only possible jump points are at the $x_1, \ldots, x_K$, and the result is easily checked. Now we may proceed by induction on m, using Rolle's theorem just as in the proof of Theorem 4.53. ∎

We give two simple examples to illustrate the difference between K odd and K even.

EXAMPLE 8.5

Let $m=2$, $\Delta = \{1/2, 3/2\}$, and let $s(x) = -x + 2(x-\frac{1}{2})_+ - 2(x-3/2)_+$, where $[a,b]=[0,2]$.

Discussion. The spline s is illustrated in Figure 20. It belongs to $\mathring{\mathcal{S}}_2(\Delta)$, has two knots, and has two zeros. ∎

EXAMPLE 8.6

Let $m=2$, $K=3$, and let $\Delta = \{1/2, 3/2, 5/2\}, [a,b]=[0,3]$.

Discussion. Every periodic spline in $\mathring{\mathcal{S}}_2(\Delta)$ has at most two zeros on $[0,3]$. A typical spline in this class is shown in Figure 21. ∎

In Section 4.8 we were able to use the basic bound on the number of zeros of a polynomial spline to establish important results about the nonsingularity of determinants formed from B-splines. We can do the

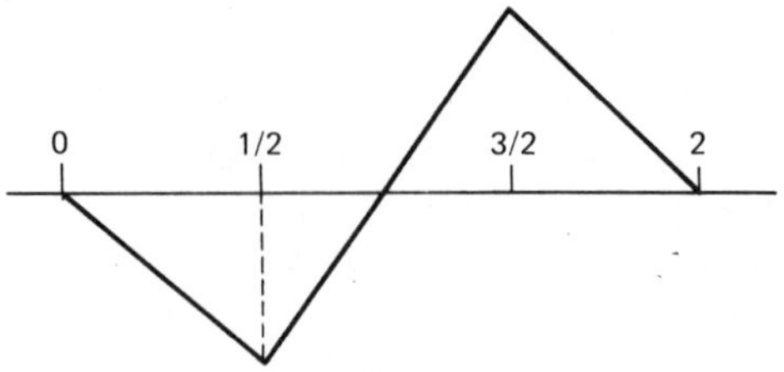

Figure 20. The spline in Example 8.5.

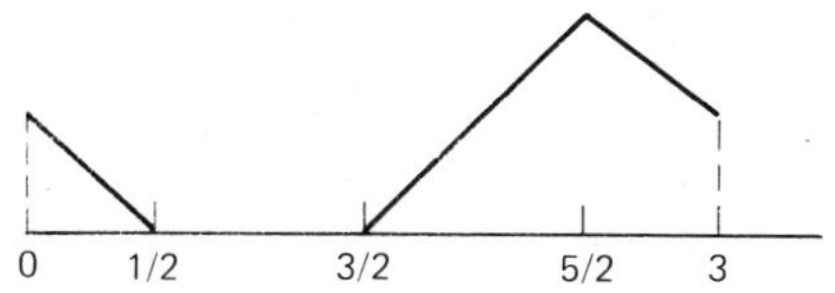

Figure 21. The spline in Example 8.6.

same for periodic splines in the case where K is odd. An analogous result cannot be established for K even because our zero bound is weaker in this case—see Example 8.10.

We begin by posing a periodic version of the Hermite interpolation Problem 4.69.

PROBLEM 8.7. Periodic Hermite Interpolation by Splines

Let $t_1 \leqslant t_2 \leqslant \cdots \leqslant t_K$ be points on the circle $[a,b)$ of circumference $L=b-a$. Suppose $t_K - t_1 < L$. Let $\theta_1,\ldots,\theta_K$ be a sequence of signs, and let $\mathring{N}_1^m,\ldots,\mathring{N}_K^m$ be the set of periodic B-splines of Theorem 8.2. For $i=1,2,\ldots,K$, let

$$e_i = \begin{cases} 0, & \text{if } t_i \notin \Delta \\ m_j, & \text{if } t_i = x_j \in \Delta. \end{cases} \tag{8.10}$$

The integer e_i counts the number of knots equal to t_i, if any. Define

$$d_i = \begin{cases} m - e_i + \max\{j:\ t_{i+j} = \cdots = t_i \text{ and } \theta_{i+j} = \cdots = \theta_i\}, & \text{if } \theta_i = - \\ \max\{j:\ t_i = \cdots = t_{i-j} \text{ and } \theta_i = \cdots = \theta_{i-j}\}, & \text{if } \theta_i = + \end{cases}. \tag{8.11}$$

We assume that the t's and θ's are chosen so that

$$0 \leqslant d_i \leqslant m-1; \tag{8.12}$$

$$\text{if } \theta_i = + \text{ and } \theta_{i+1} = -, \qquad \text{then } t_i < t_{i+1}; \tag{8.13}$$

$$\text{if } \theta_i = - \text{ and } \theta_{i+1} = +, \qquad \text{then } t_{i+m-e_i} = \cdots = t_{i+1} = t_i. \tag{8.14}$$

The Hermite interpolation problem is as follows: Given $\{v_i\}_1^K$, find $s \in \mathring{\mathcal{S}}(\mathcal{P}_m; \mathfrak{M}; \Delta)$ such that

$$D_{\theta_i}^{d_i} s(t_i) = v_i, \qquad i = 1,2,\ldots,K. \tag{8.15}$$

THEOREM 8.8

Let K be odd, and let $\mathring{B}_i = \mathring{N}_i^m$, $i = 1,2,\ldots,K$ be periodic B-splines as in Theorem 8.2. Given $\{t_i\}_1^K$ and $\{\theta_i\}_1^K$ as in Problem 8.7, let

$$\hat{D}\begin{bmatrix} t_1,\ldots,t_K \\ \theta_1,\ldots,\theta_K \\ \mathring{B}_1,\ldots,\mathring{B}_K \end{bmatrix} = \det\left[\theta_i^{d_i} D_{\theta_i}^{d_i} \mathring{B}_j(t_i)\right]_{i,j=1}^K. \tag{8.16}$$

Then, for all $1 \leqslant q \leqslant K$,

$$\hat{D}\begin{bmatrix} t_1,\ldots,t_K \\ \theta_1,\ldots,\theta_K \\ \mathring{B}_q,\ldots,\mathring{B}_K,\mathring{B}_1,\ldots,\mathring{B}_{q-1} \end{bmatrix} = \hat{D}\begin{bmatrix} t_1,\ldots,t_K \\ \theta_1,\ldots,\theta_K \\ \mathring{B}_1,\ldots,\mathring{B}_K \end{bmatrix} \geqslant 0. \tag{8.17}$$

Moreover, the determinant is positive if and only if for some choice of $1 \leqslant q \leqslant K$, the B-splines

$$\mathring{B}_1^*,\ldots,\mathring{B}_K^* = \mathring{B}_q,\ldots,\mathring{B}_K,\mathring{B}_1,\ldots,\mathring{B}_{q-1}$$

satisfy

$$t_i \in \sigma_i^* = (y_i^*, y_{i+m}^*) \cup \left\{x: D_{\theta_i}^{d_i} \mathring{B}_i^*(x) \neq 0\right\}, \qquad i = 1,2,\ldots,K, \tag{8.18}$$

where $y_i^*,\ldots,y_{i+m}^*$ are the knots (thought of as lying on the circle) associated with $\mathring{B}_i^*$, $i = 1,2,\ldots,K$.

Proof. Since K is odd, there is no change in the value of the determinant if we change the order of its columns from $1,2,\ldots,K$ to $K,1,2,\ldots,K-1$. This can be repeated as often as desired, and the equality of determinants asserted in (8.17) is established.

The proof that $\hat{D} \neq 0$ when (8.18) holds can be carried out using the same ideas as in the proof of Theorem 4.71. In particular, if $\hat{D} = 0$ and (8.18) holds, then we can find $\{c_i\}_1^K$, not all zero, so that

$$D_{\theta_i}^{d_i} s(t_i) = \sum_{j=1}^K c_j D_{\theta_i}^{d_i} \mathring{B}_j(t_i) = 0, \qquad i = 1,2,\ldots,K.$$

If s does not vanish on any subinterval of the circle, then it is a spline in $\mathring{\mathcal{S}}(\mathcal{P}_m; \mathfrak{M}; \Delta)$ with K zeros counting multiplicities, (no two zeros can be swallowed up in an interval where s vanishes in this case). This contradicts

Theorem 8.4. On the other hand, if s does vanish on some interval, then we may arrive at a contradiction using ordinary B-splines just as before.

The proof that $\hat{D}$ is positive when (8.18) holds can also be established along the lines of the proof of Theorem 4.72 via a limiting argument. We turn now to the proof of the fact that $\hat{D}=0$ whenever (8.18) fails. This is more delicate than in the nonperiodic case.

Suppose (8.18) fails. In particular, suppose $\mathring{B}_1,\ldots,\mathring{B}_K$ have been renumbered so that $t_i \in \sigma_i$, $i=1,2,\ldots,p<K$, with p maximal. Since p is maximal, t_{p+1} must either be too far right or too far left (on the circle) to belong to σ_{p+1}. We consider these two cases separately.

CASE 1. Suppose t_{p+1} is too far right for $\mathring{B}_{p+1}$. Let $\mathring{B}_r$ be the first B-spline following $\mathring{B}_{p+1}$ so that

$$t_{p+1} \in \sigma\left(D_{\theta_{p+1}}^{d_{p+1}} \mathring{B}_r\right) = (y_r, y_{r+m}) \cup \left\{x\colon D_{\theta_{p+1}}^{d_{p+1}} \mathring{B}_r(x) \neq 0\right\}.$$

Renumber the $\mathring{B}$'s as $\tilde{B}_1,\ldots,\tilde{B}_K$ so that $\tilde{B}_{p+1} = \mathring{B}_r$, and consider the matrix

$$\tilde{M} = \left[\theta_i^{d_i} D_{\theta_i}^{d_i} \tilde{B}_j(t_i)\right]_{i,j=1}^{K}.$$

Since p is maximal, $t_j \notin \sigma(D_{\theta_j}^{d_j} \tilde{B}_j)$ for some $1 \leqslant j \leqslant p$. Since the support sets of $\{\tilde{B}_j\}$ are further right than those of the $\{\mathring{B}_j\}$, this can happen only if for some $1 \leqslant j \leqslant p$, t_j is too far left for $\tilde{B}_j$. But then columns j through p of $\hat{M}$ have nonzero entries only in rows $j+1$ through p, and it follows that $\hat{D} = \det(\tilde{M}) = 0$.

CASE 2. Suppose t_{p+1} is too far left for $\mathring{B}_{p+1}$. Let $\mathring{B}_l$ be the first B-spline preceding $\mathring{B}_{p+1}$ so that $t_{p+1} \in \sigma(D_{\theta_{p+1}}^{d_{p+1}} \mathring{B})$. Renumber the $\mathring{B}_1,\ldots,\mathring{B}_K$ as $\tilde{B}_1,\ldots,\tilde{B}_K$ so that $\tilde{B}_{p+1} = \mathring{B}_l$. Since p is maximal, $t_j \notin \sigma(D_{\theta_j}^{d_j} \tilde{B}_j)$ for some $1 \leqslant j \leqslant p$. Since we have shifted the B-splines so that their supports lie further to the left, we conclude that for some $1 \leqslant j \leqslant p$, t_j is too far right for $\tilde{B}_j$. It follows that the columns 1 through j and $p+2$ through K of $\tilde{M}$ have zero entries in rows j through $p+1$. This means that this set of $K-p+j-1$ columns has nonzero entries in at most $K-p+j-2$ rows, and $\hat{D} = \det(\tilde{M}) = 0$. ■

The need to introduce the integer q in the statement of Theorem 8.8 is due to the fact that on the circle there is no natural way to choose the numbering of the t's. In fact, (8.18) can hold for more than one choice of q, as the following example shows.

EXAMPLE 8.9

Let $[a,b) = [0,3)$, $m=2$, $K=3$, and $\Delta = \{\frac{1}{2}, \frac{3}{2}, \frac{5}{2}\}$. Let $t_1 = \frac{1}{4}$, $t_2 = 1$, and $t_3 = 2$.

Discussion. In this case (8.18) holds with $q=0$. It is easily checked that it also holds for $q=1$. We have

$$\begin{vmatrix} \mathring{B}_1(t_1)\mathring{B}_2(t_1)\mathring{B}_3(t_1) \\ \mathring{B}_1(t_2)\mathring{B}_2(t_2)\mathring{B}_3(t_2) \\ \mathring{B}_1(t_3)\mathring{B}_2(t_3)\mathring{B}_3(t_3) \end{vmatrix} = \begin{vmatrix} \mathring{B}_2(t_1)\mathring{B}_3(t_1)\mathring{B}_1(t_1) \\ \mathring{B}_2(t_2)\mathring{B}_3(t_2)\mathring{B}_1(t_2) \\ \mathring{B}_2(t_3)\mathring{B}_3(t_3)\mathring{B}_1(t_3) \end{vmatrix} = \frac{1}{4}.$$

See Figure 22. ■

The following example shows that Theorem 8.8 cannot hold in the case where K is even:

EXAMPLE 8.10

Let $[a,b)=[0,2)$, $m=2$, $K=2$, and $\Delta=\{1/2,3/2\}$.

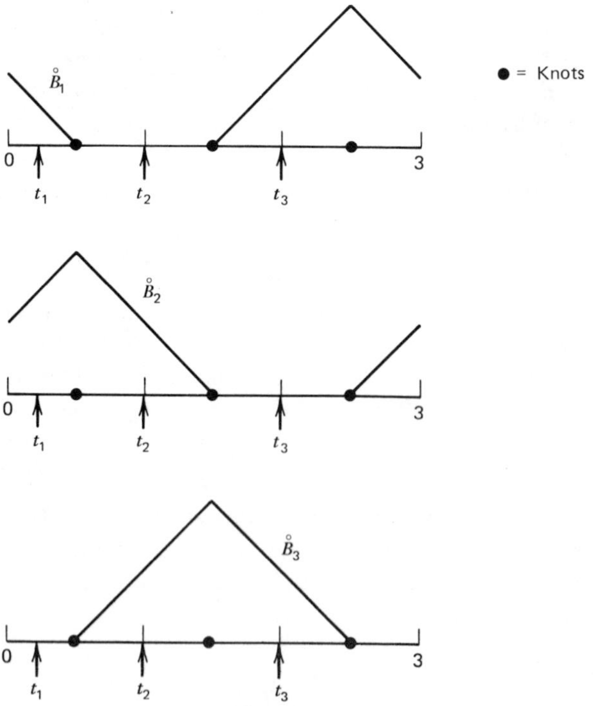

Figure 22. B-Splines for Example 8.9.

Discussion. The two periodic B-splines forming a basis for $\mathring{\mathcal{S}}_2(\Delta)$ are shown in Figure 23. It is clear that $t_1=0, t_2=1$ satisfy (8.18), but

$$\hat{D}\begin{pmatrix} t_1, t_2 \\ \mathring{B}_1, \mathring{B}_2 \end{pmatrix} = \begin{vmatrix} \frac{1}{2} & \frac{1}{2} \\ \frac{1}{2} & \frac{1}{2} \end{vmatrix} = 0. \qquad \blacksquare$$

In the remainder of this section we discuss the approximation power of periodic splines. Our main tool for obtaining direct theorems will be a periodic analog of the local spline approximation operator Q defined in Theorem 6.18. Given a partition Δ of the circle $[a,b)$ into k subintervals, it is clear that the method of Lemma 6.17 can be applied to choose a thinned-out partition Δ^* with

$$\frac{\overline{\Delta}}{2} \leqslant \underline{\Delta}^* \leqslant \overline{\Delta}^* \leqslant \frac{3\overline{\Delta}}{2},$$

where $\underline{\Delta}$ and $\overline{\Delta}$ are the lengths of the smallest and largest subintervals of the circle $[a,b)$ corresponding to Δ, while $\underline{\Delta}^*$ and $\overline{\Delta}^*$ are the analogous quantities corresponding to Δ^*.

Suppose we number the points in Δ^* as $\Delta^* = \{a < y_{m+1} < \cdots < y_{m+k^*} < b\}$. Associated with Δ^* define the extended partition $\Delta_e^* = \{y_i\}_1^{2m+k^*}$ by

$$y_i = y_{i+k^*} - L, \qquad y_{i+m+k^*} = y_{m+i} + L, \qquad i = 1, 2, \ldots, m,$$

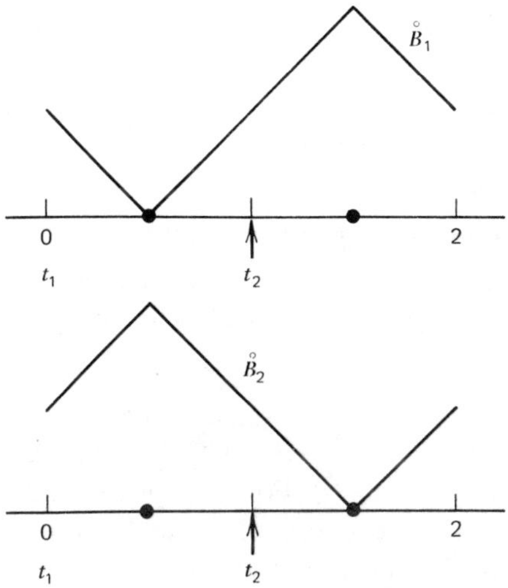

Figure 23. The B-splines for Example 8.10.

where $L=b-a$. Let $\{B_i\}_1^{m+k^*}$ be the usual B-splines associated with the extended partition Δ_e^*, and let $\{\mathring{B}_i\}_1^{k^*}$ be the corresponding periodic B-splines [cf. (8.6)–(8.7)]. Finally, let $\{\tau_{ij}\}_{j=1,i=1}^{m\ \ m+k^*}$ and $\{\alpha_{ij}\}_{j=1,i=1}^{m\ \ m+k^*}$ be defined as in (6.37) and (6.38).

We are ready to define our spline approximation operator. Given any periodic function defined on $[a,b]$, let

$$\mathring{Q}f(x)=\sum_{i=1}^{k^*}\sum_{j=1}^{m}\left(\alpha_{ij}[\tau_{i1},\ldots,\tau_{ij}]\mathring{f}\,\right)\mathring{B}_i(x), \tag{8.19}$$

where $\mathring{f}$ is the periodic extension of f to $[a-L,b+L]$ defined by

$$\mathring{f}\,(x)=\begin{cases} f(x+L), & a-L\leqslant x\leqslant a\\ f(x), & a\leqslant x\leqslant b\\ f(x-L), & b\leqslant x\leqslant b+L\end{cases}.$$

The following theorem gives several basic properties of $\mathring{Q}$:

THEOREM 8.11

$\mathring{Q}$ is a linear operator mapping the periodic functions on $[a,b]$ into $\mathring{\mathcal{S}}_m(\Delta^*)\subseteq\mathring{\mathcal{S}}_m(\Delta)$. It can also be written in the form

$$\mathring{Q}f(x)=\sum_{i=1}^{m+k^*}\sum_{j=1}^{m}\left(\alpha_{ij}[\tau_{i1},\ldots,\tau_{ij}]\mathring{f}\,\right)B_i(x),\qquad a\leqslant x\leqslant b, \tag{8.20}$$

Moreover, $\mathring{Q}$ reproduces polynomials locally. More precisely, if $\mathring{f}$ is a polynomial of order m on an interval of the form $[y_{l+1-m},y_{l+m}]$, then $\mathring{Q}f(x)=f(x)$ for all $y_l\leqslant x\leqslant y_{l+1}$, any $m+1\leqslant l\leqslant m+k^*$.

Proof. It is clear from the definition that $\mathring{Q}f\in\mathring{\mathcal{S}}_m(\Delta^*)\subseteq\mathring{\mathcal{S}}_m(\Delta)$. The alternate expression (8.20) for $\mathring{Q}f$ follows from Theorem 8.3. The reproductive property is then a consequence of Theorem 6.18 and the local character of $\mathring{Q}$. ■

It is now clear that Theorem 6.20 and Corollary 6.21 can be applied to give bounds for $D^r(f-\mathring{Q}f)$ and for $D^r\mathring{Q}f$. These results translate immediately into error bounds on $\mathring{Q}f$. For example, Corollary 6.21 yields the following theorem, where the spaces $\mathring{C},\mathring{L}_p^\sigma$ and the periodic modulus of continuity are as defined in Section 2.11:

THEOREM 8.12

Let Δ be an arbitrary partition of the interval $[a,b]$, and suppose $1 \leqslant \sigma \leqslant m$. Then for all $f \in \mathring{C}^{\sigma-1}[a,b]$,

$$\left.\begin{array}{ll} \|D^r(f-\mathring{Q}f)\|_\infty, & r=0,\ldots,\sigma-1 \\ \|D^r\mathring{Q}f\|_\infty, & r=\sigma,\ldots,m-1 \end{array}\right\} \leqslant C_1(\bar{\Delta})^{\sigma-r-1}\mathring{\omega}(D^{\sigma-1}f,\bar{\Delta}). \tag{8.21}$$

Moreover, if $f \in \mathring{L}_p^\sigma[a,b]$ and $1 \leqslant p \leqslant q \leqslant \infty$, then

$$\left.\begin{array}{ll} \|D^r(f-\mathring{Q}f)\|_q, & r=0,1,\ldots,\sigma-1 \\ \|D^r\mathring{Q}f\|_q, & r=\sigma,\ldots,m-1. \end{array}\right\} \leqslant C_2(\bar{\Delta})^{\sigma-r+1/q-1/p}\|D^\sigma f\|_p. \tag{8.22}$$

Since Theorem 8.12 involves estimates for derivatives, we can restate it in terms of the Sobolev norm (cf. Corollary 6.26).

COROLLARY 8.13

Suppose $0 \leqslant r \leqslant \sigma-1 \leqslant m-1$. Then there exists a constant C such that for all $f \in \mathring{L}_p^\sigma[a,b]$,

$$\|f-\mathring{Q}f\|_{\mathring{L}_q^r[a,b]} \leqslant C(\bar{\Delta})^{\sigma-r+1/q-1/p}\|f\|_{\mathring{L}_p^\sigma[a,b]},$$

$1 \leqslant p \leqslant q \leqslant \infty$.

We can also give an estimate for $\mathring{L}_p[a,b]$ functions.

THEOREM 8.14

Suppose $1 \leqslant p < \infty$. Then there exists a constant C such that for all $f \in \mathring{L}_p[a,b]$,

$$d\left[f, \mathring{\mathcal{S}}_m(\Delta)\right]_p \leqslant C_1\mathring{\omega}_m(f;\bar{\Delta})_p. \tag{8.23}$$

If $f \in \mathring{C}[a,b]$, the same inequality holds with $p=\infty$.

Proof. For every $g \in \mathring{L}_p^m[a,b]$, Theorem 8.12 guarantees that $s=\mathring{Q}g$ satisfies

$$\|f-s\|_p \leqslant \|f-g\|_p + \|g-s\|_p \leqslant \|f-g\|_p + C_2\bar{\Delta}^m\|D^m g\|_p.$$

Taking the infimum over all $g \in \mathring{L}_p^m[a,b]$, we obtain

$$d\big[f, \mathring{\mathcal{S}}_m(\Delta)\big]_p \leqslant C_2 \mathring{K}_{m,p}(\bar{\Delta}^m)$$

(assuming $C_2 \geqslant 1$). The assertion now follows from the equivalence of the K-functional with the modulus of smoothness (cf. §2.11). ■

We also have the following result in terms of the modulus of smoothness of a derivative:

THEOREM 8.15

Let $1 \leqslant \sigma \leqslant m$ and $1 \leqslant p \leqslant q \leqslant \infty$. Then there exists a constant C such that for all $f \in \mathring{L}_p^\sigma[a,b]$,

$$d\big[f, \mathring{\mathcal{S}}_m(\Delta)\big]_q \leqslant C_1 (\bar{\Delta})^{\sigma + 1/q - 1/p} \mathring{\omega}_{m-\sigma}\big(D^\sigma f; \bar{\Delta}\big)_p.$$

Proof. Given $g \in \mathring{L}_p^m[a,b]$, let $G = g^{(\sigma)} \in \mathring{L}_p^{m-\sigma}[a,b]$. Then

$$\|f - \mathring{Q}f\|_q \leqslant \|(f-g) - \mathring{Q}(f-g)\|_q + \|g - \mathring{Q}g\|_q.$$

Using Theorem 8.12 we obtain

$$\|f - \mathring{Q}f\|_q \leqslant \bar{\Delta}^{\sigma + 1/q - 1/p}\big(C_2\|D^\sigma f - G\|_p + C_3 \bar{\Delta}^{m-\sigma}\|D^{m-\sigma}G\|_p\big).$$

The expression in the parentheses is bounded by a constant times $\mathring{K}_{m-\sigma}(\bar{\Delta})D^\sigma f$, and the assertion follows from the equivalence of the K-functional with the modulus of smoothness. ■

The above discussion shows that the direct theorems for ordinary polynomial spline approximation can all be carried over easily to the periodic case. The ambitious reader can check that there are analogous results for periodic Besov spaces and for other intermediate spaces.

This same comment applies to lower bounds and inverse theorems. Indeed, the lower bounds established in §6.6 for ordinary polynomial spline approximation are also valid for periodic spline approximation since the functions constructed there are all locally defined, hence they can be extended trivially to be periodic. The n-width results of §6.7 also carry over directly. Finally, we also note that various classes of periodic functions (e.g. certain periodic Lipschitz spaces $\mathring{\mathrm{Lip}}_p^{k,\alpha}$ or periodic Zygmund spaces $\mathring{Z}^k$) can be characterized completely (cf. §6.8 – 6.9 for the nonperiodic case) in terms of approximation by periodic splines.

§ 8.2. NATURAL SPLINES

In this section we shall consider a linear subspace of $\mathcal{S}(\mathcal{P}_m;\mathcal{M};\Delta)$ that plays an important role in applications. Suppose $\mathcal{M}=(m_1,\ldots,m_k)$ with $1\leqslant m_i\leqslant m$, $i=1,2,\ldots,k$. We call

$$\mathcal{NS}(\mathcal{P}_{2m};\mathcal{M};\Delta)=\{s\in\mathcal{S}(\mathcal{P}_{2m};\mathcal{M};\Delta):\ s_0=s|_{[a,x_1)} \text{ and } s_k=s|_{[x_k,b]} \text{ belong to } \mathcal{P}_m\} \tag{8.24}$$

the space of *natural polynomial splines of order* $2m$ *with knots at* $x_1,\ldots,x_k$ *of multiplicities* $m_1,\ldots,m_k$.

Since the dimension of $\mathcal{S}(\mathcal{P}_{2m};\mathcal{M};\Delta)$ is $2m+K$ while we have enforced $2m$ extra conditions to define $\mathcal{NS}(\mathcal{P}_{2m};\mathcal{M};\Delta)$, it is natural to expect the dimension of $\mathcal{NS}$ to be K. The following theorem confirms this:

THEOREM 8.16

$\mathcal{NS}(\mathcal{P}_{2m};\mathcal{M};\Delta)$ is a linear space of dimension K.

Proof. Proceeding again as in Theorem 4.4, we now find that the ties between the various polynomial pieces can be described by $A\mathbf{c}=\mathbf{0}$ with

$$A=\begin{bmatrix} E & & & & & \\ A_1 & -A_1 & & & & \\ & A_2 & -A_2 & & & \\ & & \cdots & & & \\ & & & A_k & -A_k \\ & & & & E \end{bmatrix},$$

where now

$$A_i=\begin{bmatrix} 1 & x_i & \cdots & & x_i^{2m-1}/(2m-1)! \\ 0 & 1 & \cdots & & x_i^{2m-2}/(2m-2)! \\ \cdots & & & & \\ 0 & 0 & 1 & \cdots & x_i^{2m-m_i}/(2m-m_i)! \end{bmatrix},$$

and E is the m by $2m$ matrix with zeros in the first m columns and the m by m identity in the last m columns. Now, the matrix A is again seen to be of full rank, $2m+\Sigma_1^k(2m-m_i)$. Since $\mathbf{c}$ is a vector with $2m(k+1)$ components, we deduce $\mathcal{NS}$ is of dimension K. ■

In some applications it may be possible to deal with natural splines by using a basis for $\mathcal{S}(\mathcal{P}_{2m};\mathfrak{M};\Delta)$ and enforcing the end conditions. For other applications it is desirable to have a basis for $\mathfrak{N}\mathcal{S}(\mathcal{P}_{2m};\mathfrak{M};\Delta)$ itself. We now show how to construct such a basis consisting of splines with small supports.

Given $j \geqslant 1$, $n \geqslant 1$, and $y_i \leqslant y_{i+1} \leqslant \cdots \leqslant y_{i+j}$, we define

$$L_{i,j}^n(x) = \begin{cases} [y_i,\ldots,y_{i+j}](y-x)_+^{n-1}, & y_i < y_{i+j} \\ 0, & y_i = y_{i+j} \end{cases}, \tag{8.25}$$

and

$$R_{i,j}^n(x) = \begin{cases} (-1)^j[y_i,\ldots,y_{i+j}](x-y)_+^{n-1}, & y_i < y_{i+j} \\ 0, & y_i = y_{i+j} \end{cases}. \tag{8.26}$$

The usual B-spline defined in Section 4.3 is given by

$$Q_i^n(x) = L_{i,n}^n(x) = R_{i,n}^n(x) \tag{8.27}$$

except possibly when x falls at an n-tuple knot, cf. (4.16). Clearly $L_{i,j}^n$ and $R_{i,j}^n$ are polynomial splines of order n. We collect a number of their properties in the following theorem:

THEOREM 8.17

For all $0 \leqslant j \leqslant n$,

$$L_{i,j}^n(x)\begin{cases} =0 & \text{for } x > y_{i+j} \\ >0 & \text{for } x < y_{i+j} \end{cases}, \tag{8.28}$$

$$L_{i,j}^n(x) \quad \text{is a polynomial of exact order } n-j \text{ for } x < y_i; \tag{8.29}$$

$$D_+L_{i,j}^n(x) = -(n-1)L_{i,j}^{n-1}(x); \tag{8.30}$$

$$L_{i,j}^n(x) = \begin{cases} L_{i,j-1}^{n-1}(x) + (y_{i+j}-x)L_{i,j}^{n-1}(x) & \text{if } j>1 \\ (y_{i+j}-x)L_{i,j}^{n-1}(x) & \text{if } j=0 \end{cases}. \tag{8.31}$$

Similarly,

$$R_{i,j}^n(x)\begin{cases} =0 & \text{for } x < y_i \\ >0 & \text{for } x > y_i \end{cases}, \tag{8.32}$$

$$R_{i,j}^n(x) \quad \text{is a polynomial of exact order } n-j \text{ for } x > y_{i+j}, \tag{8.33}$$

$$D_+ R_{i,j}^n(x) = (n-1) R_{i,j}^{n-1}(x); \tag{8.34}$$

$$R_{i,j}^n(x) = R_{i,j-1}^{n-1}(x) + (x-y_i) R_{i,j}^{n-1}(x), \qquad j>1. \tag{8.35}$$

Proof. We concentrate on the L's; the properties of the R's can be established similarly. The first part of (8.28) is clear since $(y-x)_+^{n-1}$ vanishes identically for $x>y$. The positivity assertion follows inductively from the recursion (8.31) and the fact that $L_{i,0}^n(x) = (y_i - x)_+^{n-1}$. The differentiation formula (8.30) follows directly from the definition. Finally, to prove the recursion, we apply Leibnitz's rule to $(y-x)_+^{n-2}(y-x)$ to obtain

$$\begin{aligned} L_{i,j}^n(x) &= [y_i, \ldots, y_{i+j-1}](y-x)_+^{n-2}[y_{i+j-1}, y_{i+j}](y-x) \\ &\quad + [y_i, \ldots, y_{i+j}](y-x)_+^{n-2}[y_{i+j}](y-x) \\ &= L_{i,j-1}^{n-1}(x) + (y_{i+j} - x) L_{i,j}^{n-1}(x). \end{aligned}$$

■

THEOREM 8.18

Let $K \geqslant 2m$, and set

$$B_i(x) = \begin{cases} L_{m+1, m+i-1}^{2m}(x), & i = 1, 2, \ldots, m \\ N_i^{2m}(x), & i = m+1, \ldots, K-m \\ R_{i, m+K-i}^{2m} & i = K-m+1, \ldots, K. \end{cases} \tag{8.36}$$

Then $\{B_i\}_1^K$ is a basis for $\mathfrak{NS}(\mathcal{P}_{2m}; \mathfrak{M}; \Delta)$.

Proof. By Theorem 8.17 each of these splines belongs to $\mathfrak{NS}$. Since we already know the dimensionality of $\mathfrak{NS}$ is K, it remains only to establish the linear independence. Suppose $\Sigma_1^K c_i B_i(x) = 0$. Then for $x < y_1$ we have $\Sigma_1^m c_i B_i(x) = 0$. But each of these splines reduces to a polynomial of *exact* order $m - i + 1$, and so the c's must be zero. A similar argument on the right end shows the $c_{K-m+1}, \ldots, c_K$ to be zero. But the remaining c's must also be zero by the linear independence of the B-splines N_i^{2m}, $i = m+1, \ldots, K-m$. ■

We have introduced the restriction that $K \geqslant 2m$ in Theorem 8.18 in order that there be a complete set of m natural splines on both ends. In practice, we will normally use small m and large K, so the assumption is not unreasonable. If $K = 2m$, then none of the usual B-splines are needed. We should point out that the first and last m basis elements in Theorem 8.18 are not local support functions in the usual sense, but each of the splines does have support on at most $2m$ of the intervals defined by the partition, where $(-\infty, x_1)$ and (x_k, ∞) are included. Figure 24 shows a typical case of Theorem 8.18 with cubic natural splines.

Our next theorem shows how a B-spline expansion for a natural spline can be manipulated to obtain expansions for its various derivatives. The result is useful for numerical applications.

THEOREM 8.19

Suppose $s \in \mathfrak{N}\mathcal{S}(\mathcal{P}_{2m}; \mathfrak{M}; \Delta)$ is given by

$$s(x) = \sum_{1}^{m} c_i L_{m+1,m+i-1}^{2m}(x) + \sum_{m+1}^{K-m} c_i N_i^{2m}(x) + \sum_{K-m+1}^{K} c_i R_{i,m+K-i}^{2m}(x). \tag{8.37}$$

Then for $0 \leqslant d \leqslant m$,

$$D_+^d s(x) = \sum_{i=1}^{m-d} c_i^{(d)} L_{m+1,m+i-1}^{2m-d}(x) + \sum_{i=m-d+1}^{K-m} c_i^{(d)} N_{i+d}^{2m-d}(x) + \sum_{i=K-m+1}^{K-d} c_i^{(d)} R_{i+d,m+K-i-d}^{2m-d}(x), \tag{8.38}$$

where

$$c_i^{(0)} = c_i, \qquad i = 1, 2, \ldots, K;$$

$$c_i^{(d)} = \frac{(2m-d)\left(c_{i+1}^{(d-1)} - c_i^{(d-1)}\right)}{(y_{2m+i} - y_{i+d})}, \qquad i = m-d+1, \ldots, K-m,$$

and

$$c_i^{(d)} = \begin{cases} -(2m-d)c_i^{(d-1)}, & i = 1, 2, \ldots, m-d \\ (2m-d)c_{i+1}^{(d-1)} & i = K-m+1, \ldots, K-d \end{cases}.$$

Moreover, for $m \leqslant d \leqslant 2m-1$,

$$D_+^d s(x) = \sum_{i=1}^{K-2m+d} c_i^{(d)} N_{i+m}^{2m-d}(x), \tag{8.39}$$

where

$$c_i^{(d)} = \frac{(2m-d)\left(c_i^{(d-1)} - c_{i-1}^{(d-1)}\right)}{(y_{3m+i-d-1} - y_{i+m})}, \qquad i = 1, \ldots, K-2m+d.$$

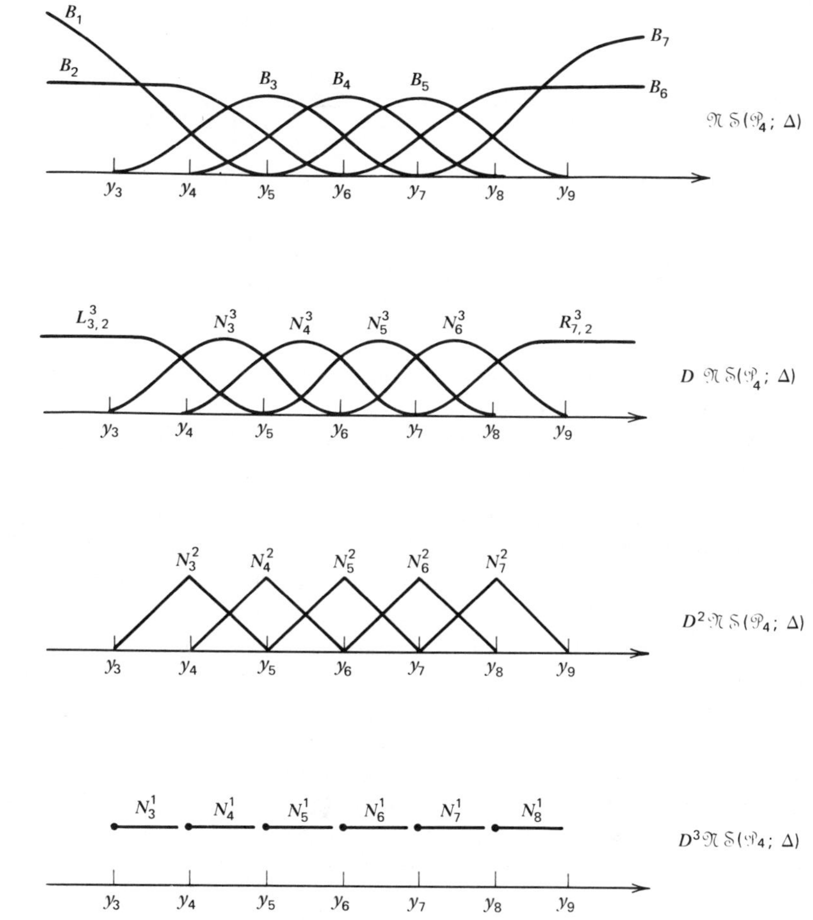

Figure 24. Bases for cubic natural splines and derivatives.

Proof. For $d=0$ there is nothing to prove. We now proceed by induction. Suppose (8.38) has been proved for $d-1$, $0<d<m$. Then differentiating, we obtain

$$\frac{D_+^d s(x)}{(2m-d)} = -\sum_{1}^{m-d+1} c_i^{(d-1)} L_{m+1,m+i-1}^{2m-d}(x) + \sum_{K-m+1}^{K-d+1} c_i^{(d-1)} R_{i+d-1,m+K-i-d+1}^{2m-d} + \sum_{m-d+2}^{K-m} c_i^{(d-1)} \left[\frac{N_{i+d-1}^{2m-d}(x)}{(y_{i+2m-1}-y_{i+d-1})} - \frac{N_{i+d}^{2m-d}(x)}{(y_{i+2m}-y_{i+d})} \right].$$

But

$$L_{m+1,2m-d}^{2m-d} = Q_{m+1}^{2m-d} = \frac{N_{m+1}^{2m-d}}{(y_{3m-d+1} - y_{m+1})},$$

and

$$R_{K-m+d,2m-d}^{2m-d} = Q_{K-m+d}^{2m-d} = \frac{N_{K-m+d}^{2m-d}}{(y_{K+m} - y_{K-m+d})}.$$

Thus by combining terms and rearranging (compare the proof of Theorem 5.9), we obtain (8.38) for d.

Equation (8.39) for $d=m$ agrees with (8.38) for $d=m$. To prove (8.39) for $m<d\leqslant 2m-1$, we again proceed by induction. ■

Figure 24 shows the shape of the various splines involved in the expansions of the derivatives of a cubic natural spline. All of the B-splines shown in Figure 24 can also be found in the array shown in Figure 25. For any given x, all of the nonzero entries in this array can be computed recursively. In particular, if $y_l \leqslant x < y_{l+1}$ with $3 \leqslant l \leqslant 7$, then we may start with $Q_l^1(x) = 1/(y_{l+1} - y_l)$ and generate the pyramid lying under this element. This pyramid is outlined in Figure 25 for the case of $l=6$. The entries of the form $L_{i,j}^n$ can be computed using the recursion (8.31). Those of the form $R_{i,j}^n$ may be computed using (8.35), and the Q_i^n can be obtained from the elements above using the recursion (4.22). The process is a stable one in that all elements of the array are nonnegative, while within the pyramid all of the coefficients in the recursions are also nonnegative.

If we are interested in $x < y_3$, then we may start with $L_{3,0}^1(x) = 1$, and generate the pyramid lying under it using the recursions for the L's. Similarly, if $x \geqslant y_7$, we start with $R_{9,0}^1(x) = 1$, and generate the pyramid lying under it.

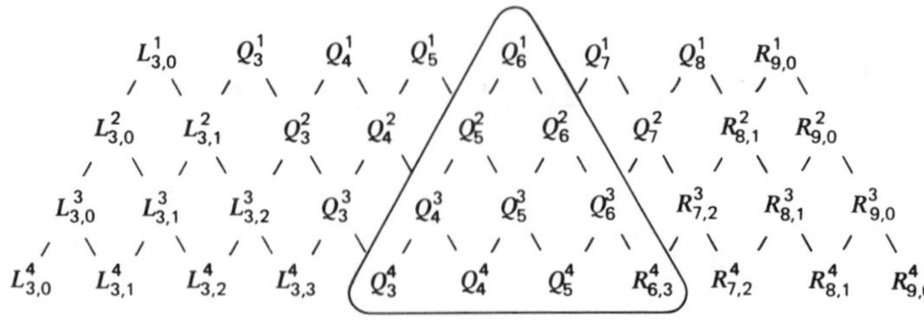

Figure 25. The B-spline array for the splines in Figure 24.

We now discuss some algorithms for dealing with natural splines numerically. Our first algorithm may be used to evaluate B-splines at a given point (cf. Algorithm 5.5).

ALGORITHM 8.20. To Compute the Pyramid of B-Splines With Values at x, given $y_l \leqslant x < y_{l+1}$

1. If $x < y_{m+1}$:
   a. Set $L^1_{m+1,0} = 1$;
   b. Compute the pyramid lying under $L^1_{m+1,0}$ using the recursion (8.31);
2. If $y_l \leqslant x < y_{l+1}$ for some $m+1 \leqslant l \leqslant m+K-1$:
   a. Set $Q_l^1 = 1/(y_{l+1} - y_l)$;
   b. Compute the pyramid lying under Q_l^1 using (8.31), (4.22), and (8.35) as necessary;
3. If $x \geqslant y_{m+K}$:
   a. Set $R^1_{m+K,0} = 1$;
   b. Compute the pyramid lying under $R^1_{m+K,0}$ using (8.35).

The next algorithm can be used to find the coefficients of the B-spline expansions of the various derivatives of a natural spline.

ALGORITHM 8.21. To Compute the Array $c_i^{(d)}$ of Theorem 8.19

1. For $j \leftarrow 1$ step 1 until K
   $cd(0,j) \leftarrow c(j)$;
2. For $d \leftarrow 1$ step 1 until m
   a. for $j \leftarrow m-d+1$ step 1 until $K-m$
      $cd(d,j) \leftarrow (2m-d)*[cd(d-1,j+1] - cd[d-1,j)]/[y(2m+j) - y(j+d)]$;
   b. for $j \leftarrow 1$ step 1 until $m-d$
      $cd(d,j) \leftarrow -(2m-d)*cd(d-1,j)$;
      $cd(d,K-m+j) \leftarrow (2m-d)*cd(d-1,K-m+j+1)$;
3. For $d \leftarrow m+1$ step 1 until $2m-1$
   for $j \leftarrow 1$ step 1 until $K-2m+d$
   $cd(d,j) \leftarrow (2m-d)*[cd(d-1,j) - cd(d-1,j-1)]/[y(3m+j-d-1) - y(i+m)]$.

Algorithms 8.20 and 8.21 can now be combined to produce an algorithm for evaluating a natural spline and its first $2m-1$ derivatives.

ALGORITHM 8.22. To Compute $s(x),\ldots,D^{2m-1}s(x)$ for given $a \leqslant x \leqslant b$

1. Find l such that $y_l \leqslant x < y_{l+1}$ by Algorithm 5.4;
2. Compute the pyramid of B-splines with value at x by Algorithm 8.20;

3. For $d=0$ step 1 until m compute $D^d s(x)$ by formula (8.38);
4. For $d=m+1$ step 1 until $2m-1$ compute $D^d s(x)$ using formula (8.39).

Discussion. It is assumed that the array $[c_i^{(d)}]$ of coefficients of the B-spline expansions of the various derivatives has already been computed by Algorithm 8.21. ■

Algorithm 8.22 can be applied in the same way as in the ordinary polynomial spline case (cf. Algorithm 5.15) to convert the B-spline expansion of a natural spline to a piecewise polynomial representation. There are some obvious simplifications that can be made in the algorithms in the case of equally spaced knots. We should also mention that it is possible to establish the analog of Theorem 5.7 and to design an algorithm based on it that is similar to Algorithm 5.8. This would produce a somewhat more efficient alternate to Algorithm 8.22.

We close this section by noting that it is possible to develop zero properties and to examine the sign structure of determinants formed from the natural B-splines in much the same way as was done for the ordinary polynomial splines. But as there is little need for these results in practice, we do not bother to work them out here. We should also emphasize that in applications of splines we would generally not *select* the space of natural splines for a particular numerical process. We have examined the space and how to handle it on a computer primarily because natural splines turn out to be the solution of various best interpolation and smoothing problems.

§ 8.3. g-SPLINES

In this section we study certain linear spaces of generalized polynomial splines (called g-splines) that arise naturally in several applications (including best interpolation, smoothing, and the construction of optimal quadrature formulae). We begin with their definition.

Let $\Delta=\{a=x_0<x_1<\cdots<x_{k+1}=b\}$ be a partition of the interval $[a,b]$ into subintervals $I_i=[x_i,x_{i+1})$, $i=0,1,\ldots,k-1$ and $I_k=[x_k,x_{k+1}]$. For each $i=1,2,\ldots,k$, suppose $1\leqslant m_i\leqslant m$ and that

$$\Gamma_i=\{\gamma_{ij}\}_{j=1}^{m-m_i} \tag{8.40}$$

is a collection of linear functionals defined on $\mathcal{P}_m$. We define the space of *g-splines* by

$$\mathcal{S}(\mathcal{P}_m;\Gamma;\Delta)=\left\{\begin{array}{l} s\colon \text{there exist } s_0,\ldots,s_k\in\mathcal{P}_m \text{ with} \\ s|_{I_i}=s_i,\ i=0,1,\ldots,k \text{ such that} \\ \gamma_{ij}s_{i-1}=\gamma_{ij}s_i \text{ for } j=1,\ldots,m-m_i \text{ and } i=1,\ldots,k \end{array}\right\}. \tag{8.41}$$

Although many of the results presented here are valid for general linear functionals (cf. Chapter 11), throughout this section we shall restrict our attention to the case where the linear functionals γ_{ij} are linear combinations of derivatives; that is, we suppose

$$\gamma_{ij} = e_{x_i} \sum_{\nu=1}^{m} \gamma_{j\nu}^{i} D^{\nu-1}, \tag{8.42}$$

where e_x denotes the usual point evaluator functional defined by $e_x f = f(x)$ If $\Gamma_i = \{\gamma_{ij}\}_{j=1}^{m-m_i}$ is a set of linear functionals of the form (8.42), then we refer to Γ_i as a set of *Extended Hermite-Birkhoff (EHB-) linear functionals*.

Before proceeding to the basic constructive properties of $\mathcal{S}(\mathcal{P}_m;\Gamma;\Delta)$, it is instructive to consider several examples.

EXAMPLE 8.23

Let $m=2$, $[a,b]=[0,5]$, $\Delta=\{1,2,3,4\}$, and $\Gamma=\{e_1',e_2',e_3',e_4'\}$, where $e_i'f=f'(i)$.

Discussion. The space $\mathcal{S}(\mathcal{P}_2;\Gamma;\Delta)$ consists of piecewise linear polynomials such that across the knots the slope is continuous, but the function itself need not be. A typical member of this space is illustrated in Figure 26. ■

EXAMPLE 8.24. Hermite-Birkhoff Linear Functionals

Suppose each Γ_i consists of linear functionals of the form

$$\gamma_{ij} = e_{x_i} D^{\nu_{ij}}, \qquad j=1,2,\ldots,m-m_i, \tag{8.43}$$

where

$$0 \leqslant \nu_{i,1} < \cdots < \nu_{i,m-m_i} \leqslant m-1,$$

$i=1,2,\ldots,k$. We call such a set Γ_i of linear functionals a *Hermite-Birkhoff (HB-) set*.

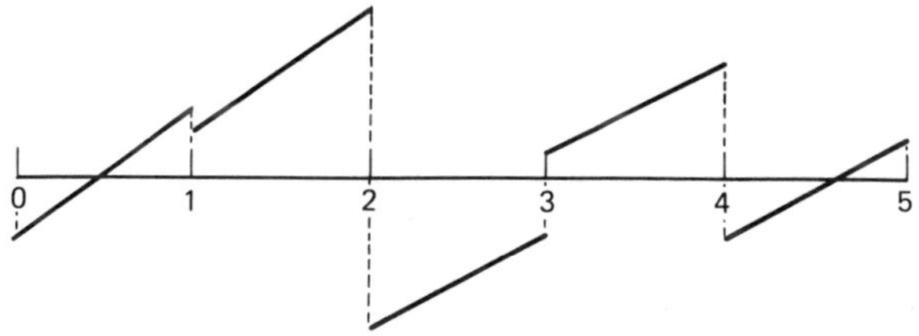

Figure 26. A g-spline (cf. Example 8.23).

Discussion. A set $\Gamma = \cup_{i=1}^{k} \Gamma_i$ of HB linear functionals can also be conveniently described in terms of a so-called *incidence matrix*

$$E = (E_{ij})_{i=1,\,j=1}^{k\quad m}, \qquad E_{ij} = \begin{cases} 0 & \text{if } e_{x_i} D^{m-j} \in \Gamma \\ 1 & \text{otherwise.} \end{cases} \tag{8.44}$$

The number of functionals in Γ is equal to the number of unit entries in E; namely, $K = \sum_{i=1}^{k} m_i$. The set Γ in Example 8.23 is an example of an HB set of functionals. The corresponding incidence matrix is given by

$$E = \begin{bmatrix} 0 & 1 \\ 0 & 1 \\ 0 & 1 \\ 0 & 1 \end{bmatrix}.$$

■

We can specialize Γ still further so that $\mathcal{S}(\mathcal{P}_m; \Gamma; \Delta)$ reduces to the space of polynomial splines $\mathcal{S}(\mathcal{P}_m; \mathfrak{M}; \Delta)$ discussed in Chapter 4.

EXAMPLE 8.25. Hermite Linear Functionals

Suppose

$$\Gamma_i = \left\{ e_{x_i}, e'_{x_i}, \ldots, e_{x_i}^{(m-m_i-1)} \right\}, \tag{8.45}$$

$i = 1, 2, \ldots, k$ where $e_{x_i}^{(\nu)} = e_{x_i} D^{\nu}$, in general. We call $\Gamma = \cup_{i=1}^{k} \Gamma_i$ a set of *Hermite linear functionals* in this case.

Discussion. Here $\mathcal{S}(\mathcal{P}_m; \Gamma; \Delta) = \mathcal{S}(\mathcal{P}_m; \mathfrak{M}; \Delta)$ with $\mathfrak{M} = (m_1, \ldots, m_k)$. In this case the ith row of the incidence matrix consists of a sequence of m_i ones, followed by $m - m_i$ zeros, $i = 1, 2, \ldots, k$. ■

It is clear that the space of g-splines $\mathcal{S}(\mathcal{P}_m; \Gamma; \Delta)$ is a linear space. The following theorem gives its dimension. We assume here and throughout the remainder of the section that for each $i = 1, 2, \ldots, k$, the linear functionals $\gamma_{i1}, \ldots, \gamma_{i,m-m_i}$ making up Γ_i are linearly independent over $\mathcal{P}_m$. This is equivalent to the assumption that the matrix

$$G_i = \left(\gamma_{j\nu}^{i}\right)_{j=1,\,\nu=1}^{m-m_i\,m} \tag{8.46}$$

is of full rank $m - m_i$.

THEOREM 8.26

The dimension of $\mathbb{S}(\mathcal{P}_m;\Gamma;\Delta)$ is $m+K$, where $K=\sum_1^k m_i$.

Proof. The proof is exactly as in Theorem 4.4, except that now the matrix A is given by

$$A=\begin{bmatrix} A_1 & -A_1 & & & & \\ & A_2 & -A_2 & & & \\ & & \cdots & & & \\ & & & & A_k & -A_k \end{bmatrix}, \qquad A_i=G_i\cdot V_i,$$

where $V_i=V(x_i,\dots,x_i)$ is the m by m VanderMonde (in this case Wronskian) matrix. Since V_i is rank m and G_i is rank $m-m_i$, each A_i is also of rank $m-m_i$. ■

We would now like to construct a basis for $\mathbb{S}(\mathcal{P}_m;\Gamma;\Delta)$. As in Chapter 4, we will begin with a basis of one-sided splines. Clearly, we should include the functions $1,x,\dots,x^{m-1}$. In addition, we will want to construct m_i splines associated with each knot x_i and in such a way that they vanish for $x<x_i$. The following lemma considers the possibility of such a construction:

LEMMA 8.27

Suppose

$$G_i\begin{bmatrix}\beta_m\\ \vdots \\ \beta_1\end{bmatrix}=0, \tag{8.47}$$

where G_i is the matrix of coefficients of the linear functionals Γ_i associated with the knot x_i—see (8.46). Then

$$\rho(x)=\sum_{\nu=1}^{m}\beta_\nu\frac{(x-x_i)_+^{m-\nu}}{(m-\nu)!}$$

is a g-spline in the space $\mathbb{S}(\mathcal{P}_m;\Gamma;\Delta)$.

Proof. Clearly ρ is a piecewise polynomial. We need only check that it satisfies the proper smoothness conditions at the knot x_i; namely,

$$\sum_{\nu=1}^{m}\beta_\nu\sum_{\mu=1}^{m}\gamma_{j\mu}^{i}D^{\mu-1}\frac{(x-x_i)_+^{m-\nu}}{(m-\nu)!}\bigg|_{x=x_i}=0,\qquad j=1,2,\dots,m-m_i.$$

But these are just the equations (8.47). ■

Since (8.47) is a system of $m-m_i$ equations with m unknowns, there exist m_i different (linearly independent) solutions. We are now ready to construct our one-sided basis for $\mathcal{S}(\mathcal{P}_m;\Gamma;\Delta)$.

THEOREM 8.28

For each $i=1,2,\ldots,k$, let

$$\beta^{ij}=(\beta_1^{ij},\ldots,\beta_m^{ij}), \qquad j=1,2,\ldots,m_i$$

be a set of m_i linearly independent solutions of (8.47). Set

$$\rho_{ij}(x)=\sum_{\nu=1}^{m}\beta_\nu^{ij}\frac{(x-x_i)_+^{m-\nu}}{(m-\nu)!}, \qquad \begin{array}{l} j=1,2,\ldots,m_i \\ i=1,2,\ldots,k \end{array}, \tag{8.48}$$

and

$$\rho_{0j}(x)=x^{j-1}, \qquad j=1,2,\ldots,m_0, \text{ with } m_0=m. \tag{8.49}$$

Then

$$\{\rho_{ij}(x)\}_{i=0,\ j=1}^{k \quad m_i}$$

is a basis for $\mathcal{S}(\mathcal{P}_m;\Gamma;\Delta)$. Moreover, each $\rho_{ij}(x)$ vanishes for $x<x_i$.

Proof. By Lemma 8.27 each of these functions belongs to $\mathcal{S}$. For their linear independence, suppose $\Sigma_0^k\Sigma_1^{m_i}c_{ij}\rho_{ij}(x)=0$. Then for x in (a,x_1) we have $\Sigma_1^m c_{0j}x^{j-1}=0$, which implies $c_{01}=\cdots=c_{0m}=0$. Now, looking at x in (x_1,x_2), we have $\Sigma_1^{m_1}c_{1j}\rho_{1j}(x)=0$. But by the linear independence of the $\beta^{11},\ldots,\beta^{1m_1}$, the splines $\rho_{11},\ldots,\rho_{1m_1}$ are linearly independent on this interval, and again the c's are zero. This process may be continued to argue that all c's are zero, so the splines $\{\rho_{ij}\}$ are linearly independent. Since there are $m+K$ of them, in view of Theorem 8.26, they form a basis. ■

We give several examples:

EXAMPLE 8.29

Suppose Γ is a HB-set of linear functionals.

Discussion. In this case the one-sided basis is somewhat easier to describe. It is given by

$$\{1,x,\ldots,x^{m-1}\}\cup\{(x-x_i)_+^{m-j}\}_{i,j} \qquad \text{such that } E_{ij}=1,$$

where E is the incidence matrix associated with Γ as in Example 8.24 In other words, at each knot we take the splines $(x-x_i)_+^j$ with

$$j\in\{0,1,\ldots,m-1\}\backslash\{\nu_{i,1},\ldots,\nu_{i,m-m_i}\}, \qquad i=1,2,\ldots,k,$$

where the ν's list the derivatives in Γ (see Example 8.24). ■

EXAMPLE 8.30

Give a one-sided basis for the space $\mathcal{S}(\mathcal{P}_2;\Gamma;\Delta)$ discussed in Example 8.23.

Discussion. Applying Example 8.29, we may take the functions

$$\{1,x\}\cup\{(x-x_i)_+^0\}_{i=1}^k. \qquad ■$$

Our next task is to construct a local support basis for the space of g-splines. This is not a trivial assignment, as the following example (where there does not even exist a local support basis) shows.

EXAMPLE 8.31

Construct a local support basis for the space of splines discussed in Examples 8.23 and 8.30.

Discussion. Any spline in $\mathcal{S}(\mathcal{P}_2;\Gamma;\Delta)$ that vanishes outside of an interval is forced by the continuity of the slopes to have zero slope everywhere. Thus the only local support splines in $\mathcal{S}(\mathcal{P}_2;\Gamma;\Delta)$ are those with zero slopes; that is, piecewise constants. But the spline $s(x)=x$ belongs to the space, and clearly it cannot be represented as a linear combination of piecewise constants. It follows that $\mathcal{S}(\mathcal{P}_2;\Gamma;\Delta)$ cannot have a basis consisting solely of local support splines. ■

As a first step toward constructing a local support basis for $\mathcal{S}(\mathcal{P}_m;\Gamma;\Delta)$, we put the one-sided splines in lexicographical order:

$$\rho_1,\ldots,\rho_{m+K} := 1,x,\ldots,x^{m-1},\rho_{11},\ldots,\rho_{1m_1},\ldots,\rho_{k1},\ldots,\rho_{km_k}. \tag{8.50}$$

If we introduce the notation

$$y_{m+1}\leqslant y_{m+2}\leqslant\cdots\leqslant y_{m+K}=\overbrace{x_1,\ldots,x_1}^{m_1},\ldots,\overbrace{x_k,\ldots,x_k}^{m_k},$$

where each x_i is repeated exactly m_i times, $i=1,2,\ldots,k$, then by the construction in Theorem 8.28, each of the splines $\rho_i(x)$ vanishes identically for $x<y_i$, $i=m+1,\ldots,m+K$.

For convenience, we write $x_0 = a$ and $m_0 = m$. Then there exists coefficients C_{ij} so that

$$\rho_j(x) = \begin{cases} 0, & x < y_i \\ \sum_{i=1}^{m} C_{ij} x^{i-1} i / (i-1)! & x \geq y_i \end{cases} \tag{8.51}$$

(with $y_1 = \ldots = y_m = a$).

The basis $\{\rho_i\}_1^{m+K}$ for $\mathcal{S}(\mathcal{P}_m; \Gamma; \Delta)$ is completely described by the set $\{y_i\}_1^{m+K}$ and the matrix $C = (C_{ij})_{i=1, j=1}^{m, m+K}$. We shall use the notation $C\langle i_1, \ldots, i_r \rangle$ to denote the submatrix of C obtained by taking only the columns $i_1, \ldots, i_r$.

The following analog of Lemma 4.7 shows when it is possible to form a linear combination of one-sided splines to produce a spline with local support:

LEMMA 8.32

Suppose $1 \leq i_1 < i_2 < \cdots < i_r \leq m + K$, and suppose $\boldsymbol{\delta} = (\delta_1, \ldots, \delta_r)^T$ is a solution of

$$C\langle i_1, \ldots, i_r \rangle \boldsymbol{\delta} = 0. \tag{8.52}$$

Then

$$B(x) = \sum_{j=1}^{r} \delta_j \rho_{i_j}(x)$$

is a spline in $\mathcal{S}$ which vanishes for $x < y_{i_1}$ (if $i_1 > m$) and for $x > y_{i_r}$.

Proof. Clearly $B \in \mathcal{S}$ as it is a linear combination of the ρ's. Since all of the $\rho_{i_1}(x), \ldots, \rho_{i_r}(x)$ vanish for $x < y_{i_1}$ by definition (if $i_1 > m$), it follows that $B(x)$ also does. Now for $x > y_{i_r}$, each of the functions ρ is given by its expansion

$$\rho_{i_j}(x) = [1, x, \ldots, x^{m-1}] C_{i_j}.$$

Thus

$$B(x) = [1, x, \ldots, x^{m-1}] C\langle i_1, \ldots, i_r \rangle \boldsymbol{\delta} = 0 \qquad \text{for } x > y_{i_j}.$$ ■

In applying Lemma 8.32 to the construction of local support splines, we may choose any value of r for which a corresponding nontrivial $\boldsymbol{\delta}$ satisfying (8.52) can be found. In some cases, local support splines can be constructed using only two one-sided basis splines. In general, however, it will be necessary to use larger values for r. (For example, in the polynomial spline case, we saw in Lemma 4.7 that it was necessary to take $r \geqslant m+1$.)

Lemma 8.32 can be used to construct local support splines as linear combinations of one-sided splines. An important question associated with this process is: When are the resulting splines linearly independent? The following lemma gives simple algebraic conditions which assure linear independence:

LEMMA 8.33

Suppose $\{\boldsymbol{\beta}_\nu = (\beta_{\nu 1}, \ldots, \beta_{\nu, m+K})\}_{\nu=1}^{q}$ is a set of q linearly independent vectors with $m+K$ components, and that

$$B_\nu(x) = \sum_{j=1}^{m+K} \beta_{\nu j} \rho_j(x), \qquad \nu = 1, 2, \ldots, q. \tag{8.53}$$

Then the splines $B_1, \ldots, B_q$ are linearly independent.

Proof. If $d_1 B_1 + \cdots + d_q B_q = 0$ on $[a,b]$, then

$$\sum_{\nu=1}^{q} d_\nu \sum_{j=1}^{m+K} \beta_{\nu j} \rho_j = \sum_{j=1}^{m+K} \rho_j \sum_{\nu=1}^{q} d_\nu \beta_{\nu j} = 0.$$

By the linear independence of the ρ's, it follows that $d_1 \boldsymbol{\beta}_1 + \cdots + d_q \boldsymbol{\beta}_q = 0$. This in turn implies that $d_1 = \cdots = d_q = 0$ since we have assumed the vectors $\boldsymbol{\beta}_1, \ldots, \boldsymbol{\beta}_q$ are linearly independent. ■

Lemmas 8.32 and 8.33 together show how a local support basis for $\mathbb{S}$ could be constructed. We must find $m+K$ linearly independent vectors in $\mathbf{R}^{m+K}$ to serve as the coefficients of local support splines. Whether we can choose such vectors will depend heavily on the properties of the matrix C. Our next theorem gives a set of conditions on the matrix C which is sufficient to guarantee the existence of a basis for $\mathbb{S}$ which consists of splines whose support intervals are not too big.

THEOREM 8.34

Let $\varepsilon_0 = 0$, and define

$$\varepsilon_i = \varepsilon_{i-1} + m_{i-1} \qquad i = 1, 2, \ldots, k+1.$$

Suppose the matrix C defining the one-sided basis $\{\rho_j\}_1^{m+K}$ has the property that

$$C\langle \varepsilon_{i+1}+1,\ldots,\varepsilon_{i+m+1}\rangle \quad \text{is of full rank } m \tag{8.54}$$

for each $i=0,1,\ldots,k-m$. Then there exists a basis $\{B_i\}_1^{m+K}$ for $\mathbb{S}$ with the properties that

$$B_{\varepsilon_i+1},\ldots,B_{\varepsilon_i+m_i} \quad \text{have support on } [x_i,x_{i+m}], \qquad i=0,1,\ldots,k-m, \tag{8.55}$$

$$B_{\varepsilon_i+1},\ldots,B_{\varepsilon_i+m_i} \quad \text{have support on } [x_i,b], \qquad i=k-m+1,\ldots,k. \tag{8.56}$$

Proof. We construct coefficient vectors $\boldsymbol{\beta}_1,\ldots,\boldsymbol{\beta}_{m+K}$. For $i=0,1,\ldots,k-m$ and $j=1,2,\ldots,m_i$, choose $\boldsymbol{\beta}_{\varepsilon_i+j}$ to be a vector in $\mathbf{R}^{m+K}$ whose ε_i+j component is equal to 1, whose $\varepsilon_{i+1}+1,\ldots,\varepsilon_{i+m+1}$ components are equal to $\boldsymbol{\delta}$, where $\boldsymbol{\delta}$ is any solution of the system

$$C\langle \varepsilon_{i+1}+1,\ldots,\varepsilon_{i+m+1}\rangle\boldsymbol{\delta}=-C\langle \varepsilon_i+j\rangle,$$

and whose remaining coefficients are zero. For $i=k-m+1,\ldots,k$ and $j=1,2,\ldots,m_i$, let $\boldsymbol{\beta}_{\varepsilon_i+j}$ be an $m+K$ vector with its ε_i+j component equal to 1, and all other components equal to zero. By the construction we see that the corresponding B-splines have the stated support properties. In particular, we note that

$$B_{\varepsilon_i+j}=\rho_{\varepsilon_i+j}, \qquad j=1,2,\ldots,m_i, \qquad i=k-m+1,\ldots,k.$$

Moreover, by the construction, these coefficient vectors are clearly linearly independent, and the result follows from Lemma 8.33. ■

Some discussion of the ε's in Theorem 8.34 may be in order. They are defined so that

$$\varepsilon_i+1=\min\{j: y_j=x_i\};$$

$$\varepsilon_{i+1}=\max\{j: y_j=x_i\}.$$

Thus the part of the matrix C involved in the hypothesis (8.54) of Theorem 8.34 is precisely those columns which are related to one-sided splines associated with the knots $x_{i+1},\ldots,x_{i+m}$. The condition (8.54) is not satisfied for the matrix C corresponding to Example 8.31.

As an application of Theorem 8.34, we now discuss one commonly occurring situation where a local support basis for a g-spline space can be constructed.

COROLLARY 8.35

Suppose none of the functionals in Γ involves the $m-1^{\text{st}}$ derivative. Then (8.54) holds, and Theorem 8.34 provides a local basis for $\mathbb{S}$.

Proof. The matrix in (8.54) contains the submatrix

$$C\langle \varepsilon_{\nu+1},\ldots,\varepsilon_{\nu+m}\rangle.$$

Since

$$\rho_{\varepsilon_{\nu+j}+1}(x)=\frac{(x-x_{\nu+j})_+^{m-1}}{(m-1)!}=\sum_{l=0}^{m-1}\frac{x^l(-x_{\nu+j})^{m-1-l}}{l!(m-1-l)!},\qquad x>x_{\nu+j},$$

this matrix is just a constant multiple of the VanderMonde matrix $V(x_{\nu+1},\ldots,x_{\nu+m})$, which, of course, is nonsingular. ■

While Corollary 8.35 does produce a basis for the space of g-splines which consists of splines with relatively small supports, we emphasize that we have *not* shown that these basis splines are positive in the interior of their support. In general, they will not be. In addition, it is also not generally the case that the local support splines constructed here have the smallest possible support sets—there may be other splines forming a basis that have smaller supports.

In the remainder of this section we discuss zero properties of g-splines. We begin with a rather weak result.

THEOREM 8.36

For each $i=1,2,\ldots,k$, let $\tilde{m}_i$ be such that the Hermite functionals e_{x_i}, $e'_{x_i},\ldots,e_{x_i}^{(m-1-\tilde{m}_i)}$ are contained in Γ_i, $i=1,2,\ldots,k$. Then for every nontrivial s in $\mathbb{S}(\mathcal{P}_m;\Gamma;\Delta)$,

$$Z(s)\leqslant m+\sum_{1}^{k}\tilde{m}_i-1,$$

where $Z(s)$ counts the number of zeros of s with multiplicities as in Definition 4.47.

Proof. Since $\mathcal{S}(\mathcal{P}_m;\Gamma;\Delta)\subseteq\mathcal{S}(\mathcal{P}_m;\tilde{\mathfrak{M}};\Delta)$ with $\tilde{\mathfrak{M}}=(\tilde{m}_1,\ldots,\tilde{m}_k)$, the assertion follows immediately from Theorem 4.53. ■

In order to get sharper bounds, we now restrict our attention to the case where Γ consists of HB-linear functionals. Given such a set of linear functionals, suppose E is the corresponding incidence matrix as in Example 8.24. Our zero bound will depend on the structure of E. To state it, we need some additional terminology. A *sequence of 1's* in E is any sequence of consecutive 1's appearing in a given row of E. There may be more than one sequence of 1's in a given row of E (separated by at least one zero). Each row contains at least one sequence of 1's. A sequence of 1's is called an *odd sequence* provided it has an odd number of 1's in it.

THEOREM 8.37

Let Γ be a set of HB-linear functionals, and let E be the associated incidence matrix. Let K be the number of 1's appearing in E, and let p be the number of odd sequences of 1's in E that do not start in the first column. Then for any $s\in\mathcal{S}(\mathcal{P}_m;\Gamma;\Delta)$ such that $D^{m-1}s$ does not vanish on any interval,

$$Z(s)\leqslant m+K+p-1, \tag{8.57}$$

where $Z(s)$ counts the number of zeros of s with multiplicites as in Definition 4.47.

Proof. The proof is similar to the proof of Theorem 4.53. First, by the assumption on $D^{m-1}s$, we know that s and all of its derivatives have only isolated zeros. On the other hand, we cannot apply the trick of pulling multiple knots apart to get a nearby continuous spline, and thus we have to deal directly with the jump zeros of s; that is, with the points where $s(t-)s(t+)\leqslant 0$ and $s(t-)\neq s(t+)$. We classify the jump zeros of s as follows:

t is type $+$ if $Ds(t-)Ds(t+)<0$;

t is type c if $Ds(t-)Ds(t+)>0$ and $s(t-)Ds(t-)<0, s(t+)Ds(t+)>0$;

t is type $-$ if $Ds(t-)Ds(t+)>0$ and $s(t-)Ds(t-)>0, s(t+)Ds(t+)<0$;

t is type 0 otherwise.

We illustrate these various types of jump zeros in Figure 27. The important point to note is that if t is of type c, then it can serve as a Rolle's point (cf. Definition 2.18) for both of the intervals with end-point t. If t is of type $+$

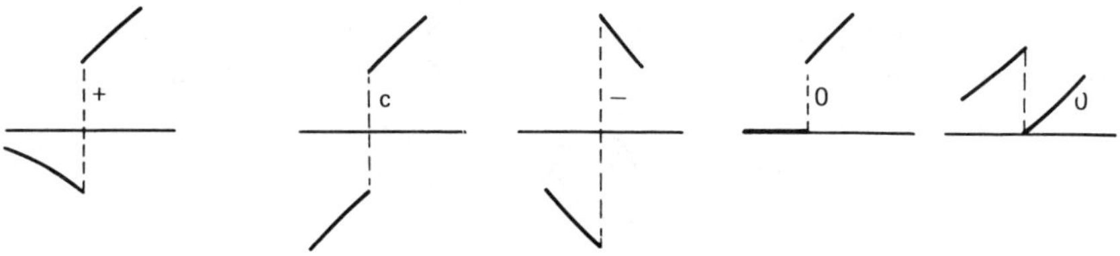

Figure 27. Types of jump zeros.

or type 0 then it is a Rolle's point for at least one of these intervals. If t is of type $-$, then it is not a Rolle's point for either interval.

Suppose now that s has zeros at the points $t_1<t_2<\cdots<t_r$ with multiplicities l_i, $i=1,2,\ldots,r$. Then Ds has a zero of multiplicity l_i-1 at t_i, $i=1,2,\ldots,r$. Now suppose that μ^-, μ^+, and μ^0 denote the number of zeros of s of type $-$, $+$, and 0, respectively. Then the t's define $r-1-\mu^--\mu^0-\mu^+$ intervals where both endpoints are either points where s has no jump discontinuity, or are points with jump discontinuity of type c. We throw out all such intervals that contain type$-$jump zeros. This leaves $r-1-2\mu^--\mu^+$ intervals on which the extended Rolle's Theorem 2.19 can be applied to conclude that Ds has a zero in the interval. We deduce that

$$Z(Ds)\geqslant \sum_{i=1}^{r}(l_i-1)+r-1-2\mu^--\mu^0-\mu^++\mu^+,$$

which can be rewritten as

$$Z(s)\leqslant Z(Ds)+1+2\mu^-+\mu^0.$$

Similarly, the number of zeros of Ds can be related to the number of zeros of D^2s, and so on.

It remains to count zeros. To this end it is convenient to introduce a companion matrix to E which describes the types of jump zeros that s and its derivatives have at the knots. We define $E^*=(E^*_{ij})_{i=1,\,j=1}^{k\quad m}$, where

$$E^*_{ij}=\begin{cases}2 & \text{if } D^{m-j}s \text{ has a zero of type}-\text{at } x_i\\ 1 & \text{if } D^{m-j} \text{ has a zero of type 0 at } x_i\\ 0 & \text{otherwise.}\end{cases}$$

Now, if we string the inequalities relating the number of zeros to the higher derivatives together, we obtain

$$Z(s)\leqslant \sum_{i=1}^{k}\sum_{j=2}^{m}E^*_{ij}+m-1+Z(D^{m-1}s).$$

Since $D^{m-1}s$ is a piecewise constant, it can have zeros only at the knots, and in particular, only at those knots where $E_{i1}^*=1$. We conclude that

$$Z(s) \leqslant \sum_{i=1}^{k} \sum_{j=1}^{m} E_{ij}^* + m - 1.$$

We now relate the sum to the original incidence matrix. First, we note that the only time $E_{ij}^*=2$ is possible is if $D^{m-j}s$ has a zero of type $-$ at x_i. But then $D^{m-j+1}s$ does not have a discontinuous zero at x_i, so it follows that $E_{ij-1}^*=0$. On the other hand, if $D^{m-j-1}s$ has a jump zero at x_i whose derivative $D^{m-j}s$ is also a jump zero of type $-$, then $D^{m-j}s$ changes sign at x_i, so $D^{m-j-1}s$ has a type $+$ zero at this x_i. We have shown

$$E_{ij}^*=2 \qquad \text{implies } j>1, E_{ij-1}^*=0, \text{ and if } j<m, E_{ij+1}^*=0.$$

Now we compare the sum $\Sigma\Sigma E_{ij}^*$ with $\Sigma\Sigma E_{ij}$. Consider the ith row of E. If it has a string of 1's with an even number in it, then the corresponding string in E^* has a sum no larger than the number of 1's. (Indeed, if any of the E_{ij}^*'s in this string is 2, then the entry in front of and behind the 2 must be a zero, so the sum is the same.) Consider now a string of 1's of length r with r odd. If this string starts in the first column, then since E_{i1}^* cannot be 2, it follows that the sum of the corresponding E^*'s is at most r. On the other hand, if this string starts in the second column or later, then the corresponding string of E^*'s has a maximal sum when it has the form $2,0,2,\ldots,0,2$, in which case it adds up to $r+1$. In summary, we see that

$$\sum_{i=1}^{k} \sum_{j=1}^{m} E_{ij}^* \leqslant \sum_{i=1}^{k} \sum_{j=1}^{m} E_{ij} + p$$

and the theorem is proved. ■

The following example shows that the hypothesis on $D^{m-1}s$ is necessary for the strong zero bound of Theorem 8.37. It is possible to give bounds without this hypothesis, but then the counting procedure must be significantly weakened.

EXAMPLE 8.38

Let $m=9$ and $\Delta=\{0,1\}$. Let

$$E=\begin{bmatrix} 1\,0\,0\,0\,0\,0\,0\,0\,1 \\ 1\,0\,0\,0\,0\,0\,0\,0\,1 \end{bmatrix},$$

and

$$s(x)=\begin{cases} -x^8, & x<0 \\ 1, & 0\leqslant x<1 \\ -(x-1)^8, & 1\leqslant x. \end{cases}$$

Discussion. Clearly s has a zero of order 9 at each of the points 0 and 1, hence $Z(s)=18$. On the other hand, $K=4$ and $p=2$, so $m+K+p-1=14$. Thus (8.57) does not hold. The reason is, of course, that Ds (and all higher derivatives) vanish identically on [0, 1]. ■

There is a slight variant of Theorem 8.37 that holds for *confined splines*; that is, for splines that vanish outside of a finite interval. To state the result we need more notation. We say that an odd sequence of 1's in the incidence matrix is *supported* provided it begins with an element $E_{ij}=1$ with $1<i<k$ and $1<j$ and provided there exist $i'<i<i''$ and $1\leqslant j',j''<j$ such that $E_{i',j'}=E_{i'',j''}=1$.

THEOREM 8.39

Let Γ and E be as in Theorem 8.37, and suppose $s\in\mathbb{S}(\mathcal{P}_m;\Gamma;\Delta)$ is a g-spline such that $D^{m-1}s$ does not vanish on any subinterval of (x_1,x_k). Suppose s vanishes identically outside of (x_1,x_k). Then

$$Z_{(x_1,x_k)}(s)\leqslant m+\tilde{K}+\tilde{p}-1,$$

where $\tilde{K}$ counts the number of 1's in rows $2,\ldots,k-1$ of E, and $\tilde{p}$ counts the number of odd supported sequences in E.

Proof. The proof is nearly identical to the proof of Theorem 8.37. Since we are counting zeros only on (x_1,x_k), we do not need to add in the values of $E^*_{i,j}$ for $i=1$ and $i=k$. Moreover, since s is identically zero outside of $[x_1,x_k]$, if $D^{m-j}s$ has a type $-$ jump zero at some x_i in (x_1,x_k), then at least one of the functions $D^{m-j+1}s,\ldots,D^{m-1}s$ must have a jump zero to the left of x_i (otherwise $D^{m-j}s$ would be identically zero to the left of x_i, and x_i could not be of type $-$). This means that $E_{i',j'}=1$ for some $i'<i$ and for some $j'<j$. A similar argument applies to the right of x_i. ■

We give only one simple example to illustrate Theorem 8.39.

EXAMPLE 8.40

Let $\Delta=\{-1,0,1\}, m=3$, and

$$E=\begin{bmatrix} 1 & 1 & 1 \\ 0 & 1 & 0 \\ 1 & 1 & 1 \end{bmatrix}.$$

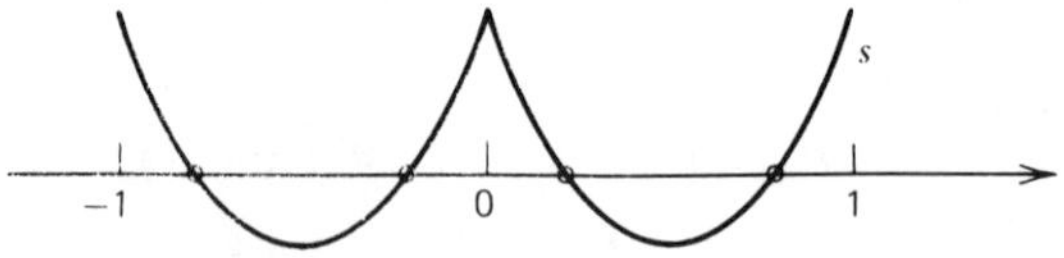

Figure 28. A g-spline with four zeros (cf. Example 8.40).

Discussion. By Theorem 8.39, if s is a g-spline in $\mathcal{S}(\mathcal{P}_3;\Gamma;\Delta)$ with Γ corresponding to this incidence matrix E, and if s vanishes outside of $[-1,1]$, then $Z_{(-1,1)}(s) \leqslant m+\tilde{K}+\tilde{p}-1=3+1+1-1=4$. In Figure 28 we illustrate a member of this spline space which has four zeros. ■

§ 8.4. MONOSPLINES

In this section we consider a class of piecewise polynomials (called monosplines) that play an interesting role in several applications including best approximation and optimal quadrature formulae.

Given a set $\Delta=\{x_1<x_2<\cdots<x_k\}$ and a multiplicity vector $\mathfrak{M}=(m_1,\ldots,m_k)$ with $1\leqslant m_i\leqslant m$, $i=1,2,\ldots,k$, we define the class of *polynomial monosplines of degree m* as

$$\mathfrak{M}\mathcal{S}(\mathcal{P}_m;\mathfrak{M};\Delta)=\left\{\frac{x^m}{m!}+s(x)\colon s\in\mathcal{S}(\mathcal{P}_m;\mathfrak{M};\Delta)\right\}. \tag{8.58}$$

While $\mathfrak{M}\mathcal{S}(\mathcal{P}_m;\mathfrak{M};\Delta)$ is obtained by translating the linear space $\mathcal{S}(\mathcal{P}_m;\mathfrak{M};\Delta)$, it is clear that $\mathfrak{M}\mathcal{S}$ itself is not a linear space (it is convex). Despite this fact, monosplines are easy to deal with on a computer—indeed, it suffices to work with their spline parts. Many of the results obtained in Chapter 4 for polynomial splines imply analogous results for monosplines. To give just one example, we state a simple property of derivatives of monosplines:

THEOREM 8.41

Let $f\in\mathfrak{M}\mathcal{S}(\mathcal{P}_m;\mathfrak{M};\Delta)$. Then

$$D_+f\in\mathfrak{M}\mathcal{S}(\mathcal{P}_{m-1};\mathfrak{M}';\Delta),$$

where $\mathfrak{M}=(m_1',\ldots,m_k')$, and $m_i'=\min(m_i,m-1)$, $i=1,2,\ldots,k$.

Proof. This is an immediate corollary of Theorem 4.49. ■

There is no need to list explicitly the many other simple properties that monosplines inherit from polynomial splines. Instead, we turn now to some results on zeros of monosplines which are very useful in applications. First we give a Budan-Fourier-type theorem that provides an upper bound on the number of zeros a monospline can have.

To state the Budan-Fourier theorem for monosplines, we must first agree on how to count multiple zeros. Because of the monomial term, monosplines cannot vanish on intervals, and we are spared that complication. We still have to deal with zeros occurring at a knot, however. We shall use the following counting procedure (cf. Definition 4.45 for the case of polynomial splines):

DEFINITION 8.42

Suppose f is a monospline of order m such that $f(t-)=D_-f(t)=\cdots=D_-^{l-1}f(t)=0\neq D_-^l f(t)$ and $f(t+)=D_+f(t)=\cdots=D_+^{r-1}f(t)=0\neq D_+^r f(t)$, some $l,r\geqslant 0$. Let $\alpha=\max(l,r)$. Then we say that f has a *zero of multiplicity* z at t, where

$$z=\begin{cases}\alpha+1, & \text{if } \alpha \text{ is even and } s \text{ changes sign at } t\\ \alpha+1, & \text{if } \alpha \text{ is odd and } s \text{ does not change sign at } t\\ \alpha, & \text{otherwise.}\end{cases}$$

Definition 8.42 counts a point where a monospline jumps through zero as a zero of multiplicity 1. The definition is such that odd-order zeros are associated with sign changes, while even-order ones are not. The maximal order a zero can possess is $m+1$. Given a monospline f, we write $Z(f)$ for the number of zeros on the entire line **R**, counting multiplicities as in Definition 8.42. Similarly, if (a,b) is any subinterval of **R**, we write $Z_{(a,b)}(f)$ for the number of zeros of f on (a,b).

THEOREM 8.43. Budan-Fourier Theorem For Monosplines

Given $\Delta=\{x_1<\cdots<x_k\}$ and multiplicities $m_1,\ldots,m_k$, define

$$\sigma_i=\begin{cases}0, & \text{if } m_i \text{ is even}\\ 1, & \text{if } m_i \text{ is odd.}\end{cases} \tag{8.59}$$

Then for any $f\in\mathfrak{M}\mathcal{S}(\mathcal{P}_m;\mathfrak{M};\Delta)$,

$$Z(f)\leqslant m+\sum_{i=1}^{k}(m_i+\sigma_i). \tag{8.60}$$

More precisely, for any $a<b$,

$$Z_{(a,b)}(f) \leqslant m + \sum_{i=1}^{k} (m_i+\sigma_i) - S^+\left[f(a), -D_+f(a),\ldots,(-1)^m D_+^m f(a)\right]$$
$$- S^+\left[f(b), D_-f(b),\ldots,D_-^m f(b)\right], \tag{8.61}$$

where S^+ counts weak sign changes (cf. Definition 2.10).

Proof. Statement (8.60) follows immediately from (8.61). To establish (8.61), we apply the Budan-Fourier theorem for polynomials (Theorem 3.9) to each subinterval $I_i = (x_i, x_{i+1})$, $i=0,1,\ldots,k$, where for convenience we set $a = x_0$ and $b = x_{k+1}$. Because of the monomial term, f is a polynomial of exact order $m+1$ on each such subinterval, and Theorem 3.9 implies

$$Z_{(x_i,x_{i+1})}(f) \leqslant m - L_{i+1} - R_i,$$

where

$$R_i = S^+\left[f(x_i+), -D_+f(x_i),\ldots,(-1)^m D_+^m f(x_i)\right], \qquad i=0,1,\ldots,k;$$
$$L_i = S^+\left[f(x_i), D_-f(x_i),\ldots,D_-^m f(x_i)\right], \qquad i=1,2,\ldots,k+1.$$

Suppose f has a zero of multiplicity z_i at x_i, $i=1,2,\ldots,k$. Then summing up the above bounds and rearranging, we obtain

$$Z_{(a,b)}(f) \leqslant m + \sum_{i=1}^{k} (m_i+\sigma_i) - R_0 - L_{k+1}$$
$$- \sum_{i=1}^{k} (\gamma_i - 1 + \sigma_i), \tag{8.62}$$

where

$$\gamma_i = L_i + R_i + m_i + 1 - m - z_i.$$

The statement (8.61) now follows if we can show that $\gamma_i \geqslant 0$ for all $i=1,2,\ldots,k$.

The proof that $\gamma_i \geqslant 0$ requires several cases. We carry out only the case of $0 \leqslant z_i \leqslant m - m_i - 1$. Taking account of the fact that f and its derivatives up to order $m - m_i - 1$ are all continuous across x_i, and using the simple equality (2.48) for weak sign changes, we have

$$\gamma_i = m_i + 1 - m - z_i + L_i + R_i = m_i + 1 - m - z_i + 2z_i$$

$$+ S^+\left[D_-^{z_i} f(x_i), \ldots, D_-^m f(x_i)\right] + S^+\left[D_+^{z_i} f(x_i), \ldots, (-1)^{m-z_i} D_+^m f(x_i)\right]$$

$$\geq m_i + 1 - m + z_i + S^+\left[D^{z_i} f(x_i), \ldots, D^{m-m_i-1} f(x_i)\right]$$

$$+ S^+\left[D^{z_i} f(x_i), \ldots, (-1)^{m-m_i-1} D^{m-m_i-1} f(x_i)\right]$$

$$\geq m_i + 1 - m + z_i + m - m_i - 1 - z_i = 0.$$

Now $\gamma_i = 0$ implies $(-1)^{m_i} D^{m-m_i-1} f(x_i) < 0$ so m_i must be odd. ■

The assertion (8.61) can also be established directly by the same kind of inductive proof used to prove the Budan-Fourier Theorem 4.58 for polynomial splines. Here we have been able to use the Budan-Fourier theorem for polynomials in each of the subintervals because in each such subinterval a monospline is a polynomial of exact degree m.

Our next two theorems present some important properties for monosplines possessing a maximal number of zeros.

THEOREM 8.44

Let $f \in \mathfrak{M}\mathcal{S}(\mathcal{P}_m; \mathfrak{M}; \Delta)$ have zeros $t_1 \leq \cdots \leq t_N$ with $N = m + \sum_{i=1}^k (m_i + \sigma_i)$. Then for $i = 1, 2, \ldots, k$,

$$t_{l_i} \leq x_i \leq t_{m+l_{i-1}+1}, \qquad \text{if } m_i < m, \tag{8.63}$$

$$x_i = t_{l_i}, \qquad \text{if } m_i = m \text{ and } m \text{ is odd}, \tag{8.64}$$

where $l_i = \sum_{j=1}^i (m_j + \sigma_j)$. The inequalities in (8.63) are strict if x_i is at most an $m - m_i$-tuple zero.

Proof. Suppose $x_i < t_{l_i}$. Then the monospline f_R which agrees with f on the interval (x_i, ∞) has at least $m + \sum_{j=i+1}^k (m_j + \sigma_j) + 1$ zeros, but only $\sum_{j=i+1}^k m_j$ knots, counting multiplicities. This contradicts Theorem 8.43. If t_{l_i} falls at x_i, it can still be counted in obtaining this contradiction for f_R provided that the multiplicity of the zero of f at x_i is at most $m - m_i$. The upper inequalities in (8.63) are established similarly. Finally, to prove (8.64) we note that if m is odd and $m_i = m$, then $l_i = l_{i-1} + m + 1$, and the upper and lower bounds in (8.63) coincide. ■

THEOREM 8.45

Let f be as in Theorem 8.44 (i.e., f has a maximal set of zeros). Suppose we expand f as

$$f(x)=\frac{x^m}{m!}+\sum_{j=0}^{m-1}\frac{a_jx^j}{j!}+\sum_{i=1}^{k}\sum_{j=1}^{m_i}c_{ij}\frac{(x-x_i)_+^{m-j}}{(m-j)!}. \tag{8.65}$$

Then for all $1\leqslant i\leqslant k$ with m_i odd

$$c_{ij}<0, \qquad \text{all odd } j \text{ with } 1\leqslant j\leqslant m_i. \tag{8.66}$$

Moreover,

$$D^jf(x)>0, \qquad j=0,1,\ldots,m-1, \qquad \text{all } x\geqslant x_k \tag{8.67}$$

$$(-1)^{m-j}D^jf(x)>0, \qquad j=0,1,\ldots,m-1, \qquad \text{all } x\leqslant x_1.$$

Proof. If f has a maximal set of $m+\Sigma_{i=1}^k(m_i+\sigma_i)$ zeros, then by the sharp form (8.62) of the Budan-Fourier theorem, we must have

$$S^+[f(a),-Df(a),\ldots,(-1)^mD^mf(a)]=0$$

$$S^+[f(b),Df(b),\ldots,D^mf(b)]=0$$

and

$$\gamma_i=1-\sigma_i \qquad i=1,2,\ldots,k.$$

Since $a<x_1$ and $b>x_k$ are arbitrary, the first two of these assertions imply (8.67).

Suppose now that $1\leqslant i\leqslant k$ and that m_i is odd (so that $\sigma_i=1$). Then we must have $\gamma_i=0$. But the string of inequalities used in the proof of Theorem 8.43 to show that $\gamma_i\geqslant 0$ shows that $\gamma_i=0$ can happen only if

$$\begin{aligned}&S^+[D_-^{z_i}f(x_i),\ldots,D_-^mf(x_i)]\\&\quad+S^+[D_+^{z_i}f(x_i),\ldots,(-1)^{m-z_i}D_+^mf(x_i)]\\&\quad=S^+[D_-^{z_i}f(x_i),\ldots,D_-^{m-m_i-1}f(x_i)]\\&\quad+S^+[D_+^{z_i}f(x_i),\ldots,(-1)^{m-m_i-1}D_+^{m-m_i-1}f(x_i)]=m-m_i-1-z_i.\end{aligned}$$

This can happen [cf. (2.48)] if and only if all components of these vectors are nonzero, and since $D_-^m f(x_i) = D_+^m f(x_i) > 0$ it follows that

$$D^\nu f(x_i) \neq 0, \qquad \nu = z_i, \dots, m - m_i - 1$$

$$D_-^\nu f(x_i) > 0, \qquad \nu = m - m_i, \dots, m$$

$$(-1)^{m-\nu} D_+^\nu f(x_i) > 0, \qquad \nu = m - m_i, \dots, m.$$

But $c_{ij} = D_+^{m-j} f(x_i) - D_-^{m-j} f(x_i)$, and assertion (8.66) follows for odd j with $1 \leqslant j \leqslant m_i$. ■

Theorem 8.43 showed that the maximal number of zeros a monospline can possess is $N = m + \sum_{i=1}^k (m_i + \sigma_i)$. In the remainder of this section we consider the problem of constructing a monospline with a given set of N zeros (counting multiplicities). This problem turns out to be more difficult than it sounds. We begin by considering the case of $m = 1$.

THEOREM 8.46

Let $t_1 \leqslant t_2 \leqslant \cdots \leqslant t_{1+2k}$ be prescribed points with $t_i < t_{i+2}$, $i = 1, 2, \dots, 2k-1$. Then there exists a unique monospline f of degree 1 with k simple knots such that f has zeros at $t_1, \dots, t_{1+2k}$.

Proof. If f is to be a monospline of degree 1 with k simple knots $x_1 < x_2 < \cdots < x_k$ and the $1+2k$ zeros $t_1 \leqslant t_2 \leqslant \cdots \leqslant t_{1+2k}$, then according to Theorem 8.44 the knots must be located at the points

$$x_i = t_{2i}, \qquad i = 1, 2, \dots, k.$$

Now we know that such a monospline can be written in the form

$$f(x) = x + \sum_{i=1}^{k} c_i (x - x_i)_+^{=} + c_0. \tag{8.68}$$

It remains to determine the coefficients $c_0, c_1, \dots, c_k$. If the t's are distinct, we have the following nonsingular system:

$$\begin{bmatrix} 1 & (t_1 - x_1)_+^0 & \cdots & (t_1 - x_k)_+^0 \\ 1 & (t_3 - x_1)_+^0 & \cdots & \\ \cdot & & & \\ 1 & (t_{2k+1} - x_1)_+^0 & \cdots & (t_{2k+1} - x_k)_+^0 \end{bmatrix} \begin{bmatrix} c_0 \\ c_1 \\ \vdots \\ c_k \end{bmatrix} = \begin{bmatrix} -t_1 \\ -t_3 \\ \vdots \\ -t_{2k+1} \end{bmatrix}.$$

The result for a general set of $t_1 \leqslant \cdots \leqslant t_{1+2k}$ follows by taking the limit of a sequence of monosplines with simple zeros. By construction, it is clear that the corresponding monospline (8.68) always vanishes at one or both of the points $t_{2i+1}-$ and $t_{2i+1}+$, $i=0,1,\ldots,k$. Thus, for example, if $t_1<t_2<t_3$, then f has a simple zero at t_1, and (since f has slope 1 almost everywhere) f jumps down through zero at t_2. Similarly, if $t_1=t_2<t_3$, then f has a double zero at t_1, and if $t_1<t_2=t_3$, then f has a simple zero at t_1 and a double zero at t_2. Figure 29 shows the various cases. This argument can be continued to show that f has precisely the zeros $\{t_i\}_1^{1+2k}$. A typical linear monospline with the various cases combined in one example is shown in Figure 30. Since both the knots and the coefficients are uniquely determined, it follows that f is also unique. ■

It should be observed that in Theorem 8.46 we were able to specify the number of knots of the desired monospline, but not their locations. This suggests that if we are seeking monosplines of order $m>1$ with a maximal set of prescribed zeros, we should be able to specify the number of knots and their multiplicities, but not their locations. The following theorem shows that this is indeed the case, at least when the prescribed zeros are all simple and distinct:

THEOREM 8.47

Let m and $1 \leqslant m_i \leqslant m$, $i=1,2,\ldots,k$ be prescribed, and let

$$\sigma_i = \begin{cases} 0, & \text{if } m_i \text{ is even} \\ 1, & \text{if } m_i \text{ is odd.} \end{cases}$$

Then for any $t_1<t_2<\cdots<t_N$ with $N=m+\sum_{i=1}^k(m_i+\sigma_i)$, there exists a set of knots $x_1<x_2<\cdots<x_k$ and a corresponding monospline f of degree m with knots at the x_i of multiplicity m_i such that

$$f(t_i)=0, \qquad i=1,2,\ldots,N.$$

If $m_1,\ldots,m_k$ are all odd, this monospline is unique.

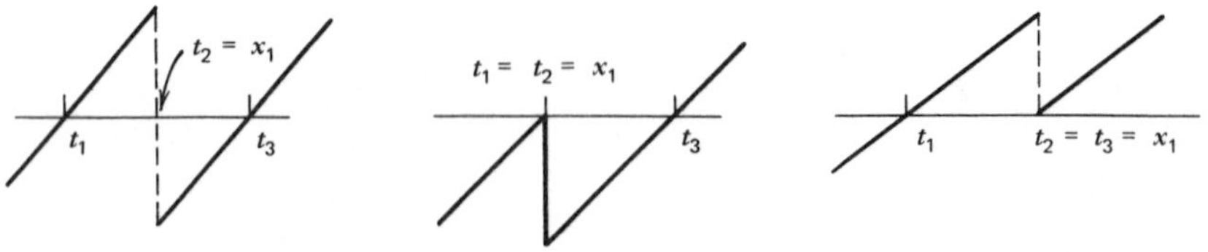

Figure 29. Zeros of a linear monospline.

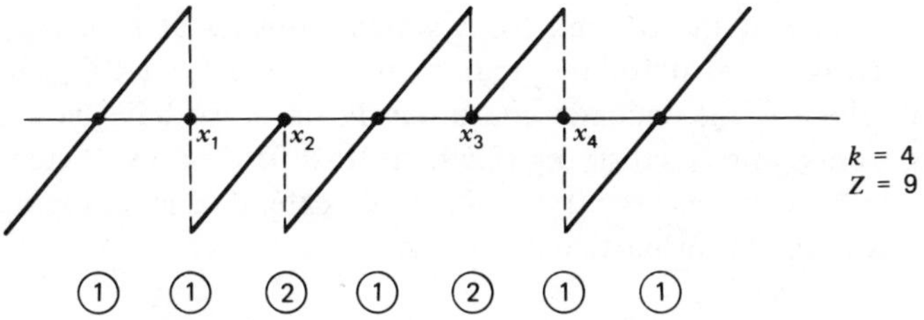

Figure 30. A linear monospline with $1+2k$ zeros.

Proof. We assume first that all $m_1,\ldots,m_k$ are odd, and furthermore that $m_i \leqslant m-2$, $i=1,2,\ldots,k$. To prove the theorem in this case we proceed by induction on k. The result is obvious for $k=0$. Now suppose it has been established for monosplines with $k-1$ knots; that is, there exists a unique monospline

$$g(x)=\frac{x^m}{m!}+\sum_{i=0}^{m-1}\frac{a_i x^i}{i!}+\sum_{i=1}^{k-1}\sum_{j=1}^{m_i}\frac{c_{ij}(x-x_i)_+^{m-j}}{(m-j)!}$$

with $g(t_i)=0$, $i=1,2,\ldots,n=m+\sum_{i=1}^{k-1}(m_i+1)$. To establish the result for k knots, we are going to show that there exists an interval $\bar{I}=[\underline{\xi},\bar{\xi}]$ such that for each $\xi\in\bar{I}$ there exists a monospline of the form

$$f_\xi(x)=\sum_{i=0}^{m-1}a_i(\xi)\frac{x^i}{i!}+\sum_{i=1}^{k}\sum_{j=1}^{m_i}c_{ij}(\xi)\frac{(x-x_i(\xi))_+^{m-j}}{(m-j)!},\tag{8.69}$$

with

$$f_\xi(t_i)=0,\qquad i=1,2,\ldots,N-1\qquad\text{and } f_\xi(t_N)\leqslant 0.\tag{8.70}$$

It will turn out that f_ξ is the monospline that vanishes at all of the $t_1,\ldots,t_N$.

To define the interval $\bar{I}$ and the mapping $\xi\mapsto f_\xi$, we first have to show that there is at least one monospline satisfying (8.70). We construct such a monospline by tacking something onto the monospline g. Let $l_1(x),\ldots,l_{m_k}(x)$ be the Lagrange polynomials of order m_k associated with Lagrange interpolation at the points $t_{n+1},\ldots,t_{n+m_k}$ (cf. Problem 2.6). Then for any given $t_n\leqslant\xi<t_{n+1}$, it is clear that

$$f_\xi(x)=g(x)-(x-\xi)_+^{m-m_k}\sum_{j=1}^{m_k}\frac{g(t_{n+j})l_j(x)}{(t_{n+j}-\xi)^{m-m_k}}\tag{8.71}$$

is a monospline of the desired form which vanishes at $t_1,\ldots,t_{N-1}$. Since $\lim_{\xi\to t_{n+1}} f_\xi(t_N) = -\infty$, it follows that there is some $t_n < \bar{\xi} < t_{n+1}$ such that $f_{\bar{\xi}}(t_N) < 0$. Since $f_\xi(t_N)$ depends continuously on ξ, we have now set up a mapping from ξ into f_ξ satisfying (8.69), at least for ξ in a sufficiently small interval with right end-point $\bar{\xi}$. We now extend this mapping as far leftward as possible. In particular, define

$$\underline{\xi} = \inf\left\{\begin{array}{l}\tau\text{: for all } \tau < \xi \leqslant \bar{\xi} \text{ there is a unique}\\ \text{monospline } f_\xi \text{ as in (8.69) satisfy-}\\ \text{ing (8.70), and the mapping from}\\ (\xi, x) \text{ into } f_\xi(x) \text{ is a } C^1 \text{ function for}\\ \text{all } (\xi, x) \in (\tau, \bar{\xi}] \times \mathbf{R}\end{array}\right\}.$$

In general, $\underline{\xi}$ may be smaller than t_n.

For each $\xi \in I = (\underline{\xi}, \bar{\xi}]$, the fact that $f_{\underline{\xi}}(t_N) < 0$ while $f_\xi(x)$ increases to ∞ as $x \to \infty$ implies that there exists $t_N(\xi) > t_N$ such that $f_\xi[t_N(\xi)] = 0$. Since f_ξ can have at most N zeros, $t_N(\xi)$ is in fact unique. We now show that it is an increasing function of ξ. To prove this, it suffices to show that for all $x > t_N$, $f_\xi(x)$ is an increasing function of ξ. To this end, we compute the partial derivative with respect to ξ:

$$\frac{\partial}{\partial\xi} f_\xi(x) = \sum_{i=0}^{m-1} \frac{a_i'(\xi) x^i}{i!} + \sum_{i=1}^{k} \sum_{j=1}^{m_i} c_{ij}'(\xi) \frac{(x - x_i(\xi))_+^{m-j}}{(m-j)!}$$
$$- \sum_{i=1}^{k} \sum_{j=1}^{m_i} c_{ij}(\xi) x_i'(\xi) \frac{(x - x_i(\xi))_+^{m-j-1}}{(m-j-1)!}$$
$$= \sum_{i=0}^{m-1} a_i'(\xi) \frac{x^i}{i!} + \sum_{i=1}^{k} \sum_{j=1}^{m_i} d_{ij}(\xi) \frac{(x - x_i(\xi))_+^{m-j}}{(m-j)!}$$
$$- \sum_{i=1}^{k} c_{im_i}(\xi) x_i'(\xi) \frac{(x - x_i(\xi))_+^{m-m_i-1}}{(m-m_i-1)!},$$

where we have written $d_{ij} = c_{ij}' - c_{ij-1} x_i'$ in order to achieve a more tractable form. Since for any $1 \leqslant \nu \leqslant N-1$, $f_\xi(t_\nu) = 0$, we also have

$$0 = \sum_{i=0}^{m-1} a_i'(\xi) \frac{t_\nu^i}{i!} + \sum_{i=1}^{k} \sum_{j=1}^{m_i} d_{ij}(\xi) \frac{(t_\nu - x_i(\xi))_+^{m-j}}{(m-j)!}$$
$$- \sum_{i=1}^{k} c_{im_i}(\xi) x_i'(\xi) \frac{(t_\nu - x_i(\xi))_+^{m-m_i-1}}{(m-m_i-1)!},$$

$\nu=1,2,\ldots,N-1$. These are N equations for the N values of $a_0',\ldots,a_{m-1}'$, $\{d_{ij}\}_{j=1,i=1}^{m_i\ \ k}$, and $\{c_{im_i}x_i'\}_{i=1}^{k}$. Solving for $c_{im_i}x_i'$, we obtain

$$c_{im_i}x_i'=(-1)^{m_i}\frac{\dfrac{\partial f_\xi}{\partial \xi}(x)g_m\begin{bmatrix} t_1,\ldots,t_{N-1},x \\ \underbrace{0,\ldots,0}_{m},\ \underbrace{x_1,\ldots,x_1}_{m_1+1},\ldots,\underbrace{x_i,\ldots,x_i}_{m_i},\ldots,\underbrace{x_k,\ldots,x_k}_{m_k+1}\end{bmatrix}}{g_m\begin{bmatrix} t_1,\ldots,t_{N-1},x \\ \underbrace{0,\ldots,0}_{m},\ \underbrace{x_1,\ldots,x_1}_{m_1+1},\ldots,\underbrace{x_k,\ldots,x_k}_{m_k+1}\end{bmatrix}},$$

where the determinants formed from the Green's function g_m are defined in Theorem 4.78. This expression is well defined since f_ξ has a maximal set of zeros, hence the conditions (8.63) hold (strictly), and thus by Theorem 4.78, the determinant in the denominator is positive. By the same theorem, the determinant in the numerator is nonnegative, while the factor $(-1)^{m_i}$ is negative since m_i is odd. On the other hand, by Theorem 8.45, $c_{im_i}<0$ for all i. We conclude that since $x_k'(\xi)=\xi'=1$, $\partial f_\xi(x)/\partial\xi\geqslant 0$. We have proved that $t_N(\xi)$ is an increasing function as asserted. As a by-product of this argument, we also observe that $x_i'(\xi)\geqslant 0$, $i=1,2,\ldots,k-1$.

We now define a monospline $f_{\bar\xi}(x)$ associated with $\bar\xi$. Since $t_N(\xi)$ is monotone increasing, it follows that for all $\xi\in I$ all N zeros of f_ξ lie in a fixed bounded interval. This implies (cf. Lemma 8.48 below) that all of the coefficients of f_ξ are uniformly bounded, hence they converge for some sequence of ξ_ν going to $\bar\xi$. As $f_{\bar\xi}$ is the pointwise limit of f_{ξ_ν}, it follows that $f=f_{\bar\xi}$ satisfies (8.70). We claim that f also vanishes at t_N, hence it is the desired monosplines.

To show that $f(t_N)=0$, we consider the mapping

$$\varphi:(a_0,\ldots,a_{m-1},\ldots,c_{11},\ldots,c_{1m_1},x_1,\ldots,c_{k1},\ldots,c_{km_k},\xi)\to[f(t_1),\ldots,f(t_N)].$$

If $f(t_N)<0$, then the Jacobian of φ with respect to its first $N-1$ variables is nonzero when evaluated at the coefficients of f. By the implicit function theorem, we could then extend the definition of f_ξ to $\xi<\bar\xi$, contradicting the definition of $\bar\xi$. We have proved the existence in the case where all knots are of odd multiplicity at most $m-2$. Concerning uniqueness in this case, suppose $\hat f$ were another monospline of the form (8.69) with the zeros $t_1,\ldots,t_N$. By the same kind of analysis just carried out, we can deform $\tilde f$ continuously such that (8.70) remains satisfied until its largest knot reaches t_n. Since uniqueness was assumed for monosplines with $k-1$ knots, the deformation must be equal to the monospline in (8.71) at $\xi = t_n$. But this monospline was uniquely deformed into f_ξ, and we conclude that $\tilde{f} = f_\xi$.

We now show that the result also holds if some knots have even multiplicity, $m_i \leqslant m-2$, all i. Let $\mathcal{E} = \{j: m_j \text{ is even}\}$ and $\mathcal{O} = \{j: m_j \text{ is odd}\}$. Let $N = m + \sum_{j=1}^k (m_j + \sigma_j)$. Then, given any $t_1 < \cdots < t_N$ there exists a monospline

$$f^*(x) = \frac{x^m}{m!} + \sum_{i=0}^{m-1} \frac{a_i x^i}{i!} + \sum_{\substack{i=1\\ i\in\mathcal{E}}}^{k} \sum_{j=1}^{m_i} c_{ij}^*(x-x_i)_+^{m-j} + \sum_{\substack{i=1\\ i\in\mathcal{O}}}^{k} \sum_{j=1}^{m_i} c_{ij}(x-x_i)_+^{m-j}$$

with zeros at $t_1, \ldots, t_N$. Let $n = m + \sum_{i=1}^k (m_i + 1) > N$, and define a mapping ψ from $\mathbf{R}^n$ into $\mathbf{R}^N$ by

$$\psi: (a_0, \ldots, a_{m-1}, c_{11}, \ldots, c_{1m_1}, x_1, \ldots, c_{k1}, \ldots, c_{km_k}, x_k) \mapsto [f(t_1), \ldots, f(t_N)],$$

where

$$f(x) = \frac{x^m}{m!} + \sum_{i=0}^{m-1} \frac{a_i x^i}{i!} + \sum_{i=1}^{k} \sum_{j=1}^{m_i} c_{ij}(x-x_i)_+^{m-j}.$$

Note that ψ vanishes if we put in the coefficients of f^*. The Jacobian of ψ with respect to all of its variables except for $\{x_i: i \in \mathcal{E}\}$ is

$$J = C \prod_{i\in\mathcal{O}} c_{im_i} \det g_m \begin{bmatrix} & & t_1, \ldots, t_N & \\ \underbrace{0, \ldots, 0}_{m}, & \underbrace{x_1, \ldots, x_1}_{m_1+\sigma_1}, & \ldots, & \underbrace{x_k, \ldots, x_k}_{m_k+\sigma_k} \end{bmatrix},$$

where C is a nonzero constant. J is a continuous function of its arguments. Since f^* has a maximal set of zeros, Theorem 4.78 asserts that J is not zero at the coefficients of f^*. By the implicit function theorem, the monospline $f(x)$ will have the same zeros for all coefficients in a neighborhood of those of f^*.

We now establish the result allowing knots with multiplicity greater than $m-2$. Suppose for the moment there is just one such knot, say x_ν, with $m_\nu = m-1$, m even. Let f_L be a monospline with $f_L(t_i) = 0$, $i = 1, 2, \ldots, n = m + \sum_{j=1}^{\nu-1}(m_j + \sigma_j)$, and f_R a monospline with $k - \nu$ knots satisfying $f_R(t_i) = 0$, $i = n+1, \ldots, N$. By (8.60), $f_L(x) > 0$ for $x > t_n$ and $f_R(x) > 0$ for $x < t_{n+1}$. Thus there exists a unique $t_n < x_\nu < t_{n+1}$ such that $f_L(x) = f_R(x)$. Clearly the monospline

$$f(x) = \begin{cases} f_L(x), & x \leqslant x_\nu \\ f_R(x), & x \geqslant x_\nu \end{cases}$$

has the desired zero set. If m is odd, then we construct f_R to vanish at $(t_n + t_{n+1})/2, t_{n+1}, \ldots, t_N$ since $\sigma_\nu = 1$ and so $N - n = m - 1 + \Sigma_{\nu+1}^k (m_j + \sigma_j)$. Again, we can piece f_L and f_R together to produce a monospline f with the desired zeros. (Clearly this construction is not unique.) A similar argument works if $m_i = m$. ■

Theorem 8.47 coupled with the bound (8.60) is usually referred to as the *fundamental theorem of algebra for monosplines*. The following lemma was used in the proof of Theorem 8.47, but since it is of interest in its own right, we state it separately:

LEMMA 8.48

Suppose f is a monospline as in (8.69) with $\{m_i\}_1^k$ odd and $N = m + \Sigma_{i=1}^k (m_i + 1)$ distinct zeros in the interval $(-K, K)$. Then there exists a constant C (depending only on m and K) such that

$$|a_i| \leqslant C, \qquad i = 0, 1, \ldots, m-1$$

$$|c_{ij}| \leqslant C, \qquad j = 1, 2, \ldots, m_i$$
$$i = 1, 2, \ldots, k.$$

Proof. The proof proceeds by double induction on m and on k. The case of $m \geqslant 1$, and $k = 0$ is obvious. The case of $m = 1$ and $k \geqslant 1$ follows directly from properties of the linear monosplines (cf. Figures 29 and 30). Now suppose the result has been established for all monosplines of degree $m - 1$ with k knots and for all monosplines of degree m with $k - 1$ knots. We now establish it for monosplines of degree m with k knots. Let f be such a monospline. Suppose that $m_i < m$, $i = 1, 2, \ldots, k$. Then by applying the analog of Rolle's theorem for monosplines (cf. Theorem 4.50), we conclude that the monospline $D_+ f$ has $N - 1$ distinct zeros in the interval $(-K, K)$. Since by Theorem 8.41 $D_+ f$ is a monospline of degree $m - 1$, the induction hypothesis assures that its coefficients are bounded. We conclude that except possibly for a_0, the coefficients of f are also bounded. But since f vanishes at least once, a_0 is also bounded.

If f has some m-tuple knot, say x_l, then it is easily seen that the monosplines f_L and f_R, which agree with f to the left and right of x_l, respectively, both have a maximum set of zeros, and by the inductive hypothesis must have bounded coefficients. The result follows. ■

It is easy to show by example that Lemma 8.48 does not hold when knots of even multiplicity are present.

We illustrate the difference between the cases where the multiplicities are even and odd in the following example. It shows clearly how non-uniqueness occurs in the even case.

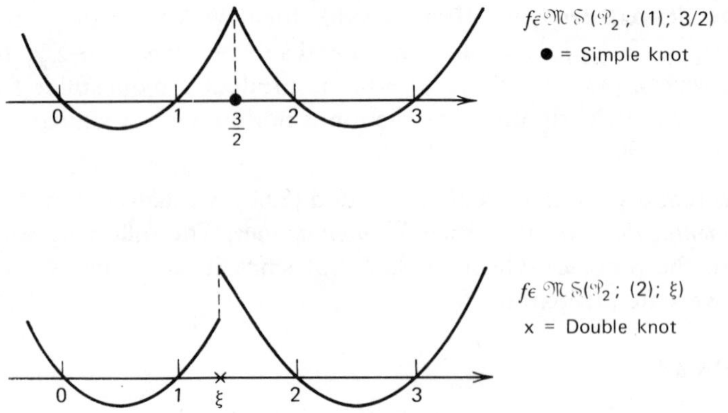

Figure 31. Monosplines of Example 8.49.

EXAMPLE 8.49

Find a monospline of degree 2 with one knot (of either multiplicity 1 or multiplicity 2) that has the zeros 0, 1, 2, and 3.

Discussion. If we want a monospline with one simple knot, then the unique solution of the problem is given by

$$f(x)=\begin{cases} \dfrac{x(x-1)}{2}, & x \leqslant \dfrac{3}{2} \\ \dfrac{(x-2)(x-3)}{2}, & x \geqslant \dfrac{3}{2} \end{cases}.$$

This monospline is shown in Figure 31. If we consider monosplines of degree 2 with one double knot, then there is a whole one-parameter family of solutions; namely,

$$f_\xi(x)=\begin{cases} \dfrac{x(x-1)}{2}, & x \leqslant \xi \\ \dfrac{(x-2)(x-3)}{2}, & x \geqslant \xi \end{cases}$$

for any $1<\xi<2$. A typical monospline in this family is also depicted in Figure 31. ■

§ 8.5. DISCRETE SPLINES

In this section we examine certain linear spaces of piecewise polynomials (called discrete splines) defined on discrete subsets of the real line **R**. In particular, given a fixed a and $h>0$, we are interested in splines defined on the discrete line

$$\mathbf{R}_{a,h} = \{\ldots, a-h, a, a+h, a+2h, \ldots\}$$

or on discrete intervals of the form

$$[a,b]_h = \{a, a+h, \ldots, a+Nh\}, \qquad b = a+Nh.$$

We begin with the definition of the space of discrete splines. Let $\Delta = \{a = x_0 < x_1 < \cdots < x_{k+1} = b\} \subseteq \mathbf{R}_{a,h}$ be a set of points partitioning the discrete interval $[a,b]_h$ into intervals $I_0, I_1, \ldots, I_k$. Let $\mathfrak{M} = (m, \ldots, m_k)$ be a vector of positive integers with $m_i \leqslant m$, $i = 1, 2, \ldots, k$. Then we define the space of *discrete polynomial splines of order m with knots at* $x_1, \ldots, x_k$ *of multiplicities* $m_1, \ldots, m_k$, by

$$\mathcal{S}(\mathcal{P}_m; \mathfrak{M}; \Delta; h) = \left\{ \begin{array}{l} s\text{: there exists } s_0, \ldots, s_k \text{ in } \mathcal{P}_m \text{ with} \\ s|_{I_i} = s_i, \quad i = 0, 1, \ldots, k \quad \text{and} \\ D_h^{j-1} s_{i-1}(x_i) = D_h^{j-1} s_i(x_i) \quad \text{for } j = \\ 1, \ldots, m - m_i, \ i = 1, 2, \ldots, k \end{array} \right\}, \tag{8.72}$$

where D_h^j is the difference operator defined by

$$D_h^j f(x) = \frac{\Delta_h^j f(x)}{h^j} = \sum_{\nu=0}^{j} \frac{\binom{j}{\nu}(-1)^{j-\nu} f(x+\nu h)}{h^j}. \tag{8.73}$$

The space $\mathcal{S}(\mathcal{P}_m; \mathfrak{M}; \Delta; h)$ is very similar to the space of polynomial splines studied in Chapter 4 except that the ties between the polynomial pieces are described here in terms of forward differences instead of derivatives. As in ordinary polynomial splines, we shall write

$$\mathcal{S}_m(\Delta; h) = \mathcal{S}(\mathcal{P}_m; \mathfrak{M}^1; \Delta; h) \qquad \text{if } \mathfrak{M}^1 = (1, \ldots, 1)$$

for the space of discrete splines with simple knots.

Although our definition of $\mathcal{S}(\mathcal{P}_m; \mathfrak{M}; \Delta; h)$ makes sense for any interval $[a,b]_h$ and any partition Δ, in order to rule out special (and uninteresting) cases, we shall henceforth assume that

$$\text{each interval } I_i \text{ contains at least } m_i \text{ points, } i = 0, 1, \ldots, k, \tag{8.74}$$

where for convenience we set $m_0 = m$. This prevents the conditions tying the polynomials together at the knot x_i from involving points that do not lie in the interval immediately to the right of x_i. It also assures that $[a,b]_k$ contains at least $m+K$ points, where $K=\sum_{i=1}^k m_i$.

It is clear that $\mathbb{S}(\mathcal{P}_m;\mathfrak{M};\Delta;h)$ is a linear space. The following theorem gives its dimension:

THEOREM 8.50

Suppose (8.74) holds. Then $\mathbb{S}(\mathcal{P}_m;\mathfrak{M};\Delta;h)$ has dimension $m+K$, where $K=\sum_{i=1}^k m_i$.

Proof. The proof follows along the same lines as the proof of Theorem 4.4 with

$$A_i=\begin{bmatrix} 1 & x_i & x_i^2 & \cdots & x_i^{m-1} \\ 0 & 1 & D_h^1 x_i^2 & \cdots & D_h^1 x_i^{m-1} \\ \cdots & & & & \\ 0 & 0 \quad \cdots & 1 & \cdots & D_h^{m-1-m_i} x_i^{m-1} \end{bmatrix}. \quad \blacksquare$$

We turn now to the task of constructing a one-sided basis for $\mathbb{S}$. As we shall see, it is convenient to introduce the *factorial functions*

$$x^{(0)_h}=1, \qquad x^{(n)_h}=x(x-h)\cdots[x-(n-1)h], \qquad n\geqslant 1. \tag{8.75}$$

The nice thing about these functions as compared to the usual powers of x is that

$$D_h^j x^{(n)_h}=\begin{cases} \dfrac{n!\,x^{(n-j)_h}}{(n-j)!}, & 1\leqslant j\leqslant n \\ 0, & n<j. \end{cases} \tag{8.76}$$

We shall also need the discrete plus function

$$(x-y)_+^{(n)_h}=(x-y)^{(n)_h}(x-y)_+^0. \tag{8.77}$$

It is clear that this function is identically zero to the left of y, is a polynomial of order n to the right of y, and vanishes at the points $y, y+h,\ldots,y+(n-1)h$. We also note that

$$D_h^j(x-y)_+^{(n)_h}\big|_{x=y}=n!\,\delta_{j,n}, \qquad j=0,1,\ldots,n. \tag{8.78}$$

The results of Chapter 4 now suggest that a one-sided basis for $\mathbb{S}$ should be given by the functions

$$\left\{\rho_{ij}(x)=(x-x_i)_+^{(m-j)_h}\right\}_{j=1,i=0}^{m_i\ \ k},$$

where $m_0=m$, and this can, in fact, be established by the same methods as used in the proof of Theorem 4.5. The following theorem gives an alternate (and somewhat simpler) one-sided basis:

THEOREM 8.51

Let $y_i=a-(m-i)h$, $i=1,2,\ldots,m$, and let

$$y_{m+1}<\cdots<y_{m+K}=x_1-(m_1-1)h,\ldots,x_1-h,x_1,\ldots,x_k-(m_k-1)h,\ldots,x_k.$$

Then the functions $\{\rho_i(x)=(x-y_i)_+^{(m-1)_h}\}_{i=1}^{m+K}$ form a basis for $\mathbb{S}(\mathcal{P}_m;\mathfrak{M};\Delta;h)$.

Proof. First, we claim that each of these functions belongs to $\mathbb{S}$. To show this, consider a typical one, say

$$\rho(x)=(x-x_i-l\cdot h)_+^{(m-1)_h}.$$

By the definition of the factorial function, ρ is zero at the points $x_i-l\cdot h,\ldots,x_i+(m-l-2)h$. It follows that for $x\in[a,b]_h$, ρ is a polynomial of order m for $x\geqslant x_i$ and $\rho(x)\equiv 0$ for $x<x_i$. In addition,

$$D_h^j\rho(x_i)=0, \qquad j=0,1,\ldots,m-l-2.$$

It follows that ρ belongs to $\mathbb{S}$ as long as $0\leqslant l\leqslant m_i-1$. The one-sided nature of the ρ's is clear, and the proof can be completed by showing their linear independence by the same kind of argument used to prove Theorem 4.5. ■

Theorem 8.51 asserts that if we take $\tilde{\Delta}=\{y_i\}_{i=m+1}^{m+K}$, then

$$\mathbb{S}(\mathcal{P}_m;\mathfrak{M};\Delta;h)=\mathbb{S}_m(\tilde{\Delta};h).$$

It should perhaps be emphasized that this equivalence holds considering the splines as functions on $[a,b]_h$. Since elements in these spaces are piecewise polynomials, they make sense for all $x\in\mathbf{R}$—the two spaces do not agree on this larger set, however.

Our next task is to construct a basis for $\mathbb{S}$ consisting of local support splines similar to the basis of B-splines constructed in Chapter 4 for the usual polynomial splines. The development there suggests that we should define basis splines as divided differences of the function $(y-x)_+^{(m-1)_h}$ over an appropriate extended partition. We do exactly that in the following theorem:

THEOREM 8.52

Let $\{y_i\}_1^{m+K}$ be as in Theorem 8.51, and set $y_{m+K+i}=b+(i-1)h$, $i=1,2,\ldots,m$. For each $i=1,2,\ldots,m+K$ define

$$B_i(x)=(-1)^m(y_{i+m}-y_i)[y_i,\ldots,y_{i+m}](y-x)_+^{(m-1)_h}.$$

Then $\{B_i\}_1^{m+K}$ is a basis for $\mathcal{S}(\mathcal{P}_m;\mathfrak{M};\Delta;h)$, and

$$B_i(x)=0 \qquad \text{for } x<y_i \text{ and } y_{i+m}<x;$$

$$B_i(x)>0 \qquad \text{for } y_i+(m-2)h<x<y_{i+m}, \qquad i=1,2,\ldots,m+K;$$

$$\sum_{i=1}^{m+K} B_i(x)=1, \qquad \text{all } a\leqslant x\leqslant b.$$

Proof. By the definition of the divided difference, for $x\in[a,b]_h$ each B_i is a linear combination of the functions $\rho_1,\ldots,\rho_{m+K}$ of Theorem 8.51, and thus is an element of $\mathcal{S}$. The linear independence of the B's as well as the other stated properties follow from general results on B-splines, to be established below. ■

We now briefly recount the main properties of discrete B-splines. Since the development follows that in §4.3 for polynomial B-splines, and since we do not need to work with multiple knots, we can suppress most of the details. We begin with a definition.

Suppose $y_i<y_{i+1}<\cdots<y_{i+m}$ are points in $\mathbf{R}_{a,h}$. Then we call

$$Q_{i,h}^m(x)=\begin{cases}(-1)^m[y_i,\ldots,y_{i+m}](x-y)_+^{(m-1)_h}, & y_i<y_{i+m}\\ 0, & \text{otherwise}\end{cases} \tag{8.79}$$

the *mth order discrete B-spline* associated with the knots $y_i,\ldots,y_{i+m}$. It is also useful to introduce the normalized version

$$N_{i,h}^m(x)=(y_{i+m}-y_i)Q_{i,h}^m(x).$$

For $m=1$ and 2 it is clear that the discrete B-splines are identical with the usual polynomial B-splines. For $m>2$ they differ slightly from the usual B-splines because we are using the factorial function $(y-x)_+^{(m-1)_h}$ instead of the Green's function $(y-x)_+^{m-1}$. In particular, if we consider

$Q_{i,h}^m(x)$ as a function of $x \in \mathbf{R}$, then we see that it has a small ripple in the interval $[y_i, y_i + (m-2)h]$ (where it even goes negative). See Figure 32. It is true (but we do not make use of it) that if $y_i, \ldots, y_{i+m}$ are held fixed and h is allowed to go to zero, then $Q_{i,h}^m(x)$ approaches $Q_i^m(x)$ for all $x \in \mathbf{R}$.

As in the usual case, discrete B-splines satisfy an important recursion relation.

THEOREM 8.53

Let $m \geqslant 2$. Then for all $x \in \mathbf{R}$,

$$Q_{i,h}^m(x) = \frac{[x - y_i - (m-2)h]\, Q_{i,h}^{m-1}(x) + [y_{i+m} - x + (m-2)h]\, Q_{i+1,h}^{m-1}(x)}{(y_{i+m} - y_i)}. \tag{8.80}$$

Proof. The proof proceeds exactly as in Theorem 4.15 by applying the Leibnitz rule for divided differences of a product of functions to

$$(x-y)^{(m-1)_h} = (x-y)^{(m-2)_h}[x - y - (m-2)h]. \qquad \blacksquare$$

The recursion (8.80) is clearly an important tool for dealing computationally with discrete B-splines. It is also a valuable theoretical tool. For example, by a simple inductive argument (cf. Theorem 4.17), we may show that

$$Q_{i,h}^m(x) > 0, \qquad y_i + (m-2)h < x < y_i, \tag{8.81}$$

while from the definition,

$$Q_{i,h}^m(x) = 0, \qquad x \leqslant y_i, \qquad y_{i+m} \leqslant x. \tag{8.82}$$

We have already noted above that $Q_{i,h}^m(x)$ takes on negative values in some subintervals of $[y_i, y_i + (m-2)h]$. On the other hand, clearly for $x \in \mathbf{R}_{a,h}$, it

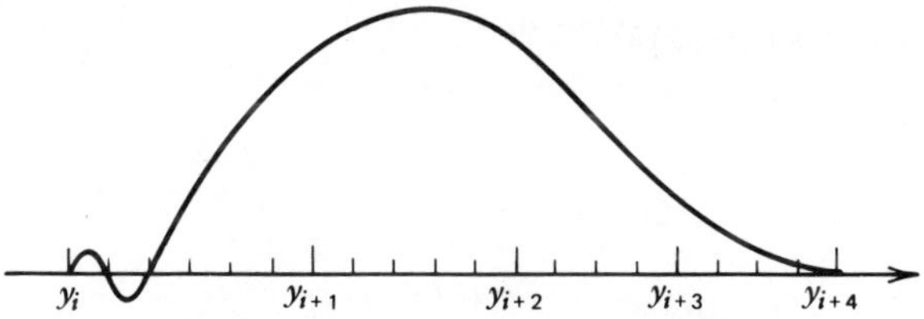

Figure 32. A discrete cubic B-spline.

is a nonnegative function; indeed,

$$Q_{i,h}^m(y_i+jh)=0, \qquad j=0,1,\ldots,m-2.$$

To discuss the linear independence of discrete B-splines, we first need an analog of Lemma 4.7.

LEMMA 8.54

Let $\tau_1<\tau_2<\cdots<\tau_m$, and suppose

$$\sum_{j=1}^{m} \alpha_j(x-\tau_j)_+^{(m-1)_h}=0 \qquad \text{for all } x>\tau_m.$$

Then $\alpha_1=\cdots=\alpha_m=0$.

Proof. The proof follows immediately after substituting

$$(x-\tau_j)^{(m-1)_h}=\sum_{\nu=0}^{m-1}\binom{m-1}{\nu}(-\tau_j)^{(m-\nu-1)_h}\cdot x^{(\nu)_h} \tag{8.83}$$

and using the linear independence of 1, $x^{(1)_h},\ldots,x^{(m-1)_h}$ (cf. the proof of Lemma 4.7). ■

THEOREM 8.55

Suppose $l<r$ and that $y_l<y_{l+1}$ and $y_{r-1}<y_r$. Then $\{Q_{i,h}^m\}_{i=l+1-m}^r$ are linearly independent on the interval $[y_l,y_r]_h$.

Proof. The proof proceeds exactly as in Theorem 4.18, using Lemma 8.54. ■

Our next result is the analog of Marsden's identity.

THEOREM 8.56

Let $l<r$ and $y_l<y_{r+1}$. Then for any $x,y\in\mathbf{R}_{a,h}$,

$$(y-x+(m-2)h)^{(m-1)_h}=\sum_{i=l+1-m}^{r}\varphi_{i,m}(y)N_{i,h}^m(x), \tag{8.84}$$

where

$$\varphi_{i,m}(y)=\prod_{\nu=1}^{m-1}(y-y_{i+\nu}).$$

Moreover, for $j=1,2,\dots,m$,

$$x^{(j-1)_h}=\sum_{i=l+1-m}^{r}\xi_{i,h}^{(j)}N_{i,h}^{m}(x), \tag{8.85}$$

where

$$\xi_{i,h}^{(j)}=(-1)^{j-1}(j-1)!\,D_h^{m-j}\varphi_{i,m}[(j-2)h], \qquad j=1,\dots,m.$$

Proof. For $m=1$, we have $\sum_{i=1}^{m+K}N_{i,h}^{1}(x)=\sum_{i=1}^{m+K}N_i^{1}(x)=1$. Now we proceed by induction on m. For convenience we write $\delta=(m-2)h$. Using (8.80), we obtain

$$\begin{aligned}
&\sum_{i=1}^{m+K}\varphi_{i,h}(y)N_{i,m}^{m}(x)\\
&\quad=\sum_{i=1}^{m+K}\varphi_{i,m}(y)\left[(x-y_i-\delta)n_{i,h}^{m-1}(x)+(y_{i+m}-x+\delta)n_{i+1,h}^{m-1}(x)\right]\\
&\quad=\sum_{i=1}^{m+K}N_{i,h}^{m-1}(x)\left[\frac{(x-y_i-\delta)\varphi_{i,m}(y)+(y_{i+m-1}-x+\delta)\varphi_{i-1,m}(y)}{(y_{i+m-1}-y_i)}\right]
\end{aligned}$$

for all $a=y_m\leqslant x<b=y_{m+K+1}$. But the quantity in brackets (cf. the proof of Theorem 4.21) is exactly $\varphi_{i,m-1}(y)(y-x+\delta)$. Hence assuming (8.84) for $m-1$, we obtain

$$\sum_{i=1}^{m+K}\varphi_{i,m}(y)N_{i,h}^{m}(x)=\sum_{i=1}^{m+K}\varphi_{i,m-1}(y)(y-x+\delta)N_{i,h}^{m-1}=(y-x+\delta)^{(m-1)_h}.$$

If we apply the operator D_h^{m-j} to (8.84) and evaluate at $y=(j-m)h$, we obtain (8.85). ■

The most important case of (8.85) is the case of $j=1$. Since $\xi_{i,h}^{(1)}=1$, $i=1,2,\dots,m+K$, we obtain

$$\sum_{i=1}^{m+K}N_{i,h}^{m}(x)\equiv 1 \qquad \text{for all } a\leqslant x\leqslant b,$$

that is, the discrete B-splines also form a partition of unity.

We now give some results on differences of discrete splines. First we observe that the difference of a polynomial discrete spline of order m is a discrete spline of order $m-1$.

THEOREM 8.57

Let $s \in \mathcal{S}(\mathcal{P}_m; \mathcal{M}; \Delta; h)$. Then $D_h s \in \mathcal{S}(\mathcal{P}_{m-1}; \mathcal{M}'; \Delta; h)$, where $\mathcal{M}' = (m_1', \ldots, m_k')$ with $m_i' = \min(m-1, m_i)$, $i = 1, 2, \ldots, k$.

Proof. The proof is nearly identical with the proof of Theorem 4.49, using the fact that if $p \in \mathcal{P}_m$, then $D_h p \in \mathcal{P}_{m-1}$. ■

The following result shows that the difference of a discrete B-spline of order m can be written as a linear combination of two discrete B-splines of order $m-1$:

THEOREM 8.58

Let $y_i < y_{i+m}$. Then

$$D_h Q_i^m = (m-1)\left(\frac{Q_{i,h}^{m-1} - Q_{i+1,h}^{m-1}}{y_{i+m} - y_i} \right). \tag{8.86}$$

Proof. The proof follows that of Theorem 4.16, using the fact that

$$D_h (x-y)_+^{(m-1)_h} = (m-1)(x-y)^{(m-2)_h}.$$ ■

Theorem 8.58 coupled with the proof of Theorem 5.9 yields the following result, which is useful in computations with discrete splines.

THEOREM 8.59

Let $s = \sum_{i=1}^n c_i N_{i,h}^m$, and suppose $1 \leq d \leq m$. Then

$$D_h^{d-1} s(x) = \sum_{i=d}^{n} c_i^{(d)} N_{i,h}^{m-d+1},$$

where $\{c_i^{(d)}\}$ are as given in Theorem 5.9.

The size of the differences $D_h^j N_{i,h}^m$ of the normalized discrete B-splines can be estimated by the same methods used in Theorem 4.22. We have the following theorem:

THEOREM 8.60

Suppose $1 \leq r \leq m$ and that $y_l \leq x < y_{l+1}$. Let Γ_{mr} and the quantities $\Delta_{i,l,m}$ be as in Theorem 4.22. Then

$$|D_h^r N_{i,h}^m(x)| \leq \frac{\Gamma_{mr}}{\Delta_{i,l,m-1} \cdots \Delta_{i,l,m-r}}.$$

Proof. We have

$$D_h^r N_{i,h}^m(x) = \frac{(-1)^m (m-1)!}{(m-r-1)!}(y_{i+m}-y_i)[y_i,\ldots,y_{i+m}](y-x)_+^{(m-r-1)_h},$$

and this can be estimated in terms of divided differences of order $m-r$, just as in the proof of Theorem 4.22. ∎

We now examine to what extent the results on integrals of polynomial splines can be carried over to discrete splines. First we need to replace the integral by an appropriate sum. Given any function f defined on $[a,b]_h$ with $b=a+Nh$, let

$$\int_a^b f(x)\,d_h x := \begin{cases} 0, & b-a<h \\ h\sum_{i=0}^{N-1} f(a+ih), & b-a \geqslant h. \end{cases} \tag{8.87}$$

We observe that

$$\int_a^b \frac{x^{(j)_h} d_h x}{j!} = \frac{b^{(j+1)_h} - a^{(j+1)_h}}{(j+1)!}. \tag{8.88}$$

The following theorem is a discrete version of the Taylor expansion (cf. Theorem 2.1):

THEOREM 8.61

For every $x \in [a,b]_h$,

$$f(x) = \sum_{i=0}^{m-1} \frac{D_h^j f(a)(x-a)^{(j)_h}}{j!} + \int_a^b \frac{(x-y-h)_+^{(m-1)_h}}{(m-1)!} D_h^m f(y)\,d_h y. \tag{8.89}$$

Proof. The formula is clear for $m=1$. Now, applying the summation by parts formula

$$\int_a^b u(y) D_h v(y)\,d_h y = [u(y)v(y)]_a^b - \int_a^b v(y+h) D_h u(y)\,d_h y, \tag{8.90}$$

we obtain

$$\int_a^b \frac{(x-y-h)_+^{(m-2)_h} D_h^{m-1} f(y)\, d_h y}{(m-2)!} = \frac{-1}{(m-1)!}\left[(x-y)^{(m-1)_h} D_h^{m-1} f(y)\right]_a^b$$

$$+\int_a^b \frac{(x-y-h)^{(m-1)_h}}{(m-1)!} D_h^m f(y)\, d_h y,$$

and the result follows by induction. ■

As in the continuous case, there is also a dual version of the Taylor expansion which is useful at times.

THEOREM 8.62

For all $y \in [a,b]_h$,

$$f(y) = \sum_{j=0}^{m-1} \frac{(-1)^j D_h^j f(b-jh)(b-y)^{(j)_h}}{j!}$$

$$+(-1)^m \int_a^{b-(m-1)h} \frac{(x-y+(m-1)h)_+^{(m-1)_h}}{(m-1)!} D_h^m f(x)\, d_h x. \tag{8.91}$$

Proof. This formula follows by integration by parts using (8.90). ■

Theorems 8.61 and 8.62 lead immediately to Peano representation theorems for linear functionals defined on $B[a,b]_h$ which annihilate the polynomials of order m. In particular, we have the following important representation for the divided difference:

THEOREM 8.63

For any f defined on $[a,b]_h$ and for any $y_i, \ldots, y_{i+m}$ chosen from this discrete interval,

$$[y_i, \ldots, y_{i+m}] f = \int_a^b \frac{Q_{i,h}^m[x+(m-1)h] D_h^m f(x)\, d_h x}{(m-1)!}. \tag{8.92}$$

Proof. We apply the divided difference $[y_i, \ldots, y_{i+m}]$ to the Taylor expansion (8.91). Since it annihilates polynomials of order m, (8.92) follows. ■

The representation (8.92) for divided differences can also be established by applying the divided difference to the Taylor expansion (8.89). In this case we would obtain

$$[y_i,\ldots,y_{i+m}]f = \int_a^b \frac{[y_i,\ldots,y_{i+m}](y-x-h)_+^{(m-1)_h}}{(m-1)!} D_h^m f(y)\,d_h y.$$

While this looks formally different, it is, in fact, the same representation since

$$(y-x-h)_+^{(m-1)_h} - (-1)^m (x-y+(m-1)h)_+^{(m-1)_h} = (y-x-h)^{(m-1)_h},$$

and so

$$Q_{i,h}^m(x+(m-1)h) = [y_i,\ldots,y_{i+m}](y-x-h)_+^{(m-1)_h}.$$

Analogous to our development in Chapter 4 of the theory of ordinary polynomial splines, we now examine zero properties of discrete splines. Since we are working on a discrete set and do not have derivatives at our disposal, we need to introduce some slightly different counting techniques.

Let f be a function defined on the discrete set $[a,b]_h$. We say that f has a zero at the point $t \in [a,b]_h$ provided

$$f(t)=0 \qquad \text{or } f(t)\cdot f(t+h)<0.$$

When f vanishes at a consecutive set of points of $[a,b]_h$, say f is 0 at $t,\ldots,t+(r-1)h$, but $f(t-h)\cdot f(t+rh) \neq 0$, then we call the set $T=\{t,t+h,\ldots,t+(r-1)h\}$ a *multiple zero* of f, and we define its multiplicity by

$$Z_T(f) = \begin{cases} r, & \text{if } f(t-h)\cdot f(t+rh)<0 \text{ and } r \text{ is odd} \\ r, & \text{if } f(t-h)\cdot f(t+rh)>0 \text{ and } r \text{ is even} \\ r+1, & \text{otherwise.} \end{cases}$$

This definition assures that f changes sign at a zero if and only if the zero is of odd multiplicity. (See Figure 33 for some examples.)

If we are dealing with discrete polynomial splines, then it is possible for such a spline to vanish identically on an interval. When this is the case, we count the multiplicity of the interval exactly as in Definition 4.46. We now define

$$Z^{\mathbb{S}}(s) = \sum_{i=1}^{p} z(T_i),$$

where $T_1,\ldots,T_p$ are the zero sets of s, counted according to their multiplicity. As in Section 4.7, we have written the superscript $\mathbb{S}$ on the symbol Z

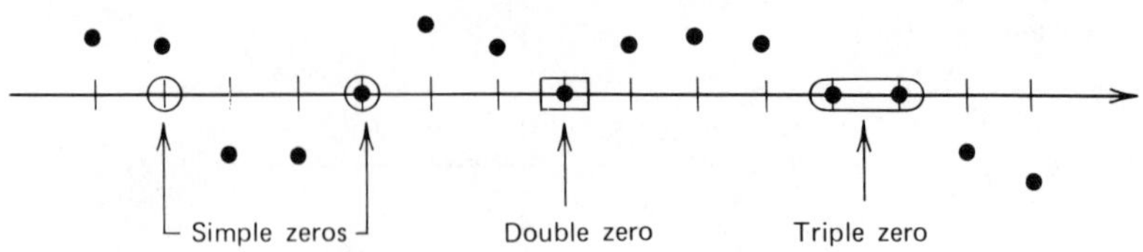

Figure 33. Zeros of a discrete spline.

to remind us that this counting procedure is relative to the space $\mathcal{S} = \mathcal{S}(\mathcal{P}_m; \mathfrak{M}; \Delta; h)$. The key to establishing bounds on $Z^{\mathcal{S}}(s)$ is the following version of Rolle's theorem for discrete splines (cf. Theorem 4.50 for the usual spline case):

THEOREM 8.64. Rolle's Theorem For Discrete Splines

For any $s \in \mathcal{S}(\mathcal{P}_m; \mathfrak{M}; \Delta; h)$,

$$Z^{D\mathcal{S}}(D_h s) \geqslant Z^{\mathcal{S}}(s) - 1, \tag{8.93}$$

where $D\mathcal{S} = \mathcal{S}(\mathcal{P}_{m-1}: \mathfrak{M}'; \Delta; h)$, with $\mathfrak{M}'$ as in Theorem 8.41.

Proof. First, if s has a z-tuple zero on the set $T = \{t, \ldots, t+(r-1)h\}$, it follows that $D_h s$ has a $z-1$-tuple zero on the set $T' = \{t, \ldots, t+(r-2)h\}$. Similarly (cf. the table in the proof of Theorem 4.50), if s has a z-tuple zero on an interval, then $D_h s$ has a $z-1$-tuple zero on the same interval. Now if T_1 and T_2 are two consecutive zero sets of s, then it is trivially true that $D_h s$ must have a sign change at some point between T_1 and T_2. Counting all of these zeros as in the case of ordinary polynomial splines, we arrive at the assertion (8.93). ■

THEOREM 8.65

For every $s \in \mathcal{S}(\mathcal{P}_m; \mathfrak{M}; \Delta; h)$ that is not identically zero,

$$Z^{\mathcal{S}}(s) \leqslant m + K - 1,$$

where $K = \sum_{i=1}^{k} m_i$. (This bound is one less than the dimension of the space.)

Proof. For $m=1$ the discrete spline is a piecewise constant, and it can have zeros only at the knots (where it jumps to or through zero). To prove the result in general, we may proceed by induction using our discrete Rolle's theorem—cf. the proof of Theorem 4.53. ■

Theorem 8.65 is an assertion about the number of zeros of a discrete spline s (counting multiplicities appropriately) in the discrete line $\mathbf{R}_{a,h}$. The result does not follow from the bounds on the number of zeros given in Section 4.7 for ordinary polynomial splines—the discrete splines have a different kind of smoothness across the knots.

We turn now to some results on determinants formed from discrete B-splines. Suppose $y_1 < y_2 < \cdots < y_{n+m}$ are given points in the discrete line $\mathbf{R}_{a,h}$, and let $B_i(x) = N_{i,h}^m(x)$, $i = 1,2,\ldots,n$ be the corresponding discrete B-splines.

THEOREM 8.66

Let $m \geqslant 2$, and suppose $t_1 < t_2 < \cdots < t_n$ are prescribed points in the discrete line $\mathbf{R}_{a,h}$. Then

$$D\begin{pmatrix} t_1,\ldots,t_n \\ B_1,\ldots,B_n \end{pmatrix} = \det\left[B_j(t_i) \right]_{i,j=1}^n \geqslant 0, \tag{8.94}$$

and strict positivity holds if and only if

$$t_i \in \sigma_i = \{ x \in \mathbf{R}_{a,h}\colon B_i(x) > 0 \} = (y_i + (m-2)h, y_{i+m})_h, \qquad i = 1,2,\ldots,n. \tag{8.95}$$

Proof. The fact that $D = 0$ whenever the conditions (8.95) fail follows directly from the support properties of the discrete B-splines, just as in the proof of Theorem 4.61. Conversely, if $D = 0$, then there exists a nontrivial spline $s = \Sigma_1^n c_i B_i$ which has zeros at each of the points $t_1,\ldots,t_n$. But if (8.95) holds, this leads to a contradiction on the number of zeros a discrete spline can have, just as in the proof of Theorem 4.61. Finally, the fact that D is positive when it is nonzero can be established by the same kind of argument used to prove Theorem 4.64. ■

Theorem 8.66 does not remain valid if we allow the t's to range over the entire real line $\mathbf{R}$ rather than over the discrete version $\mathbf{R}_{a,h}$. Indeed, for a simple example to illustrate this, we need only consider the case of $n = 1$ and recall that a discrete B-spline takes on some negative values (at points outside of the discrete set $\mathbf{R}_{a,h}$)—cf. Figure 32, page 347.

The following result shows that considered as functions on $\mathbf{R}_{a,h}$, the discrete B-splines form an OCWT system.

THEOREM 8.67

Let $\{B_i\}_1^n$ be a set of discrete B-splines associated with knots $y_1 < y_2 < \ldots < y_{n+m} \in \mathbf{R}_{a,h}$. Then for any $1 \leq \nu_1 < \nu_2 < \cdots < \nu_p \leq n$ and any $t_1 < t_2 < \cdots < t_p$ in $\mathbf{R}_{a,h}$,

$$D\begin{pmatrix} t_1, \ldots, t_p \\ B_{\nu_1}, \ldots, B_{\nu_p} \end{pmatrix} = \det\left[B_{\nu_j}(t_i) \right]_{i,j=1}^p \geq 0, \tag{8.96}$$

and strict positivity holds if and only if

$$y_{\nu_i} + (m-2)h < t_i < y_{\nu_i + m}, \qquad i = 1, 2, \ldots, p. \tag{8.97}$$

Proof. The proof proceeds along the same lines as the proof of Theorem 4.65, using Theorem 8.66. ■

The OCWT-property of the discrete B-splines leads immediately to a variation-diminishing property for discrete B-spline expansions.

THEOREM 8.68

For any nontrivial $s = \sum_{i=1}^n c_i B_i$,

$$S^-(s) \leq S^-(\mathbf{c}), \tag{8.98}$$

where S^- counts strong sign changes as in Definition 2.10, and $\mathbf{c} = (c_1, \ldots, c_n)$ is the vector of coefficients of s.

Proof. The result follows immediately from Theorem 2.42 since $\{B_i\}_1^n$ form an OCWT-system on $\mathbf{R}_{a,h}$. ■

We mention one other result connected with determinants of discrete splines, and in particular with the discrete Green's function

$$g_{m,h}(t;x) = \frac{(t-x)_+^{(m-1)_h}}{(m-1)!}. \tag{8.99}$$

THEOREM 8.69

Let $m \geq 2$. Then

$$g_{m,h}\begin{pmatrix} t_1, \ldots, t_p \\ y_1, \ldots, y_p \end{pmatrix} = \det\left[g_{m,h}(t_j; y_i) \right]_{i,j=1}^p \geq 0 \tag{8.100}$$

for all $t_1<t_2<\cdots<t_p$ and $y_1<y_2<\cdots<y_p$ in $\mathbf{R}_{a,h}$. Moreover, this determinant is strictly positive precisely when

$$t_{i-m}<y_i<t_i-(m-2)h, \qquad i=1,2,\ldots,p,$$

where the left-hand inequality is ignored when $i<m$.

Proof. The proof proceeds exactly as in the proof of Theorem 4.78, using Theorem 8.65 on the number of zeros a discrete spline can have. ∎

We devote the remainder of this section to error bounds for approximation with discrete spline functions. While it is possible to establish some results for approximation of functions on actual intervals, it is perhaps more interesting to work with functions defined only on a discrete interval $[a,b]_h$. Thus our error bounds will involve the discrete norms

$$\|f\|_{l_\infty[a,b]_h} = \max_{x\in[a,b]_h} |f(x)|, \tag{8.101}$$

and

$$\|f\|_{l_p[a,b]_h} = \left[\int_a^{b+h} |f(x)|^p d_h x\right]^{1/p} = \left[\sum_{i=0}^{N} |f(a+ih)|^p\right]^{1/p}, \tag{8.102}$$

where $b=a+Nh$. To measure the smoothness of a function defined on $[a,b]_h$, we introduce the discrete modulus of continuity defined by

$$\omega(f;t)_{l_\infty[a,b]_h} = \sup_{\substack{|y-x|\leqslant t\\ x,y\in[a,b]_h}} |f(y)-f(x)|. \tag{8.103}$$

This modulus of continuity has most of the properties of the usual one.

The key to obtaining error bounds for approximation by discrete splines is provided by a certain local spline approximation operator similar to the one constructed in Section 6.4 for ordinary polynomial splines. Let Δ be a partition of $[a,b]_h$. Using Lemma 6.17 we can select a coarser partition Δ^* such that

$$\frac{\overline{\Delta}}{2} \leqslant \underline{\Delta}^* \leqslant \overline{\Delta}^* \leqslant \frac{3\overline{\Delta}}{2}.$$

Now suppose that $y_1<y_2<\cdots<y_{m+n}$ is an extended partition associated with Δ^*, and let $\{B_i=N_{i,h}^m\}_{i=1}^n$ be the associated normalized discrete B-splines. We suppose that h is small enough so that each interval $(y_i,y_{i+1})_h$ contains at least m points of $[a,b]_h$.

For each $i=1,2,\ldots,n$, let $\tau_{i1}<\tau_{i2}<\cdots<\tau_{im}$ be points chosen from $(y_i,y_{i+m})_h$, and let $\alpha_{i1},\ldots,\alpha_{im}$ be defined as in (6.38) with $\xi_i=\xi_{i,h}$. Set

$$\lambda_i f=\sum_{j=1}^{m}\alpha_{ij}[\tau_{i1},\ldots,\tau_{ij}]f, \qquad i=1,2,\ldots,n.$$

THEOREM 8.70

For any $f\in B[a,b]_h$, define

$$Qf(x)=\sum_{i=1}^{n}(\lambda_i f)B_i(x). \tag{8.104}$$

Then Q is a linear operator mapping $B[a,b]_h$ into $\mathcal{S}_m(\Delta^*;h)\subseteq\mathcal{S}_m(\Delta;h)$. Moreover,

$$Qp=p \qquad \text{for all } p\in\mathcal{P}_m.$$

Proof. This theorem is proved in the same way as Theorem 6.18, using the discrete Marsden identity in Theorem 8.56 in place of the usual one. ■

We are ready to give a local approximation result.

THEOREM 8.71

Let $m\leqslant l\leqslant n$, $I_l=[y_l,y_{l+1})_h$, and $\tilde{I}_l=[y_{l+1-m},y_{l+m}]_h$. Then for any $1\leqslant\sigma\leqslant m$ and any $f\in B[\tilde{I}_l]$,

$$\left.\begin{matrix}\|D_h^r(f-Qf)\|_{L_q[I_l]}\\ r=0,1,\ldots,\sigma-1\\ \|D_h^rQf\|_{L_q[I_l]}\\ r=\sigma,\ldots,m-1\end{matrix}\right\}\leqslant C_1(\overline{\Delta})^{\sigma-r-1+1/q}\omega\big(D_h^{\sigma-1}f;\overline{\Delta}\big)_{L_\infty[\tilde{I}_l]},$$

where C_1 is a constant depending only on m.

Proof. The proof is very similar to the proof of Theorem 6.20 for ordinary polynomial splines. There we used a Whitney-type theorem to produce a polynomial useful for comparison purposes. For our purposes here it will be enough to take the discrete Taylor polynomial. In particular, let

$$p_f(x)=\sum_{j=0}^{\sigma-1}\frac{D_h^jf(t)}{j!}(x-t)^{(j)_h},$$

and set $R(x)=f(x)-p_f(x)$. Then defining $E_r(t)$ just as in the proof of Theorem 6.20, by the same kind of arguments used there (using the reproductive power of Q), we observe that it is enough to estimate $D_h^r QR(t)$ for each $r=0,1,\dots,m-1$. But

$$|D_h^r QR(t)| \leqslant \sum_{i=l+1-m}^{l} \sum_{j=1}^{m} |\alpha_{ij}||\lambda_{ij} R||D_h^r B_i(t)|.$$

We estimate $|\alpha_{ij}|$ as in Lemma 6.19 and the differences of the B-splines in Theorem 8.60. It remains to examine $|\lambda_{ij} R|$.

By the definition of $\lambda_{ij} R$ and the integral representation for the divided difference [cf. (8.92)], we have for $1 \leqslant j \leqslant \sigma$,

$$|\lambda_{ij} R| \leqslant |[\tau_{i1},\dots,\tau_{ij}]R| \leqslant \frac{|D_h^{j-1} R(z_{ij})|}{(j-1)!}, \qquad \text{some } z_{ij} \in \tilde{I}_l.$$

Since $D_h^i R(t)=0$, $i=0,1,\dots,\sigma-2$, expanding $D_h^{j-1} R$ in a Taylor expansion about t and applying the mean-value theorem to the integral remainder, we find that

$$D_h^{j-1} R(z_{ij}) = \theta \frac{\left[D_h^{\sigma-1} R(\tilde{z}_{ij})(z_{ij}-t)^{(\sigma-j)_h} \right]}{(\sigma-j)!},$$

where $\tilde{z}_{ij}$ is also in $\tilde{I}_l$ and $0 \leqslant \theta \leqslant 1$. But

$$|D_h^{\sigma-1} R(z_{ij})| = |D_h^{\sigma-1}[f(z_{ij}) - f(t)]| \leqslant \omega(D_h^{\sigma-1} f; |\tilde{I}_l|)_{l_\infty[\tilde{I}_l]}.$$

Since $|\tilde{I}_l| \leqslant 2m\bar{\Delta}$, combining these various facts we obtain

$$|\lambda_{ij} R| \leqslant C_1 \bar{\Delta}^{\sigma-j} \omega(D_h^{\sigma-1} f; \bar{\Delta})_{l_\infty[\tilde{I}_l]}, \qquad j=0,1,\dots,\sigma.$$

For $j=\sigma+1,\dots,m$ we use Theorem 2.56 to reduce the order of the divided difference defining λ_{ij} to a divided difference of order $\sigma-1$. This yields (cf. the proof of Theorem 6.20)

$$|\lambda_{ij} R| \leqslant \frac{C_2}{\underline{\Delta}^{j-\sigma}} \omega(D_h^{\sigma-1} R; \bar{\Delta})_{l_\infty[\tilde{I}_l]}, \qquad j=\sigma+1,\dots,m.$$

The constant depends on how we spread the τ's throughout $(y_i, y_{i+m})_h$. If we make them as equally spaced as possible, it will depend only on m, however.

The remainder of the proof now proceeds exactly as in the proof of Theorem 6.20. ∎

Theorem 8.71 has a local character—the error bound in the interval I_l depends only on the behavior of f in a somewhat larger interval $\tilde{I}_l$. We may give the following global version of this result:

THEOREM 8.72

Let $f \in B[a,b]_h$, and suppose $1 \leqslant \sigma \leqslant m$. Then

$$\left.\begin{array}{l} \|D_h^r(f-Qf)\|_{l_\infty[a,b]_h} \\ \qquad r=0,1,\ldots,\sigma-1 \\ \\ \|D_h^r Qf\|_{l_\infty[a,b]_h} \\ \qquad r=\sigma,\ldots,m-1 \end{array}\right\} \leqslant C_1(\overline{\Delta})^{\sigma-r-1}\omega\left(D_h^{\sigma-1}f;\overline{\Delta}\right)_{l_\infty[a,b]_h}.$$

Proof. We simply apply Theorem 8.71 with $q=\infty$. ∎

Theorems 8.71 and 8.72 have been stated in terms of the discrete modulus of smoothness of the difference $D_h^{\sigma-1}f$. In analogy with the continuous spline case, it is also possible to give error bounds in terms of $\|D_h^\sigma f\|_{l_p[I_l]}$ or $\|D_h^\sigma f\|_{l_p[a,b]_h}$. Lower bounds can also be established.

§ 8.6. HISTORICAL NOTES

Section 8.1

Although periodic splines have appeared in a number of papers (e.g., they arise as solutions of certain best interpolation problems in the papers of Walsh, Ahlberg, and Nilson [1962], deBoor [1963], and Schoenberg [1964a]), there seems to be very little explicit discussion of their constructive properties in the literature. The method of constructing periodic B-splines is part of the folklore. Our results on zeros and determinants of the periodic B-splines follow those of Schumaker [1976b]. The available results on the approximation power of periodic splines center around error bounds for periodic spline interpolation—see the papers quoted in the notes for Section 6.4. The construction of a local spline approximation operator used here is a simple adaptation of the similar nonperiodic version introduced in Section 6.4.

Section 8.2

Natural splines arose as solutions of best interpolation problems and also in connection with optimal quadrature formulae—see, for example, the early papers of Sard [1949], Meyers and Sard [1950a, b], Holladay [1957], Golomb and Weinberger [1959], Schoenberg [1964a], and the many papers on the subject appearing later. The construction of a local support basis given in Theorem 8.18 is credited to Greville [1969b]. Theorem 8.19 is credited to Lyche and Schumaker [1973], who used it to develop ALGOL programs for computing natural splines. For related FORTRAN programs, see Lyche, Schumaker, and Sepehrnoori [1980].

Section 8.3

Once again, the g-splines seem to have appeared first as solutions of best interpolation problems, this time with Hermite-Birkhoff interpolation conditions; see Ahlberg and Nilson [1966] and Schoenberg [1968], where the name was introduced. The problem of constructing one-sided and local support bases for classes of g-splines was first studied by Jerome and Schumaker [1971]. A more general treatment was later given by Jerome and Schumaker [1976], which we have followed here. The idea of using an incidence matrix to describe g-splines for HB-interpolation seems to have resulted from the work of Schoenberg [1968]. Zero properties of confined g-splines were treated by Ferguson [1974] and by Lorentz [1975]—see also G. D. Birkhoff [1906] and Jetter [1976].

Section 8.4

Monosplines were introduced by Schoenberg [1958], and Johnson [1960] dealt with them in connection with the best approximation of x^m by splines of order m. They are also closely connected to optimal quadrature formulae—see the articles in the text by Schoenberg [1969b]. The history of zero properties of monosplines began with Schoenberg [1958], who stated the fundamental theorem of algebra (i.e., an upper bound on the number of zeros and the existence statement of Theorem 8.47 for simple knots). The upper bound was first officially established in the article by Johnson [1960]. The existence of monosplines with simple knots and prescribed multiple zeros was established first in the article by Karlin and Schumaker [1967]—see also Schumaker [1966]. The result for multiple knots and simple zeros (and the method of proof used here) is credited to Micchelli [1972]. The Budan-Fourier theorem for monosplines is also the result of work done by Micchelli [1972]. The problem of constructing a monospline with multiple knots and with *multiple zeros* appears to be open. For results on the fundamental theorem of algebra for monosplines with boundary conditions, see Karlin and Micchelli [1972] and Micchelli and Pinkus [1977].

Section 8.5

Discrete splines were found by Mangasarian and Schumaker [1971, 1973] as solutions of certain discrete best interpolation problems. Constructive aspects of discrete splines for the case of $h=1$ were discussed by Schumaker [1973]. A more complete development of constructive aspects of discrete splines was carried out by Lyche [1976]. For a different approach to their best interpolation properties, see Astor and Duris [1974].

§ 8.7 REMARKS

Remark 8.1

In Theorem 8.15 we have examined the approximation power of periodic splines with the help of the linear spline operator $\mathring{Q}$ defined in (8.19). As in the nonperiodic case (cf. Example 6.23), $\mathring{Q}$ is not a projector. By constructing a dual basis for the periodic B-splines (following the development in Section 4.6 in the nonperiodic case), we can follow the ideas discussed in Remark 6.4 to construct a projector that produces the same error bounds as $\mathring{Q}$.

Remark 8.2

In Theorem 8.71 the approximation power of discrete splines was examined using the linear spline operator Q defined in (8.104). By constructing a dual basis for the discrete B-splines, we could obtain the same results with a projector—cf. the construction in Remark 6.4.

Remark 8.3

The discrete B-splines also turn out to be useful for expressing the divided difference of a function over a given set of points in terms of divided differences over a refinement of this set of points. These relations can be used in turn to obtain B-spline expansions of splines on one set of knots in terms of B-splines on a finer set of knots. See deBoor [1976b].

9
TCHEBYCHEFFIAN SPLINES

In the first eight chapters of this book we have dealt exclusively with polynomial splines. Here and in the following two chapters we develop the theory of similar spaces of generalized splines. We begin this development by studying Tchebycheffian splines, where, as we shall see, almost all of the results for polynomial splines can be carried over.

§ 9.1. EXTENDED COMPLETE TCHEBYCHEFF SYSTEMS

Let I be a subinterval of the real line $\mathbf{R}$, and suppose $U_m = \{u_i\}_1^m$ is a set of functions in $C^{m-1}[I]$. We call U_m an *Extended Complete Tchebycheff* (ECT-system) provided

$$D\begin{pmatrix} t_1, \ldots, t_k \\ u_1, \ldots, u_k \end{pmatrix} > 0 \qquad \text{for all } t_1 \leqslant t_2 \leqslant \cdots \leqslant t_k \text{ in } I \tag{9.1}$$

and all $k = 1, 2, \ldots, m$. In this section we show that the ECT-systems are natural generalizations of the space $\mathcal{P}_m$ of polynomials, and that many of the convenient properties of polynomials remain valid for ECT-systems. We begin with an equivalent condition for a set of functions to form an ECT-system.

THEOREM 9.1

A set of functions $u_1, \ldots, u_m$ in $C^{m-1}[I]$ form an ECT-system if and only if their *Wronskian determinants* are positive for all $x \in I$; that is,

$$W(u_1, \ldots, u_k)(x) = \det\left[D^{j-1} u_i(x)\right]_{i,j=1}^{k} > 0, \qquad \text{all } x \in I,$$

$k = 1, 2, \ldots, m$.

Proof. The necessity of this condition is clear from the definition of ECT-system. The converse is established by a standard argument for pulling apart multiple t's—see Karlin and Studden [1966], page 377. ■

If $\mathfrak{U}_m$ is an m-dimensional linear space with a basis U_m that forms an ECT-system, then we call $\mathfrak{U}_m$ an *ECT-space*. The following theorem shows that any ECT-space has an especially convenient ECT-system basis:

THEOREM 9.2

Suppose $w_i \in C^{m-i}[I]$ are positive on I, $i=1,2,\dots,m$. Then

$$\begin{aligned} u_1(x) &= w_1(x) \\ u_2(x) &= w_1(x)\int_a^x w_2(s_2)\,ds_2 \\ &\cdots \\ u_m(x) &= w_1(x)\int_a^x w_2(s_2)\int_a^{s_2}\cdots\int_a^{s_{m-1}} w_m(s_m)\,ds_m\cdots ds_2 \end{aligned} \tag{9.2}$$

form an ECT-system on I. We say it is in *canonical form*. Moreover, if $\mathfrak{U}_m$ is any ECT-space, then there is an ECT-system in canonical form that forms a basis for $\mathfrak{U}_m$.

Proof. See Karlin and Studden [1966], page 379. ■

The following two simple properties of ECT-systems are very useful:

THEOREM 9.3

If $U_m=\{u_i\}_1^m$ is an ECT-system on $[a,b]$, then we can extend each of these functions to be defined on any larger interval $[c,d]$ such that U_m is also an ECT-system on $[c,d]$.

Proof. We assume that $u_1,\dots,u_m$ are in canonical form. Then we simply extend each of the functions $w_1,\dots,w_m$ to the larger interval in such a way that they remain positive and retain their smoothness (i.e., such that $w_i \in C^{m-i}[c,d]$, $i=1,2,\dots,m$). ■

THEOREM 9.4

If $\{u_i\}_1^m$ is a canonical ECT-system on $[a,b]$, then there exists a function u_{m+1} such that $U_{m+1}=\{u_i\}_1^{m+1}$ is a canonical ECT-system of $m+1$ functions on $[a,b]$.

Proof. We simply choose $w_{m+1}=1$ and set

$$u_{m+1}(x)=w_1(x)\int_a^x w_2(s_2)\int_a^{s_2}\cdots\int_a^{s_m} w_{m+1}(s_{m+1})\,ds_{m+1}\cdots ds_2. \quad (9.3)$$

■

The most important example of an ECT-system is provided by the classical polynomials:

EXAMPLE 9.5

The space $\mathcal{P}_m$ of polynomials is an ECT-space on any interval $[a,b]$.

Discussion. A canonical basis for $\mathcal{P}_m$ on $[a,b]$ is given by the functions $u_1(x)=1$ and $u_2(x)=x-a,\ldots,u_m(x)=(x-a)^{m-1}$. These functions can be written in the canonical form (9.2) using the weight functions $w_1(x)=1, w_i(x)=i-1$, $i=2,\ldots,m$. ■

Many properties of polynomials involve working with derivatives. When dealing with ECT-systems it is convenient to replace the usual derivatives by some related differential operators. We define $D_0 f=f$, and

$$D_i f = D\left(\frac{f}{w_i}\right), \qquad i=1,2,\ldots,m. \quad (9.4)$$

Now set

$$L_i = D_i D_{i-1}\cdots D_0, \qquad i=0,1,\ldots,m. \quad (9.5)$$

The operator L_i can be regarded as a substitute for D^i. It is, in fact, just a perturbation of D^i, since by Leibniz's rule, there exist $a_{i,0},\ldots,a_{i,i-1}$ such that for all f,

$$L_i f(x) = \frac{D^i f(x)}{w_1(x)\cdots w_i(x)} + \sum_{j=0}^{i-1} a_{ij}(x) D^j f(x). \quad (9.6)$$

Clearly, if U_m is a canonical ECT-system, then $\mathcal{U}_m=\operatorname{span}(U_m)$ is the null space of L_m, and

$$L_j u_i(a) = w_i(a)\delta_{j,i-1} \qquad \begin{array}{l} j=0,1,\ldots,i-1 \\ i=1,2,\ldots,m. \end{array} \quad (9.7)$$

An important property of the polynomials is the fact that the derivatives of polynomials are again polynomials (of one order less). A similar situation persists for canonical ECT-systems. Let

$$\begin{aligned}
u_{j,1}(x) &= w_{j+1}(x) \\
u_{j,2}(x) &= w_{j+1}(x)\int_a^x w_{j+2}(s_{j+2})\,ds_{j+1} \\
&\cdots \\
u_{j,m-j}(x) &= w_{j+1}(x)\int_a^x w_{j+2}(s_{j+2})\int_a^{s_{j+2}}\cdots\int_a^{s_{m-1}} w_m(s_m)\,ds_m\cdots ds_{j+2}
\end{aligned} \tag{9.8}$$

for each $j=0,1,\ldots,m-1$. We call $U_m^{(j)}=\{u_{j,i}\}_{i=1}^{m-j}$ the *jth reduced system*. In view of Theorem 9.2, it is itself a canonical ECT-system. In addition, we have

$$L_j u_i(x)=\begin{cases} u_{j,i-j}(x), & i=j+1,\ldots,m \\ 0, & i=1,2,\ldots,j. \end{cases} \tag{9.9}$$

We write $\mathfrak{U}_m^{(j)}=\text{span}\,(U_m^{(j)})$, $j=1,2,\ldots,m-1$. It follows that if $u\in\mathfrak{U}_m$, then $L_1u\in\mathfrak{U}_m^{(1)}$, $L_2u\in\mathfrak{U}_m^{(2)}$, and so on.

The determinants D involved in the definition of ECT-systems [cf. (9.1)] are the determinants introduced in Section 2.3, where repeated t's call for taking (ordinary) derivatives. In view of the importance of the differential operators L_i introduced in (9.5), it is natural to expect that there may be some advantage to working with determinants in which D^i is replaced by L_i. We define

$$D_{U_m}(t_1,\ldots,t_m)=\det\left[L_{d_i}u_j(t_i)\right]_{i,j=1}^m, \tag{9.10}$$

where

$$d_i=\max\{j:\ t_i=\cdots=t_{i-j}\},\qquad i=1,2,\ldots,m.$$

Using (9.6), it is easy to see that by adding appropriate multiples of various rows to others, we can convert the determinant in (9.10) to the one in (9.1); that is,

$$D\begin{pmatrix} t_1,\ldots,t_m \\ u_1,\ldots,u_m \end{pmatrix}=D_{U_m}(t_1,\ldots,t_m),\qquad \text{all } t_1\leqslant t_2\leqslant\cdots\leqslant t_m\in I.$$

We conclude that the determinants (9.10) are also positive for all $t_1 \leqslant \cdots \leqslant t_m$ in I. The following result gives more precise information about the size of D_{U_m}:

LEMMA 9.6

Let

$$\begin{aligned} \underline{M}_i &= \min_{a \leqslant x \leqslant b} w_i(x) \\ \overline{M}_i &= \max_{a \leqslant x \leqslant b} w_i(x) \end{aligned}, \qquad i = 1, 2, \ldots, m. \tag{9.11}$$

Then for all $a \leqslant t_1 \leqslant t_2 \leqslant \cdots \leqslant t_m \leqslant b$,

$$C_1 V(t_1, \ldots, t_m) \leqslant D_{U_m}(t_1, \ldots, t_m) \leqslant C_2 V(t_1, \ldots, t_m), \tag{9.12}$$

where V is the VanderMonde determinant [cf. (2.65)], and where C_1 and C_2 are positive constants depending only on m and the quantities (9.11). Moreover, for all $a \leqslant x \leqslant b$,

$$\begin{aligned} C_1 |D^k V(t_1, \ldots, t_{m-1}, x)| &\leqslant |L_k D_{U_m}(t_1, \ldots, t_{m-1}, x)| \\ &\leqslant C_2 |D^k V(t_1, \ldots, t_{m-1}, x)|. \end{aligned} \tag{9.13}$$

Proof. First we claim that D_{U_m} can be written as a multiple integral (over positively oriented intervals) whose integrand is a product of the weight functions $w_1, \ldots, w_m$. To see this, suppose

$$t_1 \leqslant t_2 \leqslant \cdots \leqslant t_m = \overbrace{\tau_1, \ldots, \tau_1}^{l_1} < \cdots < \overbrace{\tau_d, \ldots, \tau_d}^{l_d}.$$

Next, by factoring $w_1(\tau_1)$ out of the first row, $w_1(\tau_2)$ out of the $l_1 + 1$st row, and so on, and then by subtracting each row with a 1 in the first column from its predecessor with a 1 in the first column, we obtain (after expanding about the first column)

$$D_{U_m} = w_1(\tau_1) \cdots w_1(\tau_d)$$

$$\cdot \int_{\tau_1}^{\tau_2} \cdots \int_{\tau_{d-1}}^{\tau_d} D_{U_m^1}\Big(\overbrace{\tau_1, \ldots, \tau_1}^{l_1 - 1}, s_1, \overbrace{\tau_2, \ldots, \tau_2}^{l_2 - 1}, s_2, \ldots, \overbrace{\tau_d, \ldots, \tau_d}^{l_d - 1}\Big) ds_1 \cdots ds_{d-1}.$$

Our claim follows by induction.

We now establish (9.12). Suppose we replace all the w's by their upper bounds $\overline{M}_1,\ldots,\overline{M}_m$. The value of the multiple integral is then larger. But with constant w's, the functions $u_1,\ldots,u_m$ reduce (modulo some factorials) to the powers $1, x,\ldots,x^{m-1}$, and thus the multiple integral is a (constant) multiple of the VanderMonde. The lower bound is obtained by replacing all w's by lower bounds on them. The proof of (9.13) is similar, taking account of the fact that in view of (9.9), the row in the determinant $L_k D_{U_m}$ corresponding to x has zeros in the first k columns. ■

In Section 2.7 we gave a detailed treatment of divided differences based on defining them as quotients of determinants. This approach suggests a natural generalization to divided differences with respect to an ECT-system. Suppose U_m is an ECT-system in canonical form, and suppose u_{m+1} is the natural extension [cf. (9.3)] to an ECT-system U_{m+1}. Then given any sufficiently differentiable function f, we define its mth *order divided difference with respect to* U_{m+1} by

$$[t_1,t_2,\ldots,t_{m+1}]_{U_{m+1}}f=\frac{D\begin{pmatrix} t_1,\ldots,t_{m+1} \\ u_1,\ldots,u_m,f \end{pmatrix}}{D\begin{pmatrix} t_1,\ldots,t_{m+1} \\ u_1,\ldots,u_{m+1} \end{pmatrix}}. \tag{9.14}$$

We suppress the subscript U_{m+1} on the divided difference symbol whenever there is no chance of confusion. The following theorem gives some elementary properties of the generalized divided differences:

THEOREM 9.7

The generalized divided difference defined in (9.14) is a linear functional such that

$$[t_1,\ldots,t_{m+1}]u=0 \qquad \text{for all } u\in\mathfrak{U}_m, \tag{9.15}$$

and

$$[t_1,\ldots,t_{m+1}]u_{m+1}=1. \tag{9.16}$$

Moreover, if

$$t_1\leqslant t_2\leqslant\cdots\leqslant t_{m+1}=\overbrace{\tau_1=\cdots=\tau_1}^{l_1}<\cdots<\overbrace{\tau_d=\cdots=\tau_d}^{l_d}, \tag{9.17}$$

then

$$[t_1,\ldots,t_{m+1}]f=\sum_{i=1}^{d}\sum_{j=1}^{l_i}\alpha_{ij}L_{j-1}f(\tau_i), \qquad \alpha_{il_i}\neq 0, \tag{9.18}$$

and

$$[t_1,\ldots,t_{m+1}]f=\sum_{i=1}^{d}\sum_{j=1}^{l_i}\beta_{ij}D^{j-1}f(\tau_i), \qquad \beta_{il_i}\neq 0,\ i=1,\ldots,d. \tag{9.19}$$

Finally, if f and g agree on the points $\{t_i\}_1^{m+1}$ in the sense that

$$D^{j-1}f(\tau_i)=D^{j-1}g(\tau_i), \qquad \begin{array}{l} j=1,2,\ldots,l_i \\ i=1,2,\ldots,d, \end{array}$$

then $[t_1,\ldots,t_{m+1}]f=[t_1,\ldots,t_{m+1}]g$.

Proof. Properties (9.15) and (9.16) follow directly from the definition. The expansion (9.18) is obtained by using the Laplace expansion on the determinant appearing in the numerator of the definition of the divided difference. The alternate form (9.19) follows from (9.6). These expansions show clearly that the divided difference is a linear functional defined for all sufficiently smooth functions, and that if f and g agree on the points $\{t_i\}_1^{m+1}$, as in (9.17), then their divided differences are equal. ■

One of the most important properties of ordinary divided differences is the recursion relation given in Theorem 2.51. The following theorem gives an analogous recursion relation for generalized divided differences:

THEOREM 9.8

Suppose $t_1\neq t_{m+1}$. Then

$$[t_1,\ldots,t_{m+1}]_{U_{m+1}}f=\frac{[t_2,\ldots,t_{m+1}]_{U_m}f-[t_1,\ldots,t_m]_{U_m}f}{[t_2,\ldots,t_{m+1}]_{U_m}u_{m+1}-[t_1,\ldots,t_m]_{U_m}u_{m+1}}. \tag{9.20}$$

Proof. Mühlbach [1973]. ■

The recursion (9.20) reduces to the usual one when $U_{m+1}=\{1,x,\ldots,x^m\}$. As an application of the generalized divided difference, we now give an exact error expression for *interpolation by functions in an* ECT-*space*.

THEOREM 9.9

Let $\mathcal{U}_m$ be an ECT-space on I, and let $\tau_1 < \tau_2 < \cdots < \tau_d$ be prescribed points in $[a,b]$. Let $l_1, \ldots, l_d$ be positive integers with $\sum_{i=1}^d l_i = m$. Then for any given real numbers $\{z_{ij}\}_{j=1,i=1}^{l_i\ \ d}$ there exists a unique $u \in \mathcal{U}_m$ such that

$$D^{j-1}u(\tau_i) = z_{ij}, \qquad \begin{array}{l} j = 1,2,\ldots,l_i \\ i = 1,2,\ldots,d. \end{array}$$

Moreover, if $\{z_{ij} = D^{j-1}f(\tau_i)\}_{j=1,i=1}^{l_i\ \ d}$ for some function f, then

$$f(x) - u(x) = \varphi(x)[t_1, \ldots, t_m, x]_{U_{m+1}} f, \tag{9.21}$$

where

$$\varphi(x) = \frac{D_{U_{m+1}}(t_1, \ldots, t_m, x)}{D_{U_m}(t_1, \ldots, t_m)}.$$

Proof. This result is the direct analog of Theorem 3.6 for polynomial interpolation. The existence of a unique interpolant follows from the fact that the interpolation conditions yield a system of m equations for the coefficients of $u = \sum_{i=1}^m c_i u_i$ which is nonsingular since $\mathcal{U}_m$ is an ECT-space. The error expansion (9.21) follows by expanding the determinants in the numerator and denominator of the divided difference. ∎

Theorem 9.9 shows that we can always perform Hermite interpolation with an ECT-system. The interpolating conditions here are described in terms of ordinary derivatives, but it is clear that the analogous interpolation problem, using the derivatives L_j in place of D^j, can also be uniquely solved for any given data, and that the error expression (9.21) also holds in this case.

An important result concerning polynomials is the Markov inequality given in Theorem 3.3. The following theorem states a similar result for functions in an ECT-space:

THEOREM 9.10. Markov Inequality

Let $\mathcal{U}_m$ be an ECT-space on I, and let $1 \leqslant p, q \leqslant \infty$. Then there exists a constant C_1 (depending only on $\mathcal{U}_m$, p, and q) such that

$$\|D^j u\|_{L_p[I]} \leqslant C_1 h^{-j+1/p-1/q} \|u\|_{L_q[I]}, \tag{9.22}$$

where h is the length of I.

Proof. We prove this inequality for a still wider class of functions in Theorem 10.2. ■

Our next aim is to establish a form of Budan-Fourier theorem, giving a sharp estimate of how many zeros a function u in an ECT-space can have in an interval. In dealing with such u, it is somewhat more convenient to define multiple zeros in terms of the derivatives L_j rather than in terms of the usual derivatives D^j. In particular, we say that u has a *zero of multiplicity z at the point t* provided

$$u(t)=L_1u(t)=\cdots=L_{z-1}u(t)=0\neq L_zu(t). \tag{9.23}$$

This definition is actually equivalent to the usual one, since in view of the connection (9.6), (9.23) holds if and only if

$$u(t)=Du(t)=\cdots=D^{z-1}u(t)=0\neq D^zu(t). \tag{9.24}$$

The advantage of defining multiple zeros as in (9.23) is that if u has a zero of multiplicity z at t, then L_1u has a zero of multiplicity $z-1$ at the same point. This observation allows the use of inductive-type arguments. In this connection, we now give a version of the extended Rolle's Theorem 2.19 in which D is replaced by L_1.

Let f be a function such that $L_1 f$ exists on the interval (a,b). We say that $a\leqslant c<b$ is a *left Rolle's point* of f provided *either* $f(c)=0$ *or* for every $\varepsilon>0$ there exists $c<t<c+\varepsilon$ with $u(t)L_1u(t)>0$. Similarly, $a<d\leqslant b$ is called a right Rolle's point for f provided *either* $u(d)=0$ *or* for every $\varepsilon>0$ there exists $d-\varepsilon<t<d$ with $u(t)L_1u(t)<0$ (cf. Definition 2.18).

THEOREM 9.11. Extended Rolle's Theorem

Suppose $L_1 f$ exists on (c,d) and that c and d are left and right Rolle's points of f, respectively. Then $L_1 f$ has at least one sign change on (c,d). If $L_1 f$ is continuous on (c,d), then it has at least one zero there.

Proof. We simply apply the extended Rolle's Theorem 2.19 to the function f/w_1. ■

We can now extend the Budan-Fourier Theorem 3.9 for polynomials to functions in an ECT-space $\mathfrak{U}_m$.

THEOREM 9.12

Let $U_m=\{u_i\}_1^m$ be an ECT-system, and suppose $u=\sum_{i=1}^m c_iu_i$ with $c_m\neq 0$.

Then

$$Z^*_{(a,b)}(u) \leq m-1-S^+\left[u(a),-L_1u(a),\ldots,(-1)^{m-1}L_{m-1}u(a)\right]$$
$$-S^+\left[u(b),L_1u(b),\ldots,L_{m-1}u(b)\right],$$

where Z^* counts multiplicities as in (9.23) and S^+ counts strong sign changes.

Proof. The proof parallels that of Theorem 3.9, using L_1 instead of the usual derivative D, and replacing the usual Rolle's theorem by Theorem 9.11. The only point that perhaps needs mentioning is that the argument that the constant $\alpha = A_0 - A_1$ with

$$A_j = S^+\left[(-1)^j L_j u(a),\ldots,(-1)^{m-1}L_{m-1}u(a)\right]$$

can be 1 only if a is a left Rolle's point is now based on the fact that

$$L_1u(t) = w_2(t)\int_a^t w_3(\xi_1)\cdots\int_a^{\xi_{r-1}} L_{r+1}u(\xi_r)\,d\xi_r\cdots d\xi_1. \qquad \blacksquare$$

The assumption that $c_m \neq 0$ in Theorem 9.12 assures that u does not lie in the smaller ECT-space $\mathcal{U}_{m-1}$. It is the analog of the hypothesis that p be of exact order in the usual Budan-Fourier theorem. It follows immediately from Theorem 9.12 that $Z^*(u) \leq m-1$ for any nontrivial function in $\mathcal{U}_m$. This fact also follows from Theorem 2.33 on ET-systems.

We close this section by introducing an important dual set of functions associated with a canonical ECT-system $\{u_i\}_1^m$. Given $u_1,\ldots,u_m$ defined as in (9.2) by the weights $w_1,\ldots,w_m$, we define the *dual canonical ECT-system* $U_m^* = \{u_i^*\}_{i=1}^m$ by

$$\begin{aligned}
&u_1^*(x) = 1;\\
&u_2^*(x) = \int_a^x w_m(s_m)\,ds_m;\\
&\cdots\\
&u_m^*(x) = \int_a^x w_m(s_m)\int_a^{s_m}\cdots\int_a^{s_2} w_2(s_2)\,ds_2\cdots ds_m.
\end{aligned} \tag{9.25}$$

Associated with this ECT-system, we have the operators

$$L_i^* = D_i^*\cdots D_0^*, \qquad i=0,1,\ldots,m, \tag{9.26}$$

where $D_0^* f = f$, and

$$D_i^* f = \frac{1}{w_{m-i+1}} Df, \qquad i = 1, 2, \ldots, m. \tag{9.27}$$

Clearly, $\mathfrak{U}_m^* = \text{span}(U_m^*)$ is the null space of L_m^*, and

$$L_j^* u_i^*(a) = 0, \qquad j = 0, 1, \ldots, i-2, \, i = 1, 2, \ldots, m. \tag{9.28}$$

The operator L_m^* is the formal adjoint of the operator L_m. On the other hand, L_i^* is not generally the adjoint of L_i for $i = 1, 2, \ldots, m-1$. Given a dual canonical ECT-system U_m^* as in (9.25), we define its jth reduced system by

$$\begin{aligned} u_{j,1}^*(x) &= 1, \\ u_{j,2}^*(x) &= \int_a^x w_{m-j}(s_{m-j})\, ds_{m-j}, \\ &\cdots \\ u_{j,m-j}^*(x) &= \int_a^x w_{m-j}(s_{m-j}) \int_a^{s_{m-j}} \cdots \int_a^{s_3} w_2(s_2)\, ds_2 \cdots ds_{m-j}, \end{aligned} \tag{9.29}$$

$j = 0, 1, \ldots, m-1$. We write $U_m^{*(j)} = \{u_{j,i}^*\}_{i=1}^{m-j}$. We observe that

$$L_j^* u_i^* = \begin{cases} 0, & i = 1, 2, \ldots, j \\ u_{j,i-j}^*, & i = j+1, \ldots, m. \end{cases} \tag{9.30}$$

§ 9.2. A GREEN'S FUNCTION

In this section we generalize the results of Section 2.2. We begin by introducing the analogs of the functions $(x-y)_+^{j-1}$. Suppose $\mathfrak{U}_m$ is an ECT-space with a canonical ECT-system basis $\{u_i\}_1^m$ corresponding to weight functions $\{w_i\}_1^m$. For each $j = 1, 2, \ldots, m$, let

$$g_j(x; y) = \begin{cases} h_j(x), & x \geqslant y \\ 0, & \text{otherwise}, \end{cases} \tag{9.31}$$

where

$$h_j(x; y) = w_1(x) \int_y^x w_2(s_2) \int_y^{s_2} \cdots \int_y^{s_{j-1}} w_j(s_j)\, ds_j \cdots ds_2. \tag{9.32}$$

The following theorem shows that $g_j(x;y)$ has the characteristic properties of a Green's function:

THEOREM 9.13

Fix $a \leqslant y < b$. Then for all x and y,

$$h_j(x;y) = \sum_{i=1}^{j} u_i(x) u^*_{m-j,j-i+1}(y)(-1)^{j-i}, \tag{9.33}$$

$j = 1,2,\ldots,m$. Moreover

$$L_i g_j(x;y)|_{x=y} = \delta_{i,j-1} w_j(y), \qquad i = 0,\ldots,j-1. \tag{9.34}$$

Proof. To prove (9.33), we establish even more; namely, that for all $r = 0,1,\ldots,j-2$,

$$\begin{aligned} w_{r+1}(x) \int_y^x \int_y^{s_{r+2}} \cdots \int_y^{s_{j-1}} w_j(s_j) \cdots w_{r+2}(s_{r+2})\, ds_j \cdots ds_{r+2} \\ = \sum_{i=r+1}^{j} (-1)^{j-i} u_{r,i-r}(x) u^*_{m-j,j-i+1}(y). \end{aligned} \tag{9.35}$$

We prove this by induction on r. For $r = j-2$ we have

$$\begin{aligned} w_{j-1}(x) \int_y^x w_j(s_j)\, ds_j &= w_{j-1}(x) \int_a^x w_j(s_j)\, ds_j - w_{j-1}(x) \int_a^y w_j(s_j)\, ds_j \\ &= u_{j-2,2}(x) = u_{j-2,2}(x) u^*_{m-j,1}(y) \\ &\quad - u_{j-2,1}(x) u^*_{m-j,2}(y), \end{aligned}$$

which is (9.35) in this case. Now we assume that (9.35) holds for $r+1,\ldots,$ $j-2$, and we prove it for r. We have

$$\begin{aligned} w_{r+1}(x) \int_y^x \cdots \int_y^{s_{j-1}} w_j \cdots w_{r+2} = w_{r+1}(x) \int_a^x \psi(s_{r+2})\, ds_{r+2} \\ - w_{r+1}(x) \int_a^y \psi(s_{r+2})\, ds_{r+2}, \end{aligned} \tag{9.36}$$

where (using the induction hypothesis)

$$\begin{aligned} \psi(s_{r+2}) &= w_{r+2}(s_{r+2}) \int_y^{s_{r+2}} \cdots \int_y^{s_{j-1}} w_j \cdots w_{r+3} \\ &= \sum_{i=r+2}^{j} (-1)^{j-i} u_{r+1,i-r-1}(s_{r+2}) u^*_{m-j,j-i+1}(y). \end{aligned}$$

Substituting this in the first term of (9.36), we see that it reduces to

$$\sum_{i=r+2}^{j} (-1)^{j-i} u_{r,i-r}(x) u_{m-j,j-i+1}^{*}(y).$$

Now, a simple induction argument shows that

$$\int_a^y \int_y^{s_{r+2}} \cdots \int_y^{s_{j-1}} w_j \cdots w_{r+2} = (-1)^{j-r} \int_a^y \int_a^{s_j} \cdots \int_a^{s_{r+3}} w_{r+2} \cdots w_j.$$

Since $u_{r,1}(x) = w_{r+1}(x)$, the second term of (9.36) is

$$(-1)^{j-r-1} u_{r,1}(x) u_{m-j,j-r}^{*}(y),$$

and (9.35) is proved for r. The assertion (9.34) is easy. ■

It follows directly from (9.30) and (9.33) that the function g_j can be obtained from g_m by differentiation:

$$g_j(x;y) = (-1)^{m-j} L_{m-j}^{*} g_m(x;y), \tag{9.37}$$

$j = 1, 2, \ldots, m-1$ (where, here, L_{m-j}^{*} operates on the y-variable). Before showing how g_j can be used to solve initial-value problems, we give a simple example to illustrate the expansion (9.33).

EXAMPLE 9.14

Let $U_m = \{1, x, \ldots, x^{m-1}\}$ on $[a,b] = [0,1]$.

Discussion. In this case $w_1(x) = 1$ and $w_i(x) = i-1$, $i = 2, \ldots, m$. A simple calculation shows that $u_1^{*}(y) = 1$, while

$$u_i^{*}(y) = \frac{(m-1)!\,y^{i-1}}{(m-i)!\,(i-1)!}, \qquad i = 2, \ldots, m.$$

It follows that the expansion for $h_j(x;y)$ reduces to the usual binomial theorem

$$h_j(x;y) = (x-y)^{j-1} = \sum_{i=1}^{j} \binom{j-1}{i-1} x^{i-1} y^{j-i} (-1)^{j-i}.$$ ■

THEOREM 9.15

Let $h \in L_1[a,b]$, and suppose $f_0,\ldots,f_{m-1}$ are given real numbers. Let u be the unique member of $\mathfrak{U}_m$ such that

$$L_i u(a) = f_i, \qquad i = 0, 1, \ldots, m-1.$$

Then

$$f(x) = u(x) + \int_a^b g_m(x;y) h(y)\, dy \tag{9.38}$$

is the unique solution of the initial-value problem

$$L_m f(x) = h(x), \qquad x \in [a,b], \tag{9.39}$$

$$L_i f(a) = f_i, \qquad i = 0, 1, \ldots, m-1. \tag{9.40}$$

Proof. It is easily checked that

$$L_i \int_a^b g_m(x;y) h(y)\, dy \big|_{x=a} = 0, \qquad i = 0, 1, \ldots, m-1,$$

and thus f satisfies the initial conditions (9.40). On the other hand, since

$$L_{m-1} g_m(x;y) = \begin{cases} w_m(x), & x \geqslant y \\ 0, & \text{otherwise,} \end{cases}$$

we have

$$L_{m-1} f(x) = L_{m-1} u(x) + \int_a^x w_m(x) h(y)\, dy.$$

Applying the differential operator D_m, we obtain (9.39). ■

As in the polynomial case, the function g_m plays an important role in a kind of Taylor expansion.

THEOREM 9.16. Generalized Taylor Expansion

Suppose $L_m f \in L_1[a,b]$. Then for all $a \leqslant x \leqslant b$,

$$f(x) = u_f(x) + \int_a^b g_m(x;y) L_m f(y)\, dy, \tag{9.41}$$

where u_f is the function in $\mathcal{U}_m$ such that

$$L_i u_f(a) = L_i f(a), \qquad i=0,1,\dots,m-1.$$

Proof. Let

$$g(x) = u_f(x) + \int_a^b g_m(x;y) L_m f(y)\,dy.$$

Theorem 9.15 implies that $L_m f(x) = L_m g(x)$, all $a \leqslant x \leqslant b$, and thus $f-g \in \mathcal{U}_m$. But Theorem 9.15 also asserts that $L_i(f-g)(a)=0$, $i=0,1,\dots,m-1$, and we conclude that $f=g$. ■

The function $g_m(x;y)$ also serves as a kind of dual Green's function for the adjoint operator L_m^*. Indeed, since [cf. (9.37)]

$$L_i^* g_m(x;y) = (-1)^i g_{m-i}(x;y),$$

we observe that for any fixed x,

$$L_m^* g_m(x;y) = 0, \qquad \text{all } x < y \leqslant b;$$

$$L_i^* g_m(x;y)|_{y=x} = (-1)^i w_1(x)\delta_{i,m-1}, \qquad i=0,1,\dots,m-1.$$

THEOREM 9.17

Let $h \in L_1[a,b]$, and let $f_0,\dots,f_{m-1}$ be given real numbers. Let u^* be the unique function in $\mathcal{U}_m$ such that

$$L_i^* u(b) = f_i, \quad i=0,1,\dots,m-1.$$

Then

$$f(y) = u^*(y) + \int_a^b (-1)^m h(x) g_m(x;y)\,dx \tag{9.42}$$

is the unique solution of the terminal-value problem

$$L_m^* f(y) = h(y), \qquad a \leqslant y \leqslant b;$$

$$L_i^* f(b) = f_i, \qquad i=0,1,\dots,m-1.$$

THEOREM 9.18. Dual Taylor Expansion

Let $L_m^* f \in L_1[a,b]$. Then for all $a \leqslant y \leqslant b$,

$$f(y) = u_f^*(y) + \int_a^b g_m(x;y)(-1)^m L_m^* f(x)\,dx, \tag{9.43}$$

where u_f^* is an element of $\mathcal{U}_m^*$ such that

$$L_i^* u_f^*(b) = L_i^* f(b), \qquad i = 0, 1, \ldots, m-1.$$

§ 9.3. TCHEBYCHEFFIAN SPLINE FUNCTIONS

Let $\Delta = \{a = x_0 < x_1 < \cdots < x_k < x_{k+1} = b\}$ be a partition of the interval $[a,b]$, and let $\mathfrak{M} = (m_1, \ldots, m_k)$ be a vector of integers with $1 \leqslant m_i \leqslant m$, $i = 1, 2, \ldots, k$. Suppose $\mathcal{U}_m$ is an m-dimensional ECT-space.

DEFINITION 9.19

We call

$$\mathcal{S}(\mathcal{U}_m; \mathfrak{M}; \Delta) = \left\{ \begin{array}{l} s\text{: there exist } s_0, \ldots, s_k \text{ in } \mathcal{U}_m \text{ with} \\ s|_{(x_i, x_{i+1})} = s_i, \qquad i = 0, 1, \ldots, k, \text{ and} \\ D^{j-1} s_{i-1}(x_i) = D^{j-1} s_i(x_i), \\ j = 1, \ldots, m - m_i,\ i = 1, 2, \ldots, k \end{array} \right\} \tag{9.44}$$

the space of Tchebycheffian spline functions with knots $x_1, \ldots, x_k$ *of multiplicities* $m_1, \ldots, m_k$. We use the notation $\mathcal{S}(\mathcal{U}_m; \Delta)$ in the case of simple knots.

This is a natural generalization of the space of polynomial splines discussed in Chapter 4. Indeed, the space $\mathcal{S}(\mathcal{U}_m; \mathfrak{M}; \Delta)$ reduces to the space of polynomial splines when we choose $\mathcal{U}_m = \mathcal{P}_m$. In defining $\mathcal{S}(\mathcal{U}_m; \mathfrak{M}; \Delta)$ we have required continuity of ordinary derivatives across the knots—we could just as well have required continuity of the derivatives L_j in view of the equivalence of (9.23) and (9.24).

It is clear that the space $\mathcal{S}(\mathcal{U}_m; \mathfrak{M}; \Delta)$ is a linear space. By arguments similar to those used in the proof of Theorem 4.4, we can easily prove that

$$\dim \mathcal{S}(\mathcal{U}_m; \mathfrak{M}; \Delta) = m + K, \qquad \text{where } K = \sum_{i=1}^{k} m_i.$$

The following theorem gives a one-sided basis for $\mathcal{S}$:

THEOREM 9.20

The functions

$$\rho_{i,j}(x) = \{g_{m-j+1}(x;x_i)\}_{j=1,i=0}^{m_i\ \ k}, \tag{9.45}$$

with $m_0 = m$, is a basis for $\mathcal{S}(\mathcal{U}_m;\mathfrak{M};\Delta)$.

Proof. From (9.33) and (9.34) we conclude that each of the functions ρ_{ij} belongs to $\mathcal{S}$ [recall the equivalence of (9.23) and (9.24)]. Since the dimension of $\mathcal{S}$ is $m+K$, it remains only to check that the ρ's are linearly independent. The proof of this is exactly like the proof of Theorem 4.5 in the polynomial case. ■

In view of the relation (9.37), we see that each of the one-sided splines in Theorem 9.20 can be defined directly in terms of the Green's function g_m; namely,

$$\rho_{i,j}(x) = (-1)^{j-1} L_{j-1}^{*} g_m(x;x_i), \qquad \begin{array}{l} j=1,2,\ldots,m_i \\ i=0,1,\ldots,k. \end{array} \tag{9.46}$$

The following analog of Lemma 4.7 discusses the possibility of constructing splines with local support:

LEMMA 9.21

Let $\tau_1 < \tau_2 < \cdots < \tau_d$ and $1 \leq l_i \leq m$, $i = 1,2,\ldots,d$ be given. Then

$$\sum_{i=1}^{d}\sum_{j=1}^{l_i} c_{ij} g_{m-j+1}(x;\tau_i) \equiv 0 \qquad \text{for all } x > \tau_d, \tag{9.47}$$

with $\sum_{i=1}^{d} l_i \leq m$, implies that the c's are all zero.

Proof. Define $u_{j,i}^{*}(y) = 0$ for $i = m-j+1,\ldots,m$. Then substituting (9.33) in (9.47) and interchanging the order of the summation, we obtain

$$\sum_{v=1}^{m} u_v(x) \sum_{i=1}^{d}\sum_{j=1}^{l_i} c_{ij} L_{j-1}^{*} u_{0,m-v+1}^{*}(\tau_i)(-1)^{j-1} \equiv 0$$

for $x > \tau_d$. By the linear independence of the $u_1,\ldots,u_m$, this implies

$$\sum_{i=1}^{d}\sum_{j=1}^{l_i} c_{ij} L_{j-1}^{*} u_{m-v+1}^{*}(\tau_i)(-1)^{j-1} = 0, \qquad v = 1,2,\ldots,m.$$

(Here we have written $u_i^* = u_{0,i}^*$, $i=1,2,\ldots,m$.) Since $U_m^* = \{u_i^*\}_1^m$ is an ECT-system in canonical form, the first $\Sigma_{i=1}^d l_i$ equations provide a nonsingular homogeneous system for the c's, and we conclude that they must all be zero. ■

Lemma 9.21 suggests that (as in the polynomial case) local support splines should be constructed as linear combinations of $m+1$ one-sided splines. By the same arguments used in Section 4.2, we are led to define Tchebycheffian B-splines as (generalized) divided differences. We study such B-splines in detail in the following section.

§ 9.4. TCHEBYCHEFFIAN B-SPLINES

Given a set of points

$$y_i \leqslant y_{i+1} \leqslant \cdots \leqslant y_{i+m} \qquad \text{with } y_i < y_{i+m},$$

we define the associated *Tchebycheffian B-spline* (*TB-splines*) as

$$Q_i(x) = (-1)^m [y_i, \ldots, y_{i+m}]_{U_{m+1}^*} g_m(x;y), \tag{9.48}$$

where the divided difference is taken with respect to the dual canonical ECT-system U_{m+1}^* associated with U_m—cf. (9.14).

The following theorem collects some of the basic properties of TB-splines:

THEOREM 9.22

Suppose

$$y_i \leqslant y_{i+1} \leqslant \cdots \leqslant y_{i+m} = \overbrace{\tau_1, \ldots, \tau_1}^{l_1} < \cdots < \overbrace{\tau_d, \ldots, \tau_d}^{l_d}.$$

Then

$$Q_i(x) = \sum_{j=1}^{d} \sum_{k=1}^{l_j} c_{jk} g_{m-k+1}(x;\tau_j), \tag{9.49}$$

where the g's are the Green's functions defined in (9.31). The B-spline has the properties

$$Q_i(x) = 0, \qquad x < y_i \text{ and } x > y_{i+m}; \tag{9.50}$$

$$Q_i(x) > 0 \qquad \text{for } y_i < x < y_{i+m}. \tag{9.51}$$

Proof. The expansion (9.49) follows by the Laplace expansion of the determinant in the numerator of (9.48), recalling that $g_{m-k}=(-1)^k L_k^* g_m$. The denominator in (9.48) is always positive. The fact that $Q_i(x)$ is zero for $x<y_i$ follows automatically from the one-sided nature of g_m. For $x>y_{i+m}$ we have the divided difference of a function in $\mathcal{U}_m^*=\text{span}\{u_i^*\}_1^m$, hence $Q_i(x)$ is also zero there. The positivity of Q_i in (y_i,y_{i+m}) follows from the normalization and the fact that it cannot have any zeros in this interval (a fact that we establish in Corollary 9.31). ■

We are now ready to give a local-support basis for the space of TB-splines.

THEOREM 9.23

Let $\Delta_e=\{y_i\}_1^{2m+K}$ be an extended partition of $[a,b]$, as in Definition 4.8, with $b<y_{2m+K}$. Let $Q_i(x)$ be the B-spline associated with $y_i,\ldots,y_{i+m}$ as in (9.48), $i=1,2,\ldots,m+K$. Then $\{Q_i\}_1^{m+K}$ is a basis for $\mathcal{S}(\mathcal{U}_m;\mathfrak{M};\Delta)$. Moreover,

$$Q_i(x)=0, \qquad x<y_i \text{ and } x>y_{i+m},$$

and

$$Q_i(x)>0, \qquad y_i<x<y_{i+m},$$

$i=1,2,\ldots,m+K$.

Proof. In view of Theorem 9.22 we see that each of the Q_i is a spline in $\mathcal{S}(\mathcal{U}_m;\mathfrak{M};\Delta)$ with the asserted properties. Since the dimension of $\mathcal{S}$ is $m+K$, it remains only to check that the Q's are linearly independent. We may establish this exactly as in the proof of Theorem 4.18, using Lemma 9.21. ■

It is also possible to define a B-spline basis for $\mathcal{S}$ in the case where the extended partition Δ_e is such that $b=y_{2m+K}$. In this case we must define

$$Q_{m+K}(b)=\lim_{x\uparrow b} Q_{m+K}(x)$$

(cf. Corollary 4.10).

We now continue with properties of the B-splines. First, we have the following simple analog of the *Peano representation of divided differences* given in Theorem 4.23:

THEOREM 9.24

Let $L_m f \in L_1[y_i, y_{i+m}]$. Then

$$[y_i, \ldots, y_{i+m}]_{U^*_{m+1}} f = \int_{y_i}^{y_{i+m}} Q_i(x) L^*_m f(x)\,dx. \tag{9.52}$$

Proof. We simply apply the divided difference to the dual Taylor expansion (9.43). ■

If we take $f = u^*_{m+1}$ in (9.52), we obtain the interesting fact that

$$1 = \int_{y_i}^{y_{i+m}} Q_i(x)\,dx. \tag{9.53}$$

As in the polynomial spline case, it is useful to have a normalized version of the TB-spline. We define

$$N_i(x) = \alpha_i Q_i(x), \tag{9.54}$$

where

$$\alpha_i = \frac{D_{U^*_{m+1}}(y_i, \ldots, y_{i+m}) D_{U^*_{m-1}}(y_{i+1}, \ldots, y_{i+m-1})}{D_{U^*_m}(y_{i+1}, \ldots, y_{i+m}) D_{U^*_m}(y_i, \ldots, y_{i+m-1})}.$$

This factor is well defined, and it is positive since U^*_{m+1} is an ECT-system. In the case where $\mathcal{U}_m = \mathcal{P}_m$, it can be shown after some calculation (cf. Example 9.14 where the u_i^*'s are calculated) that $\alpha_i = (y_{i+m} - y_i)$.

We can now establish the following version of *Marsden's identity for normalized TB-splines* (cf. Theorem 4.21 in the polynomial spline case):

THEOREM 9.25

Let $\{N_i\}_1^{m+K}$ be the normalized B-splines defined in (9.54), and let

$$\varphi_i(y) = \frac{D_{U^*_m}(y_{i+1}, \ldots, y_{i+m-1}, y)}{D_{U^*_{m-1}}(y_{i+1}, \ldots, y_{i+m-1})}, \tag{9.55}$$

for $i = 1, 2, \ldots, m+K$. Then the function h_m defined in (9.32) satisfies

$$h_m(x; y) = \sum_{i=1}^{m+K} (-1)^{m-1} \varphi_i(y) N_i(x). \tag{9.56}$$

Moreover, for all $y_m \leqslant x \leqslant y_{m+K}$,

$$u_1(x) = \sum_{i=1}^{m+K} N_i(x). \tag{9.57}$$

We also have

$$u_j(x) = \sum_{i=1}^{m+K} \eta_i^{(j)} N_i(x), \qquad j = 2,3,\ldots,m, \tag{9.58}$$

where

$$\eta_i^{(j)} = (-1)^{j-1} L_{m-j}^* \varphi_i(y)|_{y=a}. \tag{9.59}$$

In particular,

$$\eta_i^{(2)} = \frac{D\begin{pmatrix} y_{i+1},\ldots,y_{i+m-1} \\ u_1^*,\ldots,u_{m-2}^*,u_m^* \end{pmatrix}}{D_{U_{m-1}^*}(y_{i+1},\ldots,y_{i+m-1})}, \tag{9.60}$$

and

$$\eta_i^{(3)} = \frac{D\begin{pmatrix} y_{i+1},\ldots,y_{i+m-1} \\ u_1^*,\ldots,u_{m-3}^*,u_{m-1}^*,u_m^* \end{pmatrix}}{D_{U_{m-1}^*}(y_{i+1},\ldots,y_{i+m-1})}. \tag{9.61}$$

Proof. Fix $m \leqslant l < l+1 \leqslant m+K+1$ and $y_l < x \leqslant y_{l+1}$. For each i there is a unique polynomial in U_{m+1}^* (call it p_i) which interpolates $g_m(x;y)$ at the points $y = y_i,\ldots,y_{i+m}$. By the remainder theory for interpolation from U_{m+1}^* (see Theorem 9.9),

$$R_i(y) = g_m(x;y) - p_i(y) = \phi_{i,m+2}(y)[\,y_i,\ldots,y_{i+m},y\,]\,g_m(x;y),$$

where $\phi_{i,m+2}$ is the unique element of U_{m+2}^* with leading term u_{m+2}^* and zeros at $y_i,\ldots,y_{i+m}$. We observe that

$$\phi_{i,m+2}(y) = \frac{D_{U_{m+2}^*}(y_i,\ldots,y_{i+m},y)}{D_{U_{m+1}^*}(y_i,\ldots,y_{i+m})}.$$

From properties of the generalized divided difference, we note that

$$\phi_{i+1,m+1}(y)([y_{i+1},\ldots,y_{i+m},y]g_m-[y_i,\ldots,y_{i+m}]g_m)$$
$$=\phi_{i,m+2}(y)[y_i,\ldots,y_{i+m},y]g_m,$$

and

$$\phi_{i+1,m+1}(y)([y_{i+1},\ldots,y_{i+m},y]g_m-[y_{i+1},\ldots,y_{i+m+1}]g_m)$$
$$=\phi_{i+1,m+2}(y)[y_{i+1},\ldots,y_{i+m+1},y]g_m.$$

Subtracting these two expressions and setting $B_i=(-1)^mQ_i$, we obtain

$$R_i(y)-R_{i+1}(y)=\phi_{i+1,m+1}(y)(B_{i+1}(x)-B_i(x)).$$

If we sum these identities for $i=l-m,\ldots,l$ and note that $B_{l-m}(x)=B_{l+1}(x)=0$, then after rearranging we can write

$$R_{l-m}(y)-R_{l+1}(y)=\sum_{i=l+1-m}^{l}B_i(x)(\phi_{i,m+1}(y)-\phi_{i+1,m+1}(y)).$$

Since $R_{l+1}(y)=g_m(x;y)$ while $R_{l-m}(y)=g_m(x;y)-h_m(x;y)$, this gives an expansion of $h_m(x;y)$. It is clear (since they both have leading terms u_{m+1}) that the difference of $\phi_{i,m+1}(y)$ and $\phi_{i+1,m+1}(y)$ is an element in U_m^* with zeros at $y_{i+1},\ldots,y_{i+m-1}$. Thus

$$\phi_{i,m+1}(y)-\phi_{i+1,m+1}(y)=a_i\phi_{i+1,m}(y)=a_i\varphi_i(y)$$

for some constant a_i. The fact that a_i equals α_i as in (9.54) follows by letting y approach y_{i+m} from above. We have established (9.56).

To prove (9.58), we note that by (9.33)

$$h_m(x;y)=\sum_{i=1}^{m}u_i(x)u_{m-i+1}^*(y)(-1)^{m-i}.$$

Then applying L_{m-j}^* to both sides of (9.56) and evaluating at $y=a$, we are led to (9.58). The exact values of the η's can be found by evaluating $L_{m-j}^*\varphi_i(y)|_{y=a}$ carefully. ■

The expansion (9.57) shows that the normalized B-splines $N_i(x)$ are bounded by $\overline{M}_1=\max_{a\leqslant x\leqslant b}u_1(x)$. This assertion in conjunction with Markov's inequality for derivatives of an ECT-system (see Theorem 9.10) shows that

$$|D^k N_i(x)| \leqslant \frac{C}{\underline{\Delta}^k}, \qquad k=0,1,\ldots,m-1, \tag{9.62}$$

where the constant depends only on m and the weights $w_1,\ldots,w_m$ involved in the canonical basis for $\mathcal{U}_m$.

As we saw in Chapter 4, it is useful to have a dual basis for the set of B-splines. We turn now to the construction of a *dual basis for the TB-splines* of Theorem 9.25. We shall construct a set $\{\lambda_i\}_1^{m+K}$ of linear functionals defined on $L_p[a,b]$ such that

$$\lambda_i N_j = \delta_{ij}, \qquad i,j=1,2,\ldots,m+K. \tag{9.63}$$

Following the construction method for the polynomial spline case in Section 4.6, for each $i=1,\ldots,m+K$ we define

$$\lambda_i f = \int_{y_i}^{y_{i+m}} f(x) L_m^* \psi_i(x)\, dx \tag{9.64}$$

with

$$\psi_i(x) = \alpha_i \varphi_i(x)\, G_i(x),$$

where g is the transition function described in Theorem 4.37, and where φ_i is defined in (9.55) and α_i is defined in (9.54).

THEOREM 9.26

The linear functionals $\{\lambda_i\}_1^{m+K}$ defined in (9.64) form a dual basis for the B-splines $\{N_i\}_1^{m+K}$. In addition,

$$\|\lambda_i\| = \sup \frac{|\lambda_i f|}{\|f\|_{L_p[y_i, y_{i+m}]}} \leqslant C \underline{\Delta}^{-1/p}, \tag{9.65}$$

$i=1,2,\ldots,m+K$, where C is a constant depending only on m and the weights $w_1,\ldots,w_m$ involved in the basis for $\mathcal{U}_m$.

Proof. First we must show that the λ's satisfy (9.63). By the support properties of the B-splines, it is clear that

$$\lambda_i N_j = 0, \qquad j=1,2,\ldots,i-m,\ i+m,\ldots,m+K.$$

On the other hand, by the Peano representation (9.52), we have

$$\lambda_i N_j = \alpha_j [y_j, \ldots, y_{j+m}]_{U^*_{m+1}} \psi_i.$$

By construction, the function ψ_i vanishes at the points $y_{i+1}, \ldots, y_{i+m-1}$ (with multiplicities) as well as for all $x \leqslant y_i$. It follows that

$$\lambda_i N_j = 0 \qquad \text{for } j = i-m, \ldots, i-1.$$

On the other hand, for $x \geqslant y_{i+1}$ the function $\psi_i(x)$ takes on the same values as the function $\varphi_i(x)$, at least on the knots. But the divided difference of functions in $\mathfrak{U}^*_m$ is zero, hence we conclude that $\lambda_i N_j = 0$ also for $j = i+1, \ldots, i+m-1$. It remains to check the value of $\lambda_i N_i$.

On the points $y_i, \ldots, y_{i+m}$, the function $\psi_i(x)$ agrees with the function

$$\theta_i(x) = \frac{D_{U^*_{m+1}}(y_i, \ldots, y_{i+m-1}, x)}{D_{U^*_m}(y_i, \ldots, y_{i+m-1})}.$$

Indeed, both ψ_i and θ_i vanish at the points $y_i, \ldots, y_{i+m-1}$ (with multiplicities), and both have the value

$$\frac{D_{U^*_{m+1}}(y_i, \ldots, y_{i+m-1}, y_{i+m})}{D_{U^*_m}(y_i, \ldots, y_{i+m-1})}$$

at the point y_{i+m}. On the other hand,

$$\theta_i(x) = u^*_{m+1}(x) + \cdots,$$

hence

$$\lambda_i N_i = [y_i, \ldots, y_{i+m}]_{U^*_{m+1}} \theta_i = 1.$$

We now estimate the norms of the linear functionals λ_i. Let $I_i = [y_i, y_{i+m}]$. By Hölder's inequality, we have

$$|\lambda_i f| \leqslant \|f\|_{L_p[I_i]} \|L^*_m \psi_i\|_{L'_p[I_i]},$$

with $1/p + 1/p' = 1$. Using the Leibnitz rule, it is easy to show by induction on m that

$$L^*_m \psi_i(x) = \sum_{k=0}^{m} \frac{c_k(x)}{d_k(x)} D^{m-k} G_i(x) L^*_k \varphi_i(x),$$

where $c_k(x)$ depends only on the values of $\{w_i\}_1^m$ and their derivatives, while $d_k(x)$ depends only on powers of $\{w_i\}_1^m$. Since

$$\|D^{m-k}G_i(x)\|_{L_\infty[I_i]} \leqslant \frac{C_{m-k}}{(y_{i+m}-y_i)^{m-k}}, \qquad k=0,1,\ldots,m$$

and $L_m^*\varphi_i=0$, this implies

$$\frac{|\lambda_i f|}{\|f\|_{L_p[I_i]}} \leqslant (y_{i+m}-y_i)^{1-1/p} \max_{0\leqslant k\leqslant m-1} \frac{\|L_k^*\varphi_i\|_{L_\infty[I_i]}}{(y_{i+m}-y_i)^{m-k}}. \tag{9.66}$$

By Lemma 9.6, and the definition of φ_i,

$$|L_k^*\varphi_i(x)| \leqslant C_3 |D^k\Phi_i(x)|, \qquad x\in I_i,$$

where

$$\Phi_i(x) = \frac{V(y_{i+1},\ldots,y_{i+m-1},x)}{V(y_{i+1},\ldots,y_{i+m-1})}.$$

Now suppose

$$y_{i+1},\ldots,y_{i+m-1} = \overbrace{\tau_1,\ldots,\tau_1}^{l_1} < \cdots < \overbrace{\tau_d,\ldots,\tau_d}^{l_d}.$$

Then by (2.67)

$$\Phi_i(x) = \prod_{i=1}^{d} (x-\tau_i)^{l_i},$$

and it follows that

$$|D^k\Phi_i(x)| \leqslant C_4 \overline{\Delta}^{m-1-k}, \qquad k=0,1,\ldots,m-1.$$

Substituting this in (9.66) yields (9.65). ∎

The dual basis $\{\lambda_i\}_1^{m+K}$ for $\{N_i\}_1^{m+K}$ can now be used to examine the conditioning of this basis.

THEOREM 9.27

Fix $1\leqslant p\leqslant\infty$. For $i=1,2,\ldots,m+K$ let

$$B_{i,p}(x) = \overline{\Delta}^{-1/p} N_i(x). \tag{9.67}$$

Then there exist constants $0<C_1$ and $C_2<\infty$ depending only on m and the quantities $\{\underline{M}_i, \overline{M}_i\}_1^m$ in (9.11) such that

$$\left(\sum_{i=1}^{m+K}|c_i|^p\right)^{1/p} \leqslant C_1\left\|\sum_{i=1}^{m+K} c_iB_{i,p}\right\|_{L_p[a,b]} \leqslant C_2\left(\sum_{i=1}^{m+K}|c_i|^p\right)^{1/p} \tag{9.68}$$

for all sets of coefficients $c_1,\ldots,c_{m+K}$.

Proof. Let $s=\sum_{i=1}^{m+K}c_iB_i$. Then by (9.65) with $I_i=[y_i,y_{i+m}]$

$$\sum_{i=1}^{m+K}|c_i|^p=\sum_{i=1}^{m+K}|\lambda_i s|^p \leqslant C_1\sum_{i=1}^{m+K}\|s\|^p_{L_p[I_i]} \leqslant C_1\|s\|^p_{L_p[a,b]}.$$

Conversely, by our normalization of the B-splines,

$$|B_{i,p}(x)| \leqslant \overline{M}_1\cdot\underline{\Delta}^{-1/p}, \qquad i=1,2,\ldots,m+K.$$

Thus

$$\int_a^b|s(x)|^p\,dx=\sum_{j=0}^{k}\int_{x_j}^{x_{j+1}}\left|\sum_{i=j-m+1}^{j}c_iB_{i,p}\right|^p$$

$$\leqslant \sum_{j=0}^{k}\int_{x_j}^{x_{j+1}}\max_{1\leqslant i\leqslant m+K}\|B_{i,p}\|^p_{L_\infty[I_i]}m^{p-1}\sum_{i=j-m+1}^{j}|c_i|^p \leqslant C_3\sum_{i=1}^{m+K}|c_i|^p. \blacksquare$$

§ 9.5. ZEROS OF TCHEBYCHEFFIAN SPLINES

In this section we give several results on the zeros of T-splines similar to those given in Section 4.7 for polynomial splines. First we need to agree on how to count zeros. Throughout this section we shall count the zeros of a T-spline according to Definitions 4.45 to 4.47. The key to establishing bounds on the number of zeros a T-spline can possess will be an appropriate version of the Rolle's theorem for such splines. Before stating it we need the following result which shows that the derivative L_1s of a T-spline is again a T-spline (with respect to the first reduced ECT-system).

THEOREM 9.28

Let $s\in\mathcal{S}(\mathcal{U}_m;\mathfrak{M};\Delta)$, $m>1$. Then

$$L_1s\in\mathcal{S}(\mathcal{U}_m^{(1)};\mathfrak{M}^{(1)};\Delta),$$

where $\mathcal{U}_m^{(1)}$ is the space spanned by the first reduced system $U_m^{(1)}$, and where $\mathfrak{M}^{(1)} = (m_1', \ldots, m_k')$ and $m_i' = \min(m_i, m-1)$, $i = 1, 2, \ldots, k$.

Proof. Since s is piecewise in $\mathcal{U}_m$, it is clear that $L_1 s$ is piecewise in the first reduced space $\mathcal{U}_m^{(1)}$. The tie conditions on $L_1 s$ follow directly from those on s. ■

We are ready to state Rolle's theorem for T-splines.

THEOREM 9.29

Suppose $s \in \mathcal{S}(\mathcal{U}_m; \mathfrak{M}; \Delta)$ and that s is continuous. Then

$$Z^{\mathcal{D}\mathcal{S}}(L_1 s) \geqslant Z^{\mathcal{S}}(s) - 1,$$

where $\mathcal{D}\,\mathcal{S} = \mathcal{S}(\mathcal{U}_m^{(1)}; \mathfrak{M}^{(1)}; \Delta)$ is the space defined in Theorem 9.28.

Proof. The proof parallels that of Theorem 4.50 in the case of polynomial splines, using the extended Rolle's Theorem 9.11 for ECT-systems. ■

We can now prove our main result concerning the zeros of T-splines.

THEOREM 9.30

For all nontrivial $s \in \mathcal{S}(\mathcal{U}_m; \mathfrak{M}; \Delta)$,

$$Z^{\mathcal{S}}(s) \leqslant m + K - 1.$$

Proof. The proof is similar to that of Theorem 4.53. We proceed by induction on m. For $m = 1$ we are dealing with splines that are constant multiples of the positive function u_1 in each subinterval defined by the knots. We conclude that $Z^{\mathcal{S}}(s) \leqslant m + k - 1 = k$ in this case. Now if s is continuous, the same inductive argument used in proving Theorem 4.53 applies, where now we use the Rolle's Theorem 9.29 for T-splines. The extension of the result to splines that have discontinuities (i.e., some m-tuple knots) also proceeds as before. It suffices to note that the perturbation Lemmas 4.51 and 4.52 both have analogs here. ■

The following corollary of Theorem 9.30 gives bounds on the number of zeros of derivatives of B-splines. It also establishes that the TB-splines are positive on (y_i, y_{i+m}).

COROLLARY 9.31

For all $i = 1, 2, \ldots, m + K$ and all $j = 0, 1, \ldots, m-1$,

$$Z_{(y_i, y_{i+m})}(L_j Q_i) \leqslant j.$$

Proof. L_jQ_i is a T-spline of order $m-j$ with $m+1$ knots. Thus $Z(L_jQ_i) \leqslant 2m-j$. But L_jQ_i has an $m-j$th order zero on $(-\infty, y_i)$ and on (y_{i+m}, ∞), so it can have at most j zeros in (y_i, y_{i+m}). ■

Corollary 9.31 can be improved to produce a form of Budan-Fourier theorem for T-splines.

THEOREM 9.32

Suppose s is a T-spline in $\mathcal{S}(\mathfrak{U}_m; \mathfrak{M}; \Delta)$ and that $L_{d_0}s(a) \neq 0$ and $L_{d_k}s(b) \neq 0$. Then

$$Z_{(a,b)}^{\mathcal{S}}(s) \leqslant m+K-1-S^{+}\left[s(a), -L_1 s(a), \ldots, (-1)^{d_0-1} L_{d_0} s(a)\right]$$
$$- S^{+}\left[s(b), L_1 s(b), \ldots, L_{d_k} s(b)\right]. \tag{9.69}$$

Proof. The proof follows the same lines as that of Theorem 4.58. Here we must use the extended Rolle's Theorem 9.11 for ECT-systems. ■

§ 9.6. DETERMINANTS AND SIGN CHANGES

In this section we discuss various determinants formed from the B-splines and from the Green's function g_m. We also show that the TB-splines form an OCWT-system, and conclude from this that the T-splines satisfy an important variation-diminishing property.

We begin with determinants formed from B-splines. Given

$$t_1 \leqslant t_2 \leqslant \cdots \leqslant t_N,$$

let

$$d_i = \max\{j: t_i = \cdots = t_{i-j}\}, \qquad i=1,2,\ldots,N.$$

Let $\{B_i\}_1^N$ be a set of TB-splines associated with a knot sequence $y_1 \leqslant y_2 \leqslant \cdots \leqslant y_{m+N}$. Then we define the matrix

$$M\begin{pmatrix} t_1,\ldots,t_N \\ B_1,\ldots,B_N \end{pmatrix} = \left[L_{d_i} B_j(t_i)\right]_{i,j=1}^{N}, \tag{9.70}$$

cf. (4.141) for the polynomial spline case. This matrix arises if we attempt to find a linear combination of the B-splines $\{B_i\}_1^N$ to solve the Hermite interpolation problem

$$L_{d_i} s(t_i) = v_i, \qquad i=1,2,\ldots,N.$$

Let $D\begin{pmatrix} t_1,\ldots,t_N \\ B_1,\ldots,B_N \end{pmatrix}$ denote the determinant of M. The following theorem gives the precise conditions under which M is nonsingular:

THEOREM 9.33

The determinant D of the matrix M in (9.70) is nonegative. It is positive if and only if

$$t_i \in \sigma_i, \qquad i=1,2,\ldots,N \tag{9.71}$$

where

$$\sigma_i = \begin{cases} [y_i, y_{i+m}), & \text{if } y_i = \cdots = y_{i+m-y_i-1} \\ (y_i, y_{i+m}), & \text{otherwise} \end{cases}$$

Theorem 9.33 shows that the B-splines basis $\{B_i\}_1^{m+K}$ for the space $\mathcal{S}(\mathcal{U}_m; \mathcal{M}; \Delta)$ of T-splines is a WT-system. As in the polynomial case, we can prove even more; namely, that the basis is actually order complete (an OCWT-system).

THEOREM 9.34

Let $\{B_i\}_1^N$ be a set of B-splines as in Theorem 9.33. Then for any choice of $1 \leqslant \nu_1 < \cdots < \nu_p \leqslant N$ and any $t_1 < \cdots < t_p$,

$$D\begin{pmatrix} t_1,\ldots,t_p \\ B_{\nu_1},\ldots,B_{\nu_p} \end{pmatrix} \geqslant 0, \tag{9.72}$$

and strict positivity holds if and only if

$$t_i \in (y_{\nu_i}, y_{\nu_i+m}) \cup \{x: L_{d_i} B_{\nu_i}(x) \neq 0\}, \qquad i=1,2,\ldots,p. \tag{9.73}$$

Proof. The proof is a direct analog of the proof of Theorem 4.65. ■

As an immediate corollary of Theorem 9.34, we have the following variation-diminishing result for T-spline expansions.

THEOREM 9.35. Variation-Diminishing Property

$$S^-_{(a,b)}\left(\sum_{i=1}^{m+K} c_i B_i \right) \leqslant S^-(c_1,\ldots,c_{m+K}) \tag{9.74}$$

for any $c_1,\ldots,c_{m+K}$, not all zero.

Proof. In Theorem 9.34 we have shown that the B-splines form an OCWT-system. Theorem 2.42 now applies. ■

The determinants associated with certain extended-Hermite interpolation problems can also be examined for T-splines. In particular, if $\theta_1,\ldots,\theta_N$ is a sequence of signs, then we may define

$$M\begin{bmatrix} t_1,\ldots,t_N \\ \theta_1,\ldots,\theta_N \\ B_1,\ldots,B_N \end{bmatrix} = \left[\theta_i^{d_i} L_{d_i}^{\theta_i} B_j(t_i)\right]_{i,j=1}^N. \tag{9.75}$$

The analogs of Theorems 4.71 to 4.73 then hold. The proofs are nearly identical to the polynomial spline case.

The Green's function $g_m(x;y)$ also exhibits strong total positivity properties. We have the following analog of Theorem 4.78:

THEOREM 9.36

Let p be a positive integer, and suppose

$$t_1 \leqslant t_2 \leqslant \cdots \leqslant t_p$$

$$y_1 \leqslant y_2 \leqslant \cdots \leqslant y_p$$

are given with $t_i < t_{i+m}$ and $y_i < y_{i+m}$, all i. Define

$$d_i = \max\{j:\ t_{i-j} = t_i\}$$

$$q_i = \max\{j:\ y_{i-j} = y_i\}, \qquad i = 1,2,\ldots,p.$$

Then

$$\det g_m\begin{pmatrix} t_1,\ldots,t_p \\ y_1,\ldots,y_p \end{pmatrix} = \det\left[L_{d_i} L_{q_j}^* g_m(t_i;y_j)\right]_{i,j=1}^p \geqslant 0, \tag{9.76}$$

and strict positivity holds if and only if

$$y_i < t_i < y_{i+m}, \qquad i = 1,2,\ldots,p, \tag{9.77}$$

where equality is allowed on the left if $y_i = \cdots y_{i+m-d_i-1}$, and the right side is ignored if $i+m>p$. [In (9.76), in the definition of the determinant, the first operator L operates on the t-variable, and the second operates on the y-variable.]

Proof. See Theorem 4.78. ■

§ 9.7. APPROXIMATION POWER OF TCHEBYCHEFFIAN SPLINES

In this section we discuss how well smooth functions can be approximated by T-splines. We shall see that, in general, we can achieve the same orders of approximation as with polynomial splines (cf. Chapter 6). For some related inverse theorems, see Section 10.7.

Our approach to deriving estimates for the approximation power of T-splines will be to construct an explicit local spline approximation operator similar to the one constructed in Remark 6.4 for polynomial splines. Given a partition $\Delta=\{a=x_0<x_1<\cdots<x_k<x_{k+1}=b\}$ of the interval $[a,b]$, suppose Δ^* is the quasi-uniform, thinned-out partition constructed in Lemma 6.17. Associated with these points, let $\{y_i\}_1^{n+m}$ be an extended partition as in (6.36), and let $\{B_i\}_1^n$ be the associated normalized TB-splines. Finally, let $\{\lambda_i\}_1^n$ be the dual basis corresponding to $\{B_i\}_1^n$ constructed in Theorem 9.26. Now for any $f\in L_1[a,b]$, we define

$$Qf(x)=\sum_{i=1}^{n}(\lambda_i f)B_i(x). \tag{9.78}$$

It is clear that Q is a linear operator mapping $L_1[a,b]$ onto the T-spline space $\mathcal{S}(\mathcal{U}_m;\Delta^*)\subseteq\mathcal{S}(\mathcal{U}_m;\mathcal{M};\Delta)$. In fact, since the λ's form a dual basis, Q is actually a linear projector onto $\mathcal{S}(\mathcal{U}_m;\Delta^*)$ i.e., $Qs=s$ for all $s\in\mathcal{S}(\mathcal{U}_m;\Delta^*)$. The following theorem gives an error bound for approximation using this projector:

THEOREM 9.37

Fix $m\leqslant l\leqslant n$, and let $I_l=[y_l,y_{l+1}]$ and $\tilde{I}_l=[y_{l+1-m},y_{l+m}]$. Then there exists a constant C_1 such that for any $f\in L_1^m[\tilde{I}_l]$,

$$\|D^j(f-Qf)\|_{L_q[I_l]}\leqslant C_1\bar{\Delta}^{m-j+1/q-1/p}\|L_m f\|_{L_p[\tilde{I}_l]}, \tag{9.79}$$

$j=0,1,\ldots,m-1$.

Proof. Fix $0\leqslant j\leqslant m-1$, and let $y_l\leqslant t<y_{l+1}$. By the generalized Taylor expansion (9.38), we have

$$f(x)=u_f(x)+\int_t^x g_m(x;y)L_m f(y)\,dy,$$

where $u_f\in\mathcal{U}_m$ is such that

$$L_j u_f(t)=L_j f(t),\qquad j=0,1,\ldots,m-1,$$

or in view of (9.6), equivalently,

$$D^j u_f(t) = D^j f(t), \qquad j = 0, 1, \ldots, m-1.$$

Since $Qu_f = u_f$, using the boundedness of the linear functionals λ_i, we obtain

$$|D^j(f-Qf)(t)| = |D^j(u_f - Qf)(t)| = |D^j Q(u_f - f)(t)|$$

$$= \left| \sum_{i=l+1-m}^{l} \lambda_i(u_f - f) D^j B_i(t) \right| \leqslant C_5 \underline{\Delta}^{*-1/p} \sum_{i=l+1-m}^{l} \|u_f - f\|_{L_p[\tilde{I}_l]} |D^j B_i(t)|.$$

Applying Hölder's inequality to the remainder in the Taylor expansion and using the integral expansion (9.31) and (9.32) for g_m to establish

$$|g_m(x;y)| \leqslant C_4 |x-y|^{m-1}, \qquad \text{all } x \text{ and } y,$$

we obtain

$$\|f - u_f\|_{L_p[\tilde{I}_l]} \leqslant C_3 \bar{\Delta}^{*m} \|L_m f\|_{L_p[\tilde{I}_l]}.$$

On the other hand, as observed in (9.62),

$$|D^j B_i(t)| \leqslant C_2 (\underline{\Delta}^*)^{-j},$$

and combining these inequalities with the fact that $\bar{\Delta}^*/\underline{\Delta}^* \leqslant 3$ and $\bar{\Delta}^* \leqslant \bar{\Delta}$, we obtain (9.79). ■

Theorem 9.37 gives a bound on $f - Qf$ in an interval I_l which depends only on the behavior of $L_m f$ in a slightly larger interval $\tilde{I}_l$. We can give the following global version:

THEOREM 9.38

For any $f \in L_p^m[a,b]$ and any $1 \leqslant p \leqslant q \leqslant \infty$,

$$\|D^j(f - Qf)\|_{L_q[a,b]} \leqslant C_1 \bar{\Delta}^{m-j+1/q-1/p} \|L_m f\|_{L_p[a,b]}, \tag{9.80}$$

$j = 0, 1, \ldots, m-1$.

Proof. We simply add the inequalities (9.79) together for $i = m, m+1, \ldots, n$ and use the Jensen inequality (cf. Remark 6.2) on the term

$$\sum_{i=m}^{n} \|L_m f\|^q_{L_p[\tilde{I}_i]} \leqslant m \sum_{i=m}^{n} \|L_m f\|^q_{L_p[I_i]}.$$ ■

It is also of interest to give estimates for how well less-smooth classes of functions can be approximated using T-splines. The following theorem is a kind of Jackson theorem (cf. Theorem 6.27 for the polynomial spline case and Theorem 3.12 for polynomials):

THEOREM 9.39

Fix $1 \leqslant p < \infty$. Then for any $f \in L_p[a,b]$,

$$d[f, \mathbb{S}(\mathfrak{U}_m; \mathfrak{M}; \Delta)]_{L_p[a,b]} \leqslant C\left[\omega_m(f; \bar{\Delta})_{L_p[a,b]} + \bar{\Delta}^m \|f\|_{L_p[a,b]}\right], \quad (9.81)$$

where C is a constant depending only on m and $\mathfrak{U}_m$. Similarly, with $p = \infty$, (9.81) holds for all functions $f \in C[a,b]$.

Proof. The operator L_m has the form

$$L_m = \frac{D^m}{w_1 w_2 \cdots w_m} + \sum_{j=0}^{m-1} a_j D^j$$

with $a_j \in C^{m-j}[a,b]$. Thus Theorem 10.1 implies that for all $f \in L_p^m\,[a,b]$,

$$\|L_m f\|_{L_p[a,b]} \leqslant C\left(\|D^m f\|_{L_p[a,b]} + \|f\|_{L_p[a,b]}\right).$$

Inserting this in (9.80) with $j = 0$, we obtain

$$d[f, \mathbb{S}(\mathfrak{U}_m; \mathfrak{M}; \Delta)]_{L_p[a,b]} \leqslant C\bar{\Delta}^m\left(\|D^m f\|_{L_p[a,b]} + \|f\|_{L_p[a,b]}\right).$$

Now we may apply Theorem 2.68 to obtain (9.81). ■

§ 9.8. OTHER SPACES OF TCHEBYCHEFFIAN SPLINES

In this section we discuss several other spaces of T-splines useful in applications. These will include spaces of periodic T-splines, natural T-splines, T-splines with HB ties, and T-monosplines. Since the development closely parallels that in Chapter 8, where similar spaces of polynomial splines were discussed, we can afford to keep the details to a minimum.

We begin by defining the space of periodic T-splines. Let $a < b$, and let $\Delta = \{a < x_1 < \cdots < x_k < b\}$. If we think of $[a,b]$ as a circle, then Δ partitions it into k subintervals $I_i = [x_i, x_{i+1})$, $i = 1, 2, \ldots, k-1$, and $I_k = [x_k, x_1]$. Now given an ECT-space $\mathfrak{U}_m$ and a multiplicity vector $\mathfrak{M} = (m_1, \ldots, m_k)$, we define the space of *periodic T-splines* by

$$\mathring{\mathbb{S}}(\mathcal{U}_m;\mathfrak{M};\Delta)=\left\{\begin{array}{l} s\text{: there exist } s_1,\ldots,s_k \text{ in } \mathcal{U}_m \text{ with } s(x)= \\ s_i(x) \text{ on } I_i,\ i=1,2,\ldots,k, \text{ and } D^{j-1}s_{i-1}(x_i) \\ =D^{j-1}s_i(x_i),\quad j=1,\ldots,m-m_i,\quad i= \\ 1,2,\ldots,k, \text{ where } s_0=s_k \end{array}\right\} \quad (9.82).$$

It is easily seen by the same kind of argument used in Theorem 8.1 that $\mathring{\mathbb{S}}(\mathcal{U}_m;\mathfrak{M};\Delta)$ is a linear space of dimension $K=\sum_{i=1}^{k}m_i$. A basis of periodic TB-splines can be constructed from the usual TB-splines by the same method used in (8.6) and (8.7).

Concerning zero properties of periodic T-splines, the situation is very similar to the case of periodic polynomial splines. Assuming that $\mathring{Z}(s)$ counts the number of zeros of s on the circle $[a,b)$ (relative to the space $\mathring{\mathbb{S}}$), we have the following result:

THEOREM 9.40

For every nontrivial $s\in\mathring{\mathbb{S}}(\mathcal{U}_m;\mathfrak{M};\Delta)$,

$$\mathring{Z}(s)\leqslant\begin{cases} K-1, & K\text{ odd}\\ K, & K\text{ even}.\end{cases}$$

Proof. The proof proceeds by induction. The case of $m=1$ is easily verified. The remainder of the proof is exactly as in the polynomial case (cf. Theorem 8.4), using the extended Rolle's Theorem 9.29. ■

The cases where K is odd and where K is even are essentially different, as shown in Examples 8.5 and 8.6.

Just as in the polynomial spline case, Theorem 9.40 can be used to establish important results about the nonsingularity (and even the signs) of various determinants formed from the periodic TB-splines. For example, Theorem 8.8 has a direct analog for T-splines, with the only change being that the derivative $D_{\theta_i}^{d_i}$ appearing there should be replaced by the differential operator L_{d_i,θ_i}, where L_{d_i} is the differential operator defined in (9.5), and the subscript θ_i indicates that the derivatives are to be taken from the left or from the right, respectively. It is also possible to give results on the approximation power of periodic T-splines.

We now discuss the analog of the natural splines introduced in Section 8.2. Suppose U_{2m} is a canonical ECT-system, $\Delta=\{a=x_0<x_1<\cdots<x_{k+1}=b\}$ is a partition of $[a,b]$, and $\mathfrak{M}=(m_1,\ldots,m_k)$ with $1\leqslant m_i\leqslant m$, $i=1,2,\ldots,k$. Then we define the space of *natural T-splines* by

$$\mathfrak{N}\mathbb{S}(\mathcal{U}_{2m};\mathfrak{M};\Delta)=\{s\in\mathbb{S}(\mathcal{U}_{2m};\mathfrak{M};\Delta)\colon s_0,s_k\in\mathcal{U}_m\}, \quad (9.83)$$

where s_0 and s_k are the pieces of the spline in the first and last intervals, respectively.

Arguing as in Theorem 8.16 for natural polynomial splines, it is easy to see that $\mathfrak{N}\mathcal{S}(\mathcal{U}_{2m};\mathfrak{M};\Delta)$ is a linear space of dimension $K=\sum_{i=1}^{k}$. We now construct a basis with small supports. In analogy with the polynomial spline case, we introduce

$$L_{i,j}^{2m}(x)=[y_i,\ldots,y_{i+j}]_{U_{j+1}^*}g_{2m}^*(y;x), \tag{9.84}$$

and

$$R_{i,j}^{2m}(x)=(-1)^j[y_i,\ldots,y_{i+j}]_{U_{j+1}^*}g_{2m}(x;y), \tag{9.85}$$

where g_{2m} and g_{2m}^* are the Green's functions associated with the ECT-systems $\mathcal{U}_{2m}$ and $\mathcal{U}_{2m}^*$, respectively. The following theorem summarizes the properties of $L_{i,j}^{2m}$ and $R_{i,j}^{2m}$:

THEOREM 9.41

Suppose $y_i<y_{i+j}$ and $0\leqslant j<2m$. Then

$$L_{i,j}^{2m}(x)=0 \quad \text{for } x>y_{i+j}$$
$$L_{i,j}^{2m}(x)>0 \quad \text{for } x<y_{i+j}.$$

Moreover, for $x<y_i$, $L_{i,j}^{2m}(x)=(-1)^{2m-j}u_{2m-j}(x)+\sum_{\nu=1}^{2m-j-1}\alpha_\nu u_\nu(x)$. Similarly,

$$R_{i,j}^{2m}(x)=0 \quad \text{for } x<y_i$$
$$R_{i,j}^{2m}(x)>0 \quad \text{for } x>y_i,$$

and for $x>y_{i+j}$, $R_{i,j}^{2m}(x)=u_{2m-j}(x)+\sum_{\nu=1}^{2m-j-1}\beta_\nu u_\nu(x)$.

Proof. We discuss only $L_{i,j}^{2m}$ as the properties of $R_{i,j}^{2m}$ are established similarly. Since the divided difference involves a combination of $g_{2m}^*(y;x)$ for $y=y_i,\ldots,y_{i+j}$, it is apparent that it vanishes for all $x>y_{i+j}$. Clearly $L_{i,j}^{2m}$ is a T-spline relative to the space $\mathcal{U}_{2m}$, and the positivity assertion follows by counting zeros and using Theorem 9.30. Finally, to establish the nature of $L_{i,j}^{2m}(x)$ for $x<y_i$, we note that by the expansion of g_{2m}^* [cf. (9.33)],

$$L_{i,j}^{2m}(x)=\sum_{\nu=1}^{2m}[y_i,\ldots,y_{i+j}]_{U_{j+1}^*}u_\nu^*(y)u_{2m-\nu+1}(x)(-1)^{2m-\nu},$$

for all such x. Now the divided difference of u_ν^* is zero for $\nu=1,2,\ldots,j$, and is 1 for $\nu=j+1$. It follows that $L_{i,j}^{2m}$ has the asserted expansion for $x<y_i$. ■

We can now give a local support basis for the space of natural T-splines.

THEOREM 9.42

Let $K \geqslant 2m$, and let $y_{m+1} \leqslant \cdots \leqslant y_{m+K} = \{\overbrace{x_1,\ldots,x_1}^{m_1},\ldots,\overbrace{x_k,\ldots,x_k}^{m_k}\}$ in natural order. Define

$$B_i(x) = \begin{cases} L_{m+1,m+i-1}^{2m}(x), & i=1,2,\ldots,m \\ N_i^{2m}(x), & i=m+1,\ldots,K-m, \\ R_{i,m+K-i}^{2m}(x), & i=K-m+1,\ldots,K \end{cases}$$

where N_i^{2m} is the normalized TB-spline associated with the knots $y_i,\ldots,y_{i+2m}$, $i=m,\ldots,K-m$. Then $\{B_i\}_1^K$ form a basis for $\mathfrak{NS}(\mathfrak{U}_{2m};\mathfrak{M};\Delta)$.

Proof. The proof is very much like the proof of Theorem 8.18 in the polynomial spline case. The linear independence argument makes use of the fact that we know the exact structure of each of the L's for $x<x_1$ and each of the R's for $x>x_k$. ■

As pointed out in the historical notes for Section 8.2, the natural splines arise as solutions of certain best interpolation problems. Thus, although we have been able to define natural-splines associated with an arbitrary ECT-space $\mathfrak{U}_{2m}$, it turns out that the more important spline spaces arise when $\mathfrak{U}_{2m}$ is the null space of a self-adjoint differential operator. The following lemma shows how an arbitrary canonical ECT-system U_m can be extended to a canonical ECT-system U_{2m} in such a way that the corresponding space $\mathfrak{U}_{2m}$ has this property.

LEMMA 9.43

Let $U_m=\{u_i\}_1^m$ be a canonical ECT-system associated with the weights $w_1,\ldots,w_m$. Suppose $w_i \in C^{2m-i}[a,b]$, $i=1,2,\ldots,m$. Let $w_{m+1} \in C^{m-1}[a,b]$ be a positive function on $[a,b]$, and define $w_{m+i}=w_{m-i+2}$, $i=2,3,\ldots,m$. Then the canonical ECT-system U_{2m} defined as in (9.2) with the weights $w_1,\ldots,w_{2m}$ is the null space of the self-adjoint differential operator L^*L,

where

$$L=\frac{1}{\sqrt{w_{m+1}}}D\frac{1}{w_m}D\frac{1}{w_{m-1}}\cdots D\frac{1}{w_1}, \tag{9.86}$$

and

$$L^*=\frac{1}{w_1}D\frac{1}{w_2}D\cdots\frac{1}{w_m}D\frac{1}{\sqrt{w_{m+1}}}. \tag{9.87}$$

Proof. The operators associated with U_{2m} are given by

$$D_i=D\frac{1}{w_i}, \qquad i=1,2,\ldots,m;$$

$$D_{m+1}=D\frac{1}{w_{m+1}};$$

$$D_{i+m}=D\frac{1}{w_{m-i+2}}, \qquad i=2,\ldots,m.$$

It follows that $\mathfrak{U}_{2m}$ is the null space of $L^*L=D_{2m}\cdots D_2D_1$, while $\mathfrak{U}_m$ is the null space of L. ■

So far we have been discussing T-splines where the ties between the various pieces are described in terms of the continuity of a sequence of consecutive derivatives. As in the polynomial case, it is also possible to define linear spaces of T-splines where the ties are described in terms of the continuity of EHB-linear functionals (cf. Section 8.3). In particular, if

$$\Gamma=\{\gamma_{ij}\}_{j=1,i=1}^{m-m_i,k}$$

is a set of EHB-linear functionals as in (8.42), then we define the space of *Tchebycheffian g-splines* (*Tg-splines*) by

$$\mathbb{S}(\mathfrak{U}_m;\Gamma;\Delta)=\left\{\begin{array}{l} s\text{: there exist } s_0,\ldots,s_k \text{ in } \mathfrak{U}_m \text{ with } s|_{I_i}= \\ s_i,\ i=0,1,\ldots,k, \text{ and } \gamma_{ij}s_{i-1}=\gamma_{ij}s_i, \\ j=1,2,\ldots,m-m_i \text{ and } i=1,2,\ldots,k \end{array}\right\}. \tag{9.88}$$

When Γ consists of a set of Hermite linear functionals (cf. Example 8.25), the space $\mathbb{S}(\mathfrak{U}_m;\Gamma;\Delta)$ reduces to the space of T-splines studied in Sections 1 to 7 of this chapter.

The constructive theory of Tg-splines parallels that of the polynomial g-splines. For example, by the same arguments used in Theorem 8.26, we can show that $\mathcal{S}(\mathfrak{A}_m;\Gamma;\Delta)$ is a linear space of dimension $m+K$, where $K=\sum_{i=1}^{k} m_i$. As in the polynomial case, the construction of a basis for the Tg-spline space is somewhat complicated. A useful tool is provided by the following analog of Lemma 8.27:

LEMMA 9.44

Suppose $\boldsymbol{\beta}$ is a solution of the $(m-m_i)$ by m system

$$G_i\begin{bmatrix}\beta_m\\ \vdots\\ \beta_1\end{bmatrix}=0, \qquad G_i=\left(\gamma_{j\nu}^{i}\right)_{j=1,\nu=1}^{m-m_i,m}, \tag{9.89}$$

where $\gamma_{j1}^{i},\ldots,\gamma_{jm}^{i}$ are the coefficients of the linear functional γ_{ij} as in (8.42). Then

$$\rho(x)=\sum_{\mu=1}^{m}\beta_\mu g_{m-\mu}(x;x_i)$$

is a one-sided Tg-spline in the space $\mathcal{S}(\mathfrak{A}_m;\Gamma;\Delta)$.

Proof. It is clear that ρ is zero for $x<x_i$, and that it is an element of $\mathfrak{A}_m$ for $x\geqslant x_i$. To prove it belongs to $\mathcal{S}$, we need only check that it satisfies the required continuity conditions at the knot x_i, namely,

$$\sum_{\mu=1}^{m}\beta_\mu\sum_{\nu=1}^{m}\gamma_{j\nu}^{i}D^{\nu-1}g_{m-\mu}(x;x_i)\Big|_{x=x_i}=0, \qquad j=1,2,\ldots,m-m_i.$$

But since [cf. (9.34)]

$$\begin{aligned} D^{\nu-1}g_{m-\mu}(x;x_i)|_{x=x_i}&\neq 0, && \nu-1=m-\mu\\ &=0, && \text{otherwise,}\end{aligned}$$

this is precisely the system (9.89). ■

Since (9.89) is a system of $m-m_i$ equations in m unknowns, there always exist m_i linearly independent solutions. We can now exploit this fact to construct a one-sided basis for $\mathcal{S}(\mathfrak{A}_m;\Gamma;\Delta)$.

THEOREM 9.45

For each $i=1,2,\ldots,k$ let

$$\beta^{ij}=(\beta_1^{ij},\ldots,\beta_m^{ij}),\qquad j=1,2,\ldots,m_i,$$

be a set of m_i linearly independent solutions of (9.89). Set

$$\rho_{ij}(x)=\sum_{\nu=1}^{m}\beta_\nu^{ij}g_{m-\nu}(x;x_i),\qquad \begin{array}{l} j=1,2,\ldots,m_i \\ i=1,2,\ldots,k, \end{array} \tag{9.90}$$

and

$$\rho_{0j}(x)=u_j(x),\qquad j=1,2,\ldots,m_0=m. \tag{9.91}$$

Then $\{\rho_{ij}\}_{i=0,j=1}^{k\ \ m}$ is a basis for $\mathcal{S}(\mathcal{U}_m;\Gamma;\Delta)$ with the property that $\rho_{ij}(x)$ vanishes for $x<x_i$, all i and j.

Proof. The proof of this result parallels that for Theorem 8.28. In particular, the fact that these functions lie in $\mathcal{S}$ follows directly from Lemma 9.44, as does their one-sided nature. Their linear independence is proved exactly as in the polynomial g-spline case. ■

As in the polynomial g-spline case, the one-sided basis constructed in Theorem 9.45 for the space $\mathcal{S}(\mathcal{U}_m;\Gamma;\Delta)$ of Tg-splines takes a somewhat simpler form in the case where Γ consists of HB-linear functionals as defined in Example 8.24. In particular, if $E=(E_{ij})_{i=1,\ j=1}^{k\ \ m}$ is the incidence matrix defining the HB-set Γ, then in this case a one-sided basis for $\mathcal{S}(\mathcal{U}_m;\Gamma;\Delta)$ is given by the functions

$$\{u_i\}_1^m\cup\{g_{m-j+1}(x;x_i)\}_{i,j\text{ such that }E_{ij}=1} \tag{9.92}$$

(cf. Example 8.29 for the polynomial g-spline case).

Following the development in Section 8.3, our next task is to construct a local support basis for $\mathcal{S}(\mathcal{U}_m;\Gamma;\Delta)$. As shown there, this will not be possible for arbitrary sets Γ. As a first step to seeing when local support bases can be constructed, we begin by putting the one-sided splines (9.90) in lexicographical order:

$$\rho_1,\ldots,\rho_{m+K}=u_1,\ldots,u_m,\rho_{11},\ldots,\rho_{1m_1},\ldots,\rho_{k1},\ldots,\rho_{km_k}. \tag{9.93}$$

Introducing the notation

$$y_{m+1} \leqslant \cdots \leqslant y_{m+K} = \overbrace{x_1,\ldots,x_1}^{m_1},\ldots,\overbrace{x_k,\ldots,x_k}^{m_k},$$

it follows from the construction of the ρ's that $\rho_i(x)$ vanishes for $x<y_i$, $i=m+1,\ldots,m+K$. For convenience we write $x_0=a$, $m_0=m$, and $y_1 = \cdots = y_m = a$. Then we may write

$$\rho_j(x)=\begin{cases} 0 & x<y_i \\ \sum_{i=1}^{m} C_{ij}u_i(x), & x\geqslant y_i, \end{cases} \tag{9.94}$$

$i=1,2,\ldots,m+K$.

As in the polynomial g-spline case, the basis $\{\rho_i\}_1^{m+K}$ of one-sided splines for $\mathcal{S}(\mathfrak{U}_m;\Gamma;\Delta)$ is completely described in terms of the set $\{y_i\}_1^{m+K}$ and the matrix $C=(C_{ij})_{i=1,j=1}^{m,m+k}$. At this point we are in a position to establish the exact analogs of Lemmas 8.32 and 8.33 and Theorem 8.34. Indeed, these results now hold for Tg-splines, with the only change being that the functions $1,x,\ldots,x^{m-1}$ are to be replaced by the functions $u_1,\ldots,u_m$. In particular, Theorem 8.34 now gives conditions on the matrix C under which a basis for $\mathcal{S}(\mathfrak{U}_m;\Gamma;\Delta)$ can be constructed with relatively small supports. We state an analog of Corollary 8.35 showing that Theorem 8.34 can be applied whenever Γ contains all of the point evaluators $e_{x_1},\ldots,e_{x_k}$.

THEOREM 9.46

Suppose none of the functionals in Γ involves the $m-1^{\text{st}}$ derivative. Then $\mathcal{S}(\mathfrak{U}_m;\Gamma;\Delta)$ has a basis of local support splines as in Theorem 8.34.

Proof. As in Corollary 8.35 it suffices to show that the matrix $C<\varepsilon_{\nu+1},\ldots,\varepsilon_{\nu+m}>$ is nonsingular. But since for $x>x_{\nu+m}$,

$$\rho_{\varepsilon_{\nu+j}+1}(x)=g_m(x;x_{\nu+j})=\sum_{l=1}^{m}(-1)^{m-l}u_l(x)u^*_{m-l+1}(x_{\nu+j}),$$

this matrix is a constant multiple of the nonsingular matrix $D_{U^*_m}(x_\nu,\ldots,x_{\nu+m})$. ■

We conclude our discussion of Tg-splines with a few remarks about the number of zeros they can have. First, for general Γ it is clear that we can apply Theorem 9.30 to establish the exact analog of Theorem 8.36. To get better results, we have to restrict ourselves to the case where Γ consists of a set of HB-linear functionals. In this case we can establish direct analogs of Theorems 8.37 and 8.39. The only change necessary in their statement is that we should replace the hypothesis that $D^{m-1}s$ does not vanish on any interval by the hypothesis that $L_{m-1}s$ does not vanish on any interval. The proofs also carry over practically unaltered. Again, we need to replace the ordinary derivatives D^j by the operators L_j, and we must apply the extended Rolle's Theorem 9.11 for ECT-systems in place of the usual one.

We turn now to a brief discussion of T-monosplines. Suppose $U_{m+1} = \{u_i\}_1^{m+1}$ is an ECT-system. Then given $\Delta = \{x_1 < x_2 < \cdots < x_k\}$ and $\mathfrak{M} = (m_1, \ldots, m_k)$ with $1 \leq m_i \leq m$, $i = 1, 2, \ldots, k$, we define the space of *T-monosplines* by

$$\mathfrak{M}\mathcal{S}(\mathfrak{U}_{m+1}; \mathfrak{M}; \Delta) = \{u_{m+1} + s : s \in \mathcal{S}(\mathfrak{U}_m; \mathfrak{M}; \Delta)\}, \tag{9.95}$$

where $U_m = \{u_i\}_1^m$.

As in the polynomial monospline case, $\mathfrak{M}\mathcal{S}(\mathfrak{U}_{m+1}; \mathfrak{M}; \Delta)$ is not a linear space, but since it is the translation (by u_{m+1}) of the linear space $\mathcal{S}(\mathfrak{U}_m; \mathfrak{M}; \Delta)$ of T-splines, its basic properties are clear. Thus it suffices here to describe briefly the analogs of the results of Section 8.4 on zeros. The key to the kinds of inductive proofs used there to establish bounds on the number of zeros a polynomial monospline can have is the fact that the derivative of a monospline of degree m is a monospline of degree $m-1$. Our first theorem is an appropriate version of this result for T-monosplines.

THEOREM 9.47

Let $f \in \mathfrak{M}\mathcal{S}(\mathfrak{U}_{m+1}; \mathfrak{M}; \Delta)$. Then $L_1 f \in \mathfrak{M}\mathcal{S}(\mathfrak{U}_m; \mathfrak{M}^{(1)}; \Delta)$, where $\mathfrak{M}^{(1)}$ is defined in Theorem 9.28.

Proof. This follows immediately from Theorem 9.28. ■

We can now state the following analog of Theorem 8.43:

THEOREM 9.48. Budan-Fourier Theorem for Tchebycheffian Monosplines

Given $\Delta = \{x_1 < \cdots < x_k\}$ and a multiplicity vector $\mathfrak{M} = (m_1, \ldots, m_k)$, let $\sigma_1, \ldots, \sigma_k$ be defined as in (8.59). Then for any $f \in \mathfrak{M}\mathcal{S}(\mathfrak{U}_{m+1}; \mathfrak{M}; \Delta)$,

$$Z(f) \leq m + \sum_{i=1}^{k} (m_i + \sigma_i), \tag{9.96}$$

where Z counts the zeros of f on $\mathbf{R}$ with multiplicities as in Definition 8.42. More precisely, for any $a<b$,

$$Z_{(a,b)}(f) \leqslant m + \sum_{i=1}^{k} (m_i + \sigma_i) - S^{+}\left[f(b), L_1 f(b), \ldots, L_m f(b) \right]$$

$$- S^{+}\left[f(a), -L_1 f(a), \ldots, (-1)^m L_m f(a) \right]. \tag{9.97}$$

Proof. It suffices to prove the stronger version (9.97). To this end we use the Budan-Fourier Theorem 9.12 for ECT-systems on each interval in the same way as was done in the proof of Theorem 8.43. Here we should remember the equivalence of the definitions (9.23) and (9.24) of a multiple zero of a function in an ECT-space. ■

As in the polynomial monospline case, for several applications it is important to establish the existence of T-monosplines possessing a maximal set of zeros. Such a result together with the bound (9.96) would then constitute a kind of fundamental theorem of algebra for T-monosplines. As a first step toward establishing an existence theorem of this kind, we note that Theorem 8.44 immediately generalizes to T-monosplines. In addition, we have the following analog of Theorem 8.45:

THEOREM 9.49

Suppose $f \in \mathfrak{M}\mathcal{S}(\mathfrak{U}_{m+1}; \mathfrak{M}; \Delta)$ has zeros $t_1 \leqslant \cdots \leqslant t_N$ with $N = m + \Sigma_{i=1}^{k}(m_i + \sigma_i)$. Then

$$L_j f(x) > 0, \qquad j = 0, 1, \ldots, m-1, \qquad \text{all } x \geqslant x_k;$$

$$(-1)^{m-j} L_j f(x) > 0, \qquad j = 0, 1, \ldots, m-1, \qquad \text{all } x \leqslant x_1. \tag{9.98}$$

Moreover, if f is expanded in the form

$$f(x) = u_{m+1}(x) + \sum_{j=1}^{m} a_j u_j(x) + \sum_{i=1}^{k} \sum_{j=1}^{m_i} c_{ij} g_{m-j}(x; x_i), \tag{9.99}$$

then for all $1 \leqslant i \leqslant k$ with m_i odd,

$$c_{ij} < 0, \qquad \text{all odd } j,\ 1 \leqslant j \leqslant m_i. \tag{9.100}$$

Proof. These results follow from the Budan-Fourier Theorem 9.48 in the same way as in Theorem 8.45 for polynomial monosplines. ■

With these results in hand, we can now deal with the existence of T-monosplines with a prescribed maximal set of zeros.

THEOREM 9.50

Fix m and $1 \leqslant m_i \leqslant m$, and let σ_i be as in (8.59), $i=1,2,\ldots,k$. Then for any given $t_1 < t_2 < \cdots < t_N$ with $N = m + \sum_{i=1}^{k}(m_i + \sigma_i)$, there exists a set of knots $x_1 < \cdots < x_k$ and a corresponding T-monospline $f \in \mathfrak{M}\mathcal{S}(\mathfrak{U}_{m+1}; \mathfrak{M}; \Delta)$ with $f(t_i)=0$, $i=1,2,\ldots,N$. If $m_1,\ldots,m_k$ are all odd, then this monospline is unique.

Proof. The proof closely parallels that of Theorem 8.47. Some points to keep in mind in carrying out the details are the following: Since we are dealing with ECT-systems, we have the direct analogs of the Lagrange polynomials associated with an interpolation problem. Because of the form of the functions $u_1,\ldots,u_{m+1}$ in a canonical ECT-system, it follows that any function of the form $u = u_{m+1} + \sum_{j=1}^{m} c_j u_j$ behaves like t^m for large t—in particular, $u(t) \to \infty$ as $t \to \infty$. ∎

As in the polynomial monospline case, we have stated here only the case of simple zeros and possible multiple knots. It is also possible to establish a version of this theorem for multiple zeros and simple knots.

§ 9.9. EXPONENTIAL AND HYPERBOLIC SPLINES

In this section we define two classes of T-splines which arise in practice. We begin with the so-called exponential splines. Given any $\alpha_1 < \alpha_2 < \cdots < \alpha_m$, let $\mathfrak{U}_m = \text{span}\{e^{\alpha_1 x},\ldots,e^{\alpha_m x}\}$. Since elements of the space $\mathfrak{U}_m$ are usually referred to as *exponential polynomials*, it is natural to call $\mathcal{S}(\mathfrak{U}_m; \mathfrak{M}; \Delta)$ the space of *exponential splines*.

The basic properties of exponential splines all follow from the general results developed in this chapter once we show that $\mathfrak{U}_m$ is an ECT-space. To show this, we need only display a canonical ECT-system that forms a basis for it. Let $U_m = \{u_i\}_1^m$ be the canonical ECT-system defined in (9.2) with respect to the weight functions

$$w_1(x) = e^{\alpha_1 x};$$

$$w_i(x) = e^{(\alpha_i - \alpha_{i-1})x}, \qquad i = 2,\ldots,m. \tag{9.101}$$

It is easy to see that

$$u_i(x) = \sum_{\nu=1}^{i} c_{i\nu} e^{\alpha_\nu x}, \qquad i = 1,2,\ldots,m \qquad c_{ii} \neq 0,$$

and thus that $\mathfrak{U}_m = \text{span}(U_m)$. We have shown that $\mathfrak{U}_m$ is an ECT-space.

There is no need to go into detail on the properties of exponential splines since they are all direct translations of our general results on

T-splines. On the other hand, it may be useful to say a little more about the differential operators, dual ECT-system, and Green's function associated with $\mathfrak{U}_m$. The operators D_i defined in (9.4) are given here by

$$D_i f(x) = D\left[e^{(\alpha_{i-1}-\alpha_i)x} f(x)\right], \qquad i=1,2,\ldots,m.$$

After some calculation, it is easily seen that the operator $L_m = D_m \cdots D_1$ is a nonzero multiple of the differential operator

$$L = (D-\alpha_1)\cdots(D-\alpha_m)$$

(which is to be expected since $\mathfrak{U}_m$ is clearly the null space of L).

In view of the form of $u_1,\ldots,u_m$, it is clear that the associated jth reduced system is given by

$$u_{j,i}(x) = \sum_{\nu=1}^{i} d_{j,\nu} e^{(\alpha_{\nu+j}-\alpha_j)x},$$

$i=1,2,\ldots,m-j$. It is also an easy matter to compute the dual canonical ECT-system (9.25) associated with $w_1,\ldots,w_m$. We obtain

$$u_1^*(x) = 1;$$

$$u_i^*(x) = \sum_{\nu=1}^{i-1} b_{i,\nu} e^{(\alpha_m-\alpha_{m-\nu})x}, \qquad i=2,\ldots,m.$$

The Green's functions $g_j(x;y)$ associated with the operators L_j and the initial conditions $L_{\nu-1}f(a)=0$, $\nu=1,\ldots,j-1$, can be found by evaluating the integrals in (9.32). For $j>2$ we have

$$g_j(x;y) = \begin{cases} \displaystyle\sum_{\nu=1}^{j} a_{j,\nu}(y) e^{\alpha_\nu(y-x)}, & x \geq y \\ 0, & x<y, \end{cases} \tag{9.102}$$

with appropriate functions $\{a_{j,\nu}(y)\}$ such that $a_{j,j}(y)>0$.

To close this section, we now consider a related space of natural T-splines that arises in connection with a certain best interpolation problem. Suppose $0=\alpha_1<\alpha_2<\cdots<\alpha_m$. Now let $w_{m+1}(x)=e^{-2\alpha_m x}$, and choose $w_{m+i}=w_{m-i+2}$, $i=2,3,\ldots,m$. Lemma 9.43 asserts that the canonical ECT-system $U_{2m}=\{u_i\}_1^{2m}$ associated with these weights spans the null space of L^*L, where L is given in (9.86). A simple calculation shows that

$$\mathfrak{U}_{2m} = \text{span}(U_{2m}) = \text{span}\{1, x, e^{-\alpha_2 x}, e^{\alpha_2 x}, \ldots, e^{-\alpha_m x}, e^{\alpha_m x}\}.$$

This is the null space of L^*L, where

$$L = D(D-\alpha_2)\cdots(D-\alpha_m)$$

$$L^* = D(D+\alpha_2)\cdots(D+\alpha_m).$$

It is clear that the space $\mathcal{U}_{2m}$ is also spanned by the functions

$$1, x, \sinh(\alpha_2 x), \cosh(\alpha_2 x), \ldots, \sinh(\alpha_m x), \cosh(\alpha_m x).$$

In view of this fact, we call the space $\mathfrak{N}\mathcal{S}(\mathcal{U}_{2m}; \mathfrak{M}; \Delta)$ of natural T-splines associated with $\mathcal{U}_{2m}$ the space of *hyperbolic splines*.

§ 9.10. CANONICAL COMPLETE TCHEBYCHEFF SYSTEMS

So far we have been working with spaces of generalized splines where the pieces are required to be elements of a given ECT-system. Our motivation for working with ECT-systems was the fact that they closely mirror the properties of ordinary polynomials. In this section we show that there is a still larger class of functions that retain most of the features of the ECT-systems.

DEFINITION 9.51

Let u_1 be a bounded positive function on the interval $[a,b]$, and suppose $\sigma_2, \ldots, \sigma_m$ are bounded, right continuous, monotone-increasing functions on $[a,b]$. Let

$$u_2(x) = u_1(x)\int_a^x d\sigma_2(s_2), \tag{9.103}$$

$$\cdots$$

$$u_m(x) = u_1(x)\int_a^x \int_a^{s_2} \cdots \int_a^{s_{m-1}} d\sigma_m(s_m)\cdots d\sigma_2(s_2).$$

We call $U_m = \{u_i\}_1^m$ a *Canonical Complete Tchebycheff (CCT-) system*.

The set of CCT-systems is quite large. It contains all ECT-systems, for example, as is clear if the σ's have the form

$$\sigma_i(t) = \int_a^t w_i(s)\,ds, \qquad i = 2, \ldots, m$$

with $w_2, \ldots, w_m$ positive functions with $w_i \in C^{m-i+1}[a,b]$. The following example presents a CCT-system that is not an ECT-system:

EXAMPLE 9.52

Let $u_1(x)=1$, $u_2(x)=2x^{1/2}$, $u_3(x)=\frac{2}{3}x^{3/2}$, and $u_4(x)=\frac{1}{5}x^{5/2}$, and let $I=[0,1]$.

Discussion. This set of functions is a CCT-system corresponding to the functions $\sigma_2(t)=2t^{1/2}$, $\sigma_3(t)=t$, and $\sigma_4(t)=t$. These functions do *not* form an ECT-system on $[0,1]$ since the function u_2 does not have enough derivatives at $t=0$. ■

Because of the close relationship between the definition of a CCT-system and the canonical expansions (9.2) of an ECT-system, it is not surprising that many of the results for ECT-systems can be carried over without difficulty. For example, it is clear that the analogs of both Theorems 9.3 and 9.4 hold for CCT-systems. In the remainder of this section we discuss analogs for the other results of Section 9.1.

We begin by defining the analogs of the differential operators D_i and L_i defined in (9.4) and (9.5). Given any function ψ defined on $[a,b]$, we define

$$D_0\psi(t)=\frac{\psi(t)}{u_1(t)};$$

$$D_j^+\psi(t)=\lim_{\delta\downarrow 0}\frac{\psi(t+\delta)-\psi(t)}{\sigma_{j+1}(t+\delta)-\sigma_{j+1}(t)},\qquad j=1,2,\dots,m-1;\quad (9.104)$$

and

$$L_j^+=D_j^+\cdots D_1^+D_0,\qquad j=0,1,\dots,m-1.\qquad (9.105)$$

It is clear that

$$L_j^+u_i(t)|_{t=a}=\delta_{j,i-1},\qquad \begin{array}{l} j=0,1,\dots,i-1\\ i=1,2,\dots,m. \end{array}\qquad (9.106)$$

It is also of interest to define reduced systems associated with U_m. For each $j=1,2,\dots,m-1$ we define the jth reduced system associated with U_m by

$$u_{j,1}(t)=1;$$

$$u_{j,2}(t)=\int_a^t d\sigma_{j+2}(s_{j+2});$$

$$\cdots$$

$$u_{j,m-j}(t)=\int_a^t\cdots\int_a^{s_{m-1}}d\sigma_m(s_m)\cdots d\sigma_{j+2}(s_{j+2}).$$

We write $U_m^{(j)}=\{u_{j,i}\}_{i=1}^{m-j}$. These reduced systems arise when we take "derivatives" of the CCT-system U_m. In particular,

$$L_j^+u_i=\begin{cases}0, & i=1,2,\ldots,j\\ u_{j,i-j}, & i=j+1,\ldots,m.\end{cases} \tag{9.107}$$

Thus if $u\in\mathcal{U}_m=\text{span}(U_m)$, then $L_1^+u\in\mathcal{U}_m^{(1)}$, $L_2^+u\in\mathcal{U}_m^{(2)}$, and so on.

An important property of the ECT-systems is the fact that determinants formed from them are always nonnegative. A similar situation persists for the CCT-systems. Suppose we define

$$D\begin{pmatrix} t_1,\ldots,t_m\\ u_1,\ldots,u_m\end{pmatrix}=\left[L_{d_i}u_j(t_i)\right]_{i,j=1}^m, \tag{9.108}$$

where $t_1\leqslant t_2\leqslant\cdots\leqslant t_m$ and

$$d_i=\max\{j:t_i=\cdots=t_{i-j}\},\qquad i=1,2,\ldots,m.$$

We have the following result:

THEOREM 9.53

Let $\{u_i\}_1^m$ be a CCT-system on the interval I. Then

$$D\begin{pmatrix} t_1,\ldots,t_m\\ u_1,\ldots,u_m\end{pmatrix}>0 \qquad \text{for all } t_1\leqslant t_2\leqslant\cdots\leqslant t_m \text{ in } I.$$

Proof. We proceed by induction on m. For $m=1$ the result is trivial. Now suppose the result has been proved for all CCT-systems of $m-1$ functions. We shall reduce the determinant to one involving $m-1$ functions in the first reduced system. The key to this is the relation

$$\frac{u_j(\tau_2)}{u_1(\tau_2)}-\frac{u_j(\tau_1)}{u_1(\tau_1)}=\int_{\tau_1}^{\tau_2}L_1^+u_j(s)\,d\sigma_2(s),\qquad 2\leqslant j\leqslant m. \tag{9.109}$$

Now suppose

$$t_1\leqslant t_2\leqslant\cdots\leqslant t_m=\overbrace{\tau_1,\ldots,\tau_1}^{l_1}<\cdots<\overbrace{\tau_d,\ldots,\tau_d}^{l_d}.$$

Then the only nonzero entries in the first column correspond to the d rows starting with $u_1(\tau_1),\ldots,u_1(\tau_d)$. If we factor these expressions out of their rows, we are left with 1's in these positions. Now by subtracting the last of these rows from the preceeding row, then the next-to-last row from its predecessor with a 1 in column 1, we are left with a determinant with a 1 in the upper-left-hand corner and zeros in the remainder of column 1. Expanding by the first column and using (9.109), we obtain

$$D\begin{pmatrix} t_1,\ldots,t_m \\ u_1,\ldots,u_m \end{pmatrix} = \int_{\tau_1}^{\tau_2} \cdots \int_{\tau_{d-1}}^{\tau_d} \psi(s_1,\ldots,s_{d-1})\,d\sigma_2(s_1)\cdots d\sigma_2(s_{d-1}),$$

where

$$\psi(s_1,\ldots,s_{d-1}) = D\begin{pmatrix} \tau_1,\ldots,\tau_1,s_1,\tau_2,\ldots,\tau_2,s_2,\ldots,s_{d-1},\tau_d,\ldots,\tau_d \\ L_1^+ u_2,\ldots,L_1^+ u_m \end{pmatrix},$$

where τ_i appears $l_i - 1$ times, $i = 1,2,\ldots,d$. Since the σ_i are monotone increasing, there is mass in each of the intervals $I_i = [\tau_{i-1}, \tau_i]$, and since the integrand is positive throughout the interior of the region over which the integral is being taken, it follows that the integral itself is positive. ■

With this concept of determinant formed from a CCT-system, it is now clear that we can define divided differences of functions with respect to U_m in the same way as was done in (9.14) for ECT-systems. By the same arguments used before, this new divided difference has all the properties listed in Theorem 9.7 (where D^j is replaced by L_j^+).

We now make some remarks about zeros of elements $u \in U_m$. We say such u has a z-tuple zero at the point t in $[a,b]$ provided

$$u(t) = L_1^+ u(t) = \cdots = L_{z-1}^+ u(t) = 0 \neq L_z^+ u(t), \tag{9.110}$$

where $1 \leq z \leq m-1$. We say u has an m-tuple zero at t if $L_1^+ u,\ldots,L_{m-1}^+ u$ all vanish at t. The following is the analog of Theorem 3.4 for polynomials:

THEOREM 9.54

Let U_m be spanned by a CCT-system, and let $Z(u)$ denote the number of zeros of $u \in U_m$, counting multiplicities as above. Then for all u not identically zero,

$$Z(u) \leq m-1.$$

Proof. If $u = \sum_{i=1}^m c_i u_i$ has m zeros, say at points $t_1 \leq \cdots \leq t_m$, then the vector $\mathbf{c} = (c_1,\ldots,c_m)$ must satisfy the system

$$M\begin{pmatrix} t_1,\ldots,t_m \\ u_1,\ldots,u_m \end{pmatrix}\mathbf{c} = 0.$$

Since this is a nonsingular system, it follows that $\mathbf{c} = 0$ and $u = 0$. ■

It is also possible to improve this result to obtain a form of Budan-Fourier theorem for functions in the CCT-space $\mathcal{U}_m = \text{span}\,(U_m)$. To this end we need a version of the extended Rolle's theorem for such functions. Defining left and right Rolle's points exactly as in Section 9.1 (but with L_1 replaced by L_1^+), we may state it as follows:

THEOREM 9.55

Suppose $u \in \mathcal{U}_m$ and that for some $a \leqslant c < d \leqslant b$ the points c and d are left and right Rolle's points of u, respectively. Then $L_1^+ f$ has at least one sign change on (c,d). If $L_1^+ f$ is continuous on (c,d), then it has at least one zero there.

Proof. We may use the same arguments used in the proof of Theorem 2.19 in the polynomial case. ■

We emphasize that Theorem 9.55 applies only to functions u in the CCT-space $\mathcal{U}_m$, whereas the extended Rolle's Theorem 9.11 connected with ECT-systems is valid for any function such that $L_1 f$ exists.

We can now use Theorem 9.55 and the same arguments as before to prove a direct analog of the Budan-Fourier Theorem 9.12 for functions in an ECT-space. The only change required is that the operators $L_1, \ldots, L_m$ should be replaced by $L_1^+, \ldots, L_m^+$.

As we saw above, an important tool in dealing with T-splines was the dual canonical ECT-system defined in (9.25). Clearly we can define the analog here with the weights $w_2, \ldots, w_m$ replaced by the measures $\sigma_2, \ldots, \sigma_m$. Carrying the analogy further, we may define the Green's functions associated with U_m by (9.31) with

$$h_j(x;y) = u_1(x) \int_y^x d\sigma_2(s_2) \cdots \int_y^x d\sigma_m(s_m), \tag{9.111}$$

$j = 1, 2, \ldots, m$. With this definition, it is an easy matter to establish the analog of Theorem 9.13 giving an explicit expansion for h_j.

§ 9.11. DISCRETE TCHEBYCHEFFIAN SPLINES

In this section we briefly indicate how the ideas of Section 8.5 can be combined with the ideas developed in this chapter to produce a reasonable class of discrete T-splines. Throughout this section we follow the notation of Section 8.5. In particular, we write $\mathbf{R}_{a,h} = \{\ldots, a-h, a, a+h, \ldots\}$ for the discrete line and $[a,b]_h = \{a, a+h, \ldots, a+Nh\}$, $b = a+Nh$, for a discrete subinterval of it. We also use the discrete "derivatives" D_h^j defined in (8.73), and the discrete "integral" defined in (8.87).

The starting point for developing a space of generalized discrete splines is to find a suitable substitute for the space of polynomials. Guided by the discussion of CCT-systems, we suppose that $U_m = \{u_i\}_1^m$ is a set of functions defined on the discrete interval $[a,b]_h$ by

$$\begin{aligned}
u_1(x) &= w_1(x), \\
u_2(x) &= w_1(x)\int_a^x w_2(s_2)\,d_h s_2, \\
&\cdots \\
u_m(x) &= w_1(x)\int_a^x w_2(s_2)\cdots\int_a^{s_{m-1}} w_m(s_m)\,d_h s_m\cdots d_h s_2,
\end{aligned} \tag{9.112}$$

where $w_1,\dots,w_m$ are arbitrary positive functions on $[a,b]_h$.

An example of a set of functions of the form (9.112) is provided by the polynomials. Indeed, if we set $w_1=1$ and $w_i=(i-1)$, $i=2,3,\dots,m$, then we obtain the factorial functions $u_1(x)=1$ and

$$u_i(x) = x^{(i-1)_h}, \qquad i=2,3,\dots,m,$$

defined in (8.75).

We can now define the space of splines of interest in this section. Given a set of functions U_m as in (9.112), let $\mathfrak{U}_m = \operatorname{span}(U_m)$, and suppose $\Delta = \{a = x_0 < x_1 < \cdots < x_{k+1} = b\} \subseteq [a,b]_h$ and $\mathfrak{M} = (m_1,\dots,m_k)$, with $1 \leqslant m_i \leqslant m$, $i=1,2,\dots,k$, are given. Then we define the space of *discrete Tchebycheffian-splines* by

$$\mathcal{S}(\mathfrak{U}_m;\mathfrak{M};\Delta;h) = \left\{\begin{array}{l} s\text{: there exist } s_0,\dots,s_k \text{ in } \mathfrak{U}_m \text{ with } s|_{(x_i,x_{i+1})_h} \\ = s_i,\quad i=0,1,\dots,k, \quad\text{and}\quad D_h^{j-1}s_{i-1}(x_i) = \\ D_h^{j-1}s_i(x_i),\ j=1,2,\dots,m-m_i,\ i=1,2,\dots,k \end{array}\right\}. \tag{9.113}$$

This is clearly a linear space of functions defined on $[a,b]_h$. Before developing its properties further, we need to say a little more about U_m. In view of the way in which U_m is defined, it should be apparent that many of the properties developed in Section 9.10 for CCT-systems have analogs in this setting. To describe some of them, we start by introducing certain difference operators associated with U_m. Let

$$D_j^h f = D_h\left(\frac{f}{w_j}\right), \qquad j=1,2,\dots,m, \tag{9.114}$$

and define $L_0 f = f$ and

$$L_j^h = D_j^h D_{j-1}^h \cdots D_1^h, \qquad j = 1, 2, \ldots, m. \tag{9.115}$$

Then clearly

$$L_j^h u_i(x)|_{x=a} = \delta_{j,i-1} w_i(a), \qquad \begin{aligned} j &= 0, 1, \ldots, i-1 \\ i &= 1, 2, \ldots, m. \end{aligned} \tag{9.116}$$

Associated with U_m, we define the reduced systems $U_m^{(j)} = \{u_{j,i}\}_{i=1}^{m-j}$ exactly as in (9.8) (but using the discrete integral instead of the usual one). It then follows that

$$L_j^h u_i = \begin{cases} u_{j,i-j}, & i = j+1, \ldots, m \\ 0, & i = 1, 2, \ldots, j. \end{cases} \tag{9.117}$$

Thus if $u \in \mathcal{U}_m$, then $L_1^h u \in \mathcal{U}_m^{(1)} = \text{span}[U_m^{(1)}]$, and so on.

We turn now to some properties of determinants formed from $u_1, \ldots, u_m$. It will be enough to consider determinants associated with a set of distinct points $t_1 < t_2 < \cdots < t_m$ chosen from $[a,b]_h$. We have following analog of Theorem 9.53:

THEOREM 9.56

Let $U_m = \{u_i\}_1^m$ be as in (9.112). Then

$$D\begin{pmatrix} t_1, \ldots, t_m \\ u_1, \ldots, u_m \end{pmatrix} = D_{U_m}(t_1, \ldots, t_m) = \left[u_i(t_j) \right]_{i,j=1}^m > 0$$

for all $t_1 < t_2 < \cdots < t_m$ in $[a,b]_h$.

Proof. The proof proceeds along the same lines as the proof of Theorem 9.53, where now we use the following analog of the fundamental theorem of calculus:

$$\frac{f}{u_1}(d) - \frac{f}{u_1}(c) = \int_c^d L_1^h f(t)\, d_h t,$$

any c, d in $[a,b]_h$. ∎

Clearly these determinants can be used to define generalized divided differences with respect to U_m, and the analog of Theorem 9.7 can be established with D^j replaced by L_j^h.

In order to discuss an appropriate kind of Green's function (which will be useful in constructing a one-sided basis as well as in constructing local-support versions of the B-splines), we need to introduce the dual system associated with U_m. We define $U_m^* = \{u_i^*\}_{i=1}^m$ by the same formulae used in (9.25), but by using the discrete integral instead. Associated reduced systems and dual difference operators can then be defined in the same way as before.

For each $j \geqslant 1$, let $g_j(x;y) = h_j(x;y)(x-y)_+^0$, where

$$h_j(x;y) = u_1(x)\int_y^x w_2(x_2)\cdots\int_y^x w_j(s_j)d_h s_j,\ldots,d_h s_2. \tag{9.118}$$

It is easy to carry over the arguments of Theorem 9.13 to show that the expansion (9.33) holds for all x,y in $[a,b]_h$ and that the initial conditions

$$L_i^h g_j(x;y)|_{x=y} = \delta_{i,j-1} w_j(y), \qquad i=0,1,\ldots,j-1 \tag{9.119}$$

hold. The following two results highlight the importance of g_m:

THEOREM 9.57

Let ψ be any function defined on $J=[a,b-mh]_h$, and suppose $r_0,\ldots,r_{m-1}$ are prescribed real numbers. Then

$$f(x) = u_f(x) + \int_a^x g_m(x;y)\psi(y)d_h y \tag{9.120}$$

is the unique solution of the initial-value problem

$$L_m^h f(x) = \psi(x), \qquad \text{all } x \in J; \tag{9.121}$$

$$L_j^h f(a) = r_j, \qquad j=0,1,\ldots,m-1. \tag{9.122}$$

Proof. Compare the proof of Theorem 9.15. ∎

THEOREM 9.58. Discrete Generalized Taylor Expansion

For any f defined on $[a,b]_h$,

$$f(x) = u_f(x) + \int_a^x g_m(x;y)L_m^h f(y)d_h y \tag{9.123}$$

all $x \in [a,b]_h$, where $u_f \in \mathfrak{U}_m$ is such that

$$L_j^h u_f(a) = L_j^h f(a), \qquad j=0,1,\ldots,m-1. \tag{9.124}$$

Proof. The proof proceeds along the same lines as the proofs of Theorems 8.61 and 9.16, using the discrete integration by parts formula (8.90). ■

There are also dual versions of these two theorems—cf. Theorems 8.62, 9.17, and 9.18. Using the Green's function $g_m(x;y)$, we can now define a one-sided basis for the space of discrete T-splines.

THEOREM 9.59

Let $\{y_i\}_1^{m+K}$ be defined as in Theorem 8.51. Then

$$\rho_i(x) = g_m(x;y_i), \qquad i = 1,2,\ldots,m+K$$

form a basis for $\mathcal{S}(\mathcal{U}_m;\mathfrak{M};\Delta;h)$ such that $\rho_i(x)=0$ for $x<y_i$, all i.

Proof. The fact that each of these functions is a discrete T-spline follows from the fact that $g_m(x;y_i)$ can be expanded as a linear combination of $u_1(x),\ldots,u_m(x)$ for $x \geqslant y_i$ while it satisfies the appropriate smoothness conditions at y_i—cf. (9.119). The linear independence of the ρ's follows by the same arguments used in proving Theorem 8.51. ■

It is now possible to construct appropriate B-splines forming a basis for $\mathcal{S}(\mathcal{U}_m;\mathfrak{M};\Delta;h)$. The following theorem is a synthesis of Theorems 8.52 and 9.23:

THEOREM 9.60

Let $\{y_i\}_1^{m+K}$ be as in Theorem 9.59, and let $y_{m+K+i} = b+(i-1)h$, $i = 1,2,\ldots,m$. For each $i = 1,2,\ldots,m+K$ define

$$Q_i(x) = (-1)^m [y_i,\ldots,y_{i+m}]_{U_{m+1}^*}\, g_m(x;y), \tag{9.125}$$

where the divided difference is to be taken with respect to the expanded dual system U_{m+1}^*. Then $\{Q_i\}_1^{m+K}$ is a basis for $\mathcal{S}(\mathcal{U}_m;\mathfrak{M};\Delta;h)$. Moreover, if $x \in [a,b]_h$,

$$Q_i(x) = 0, \qquad x < y_i\,,\, y_{i+m} < x, \tag{9.126}$$

and

$$Q_i(x) > 0, \qquad y_i + (m-2)h < x < y_{i+m}. \tag{9.127}$$

Proof. Compare the proofs of Theorems 8.52 and 9.23. ■

Having gotten this far with the idea of discrete T-splines, it seems safe to assert that a number of other common features of discrete and

Tchebycheffian splines can be carried over to the present setting. In particular, we mention that there is a direct analog of the Peano representation of Theorems 8.63 and 9.24 giving an integral representation of the divided difference of a function with respect to U_m. In addition, by the same methods used in Theorems 8.56 and 9.25, we can show that the normalized discrete B-splines $N_i = \alpha_i Q_i$ with α_i as defined in (9.54) satisfy

$$\sum_{i=1}^{m+K} N_i(x) = u_1(x). \tag{9.128}$$

The methods of these theorems can also be used to give a version of Mardsen's identity expressing the functions $u_1, \dots, u_m$ in explicit B-spline expansions.

Another common feature of the discrete and Tchebycheffian splines was the existence of a dual basis of linear functionals for the normalized B-splines. Following the construction of Theorem 9.26, it is quite easy to construct a similar dual basis here. It can, in turn, be used to establish discrete error bounds.

We conclude our discussion of discrete T-splines with a brief outline of available results on zeros and on determinants. Here we need an amalgamation of the ideas of Sections 8.5 and 9.5. As the proofs are again inductive, the key observation is that the analog of Theorems 8.57 and 9.28 holds. It asserts that if $s \in \mathcal{S}(\mathcal{U}_m; \mathfrak{M}; \Delta; h)$, then $L_1^h s$ is in an appropriate discrete T-spline space (associated with the first reduced system).

Before dealing with zeros of discrete T-splines, it is necessary to define what we mean by zeros and multiple zeros. Here we take precisely the definitions used in Section 8.5 for discrete polynomial splines. With this definition, it is now an easy task to follow the proofs of Theorems 8.64 and 9.29 to obtain a version of Rolle's theorem. Once we have this, the proofs of Theorems 8.65 and 9.30 carry over with no difficulty to show that

$$Z^{\mathcal{S}}(s) \leqslant m + K - 1, \qquad \text{all nontrivial } s \in \mathcal{S}(\mathcal{U}_m; \mathfrak{M}; \Delta; h). \tag{9.129}$$

It is also possible to give a Budan-Fourier theorem for discrete T-Splines.

The results on determinants of discrete T-splines are also direct analogs of those for discrete and Tchebycheffian splines given in Sections 8.5 and 9.6, respectively. For example, it can be shown that

$$D\begin{pmatrix} t_1, \dots, t_n \\ N_1, \dots, N_n \end{pmatrix} \geqslant 0, \tag{9.130}$$

and strict positivity holds precisely when

$$t_i \in \sigma_i = \{x \in \mathbf{R}_{a,h}: N_i(x) > 0\} = (y_i + (m-2)h, y_{i+m})_h, \qquad (9.131)$$

$i = 1, 2, \ldots, n$. The proof is based on the results on zeros—cf. Theorems 8.66 and 9.33. This result can then be extended to show that the B-splines $\{N_i\}_1^{m+K}$ form an OCWT-system, and thus that the B-spline expansions satisfy an appropriate variation-diminishing property (cf. Theorems 8.67 and 8.68). Finally, the analog of Theorem 8.69 on determinants formed from the Green's function can also be established.

We close this section by observing that it is also possible to define spaces of discrete T-splines which are periodic or natural and to define discrete T-monosplines.

§ 9.12. HISTORICAL NOTES

Section 9.1

Extended complete Tchebycheff systems are treated in considerable detail in the books by Karlin and Studden [1966] and Karlin [1968], and most of the results of this section can be found there. Lemma 9.6 is taken from the work of Scherer and Schumaker [1980]. The recursion formula (9.20) for generalized divided differences was established by Mühlbach [1973]. We have not been able to find the Budan-Fourier Theorem 9.12 for ECT-systems in the literature.

Section 9.2

The Green's functions $g_j(x;y)$ are also studied in the texts by Karlin and Studden [1966] and Karlin [1968]. The explicit expansion for h_j given in Theorem 9.13 is taken from the work of Schumaker [1976a].

Section 9.3

We outline the history of the development of nonpolynomial splines in the historical notes for Section 11.1. Tchebycheffian splines were introduced by Karlin [1968]. They were studied in the dissertation by Schumaker [1966], in the articles by Karlin and Ziegler [1966], Karlin and Schumaker [1967], and in numerous later papers.

Section 9.4

Tchebycheffian B-splines were also introduced by Karlin [1968]. The normalization given in (9.54) is credited to Marsden [1970] where the identities of Theorem 9.25 were first established. The dual basis constructed in Theorem 9.26 is credited to Scherer and Schumaker [1980]. The main missing property of the TB-splines is an appropriate stable recursion for computing them.

Section 9.5

For some early results on the zeros of T-splines, see the dissertation by Schumaker [1966]. The method and results presented here are based on the work of Schumaker [1976a, c].

Section 9.6

The early results on determinants associated with T-splines deal with the determinants formed from the Green's function. In particular, Karlin established Theorem 9.36 with distinct y's and t's in the mid-1960s. This result was generalized to the case of multiple y's in the dissertation by Schumaker [1966]. The full result with multiple y's and t's was first established by Karlin and Ziegler [1966] by a complicated triple induction. Results on determinants formed from the B-splines themselves first appeared in the work of Karlin [1968] and Karlin and Karon [1970]. See also the dissertation by Burchard [1968]. The case of determinants where left and right derivatives are distinguished as in (9.75) was discussed in the paper by Lyche and Schumaker [1976].

Section 9.7

Bounds on the approximation power of generalized splines were first given in an abstract setting by deBoor [1968c]. Jerome [1973a] obtained results for L-splines by constructing an explicit approximation operator—see also Johnen and Scherer [1976]. Here we have followed Scherer and Schumaker [1980].

Section 9.8

Periodic and natural T-splines first arose in connection with the solution of best interpolation problems—see Karlin and Ziegler [1966]. Tchebycheffian monosplines were first discussed in the dissertation by Schumaker [1966], and in the article by Karlin and Schumaker [1967] where the fundamental theorem of algebra was established for simple knots with multiple zeros. The result presented here for multiple knots and simple zeros is credited to Micchelli [1972]. For some related results on monosplines with boundary conditions, see Karlin [1976a, b].

Section 9.9

The hyperbolic splines first arose as solutions of best interpolation problems—see Schweikert [1966a, b], Young [1968, 1969] and Baum [1976a, b].

Section 9.10

Canonical Complete Tchebycheff systems were first introduced by Mühlbach [1973]. They were used in connection with splines by Schumaker [1976a, c]. Tchebycheffian splines associated with a CCT-system of fractional powers (cf. Example 9.52) were used by Reddien and Schumaker [1976] in the numerical solution of singular boundary-value problems.

Section 9.11

Discrete nonpolynomial splines arose already in the paper by Mangasarian and Schumaker [1971] in connection with certain best interpolation problems. For an alternate treatment of such problems, see Astor and Duris [1974]. The constructive properties given here are modeled on those given in Section 8.5 for discrete polynomial splines.

10
L-SPLINES

In the previous chapter we have examined spaces of generalized splines where the pieces were drawn from a given Tchebycheff space. In this chapter we go one step further and consider spaces of functions that are defined piecewise as elements from the null space of a differential operator. We illustrate the development by including a detailed discussion of trigonometric splines.

§ 10.1. LINEAR DIFFERENTIAL OPERATORS

We devote the first two sections of this chapter to background material on differential operators. The material on L-splines proper begins in Section 10.3.

Throughout this chapter we suppose that L is a linear differential operator of the form

$$L = D^m + \sum_{j=0}^{m-1} a_j(x) D^j, \qquad a \leqslant x \leqslant b. \tag{10.1}$$

The coefficients $a_0, \ldots, a_{m-1}$ are allowed to depend on x, but they should be smooth. More precisely, we assume throughout that

$$a_j \in C^{m-j}[a,b], \qquad j = 0, 1, \ldots, m-1.$$

This assumption assures that L is not too far away from the derivative operator D^m. We have the following theorem:

THEOREM 10.1

Let $1 \leqslant p \leqslant \infty$. Then there exist constants C_1 and C_2 such that for all $f \in L_p^m[a,b]$,

$$\|Lf\|_{L_p[a,b]} \leqslant C_1 \big[\|f\|_{L_p[a,b]} + \|D^m f\|_{L_p[a,b]} \big], \tag{10.2}$$

and

$$\|D^m f\|_{L_p[a,b]} \leqslant C_2\big[\|Lf\|_{L_p[a,b]} + \|f\|_{L_p[a,b]}\big]. \tag{10.3}$$

Proof. Clearly

$$\|Lf\|_{L_p[a,b]} \leqslant \|D^m f\|_{L_p[a,b]} + \sum_{j=0}^{m-1} \|a_j\|_{L_\infty[a,b]} \|D^j f\|_{L_p[a,b]}.$$

The inequality (10.2) follows on application of the estimate (2.34) for the jth derivative in terms of $D^m f$ and f. On the other hand,

$$\|D^m f\|_{L_p[a,b]} \leqslant \|Lf\|_{L_p[a,b]} + \left\| \sum_{j=0}^{m-1} a_j D^j f \right\|_{L_p[a,b]},$$

and by (2.34)

$$\|D^m f\|_{L_p[a,b]} \leqslant \|Lf\|_{L_p[a,b]} + C_3 \sum_{j=0}^{m-1} \left(\varepsilon^{-j}\|f\|_p + \varepsilon^{m-j}\|D^m f\|_p\right)$$

for all $0 < \varepsilon < (b-a)/2$. Now taking ε such that $C_3 \sum_{j=0}^{m-1} \varepsilon^{m-j} \leqslant 1/2$, and combining the terms involving $D^m f$, we obtain (10.3). ■

Given L, we define its *null space* by

$$N_L = \{f \in L_1^m[a,b] : Lf(x) = 0, \qquad a \leqslant x \leqslant b\}. \tag{10.4}$$

Under the assumptions above on the coefficients of L, it is a well-known fact from the theory of ordinary differential equations that N_L is an m-dimensional linear subspace of $C^\infty[a,b]$. Any set of functions $u_1, \ldots, u_m \in C^\infty[a,b]$ spanning N_L is called a *fundamental solution set* for L.

The following theorem gives estimates on the size of derivatives of elements in N_L. It may be regarded as a natural generalization of the Markov inequality for polynomials (cf. Theorem 3.3).

THEOREM 10.2

There exists a constant C (depending only on N_L) such that for all intervals $I \subseteq [a,b]$ of length $h < (b-a)/2$,

$$\|D^j u\|_{L_p[I]} \leqslant C h^{-j+1/p-1/q} \|u\|_{L_q[I]}, \tag{10.5}$$

for all $u \in N_L$ and all $1 \leqslant p, q \leqslant \infty$.

Proof. If $p=q$, then since $Lu=0$, (2.34) implies

$$\|D^j u\|_{L_p[I]} \leqslant C_3\left(h^{-j}\|u\|_{L_p[I]}+h^{m-j}\|D^m u\|_{L_p[I]}\right)$$

$$\leqslant C_2 h^{-j}\|u\|_{L_p[I]}, \tag{10.6}$$

where we have used (10.3) to estimate the derivative $D^m u$. If $1 \leqslant p < q \leqslant \infty$, then by Hölder's inequality,

$$\|u\|_{L_p[I]} \leqslant h^{-1/q+1/p}\|u\|_{L_q[I]},$$

and (10.5) follows in this case. Now suppose $1 \leqslant q < p \leqslant \infty$, and let $\xi \in I$ be such that $|u(\xi)| = h^{-1/q}\|u\|_{L_q[I]}$. Define

$$\tilde{u}(x) = u(x) - \operatorname{sgn}(u(\xi)) h^{-1/q}\|u\|_{L_q[I]}.$$

Then $\tilde{u}(\xi)=0$, and using Hölder's inequality together with (10.6) for $j=1$, we obtain

$$\|\tilde{u}\|_{L_p[I]} = \left\| \int_x^{\xi} Du(t)\,dt \right\|_{L_p[I]} \leqslant h^{1-1/q+1/p}\|Du\|_{L_q[I]}$$

$$\leqslant C h^{-1/q+1/p}\|u\|_{L_q[I]}.$$

But

$$\|u\|_{L_p[I]} \leqslant \|\tilde{u}\|_{L_p[I]} + h^{-1/q+1/p}\|u\|_{L_q[I]}$$

and (10.5) follows. ■

An important tool in studying properties of the null space N_L of a differential operator L is the *Wronskian matrix*

$$WM(u_1,\ldots,u_m)(x) = M\begin{pmatrix} u_1,\ldots,u_m \\ x,\ldots,x \end{pmatrix} = \left[D^{j-1}u_i(x)\right]_{i,j=1}^m \tag{10.7}$$

and the associated *Wronskian determinant*

$$W(u_1,\ldots,u_m)(x) = \det\left[D^{i-1}u_j(x)\right]_{i,j=1}^m. \tag{10.8}$$

It is known that if $\{u_i\}_1^m$ is a fundamental solution set for L, then the associated Wronskian determinant does not vanish for any x in $[a,b]$.

In view of the importance of ECT-systems (see Chapter 9), it is of interest to ask when the null space N_L is an ECT-space. We have the following important result:

THEOREM 10.3

Suppose N_L is spanned by functions $u_1,\dots,u_m$ such that

$$W(u_1,\dots,u_k)(x)>0 \qquad \text{for all } a\leqslant x\leqslant b \text{ and all } 1\leqslant k\leqslant m.$$

(In this case we say that L has *property W of Pólya.*) Then $\{u_i\}_1^m$ form an ECT-system and N_L is an ECT-space.

Proof. This is just a restatement of Theorem 9.1. ■

All of the operators $L=L_m$ constructed in Section 9.1 trivially possess property W of Polya. On the other hand, there are many operators L that do not possess it. For such operators, the corresponding null space is not generally an ECT-space, and in fact, may not even be a T-space as the following example shows:

EXAMPLE 10.4

Let $L=D^2+1$ on $[0,2\pi]$.

Discussion. Here the null space of L is spanned by the functions $u_1(x)=\sin(x)$ and $u_2(x)=\cos(x)$. This does not form a T-system on the interval $[0,2\pi]$. It does form a T-system on the interval $[0,\pi]$, however, and, in fact, forms an ECT-system if we work on an interval of length less than π. ■

As the above example tends to indicate, it may be possible to say more about the space N_L if we restrict it to small enough intervals. This is indeed the case as we prove in our next theorem.

THEOREM 10.5

Suppose L is a differential operator as in (10.1) that is defined on an interval $[a,b]$. Then there exists a constant $\delta>0$ such that for any subinterval I of $[a,b]$ of length $|I|<\delta$, N_L is an ECT-space on I.

Proof. Suppose $\{u_i\}_1^m$ is a basis for N_L, and let $a\leqslant\xi\leqslant b$. For each $i=1,2,\dots,m$ we define

$$\tilde{u}_i(x)=\left[u_1(x),\dots,u_m(x)\right]WM^{-1}(\xi)\varepsilon(i,m),$$

where WM^{-1} is the inverse of the Wronskian matrix (10.7) of $\{u_i\}_1^m$, and where $\varepsilon(i,m)$ is an m-vector with 1 in its ith component and zeros everywhere else. It follows immediately from the definition of the $\tilde{u}_i$'s that

$$W(\tilde{u}_1,\dots,\tilde{u}_k)(\xi)=1, \qquad k=1,2,\dots,m.$$

Since each of these Wronskian determinants is a continuous function, there exists an open interval J_ξ around ξ such that they all have value of at least $1/4$. Now since $[a,b]$ is compact, it can be covered by a finite number of such intervals. The result follows with δ the length of the largest subinterval which does not contain any of the endpoints of the J's. Indeed, if I is any interval of length less than δ, then all the Wronskians $W(\tilde{u}_1,\ldots,\tilde{u}_k)(x)$ are positive for x throughout I, and Theorem 10.3 implies $\{u_i\}_1^m$ is an ECT-system on I. ∎

§ 10.2. A GREEN'S FUNCTION

Our aim in this section is to construct the Green's function associated with the differential operator L.

DEFINITION 10.6

Let $G_L(x;y)$ be a function defined on $[a,b]\times[a,b]$ such that for all fixed $y\in[a,b]$,

$$G_L(x;y)=0, \qquad \text{all } a\leqslant x\leqslant y; \tag{10.9}$$

$$LG_L(x;y)=0 \qquad \text{all } y\leqslant x\leqslant b; \tag{10.10}$$

and

$$D^jG_L(x;y)|_{y=x}=\delta_{j,m-1}, \qquad j=0,1,\ldots,m-1. \tag{10.11}$$

Then we call G_L the *Green's function associated with L.*

The following theorem shows how G_L can be used to solve an initial-value problem:

THEOREM 10.7

Let $h\in L_1[a,b]$ and real numbers $r_1,\ldots,r_m$ be given, and suppose u_f is the unique element in N_L such that

$$D^{j-1}u_f(a)=r_j, \qquad j=1,2,\ldots,m. \tag{10.12}$$

Then

$$f(x)=u_f(x)+\int_a^b G_L(x;y)h(y)\,dy \tag{10.13}$$

is the unique solution of the *initial-value problem*

$$Lf(x)=h(x), \qquad x\in[a,b]; \tag{10.14}$$

$$D^{j-1}f(a)=r_j, \qquad j=1,2,\ldots,m. \tag{10.15}$$

Proof. There is a unique u_f solving the interpolation problem (10.12) since the Wronskian $W(u_1,\ldots,u_m)(a)\neq 0$. Now by properties (10.9) and (10.11),

$$D^\nu n \int_a^b G_L(x;y)h(y)dy=\int_a^x D^\nu G_L(x;y)h(y)dy$$

and thus

$$D^\nu f(x)=D^\nu u_f(x)+\int_a^x D^\nu G_L(x;y)h(y)\,dy,$$

$\nu=0,1,\ldots,m-1$. It follows that f satisfies the initial conditions (10.15). Differentiating this formula with $\nu=m-1$, we obtain

$$\begin{aligned} D^m f(x)&=D^m u_f(x)+\int_a^x D^m G_L(x;y)h(y)\,dy+D^{m-1}G(x;y)h(y)|_{y=x} \\ &=D^m u_f(x)+\int_a^x D^m G_L(x;y)h(y)dy+h(x). \end{aligned}$$

Now by multiplying the νth equation by $a_\nu(x)$, $\nu=0,1,\ldots,m$ with $a_m(x)=1$ and adding, we obtain

$$Lf(x)=Lu_f(x)+\int_a^x LG_L(x;y)h(y)dy+h(x)=h(x). \qquad \blacksquare$$

An immediate corollary of this result is the following important Taylor expansion:

THEOREM 10.8. Generalized Taylor Expansion

Let $f\in L_1^m[a,b]$. Then

$$f(x)=u_f(x)+\int_a^b G_L(x;y)Lf(y)\,dy, \qquad \text{all } a\leqslant x\leqslant b, \tag{10.16}$$

where u_f is the unique element in N_L such that

$$D^{j-1}u_f(a)=D^{j-1}f(a), \qquad j=1,\ldots,m. \tag{10.17}$$

Proof. Let

$$g(x)=u_f(x)+\int_a^x G_L(x;y)Lf(y)\,dy.$$

Then by Theorem 10.7, $Lg=Lf$ and so $f-g\in N_L$. But $D^{j-1}g(a)=D^{j-1}f(a), j=0,1,\ldots,m-1$, and thus $f=g$. ■

Before constructing an explicit formula for the Green's function G_L associated with a given operator L, we give a simple application of the Taylor expansion which is useful in estimating how well smooth functions can be approximated by elements of N_L on small intervals. It is the analog of the Whitney-type theorems given in Section 3.5.

THEOREM 10.9

Let $f\in L_p^m[a,b]$, and let u_f be the function in N_L defined in (10.17). Then for any $1\leqslant p,\ q\leqslant\infty$,

$$\|f-u_f\|_{L_q[a,b]}\leqslant(b-a)^{1/q}\|G_L\|_{L_{p'}[a,b]}\|Lf\|_{L_p[a,b]},\tag{10.18}$$

where $1/p+1/p'=1$.

Proof. Applying the Hölder inequality to $f-u_f$ as given by (10.16), we obtain

$$|f(x)-u_f(x)|\leqslant\|G_L\|_{L_{p'}[a,b]}\|Lf\|_{L_p[a,b]}.$$

Integrating the qth power over $[a,b]$ and taking the qth root yields (10.18). ■

We turn now to the problem of constructing the Green's function G_L explicitly. For each $j=1,2,\ldots,m$ let

$$L_j^*f=(-1)^jD^jf+(-1)^{j-1}D^{j-1}(a_{m-1}f)+\cdots+a_{m-j}f.\tag{10.19}$$

For convenience we suppose that L_0^* denotes the identity operator. The L_j^*'s are called the *partial adjoints of L*. The operator $L^*=L_m^*$ is called the *adjoint of L*.

Suppose now that $u_1,\ldots,u_m$ is a basis for N_L, and define

$$\begin{bmatrix}u_1^*(x)\\ \vdots\\ u_m^*(x)\end{bmatrix}=WM^{-1}(u_1,\ldots,u_m)(x)\begin{bmatrix}0\\0\\0\\1\end{bmatrix}\qquad\text{all } a\leqslant x\leqslant b,\tag{10.20}$$

where $WM(u_1,\ldots,u_m)(x)$ is the Wronskian matrix defined in (10.7). The $u_1^*,\ldots,u_m^*$ are the functions in the last column of WM^{-1}. We call them the *adjunct functions*. We can now express the entire inverse of the Wronskian matrix $WM(u_1,\ldots,u_m)$ in terms of the adjunct functions.

THEOREM 10.10

The inverse of the Wronskian matrix of $u_1,\ldots,u_m$ is given by

$$WM^{-1}(u_1,\ldots,u_m)(x)=\left(WM_{ij}^{-1}\right)_{i,j=1}^{m}=\left[L_{m-j}^*u_i^*(x)\right]_{i,j=1}^{m} \tag{10.21}$$

for all $a\leqslant x\leqslant b$.

Proof. We proceed by induction. We need to show that

$$\sum_{i=1}^{m} D^k u_i(x)L_{m-j-1}^*u_i^*(x)=\delta_{k,j}, \qquad \begin{array}{l}k=0,1,\ldots,m-1\\ j=0,1,\ldots,m-1.\end{array} \tag{10.22}$$

For $j=m-1$ this follows directly from the definition of the u^*'s. We now suppose that (10.22) holds for $m-1,\ldots,\ j\geqslant 1$, and we prove it for $j-1$. Differentiating the identity (10.22), we obtain

$$-\sum_{i=1}^{m} D^k u_i(x)DL_{m-j-1}^*u_i^*(x)=\sum_{i=1}^{m} D^{k+1}u_i(x)L_{m-j-1}^*u_i^*(x).$$

Adding

$$\sum_{i=1}^{m} D^k u_i(x)u_i^*(x)a_j(x)=\delta_{k,m-1}a_j(x),$$

gives

$$\sum_{i=1}^{m} D^k u_i(x)L_{m-j}^*u_i^*(x)=\sum_{i=1}^{m} D^{k+1}u_i(x)L_{m-j-1}^*u_i^*(x)+a_j(x)\delta_{k,m-1}. \tag{10.23}$$

Since $\delta_{k,j-1}=\delta_{k+1,j}$, we have proved (10.22) for $j-1$ and all $k=0,1,\ldots,m-2$. Now for $k=m-1$, by substituting

$$D^m u_i(x)=-\sum_{\nu=0}^{m-1} a_\nu(x)D^\nu u_i(x)$$

in (10.23), we obtain

$$\sum_{i=1}^{m} D^{m-1}u_i(x)L_{m-j}^{*}u_i^{*}(x) = -\sum_{i=1}^{m}\sum_{\nu=0}^{m-1} a_\nu(x)D^\nu u_i(x)L_{m-j-1}^{*}u_i^{*}(x) + a_j(x)$$

$$= -a_j(x) + a_j(x) = 0,$$

and the induction step is complete. ■

THEOREM 10.11

The function

$$G_L(x;y) = \begin{cases} \sum_{i=1}^{m} u_i(x)u_i^{*}(y), & x \geqslant y \\ 0, & x \leqslant y \end{cases} \tag{10.24}$$

is the Green's function associated with L.

Proof. It is clear that $LG_L(x;y)=0$ for $x \geqslant y$ since $u_1,\ldots,u_m$ are in N_L. The condition (10.11) on the derivatives follows from (10.22) with $j=m-1$. ■

The function G_L also has the following properties for all fixed x in $[a,b]$:

$$G_L(x;y)=0, \qquad x \leqslant y \leqslant b; \tag{10.25}$$

$$L^{*}G_L(x;y)=0, \qquad a \leqslant y \leqslant x; \tag{10.26}$$

$$L_j^{*}G_L(x;y)|_{x=y} = \delta_{j,m-1}, \qquad j=0,1,\ldots,m-1. \tag{10.27}$$

Here the operators L^{*} and L_j^{*} operate on the y-variable. These properties suggest that G_L is a kind of dual Green's function for the adjoint operator L^{*}. Indeed, we have the following:

THEOREM 10.12

Let $h \in L_1[a,b]$ and real numbers $r_1,\ldots,r_m$ be given, and suppose u^{*} is the unique element of $N_{L^{*}}$ such that

$$L_{j-1}^{*}u^{*}(b) = r_j, \qquad j=1,2,\ldots,m. \tag{10.28}$$

Then

$$f(y) = u^{*}(y) + \int_a^b G_L(x;y)h(x)\,dx \tag{10.29}$$

is the unique solution of the *terminal-value problem*

$$L^*f(y)=h(y), \qquad y\in[a,b]; \tag{10.30}$$

$$L^*_{j-1}f(b)=r_j, \qquad j=1,2,\ldots,m. \tag{10.31}$$

Proof. The proof proceeds along the same lines as the proof of Theorem 10.7. ■

Theorem 10.12 can now be used to establish a dual Taylor expansion for smooth functions.

THEOREM 10.13. Dual Taylor Expansion

Let $L^*f\in L_1[a,b]$. Then

$$f(y)=u_f^*(y)+\int_a^b G_L(x;y)L^*f(x)\,dx, \tag{10.32}$$

where u_f^* is the unique element in $N_{L^*}=\text{span}\{u_i^*\}_{i=1}^m$ such that

$$L_j^*u_f^*(b)=L_j^*f(b), \qquad j=0,1,\ldots,m-1. \tag{10.33}$$

Proof. We simply apply Theorem 10.12 in the same way as in the proof of Theorem 10.8. ■

§ 10.3. L-SPLINES

Let L be a linear differential operator of order m as in (10.1), and let N_L be its null space.

DEFINITION 10.14

Suppose $\mathfrak{M}=(m_1,\ldots,m_k)$ is a vector of integers with $1\leqslant m_i\leqslant m$, $i=1,2,\ldots,k$. Given a partition $\Delta=\{a=x_0<x_1<\cdots<x_k<x_{k+1}=b\}$ of the interval $[a,b]$, let

$$\mathcal{S}(N_L;\mathfrak{M};\Delta)=\left\{\begin{matrix} s\in B[a,b]\text{: } s|_{I_i}\in N_L,\ i=0,1,\ldots,k, \textit{ and} \\ D_-^{j-1}s(x_i)=D_+^{j-1}s(x_i),\ \ j=1,2,\ldots,m- \\ m_i,\ i=1,2,\ldots,k \end{matrix}\right\}.$$

We call $\mathcal{S}$ the space of *L-splines* with *knots* at $x_1,\ldots,x_k$ of *multiplicities* $m_1,\ldots,m_k$.

The space of L-splines $\mathcal{S}$ consists of functions whose pieces belong to N_L. The smoothness with which these pieces tie together is controlled by the multiplicity vector $\mathfrak{M}$, just as in the polynomial spline case (which, of course, is just the special case where $L=D^m$). When L has property W of Pólya (cf. Theorem 10.3), then we have the special case of T-splines.

By the same kind of argument used in Theorem 4.4 for polynomial splines, we deduce that

$$\dim \mathcal{S}(N_L;\mathfrak{M};\Delta)=m+K, \qquad K=\sum_{i=1}^{k} m_i.$$

We now construct a one-sided basis for $\mathcal{S}$, using the Green's function associated with L.

THEOREM 10.15

A one-sided basis for the space of L-splines $\mathcal{S}(N_L;\mathfrak{M};\Delta)$ is given by

$$B_{ij}(x)=L^*_{j-1}G_L(x;x_i), \qquad \begin{array}{l} j=1,2,\ldots,m_i \\ i=0,1,\ldots,k, \end{array} \tag{10.34}$$

where G_L is the Green's function associated with L defined in (10.24), and the L^*'s are the partial adjoint operators defined in (10.19).

Proof. First we observe that for each $0\leqslant i\leqslant k$ and $1\leqslant j\leqslant m_i$,

$$B_{ij}(x)=\begin{cases} 0, & x\leqslant x_i \\ \sum_{\nu=1}^{m} u_\nu(x)L^*_{j-1}u^*_\nu(x_i), & x>x_i. \end{cases}$$

Thus each such function is clearly a linear combination of the u_ν's, hence it has the desired piecewise structure. Now in view of (10.22), we also see that

$$D_+^{\mu-1}B_{ij}(x_i)=\sum_{\nu=1}^{m} D_+^{\mu-1}u_\nu(x)L^*_{j-1}u^*_\nu(x_i)=\delta_{\mu,m-j+1},$$

$\mu,j=1,2,\ldots,m$, and thus the B's also have the correct continuity at the knots to belong to $\mathcal{S}$.

We already know that the dimension of $\mathcal{S}$ is $m+\sum_1^k m_i$, hence it remains only to check that these one-sided splines are linearly independent. It suffices to prove that for each $0\leqslant i\leqslant k$, the set $\{B_{i1},\ldots,B_{i,m_i}\}$ is linearly independent on $I_i=[x_i,x_{i+1}]$, since then the argument used in proving Theorem 4.5 can be applied. But it is clear that these m_i splines are linearly independent since if $\sum_{j=1}^{m_i} c_j B_{i,j}=0$, then

$$c_\mu = D_+^{m-\mu}\left(\sum_{j=1}^{m_i} c_j B_{i,j}\right)(x_i)=0, \qquad \mu=1,2,\ldots,m_i. \qquad \blacksquare$$

Our next task is to construct a basis of local support splines for $\mathcal{S}(N_L; \mathfrak{M}; \Delta)$. One approach is to make use of the results on T-splines given in Chapter 9 coupled with the fact that N_L spans an ECT-system on all sufficiently small intervals. We carry this out in detail in the following section. Another approach is to follow the ideas used in Section 4.2, where the local support splines were obtained as linear combinations of the one-sided splines. We do this now.

Throughout the remainder of this section we suppose that $\{u_i\}_1^m$ are functions forming a basis for N_L and that $\{u_i^*\}_1^m$ are the adjunct functions defined in (10.20). The following is the analog of Lemma 4.7:

LEMMA 10.16

Let $\tau_1<\tau_2<\cdots<\tau_d$ and $1\leqslant l_i\leqslant m$, $i=1,2,\ldots,d$ be given. Suppose $\mathfrak{U}^*=\text{span}\{u_1^*,\ldots,u_m^*\}$ is an ET-space on some interval (α,β) containing the τ's. Then if $\sum_{i=1}^d l_i>m$, there exists a nontrivial

$$B(x)=\sum_{i=1}^{d}\sum_{j=1}^{l_i} c_{ij}L_{j-1}^* G_L(x;\tau_i)$$

with

$$B(x)=0 \quad \text{for} \quad x<\tau_1 \text{ and } x>\tau_d.$$

On the other hand, if $\sum_{i=1}^k l_i\leqslant m$, then no such nontrivial B exists.

Proof. Any B of this form automatically vanishes for $x<\tau_1$ by the one-sided nature of G_L. If $B(x)$ is to vanish for $x>\tau_d$, then using the expansion (10.24), we must have

$$B(x)=\sum_{i=1}^{d}\sum_{j=1}^{l_i} c_{ij}\sum_{\nu=1}^{m} u_\nu(x)L_{j-1}^* u_\nu^*(\tau_i)$$

$$=\sum_{\nu=1}^{m} u_\nu(x)\sum_{i=1}^{d}\sum_{j=1}^{l_i} c_{ij}L_{j-1}^* u_\nu^*(\tau_i)=0,$$

all $\tau_d < x < \beta$. Since $u_1,\ldots,u_m$ are linearly independent, this is equivalent to

$$\sum_{i=1}^{d} \sum_{j=1}^{l_i} c_{ij} L_{j-1}^* u_\nu^*(\tau_i) = 0, \qquad \nu = 1,2,\ldots,m.$$

This is a homogeneous system of m equations for the $\Sigma_{i=1}^d l_i$ coefficients, and thus it always has a nontrivial solution if $\Sigma_{i=1}^d l_i > m$. On the other hand, if $\Sigma_{i=1}^d l_i = m$, this system is nonsingular since its determinant is given by

$$D\begin{bmatrix} \overbrace{\tau_1,\ldots,\tau_1}^{l_1},\ldots,\overbrace{\tau_d,\ldots,\tau_d}^{l_d} \\ u_1^*,\ldots,u_m^* \end{bmatrix},$$

which is nonzero in view of the assumption that $\mathfrak{U}^*$ is an ET-space. Note here that $L_{j-1}^* = (-1)^{j-1}D^{j-1} +$ lower order terms, so using ordinary derivatives in dealing with repeated τ's is equivalent to using the L^*'s. ■

As in the polynomial spline case, we now consider the case where $\Sigma_{i=1}^d l_i = m+1$. Arguing as before, it follows that the B constructed in Lemma 10.16 must have the form

$$B(x) = C_1 D\begin{bmatrix} \overbrace{\tau_1,\ldots,\tau_1}^{l_1},\ldots,\overbrace{\tau_d,\ldots,\tau_d}^{l_d} \\ u_1^*,\ldots,u_m^*, G_L(x;\cdot) \end{bmatrix}.$$

In the absence of any good choice for C_1, we take it to be 1. The following is the analog of Theorem 4.9:

THEOREM 10.17

Let $\Delta_e = \{y_i\}_1^{2m+K}$ be an extended partition as in Definition 4.8. For each $i = 1,2,\ldots,m+K$ suppose $\mathfrak{U}^*$ is an ET-space on an interval containing $[y_i, y_{i+m}]$, and define

$$B_i(x) = D\begin{pmatrix} y_i,\ldots,y_{i+m} \\ u_1^*,\ldots,u_m^*, G_L(x;\cdot) \end{pmatrix}. \tag{10.35}$$

Then $\{B_i\}_1^{m+K}$ form a basis for $\mathcal{S}(N_L; \mathfrak{M}; \Delta)$ and

$$B_i(x) = 0 \qquad \text{for } x < y_i \text{ and } x > y_{i+m}. \tag{10.36}$$

Proof. The Laplace expansion of the determinant defining $B_i(x)$ shows that if

$$y_i \leqslant y_{i+1} \leqslant \cdots \leqslant y_{i+m} = \overbrace{\tau_1,\ldots,\tau_1}^{l_1},\ldots,\overbrace{\tau_d,\ldots,\tau_d}^{l_d}$$

with $\tau_1 < \tau_2 < \cdots < \tau_d$, then

$$B_i(x) = \sum_{\nu=1}^{d} \sum_{j=1}^{l_\nu} \alpha_{\nu j} L_{j-1}^* G_L(x;\tau_\nu).$$

Since each of these is an L-spline (cf. Theorem 10.15), it follows that B_i is also an L-spline. The fact that $B_i(x)$ is zero for $x < y_i$ is clear from the one-sided nature of G_L. On the other hand, for $x > y_{i+m}$ we know by (10.24) that $G_L(x;\cdot)$ is a member of $\mathcal{U}^*$, and thus the determinant vanishes identically. The linear independence of the $B_1,\ldots,B_{m+K}$ is established in exactly the same way as was done in Theorem 4.18, but Lemma 10.16 is used in place of Lemma 4.7. ■

When Δ_e is chosen as in (4.14) with $y_{m+K+1} = \cdots = y_{2m+K} = b$, the definition of the B-spline basis in Theorem 10.17 requires the same kind of modification as discussed in Corollary 4.10 for polynomial splines.

§ 10.4. A BASIS OF TCHEBYCHEFFIAN B-SPLINES

The construction of a basis of local support splines for the space $\mathcal{S}(N_L;\mathfrak{M};\Delta)$ of L-splines carried out in the previous section is not, in general, completely satisfactory. Perhaps the most serious deficiency is that there is no natural way to normalize the B-splines constructed there. In this section we shall show that when $\bar{\Delta}$ is sufficiently small, then it is possible to construct a basis for $\mathcal{S}$ consisting of TB-splines.

We begin by finding canonical ECT-systems spanning N_L on various subintervals of $[a,b]$. Let n be such that $h = (b-a)/n < \delta/4$, where δ is the constant in Theorem 10.5. Let $a = z_{-1}$, $b = z_{n+1} = z_{n+2}$, $z_\nu = a + \nu \cdot h$, $\nu = 0,1,\ldots,n$, and $J_\nu = [z_{\nu-1}, z_{\nu+3}]$, $\nu = 0,1,\ldots,n-1$. Since $|J_\nu| < \delta$, it follows from Theorem 10.5 that there exists a canonical ECT-system $U_{m,\nu} = \{u_i^\nu\}_{i=1}^m$ which spans N_L on J_ν. Let $U_{m,\nu}^*$ be the associated dual canonical ECT-system, and let $g_{m,\nu}(x;y)$ be the associated Green's function.

We are ready to define some local support splines in $\mathcal{S}(N_L;\mathfrak{M};\Delta)$. Let $\Delta_e = \{y_i\}_1^{2m+K}$ be an extended partition associated with the Δ (cf. Definition 4.8) with $y_1 = \cdots = y_m = a$ and $b = y_{m+K+1} = \cdots = y_{2m+K}$. For fixed

$0 \leqslant \nu \leqslant n-1$ and all i such that $z_\nu \leqslant y_i < z_{\nu+1}$, we define

$$N_i(x) = (-1)^m \alpha_i [y_i, \ldots, y_{i+m}]_{U^*_{m,\nu}} g_{m,\nu}(x;y), \tag{10.37}$$

where α_i is the constant in (9.54). For each such i, N_i is the usual normalized TB-spline associated with knots $y_i, \ldots, y_{i+m}$. In order to ensure that it makes sense, we must demand that all of the points $y_i, \ldots, y_{i+m}$ lie in the interval J_ν (where the νth ECT-system is defined). Hence from now on we assume that

$$\bar{\Delta} = \max_{0 \leqslant i \leqslant k} (x_{i+1} - x_i) < h/m. \tag{10.38}$$

The following theorem summarizes some of the properties of the B-splines N_i:

THEOREM 10.18

The spline N_i defined in (10.37) satisfies

$$N_i(x) = 0, \qquad x < y_i, \qquad y_{i+m} < x, \tag{10.39}$$

and

$$N_i(x) > 0, \qquad y_i < x < y_{i+m}. \tag{10.40}$$

Aside from a constant multiplier, it is the unique spline in $\mathcal{S}(N_L; \mathfrak{M}; \Delta)$ that has these properties.

Proof. It is clear that N_i has properties (10.39) and (10.40) since it is a T-spline. But since $U_{m,\nu}$ spans N_L throughout J_ν, it follows that N_i is in fact an L-spline. Now if $\tilde{N}_i$ were another L-spline with the same two properties, then for some choice of β, the function $g = N_i - \beta \tilde{N}_i$ would have a zero in the interval (y_i, y_{i+m}). Since g is a T-spline with only $m+1$ knots (and an m-tuple zero to the left of y_i and another to the right of y_{i+m}), Theorem 9.30 asserts that $g = 0$, and the uniqueness assertion is established. ■

The above construction can be repeated for each $\nu = 0, 1, \ldots, n-1$ to construct $m+K$ local-support B-splines in $\mathcal{S}(N_L; \mathfrak{M}; \Delta)$. To illustrate that they actually form a basis for this space of L-splines, it remains to show their linear independence. To this end, it is useful to construct a dual basis.

THEOREM 10.19

For each $i = 1, 2, \ldots, m+K$, let λ_i be the linear functional defined in (9.64). Then $\{\lambda_i\}_1^{m+K}$ form a dual basis for $\{N_i\}_1^{m+K}$; that is,

$$\lambda_i N_j = \delta_{ij}, \qquad i,j = 1, 2, \ldots, m+K. \tag{10.41}$$

Moreover, there exists a constant C (depending only on $[a,b]$ and L) such that

$$|\lambda_i f| \leqslant C\underline{\Delta}^{-1/p} \| f \|_{L_p[y_i, y_{i+m}]}, \qquad i=1,2,\ldots,m+K, \tag{10.42}$$

where $\underline{\Delta} = \min_{0 \leqslant i \leqslant k}(x_{i+1} - x_i)$.

Proof. First we note that the λ_i are well defined since the construction in (9.64) is a local process and can be carried out using the ECT-system $U_{m,\nu}$ used in constructing N_i. Property (10.41) holds since for any given λ_i, all the B-splines that have values in (y_i, y_{i+m}) can be regarded as TB-splines associated with one fixed ECT-system $U_{m,\nu}$, and we know that (10.41) holds for the TB-splines. (This is where we use the assumption that $h<\delta/4$ —cf. Figure 34.) The bound (10.42) has been established in Theorem 9.26. The constant in the bounds on the various λ_i depends on the ECT-system being used in the interval where the support of λ_i lies. But since we have only a finite number of such intervals to consider, the final constant depends only on $[a, b]$ and L. ■

THEOREM 10.20

The LB-splines $\{N_i\}_1^{m+K}$ constructed in (10.37) form a basis for $\mathcal{S}(N_L; \mathfrak{M}; \Delta)$. Moreover,

$$\sum_{i=1}^{m+K} N_i(x) = u_1(x), \qquad z_0 \leqslant x \leqslant z_1, \tag{10.43}$$

and

$$\sum_{i=1}^{m+K} N_i(x) \leqslant u_1^{\nu}(x) + u_1^{\nu-1}(x), \qquad z_\nu \leqslant x \leqslant z_{\nu+1}, \qquad \nu = 1,\ldots,n. \tag{10.44}$$

Proof. The linear independence of the LB-splines $N_1,\ldots,N_{m+K}$ follows immediately from the existence of a dual basis, and since $\mathcal{S}(N_L; \mathfrak{M}; \Delta)$ has

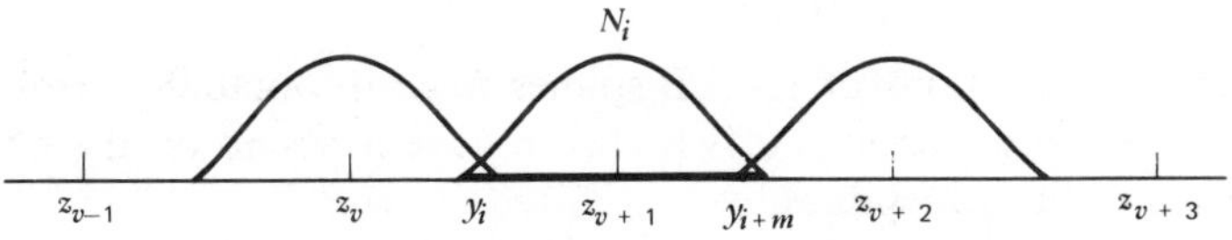

Figure 34. Construction of LB-splines.

dimension $m+K$, we have shown they form a basis. Property (10.43) follows from the analogous property of the TB-splines, cf. (9.57). On the other hand, for $\nu>0$, the B-splines with nonzero values in $[z_\nu, z_{\nu+1}]$ may be associated with two different ECT-systems $U_{m,\nu-1}$ and $U_{m,\nu}$. Since those associated with the first add up to no more than $u_1^{\nu-1}(x)$, while those associated with the second add up to no more than $u_1^\nu(x)$, the assertion (10.44) follows. ■

Our next result makes use of the boundedness of the dual basis in Theorem 10.19 to examine the conditioning of the LB-spline basis for $\mathcal{S}(N_L;\mathfrak{M};\Delta)$.

THEOREM 10.21

Given $1\leqslant p\leqslant\infty$, let

$$B_{i,p}(x)=\underline{\Delta}^{-1/p}N_i(x), \qquad i=1,2,\dots,m+K. \tag{10.45}$$

Then there exist constants $0<C_1$ and $C_2<\infty$ (depending only on $[a,b]$ and L) such that

$$\left(\sum_{i=1}^{m+K}|c_i|^p\right)^{1/p}\leqslant C_1\left\|\sum_{i=1}^{m+K}c_iB_{i,p}\right\|_{L_p[a,b]}\leqslant C_2\left(\sum_{i=1}^{m+K}|c_i|^p\right)^{1/p} \tag{10.46}$$

for all sets of coefficients $c_1,\dots,c_{m+K}$.

Proof. Let $s=\sum_{i=1}^{m+K}c_iB_{i,p}$. Then with $I_i=(y_i,y_{i+m})$,

$$\sum_{i=1}^{m+K}|c_i|^p=\sum_{i=1}^{m+K}|\lambda_i s|^p\leqslant C\sum_{i=1}^{m+K}\|s\|_{L_p[I_i]}^p\leqslant C\|s\|_{L_p[a,b]}^p.$$

This proves the first inequality. For the second, we note that

$$\int_a^b|s(x)|^p\,dx=\sum_{j=0}^{k}\int_{x_j}^{x_{j+1}}\sum_{i=j-m+1}^{j}|c_iB_{i,p}(x)|^p\,dx$$

$$\leqslant\sum_{j=0}^{k}\int_{x_j}^{x_{j+1}}\max_{1\leqslant i\leqslant m+K}\|B_{i,p}\|_{L_\infty[I_i]}^p m^{p-1}$$

$$\times\sum_{i=j-m+1}^{j}|c_i|^p\leqslant C\sum_{i=1}^{m+K}|c_i|^p.$$ ■

In many applications of the LB-splines (e.g., in obtaining error bounds for L-spline approximation) it is useful to have a bound on the size of the derivatives of the LB-splines.

THEOREM 10.22

There exists a constant depending only on $[a,b]$ and L such that

$$|D^jN_i(x)| \leqslant C_1\underline{\Delta}^{-j}, \qquad j=0,1,\ldots,m-1, \tag{10.47}$$

where $\underline{\Delta}=\min_{0\leqslant i\leqslant k}(x_{i+1}-x_i)$.

Proof. Since N_i is a piecewise element from N_L, we can apply the Markov inequality of Theorem 10.2 to each subinterval $[x_\nu, x_{\nu+1}]$ to assert that $\|D^jN_i\|_{L_\infty[x_\nu,x_{\nu+1}]} \leqslant C_2\underline{\Delta}^{-j}\|N_i\|_{L_\infty[x_\nu,x_{\nu+1}]}$. But (10.43) and (10.44) guarantee that the uniform norm of N_i is bounded, and (10.47) follows. ■

In view of the observation in Section 10.1 that L can be thought of as a perturbation of the differential operator D^m, it is reasonable to think of the LB-splines as perturbations of the usual B-splines. Our next theorem shows that this is indeed the case. It helps to provide us with some feeling for what LB-splines look like, although for a given L this theorem cannot be used to get very precise information about them.

THEOREM 10.23

Let $\{N_i\}_1^{m+K}$ be the normalized polynomial B-splines associated with the extended partition $\{y_i\}_1^{2m+K}$. Suppose $L^{(n)}$ is a sequence of linear operators defined as in (10.1) with coefficients $a_0^{(n)},\ldots,a_{m-1}^{(n)}$, and suppose $B_1^{(n)},\ldots,B_{m+K}^{(n)}$ are the corresponding normalized LB-splines. Then

$$\lim_{n\to\infty} \|a_i^{(n)}\|_{L_\infty[a,b]}=0, \qquad i=0,1,\ldots,m-1$$

implies

$$\lim_{n\to\infty} \|B_i^{(n)}-N_i\|_{L_\infty[a,b]}=0, \qquad i=1,2,\ldots,m+K.$$

Proof. From the theory of ordinary differential equations, it is known that the assumption on the a's implies that the null spaces of the corresponding differential operators $L^{(n)}$ tend to $V_m=\text{span}\{1,x,\ldots,x^{m-1}\}$. In fact, it can be shown that for n large enough these null spaces are spanned by an ECT-system, and the associated weights in its canonical expansion tend to the values associated with the functions in V_m. It follows that the Green's function $g_m(x;y)$ associated with $L^{(n)}$ tends to $(y-x)_+^{m-1}$, while by Lemma 9.6 the divided differences used in the construction of the LB-splines tend to the usual divided difference. We conclude that the $B_i^{(n)}$ converge to the N_i as asserted. ■

§ 10.5. APPROXIMATION POWER OF L-SPLINES

In this section we give some estimates for how well smooth functions can be approximated by splines in the space $\mathcal{S}(N_L;\mathfrak{M};\Delta)$. Since the T-splines are special cases of L-splines, our results will apply to them as well. For some companion lower bounds, see the following section. Inverse and saturation results are given in Section 10.7.

We begin by constructing a useful approximation operator. Suppose $\{B_i\}_1^{m+K}$ are the normalized LB-splines forming a basis for $\mathcal{S}(N_L;\mathfrak{M};\Delta)$, and suppose $\{\lambda_i\}_1^{m+K}$ is the dual basis discussed in Theorem 10.19. Then given any function $f\in L_1[a,b]$, we define

$$Qf(x)=\sum_{i=1}^{m+K}(\lambda_i f)B_i(x). \tag{10.48}$$

Clearly Q is a linear mapping of $L_1[a,b]$ into $\mathcal{S}(N_L;\mathfrak{M};\Delta)$. Q is, in fact, a projector onto this spline space since if $s=\sum_{i=1}^{m+K}c_iB_i$ is any spline in $\mathcal{S}$, then since $\lambda_i s=c_i$, it follows that $Qs=\sum_{i=1}^{m+K}c_iB_i=s$.

THEOREM 10.24

Let $1\leqslant p\leqslant q\leqslant\infty$, and suppose $f\in L_p^m[a,b]$. Then for all $j=0,1,\ldots,m-1$,

$$\|D^j(f-Qf)\|_{L_q[a,b]}\leqslant C\frac{\bar{\Delta}^{m+1/q}}{\underline{\Delta}^{j+1/p}}\|Lf\|_{L_p[a,b]}. \tag{10.49}$$

Here, C is a constant that is independent of both f and Δ.

Proof. Fix $0\leqslant j\leqslant m-1$, and suppose $t\in I_l=[y_l,y_{l+1})$. Let J_ν be the interval used in the construction of B_l (so that N_L is spanned by an ECT-system throughout J_ν). Let u_f be the unique function in N_L such that

$$D^ju_f(t)=D^jf(t),\qquad j=0,1,\ldots,m-1.$$

Since $N_L\in\mathcal{S}(N_L;\mathfrak{M};\Delta)$, we have $Qu_f=u_f$, and thus using the bound (10.42) on $|\lambda_i|$,

$$\begin{aligned}|D^j(f-Qf)(t)|&=|D^j(u_f-Qf)(t)|=|D^jQ(u_f-f)(t)|\\&\leqslant\sum_{i=l+1-m}^{l}|\lambda_i(u_f-f)||D^jB_i(t)|\\&\leqslant C\underline{\Delta}^{-1/p}\sum_{i=l+1-m}^{l}\|u_f-f\|_{L_p[\hat{I}_i]}|D^jB_i(t)|,\end{aligned}$$

where $\hat{I}_i=[y_i,y_{i+m}]$ in general. We have a bound on $D^jB_i(t)$ in Theorem 10.22. It remains to estimate u_f-f. Applying the Hölder inequality to the Taylor expansion

$$f(x)-u_f(x)=\int_t^x g_m(x;y)L_mf(y)\,dy,$$

where g_m is the Green's function associated with the ECT-system spanning N_L on J_ν, we obtain

$$\|f-u_f\|_{L_p[\hat{I}_i]}\leqslant C_2\bar{\Delta}^{1/p}\|g_m(x;\cdot)\|_{L_{p'}[t,x]}\|L_mf\|_{L_p[t,x]},$$

where $1/p'+1/p=1$.

To estimate the norm of g_m, note that by (9.31) and (9.32),

$$|g_m(x;y)|\leqslant C|x-y|^{m-1},\qquad \text{all } x,y \text{ in } J_\nu.$$

It follows that

$$\|g_m(x;\cdot)\|_{L_{p'}[t,x]}\leqslant C\bar{\Delta}^{m-1/p} \tag{10.50}$$

for all $t\in I_l$ and all $x\in\tilde{I}_l$. On the other hand,

$$L_mf(x)=W(x)Lf(x),$$

where $W(x)$ is a function that depends only on the weight functions $w_1,\ldots,w_m$ used in the canonical representation of $U_{m,\nu}$. Thus

$$\|f-u_f\|_{L_p[\hat{I}_i]}\leqslant C\bar{\Delta}^m,\qquad i=l+1-m,\ldots,l,$$

hence

$$|D^j(f-Qf)(t)|\leqslant\frac{C\bar{\Delta}^m}{\underline{\Delta}^{j+1/p}}\sum_{i=l+1-m}^{l}\|Lf\|_{L_p[\hat{I}_i]}.$$

This implies

$$\|D^j(f-Qf)\|_{L_q[I_l]}\leqslant\frac{C\bar{\Delta}^{m+1/q}}{\underline{\Delta}^{j+1/p}}\sum_{i=l+1-m}^{l}\|Lf\|_{L_p[\hat{I}_i]}.$$

Now, summing over $l=m,\ldots,m+K$ and applying the Jensen inequality (see Remark 6.2), we finally arrive at (10.49). ■

The bound on the operator Q defined in (10.48) and the error bounds (10.49) both depend on $\underline{\Delta}$ as well as $\bar{\Delta}$. It is a simple matter to construct a bounded linear operator Q mapping $L_p[a,b]$ into $\mathcal{S}(N_L; \mathfrak{M}; \Delta)$ whose norm and associated error bounds do not depend on $\underline{\Delta}$. Indeed, given Δ, suppose Δ^* is the thinned out 3-quasi-uniform partition constructed in Lemma 6.17. Then we can construct Q^* associated with the L-spline space $\mathcal{S}(N_L; \mathfrak{M}; \Delta^*)$ exactly as above. Since $\bar{\Delta}^*/\underline{\Delta}^* \leqslant 3$, the bound on $\|Q^*\|_p$ and the analogs of (10.49) are all independent of $\underline{\Delta}^*$. But since $\mathcal{S}(N_L; \mathfrak{M}; \Delta^*) \subseteq \mathcal{S} = \mathcal{S}(N_L; \mathfrak{M}; \Delta)$, Q^* maps $L_p[a,b]$ into $\mathcal{S}$. In general, Q^* will not reproduce all of $\mathcal{S}$; that is, it is not a projector onto this larger space.

Theorem 10.24 gives an estimate on the distance of a given smooth function f to the space of splines $\mathcal{S}(N_L; \mathfrak{M}; \Delta)$. This bound involves $\|Lf\|$. In our next theorem we restate this result in terms of the traditional modulus of smoothness.

THEOREM 10.25

Let $1 \leqslant p < \infty$. Then for any $f \in L_p[a,b]$,

$$d\left[f, \mathcal{S}(N_L; \mathfrak{M}; \Delta)\right]_{L_p[a,b]} \leqslant C\left[\omega_m(f; \bar{\Delta})_{L_p[a,b]} + \bar{\Delta}^m \|f\|_{L_p[a,b]}\right]. \tag{10.51}$$

where C is a constant independent of f and Δ. A similar bound holds with $p = \infty$ whenever $f \in C[a,b]$.

Proof. Coupling (10.49) for $j = 0$ with the estimate (10.2) for Lf in terms of $D^m f$ and f, we obtain

$$d(f, \mathcal{S}) \leqslant C\bar{\Delta}^m\left[\|D^m f\|_{L_p[a,b]} + \|f\|_{L_p[a,b]}\right], \qquad \text{all } f \in L_p^m[a,b].$$

The result then follows upon applying Theorem 2.68. ■

§ 10.6. LOWER BOUNDS

The general results on n-widths given in Section 2.10 show that the approximation orders given in the previous section for approximation of smooth functions by L-splines are the best possible. It is, nevertheless, of some interest to establish this directly by giving some lower bounds that are the companions of these upper bounds. In order to be able to state our results without having to pay attention to the multiplicity vector $\mathfrak{M}$, we shall work with the largest possible space of L-splines associated with a given partition; namely,

$$\mathcal{P}\mathcal{W}(N_L;\Delta)=\left\{\begin{array}{l} s\colon \text{there exist } s_0,\dots,s_k \text{ in } N_L \text{ such} \\ \text{that } s|_{(x_i,x_{i+1})}=s_i,\ i=0,1,\dots,k \end{array}\right\}. \tag{10.52}$$

This space is the analog of the space of piecewise polynomials $\mathcal{P}\mathcal{P}_m(\Delta)$. It corresponds to the space of L-splines where all knots have multiplicity m. We note that

$$\mathcal{S}(N_L;\mathfrak{M};\Delta)\subseteq\mathcal{P}\mathcal{W}(N_L;\Delta), \qquad \text{all } \mathfrak{M},$$

and thus lower bounds for $\mathcal{P}\mathcal{W}(N_L;\Delta)$ will apply to all L-spline spaces.

We begin with the companion lower bound for the case of $p=\infty$ and $j=0$ in Theorem 10.24.

THEOREM 10.26

Fix $1\leqslant p\leqslant\infty$. Then given any partition Δ, there exists a function $F\in L_p^m[a,b]$ such that

$$d[F,\mathcal{P}\mathcal{W}(N_L;\Delta)]_{L_\infty[a,b]}\geqslant C\bar{\Delta}^{m-1/p}\|LF\|_{L_p[a,b]}. \tag{10.53}$$

Proof. We construct F explicitly. Given any partition Δ of $[a,b]$, let ν be such that $x_{\nu+1}-x_\nu=\bar{\Delta}$. Let $I_\nu=[x_\nu,x_{\nu+1}]$. We now subdivide I_ν into $m+1$ equal subintervals

$$I_{\nu,i}=\left[x_\nu^i,x_\nu^{i+1}\right], \qquad i=0,1,\dots,m,$$

where, in general, $x_\nu^i=x_\nu+i\bar{\Delta}/(m+1)$. Define

$$F(x)=\begin{cases} (-1)^i\bar{\Delta}^m B_{m+1}^*\left(\dfrac{2(x-x_\nu^i)(m+1)}{\bar{\Delta}}-1\right), & x\in I_{\nu,i}, \qquad i=0,1,\dots,m \\ 0, & \text{otherwise,} \end{cases} \tag{10.54}$$

where B_{m+1}^* is the perfect (polynomial) B-spline of order $m+1$ defined on $[-1,1]$. (See Theorem 4.34.) By the properties of this B-spline, it follows that $F\in C^{m-1}[a,b]$ and

$$\|D^mF\|_{L_p[a,b]}=2^{2m-1}m!(m+1)^m\bar{\Delta}^{1/p}. \tag{10.55}$$

This shows that $F \in L_p^m[a,b]$. The perfect B-spline is normalized such that $B_{m+1}^*(0) \geqslant 1/2$, and thus by the construction of F, we also know that

$$F\left[x_\nu + \frac{(i+1/2)}{(m+1)}\right] \geqslant \frac{\bar{\Delta}^m}{2}, \qquad i=0,1,\ldots,m. \tag{10.56}$$

Now, Theorem 10.5 asserts that if $\bar{\Delta}$ is sufficiently small, then N_L is spanned by an ECT-system on I_ν. But then by the well-known alternation theorem of Tchebycheff (see Remark 7.5), it follows that zero is the best approximation of F, and thus

$$d(F, N_L)_{L_\infty[I_\nu]} = \|F\|_{L_\infty[I_\nu]} \geqslant \frac{\bar{\Delta}^m}{2}.$$

On the other hand, using Theorem 10.1 and (10.55) we see that

$$\|LF\|_{L_p[a,b]} \leqslant C\bar{\Delta}^{1/p} \tag{10.57}$$

and (10.53) follows. ∎

To further illustrate how lower bounds companion to the upper bounds of Section 10.5 can be established, we now prove a version of Theorem 10.26 in which the distance is measured in the L_q-norm, $1 \leqslant q < \infty$. Our approach will be the same as before: we attempt to construct a function F lying in $L_p^m[a,b]$ such that on some interval $I_\nu = [x_\nu, x_{\nu+1}]$ of length $\bar{\Delta}$ the best approximation of F from N_L is given by zero. The only essential problem is that we no longer have the Tchebycheff alternation theorem at our disposal to characterize best approximations. We must replace it by the orthogonality condition (see Remark 7.6)

$$\int_{x_\nu}^{x_{\nu+1}} |F(t)|^{q-1} \operatorname{sgn}[F(t)] u_i(t)\,dt = 0, \qquad i=1,2,\ldots,m. \tag{10.58}$$

The following lemma is a useful tool in constructing such functions:

LEMMA 10.27

Let $\{u_i\}_1^m$ be a set of m linearly independent functions in $L_1[c,d]$, and let θ and φ belong to $C[0,1]$. Then there exist points $0 = t_0 \leqslant t_1 \leqslant \cdots \leqslant t_{m+1} = 1$ and signs $\varepsilon_1, \ldots, \varepsilon_m$ such that the function

$$G(t) = \left\{ \varepsilon_i \theta(t_{i+1} - t_i)\varphi\left(\frac{t - t_i}{t_{i+1} - t_i}\right), \qquad t_i \leqslant t < t_{i+1}, \qquad i = 0,1,\ldots,m \right. \tag{10.59}$$

satisfies

$$\int_c^d G(t)u_i(t)\,dt=0, \qquad i=1,2,\ldots,m. \tag{10.60}$$

Proof. Let $S=\{\xi=(\xi_0,\xi_1,\ldots,\xi_m)\in\mathbf{R}^{m+1}: \Sigma_{i=0}^m|\xi_i|=d-c\}$. For each $\xi\in S$, define $\tau_i=c+|\xi_0|+|\xi_1|+\cdots+|\xi_{i-1}|$, $i=1,2,\ldots,m$, and

$$G_\xi(t)=\begin{cases}\operatorname{sgn}(\xi_0)\theta(|\xi_0|)\varphi\left(\dfrac{t-c}{|\xi_0|}\right), & c\leqslant t<\tau_1\\ \cdots & \\ \operatorname{sgn}(\xi_m)\theta(|\xi_m|)\varphi\left(\dfrac{t-\tau_m}{|\xi_m|}\right), & \tau_m\leqslant t\leqslant d.\end{cases}$$

Now define a function $\psi: S\mapsto\mathbf{R}^m\subseteq\mathbf{R}^{m+1}$ by

$$[\psi(\xi)]_i=\int_c^d G_\xi(t)u_i(t)\,dt, \qquad i=1,2,\ldots,m.$$

Clearly ψ is continuous on S and is odd; that is, $\psi(\xi)=-\psi(-\xi)$. Since S is the boundary of an open, bounded, symmetric set in $\mathbf{R}^{m+1}$ while $\psi(S)$ is contained in the proper subspace $\mathbf{R}^m$ of $\mathbf{R}^{m+1}$, it follows from Corollary 3.29 in the text by Schwartz [1969] that ψ must take on the value zero for $\xi^*\in S$. Then we may take $t_i=\tau_i^*$, $i=1,2,\ldots,m$. ∎

We are ready to give a companion for (10.49) in the case of $1\leqslant q<\infty$ and $j=0$.

THEOREM 10.28

Fix $1\leqslant p\leqslant\infty$, $1\leqslant q<\infty$. Then for any partition Δ of $[a,b]$ there exists a function $F\in L_p^m[a,b]$ such that

$$d[F,\mathcal{P}\mathcal{W}(N_L;\Delta)]_{L_q[a,b]}\geqslant C\bar{\Delta}^{m+1/q-1/p}\|LF\|_{L_p[a,b]}. \tag{10.61}$$

Proof. Let ν be such that $\bar{\Delta}=x_{\nu+1}-x_\nu$, and let $\theta(t)=t^{m(q-1)}$ and $\varphi(t)=[B_{m+1}^*(2t-1)]^{q-1}$, where B_{m+1}^* is the perfect B-spline of order $m+1$ defined on $[-1,1]$ in Theorem 4.34. By Lemma 10.27 there exist $x_\nu=t_0\leqslant t_1\leqslant\cdots\leqslant t_{m+1}=x_{\nu+1}$ and signs $\varepsilon_1,\ldots,\varepsilon_m$ such that the function

$$F(t)=\begin{cases}\varepsilon_j(t_{j+1}-t_j)^m B_{m+1}^*\left[2\left(\dfrac{t-t_j}{t_{j+1}-t_j}\right)-1\right], & t_j\leqslant t<t_{j+1},\\ & j=0,1,\ldots,m\\ 0, & \text{otherwise}\end{cases} \tag{10.62}$$

satisfies (10.58). This implies that zero is the best approximation of F from $\mathcal{PW}(N_L;\Delta)$; that is, $d[F,\mathcal{PW}(N_L;\Delta)]_{L_q[a,b]}=\|F\|_{L_q[a,b]}$. Since at least one of the subintervals $I_{\nu j}=[t_j,t_{j+1}]$ is of length at least $\bar{\Delta}/(m+1)$, we find that

$$d[F,\mathcal{PW}(N_L;\Delta)]_{L_q[a,b]}\geqslant\left[\int_{t_j}^{t_{j+1}}|F(t)|^q\,dt\right]^{1/q}$$

$$\geqslant\left(\frac{\bar{\Delta}}{m+1}\right)^{m+1/q}2^{-1/q}\|B_{m+1}^*\|_{L_q[-1,1]}.$$

Combining this with (10.57) in the same way as in the proof of Theorem 10.26 leads to (10.61). ■

§ 10.7. INVERSE THEOREMS AND SATURATION

In this section we discuss inverse, saturation, and characterization theorems for approximation by L-splines. Since the results given here closely resemble those given in Sections 6.8 and 6.9 for polynomial splines, we will keep the proofs short, emphasizing only the major changes required.

The key to obtaining inverse theorems for L-splines is to have an appropriate way of measuring the smoothness of a function. Let L be a linear differential operator as in (10.1), and let $1\leqslant p<\infty$. Then we define the *K-functional associated with L* by

$$K_{L,p}(t)f=\inf_{g\in L_p^m[a,b]}\left(\|f-g\|_{L_p[a,b]}+t^m\|Lg\|_{L_p[a,b]}\right). \tag{10.63}$$

Clearly $K_{L,p}(t)$ is a nonlinear functional defined on the space $L_p[a,b]$. We define a similar functional on $C[a,b]$ by choosing $p=\infty$ and replacing $L_p^m[a,b]$ by $C^m[a,b]$ in (10.63). It is clear that $K_{L,p}(t)$ reduces to the K-functional $K_{m,p}(t)$ introduced in Definition 2.64 if we take $L=D^m$. The following theorem shows that $K_{L,p}(t)$ is closely related to $K_{m,p}(t)$:

THEOREM 10.29

There exist positive constants C_1 and C_2 (depending only on L and $[a,b]$) such that

$$K_{L,p}(t)f\leqslant C_1\left(K_{m,p}(t)f+t^m\|f\|_{L_p[a,b]}\right) \tag{10.64}$$

$$K_{m,p}(t)f\leqslant C_2\left(K_{L,p}(t)f+t^m\|f\|_{L_p[a,b]}\right) \tag{10.65}$$

for all $f\in L_p[a,b]$, $1\leqslant p<\infty$, and for all $f\in C[a,b]$, $p=\infty$.

Proof. We simply apply the inequalities (10.2) and (10.3). ■

It is clear from the definition of $K_{L,p}(t)$ that it possesses properties similar to those listed in Theorem 2.65 for $K_{m,p}(t)$. Because of its importance for saturation theorems, we explicitly prove the analog of (2.138).

THEOREM 10.30

Suppose $1 \leqslant p < \infty$ and that $f \in L_p[a,b]$ is such that

$$\liminf_{t \to 0} t^{-m} K_{L,p}(t) f = 0. \tag{10.66}$$

Then there exists $g \in N_L$ so that $f = g$ almost everywhere. Similarly, if $f \in C[a,b]$ and (10.66) holds with $p = \infty$, then $f \in N_L$.

Proof. For any $g \in L_p^m[a,b]$ and any $0 < t \leqslant 1$,

$$K_{L,p}(1) f \leqslant t^{-m} \{ \|f - g\|_p + t^m \|Lg\|_p \} \leqslant t^{-m} K_{L,p}(t) f.$$

We conclude that $K_{L,p}(1) f = 0$, and thus there exists a sequence $g_1, g_2, \ldots$ of functions in $L_p^m[a,b]$ with

$$\lim_{n \to \infty} \|f - g_n\|_p = \lim_{n \to \infty} \|Lg_n\|_p = 0. \tag{10.67}$$

But $L_p^m[a,b]$ with the norm

$$\|g\|_L = \|g\|_{L_p[a,b]} + \|Lg\|_{L_p[a,b]}$$

is a Banach space, and thus (10.67) implies that $g_1, g_2, \ldots$ is a Cauchy sequence in $L_p^m[a,b]$. Hence it must converge to some $g \in L_p^m[a,b]$, which by (10.67) must satisfy $\|Lg\|_p = 0$ and $\|f - g\|_p = 0$. Thus $f = g$ almost everywhere, and $g \in N_L$ as was to be shown. ■

In preparation for our development of inverse theorems for approximation by L-splines, we now give an estimate for the smoothness (measured in terms of the K-functional) of a given spline in the space $\mathcal{PW}(N_L; \Delta)$ defined in (10.52).

THEOREM 10.31

Let Δ be a partition of $[a,b]$, and suppose $0 < \varepsilon \leqslant \underline{\Delta}$. Then for any $s \in \mathcal{PW}(N_L; \Delta)$,

$$K_{L,\infty}(\varepsilon) s \leqslant C_1 \sum_{j=0}^{m-1} \varepsilon^j J(D^j s), \tag{10.68}$$

where

$$J(D^j s) = \max_{1 \leq i \leq k} |\text{jump}[D^j s]_{x_i}|.$$

Proof. We define

$$g = \begin{cases} s + p_i \text{ on } J_i = [x_i - \varepsilon/2, x_i], & i = 1, 2, \ldots, k \\ s, & \text{otherwise,} \end{cases} \tag{10.69}$$

where $p_i \in \mathcal{P}_{2m}$ is chosen so that

$$D^j p_i(x_i - \varepsilon/2) = 0, \qquad D^j p_i(x_i) = \text{jump}[D^j s]_{x_i}, \qquad j = 0, 1, \ldots, m-1.$$

By construction $g \in L_\infty^m[a, b]$, and thus

$$\begin{aligned} K_{L,\infty}(\varepsilon)s &\leq \|s - g\|_{L_\infty[a,b]} + \varepsilon^m \|Lg\|_{L_\infty[a,b]} \\ &\leq \max_{1 \leq i \leq k} \|p_i\|_{L_\infty[J_i]} + \varepsilon^m \max_{1 \leq i \leq k} \|Lp_i\|_{L_\infty[J_i]} \\ &\leq \max_{1 \leq i \leq k} \left(\|p_i\|_{L_\infty[J_i]} + \varepsilon^m \sum_{\nu=0}^{m} \|a_\nu\|_{L_\infty[J_i]} \|D^\nu p_i\|_{L_\infty[J_i]} \right). \end{aligned}$$

But by a simple result on the size of Hermite interpolating polynomials (cf. Remark 10.1),

$$\|D^\nu p_i\|_{L_\infty[J_i]} \leq C\varepsilon^{-\nu} \sum_{j=0}^{m-1} \varepsilon^j J(D^j s),$$

and (10.68) follows. ■

Our first inverse result is the analog of Theorem 6.39.

THEOREM 10.32

Let Δ_ν be a sequence of partitions of $[a,b]$ satisfying the mixing condition of Definition 6.37. Then there exists a constant C_1 such that for all $f \in C[a,b]$,

$$K_{L,\infty}(\underline{\Delta}_\nu)f \leq C_1 \sup_{n \geq \nu} d[f, \mathcal{P}\mathcal{W}(N_L; \Delta_\nu)] \tag{10.70}$$

for all sufficiently large ν.

Proof. For each ν, let $s_\nu \in \mathcal{P}\mathcal{W}(N_L; \Delta_\nu)$ be such that

$$\|f - s_\nu\|_\infty \leq 2d[f, \mathcal{P}\mathcal{W}(N_L; \Delta_\nu)]_\infty.$$

Then

$$K_{L,\infty}(\underline{\Delta}_\nu)f \leqslant \|f - s_\nu\|_{L_\infty[a,b]} + K_{L,\infty}(\underline{\Delta}_\nu)s_\nu$$

$$\leqslant 2d[f, \mathcal{PW}(N_L; \underline{\Delta}_\nu)] + K_{L,\infty}(\underline{\Delta}_\nu)s_\nu.$$

Thus it suffices to estimate $K_{L,\infty}(\underline{\Delta}_\nu)s_\nu$. Here we may apply Theorem 10.31 with $\varepsilon = \underline{\Delta}_\nu$. By the mixing condition there exists $\rho > 0$ such that for each $0 \leqslant i \leqslant k_\nu$,

$$x_r^n \leqslant x_i^\nu - \rho\underline{\Delta}_\nu < x_i^\nu + \rho\underline{\Delta}_\nu \leqslant x_{r+1}^n$$

for some $n(i,\nu) \geqslant \nu$ and some $0 \leqslant r \leqslant k_n$. But then using Markov's inequality again, we see that for each $0 \leqslant j \leqslant m-1$,

$$\text{jump}[D^j s_\nu]_{x_i^\nu} \leqslant \|D^j(s_\nu - s_n)\|_{L_\infty[x_r^n, x_{r+1}^n]}$$

$$\leqslant C\underline{\Delta}_\nu^{-j}\|s_\nu - s_n\|_{L_\infty[x_r^n, x_{r+1}^n]} \leqslant 4C\underline{\Delta}_\nu^{-j} \sup_{n \geqslant \nu} d[f, \mathcal{PW}(N_L; \Delta_n)]_\infty.$$

Putting these estimates in (10.68) leads to (10.70). ■

We now give a characterization theorem that relates the order of approximation of a function by a sequence of L-spline spaces $\mathcal{PW}(N_L; \Delta_\nu)$ to its smoothness.

THEOREM 10.33

Let Δ_ν be a sequence of partitions going steadily to zero (cf. Definition 6.5) and satisfying the mixing condition. Suppose ϕ is a monotone-increasing function on $(0,\tau)$ such that $ct^m \leqslant \phi(t)$ for some $c > 0$. Then $f \in C[a,b]$ satisfies

$$d[f, \mathcal{PW}(N_L; \Delta_\nu)]_\infty \leqslant \phi(\overline{\Delta}_\nu) \tag{10.71}$$

if and only if

$$\omega_m(f;t) \leqslant C_1\phi(t), \qquad \text{all } 0 < t < \tau. \tag{10.72}$$

Proof. If f satisfies (10.71), then by Theorem 10.32,

$$K_{L,\infty}(\underline{\Delta}_\nu)f \leqslant C_5\phi(\overline{\Delta}_\nu).$$

But then

$$\omega_m(f;\underline{\Delta}_\nu) \leqslant C_4 K_{m,\infty}(\underline{\Delta}_\nu)f \leqslant C_3[K_{L,\infty}(\underline{\Delta}_\nu)f + \underline{\Delta}_\nu^m\|f\|_\infty]$$

$$\leqslant C_2\phi(\overline{\Delta}_\nu),$$

where we have used Theorem 2.67 for the first inequality, (10.65) for the second, and our assumption on ϕ for the third. Now the same arguments that were used in the proof of Theorem 6.41 can be applied, using the quasi-uniformity of Δ_ν to estimate $\omega_m(f;\overline{\Delta}_\nu)$ in terms of $\omega_m(f;\underline{\Delta}_\nu)$. The converse assertion follows immediately from our direct approximation Theorem 10.25. ■

Theorem 10.33 completely characterizes the approximation order obtainable with L-splines in terms of the smoothness of f. Using it we immediately obtain characterizations of the classical Lipschitz and Zygmund spaces in terms of their approximability by L-splines—cf. Theorem 6.43. Theorem 10.33 applies as long as $\phi(t)$ does not go to zero faster than $\mathcal{O}(t^m)$. The following theorem shows that (except for functions in N_L which are approximated exactly) m is the maximal order of convergence obtainable no matter how smooth f may be. It is the analog of the saturation Theorem 6.42 for polynomial splines.

THEOREM 10.34

Suppose Δ_ν is a sequence of partitions as in Theorem 10.33 and that $f \in C[a,b]$ is such that

$$\liminf_{\nu\to\infty} \overline{\Delta}_\nu^{-m} d[f, \mathcal{PW}(N_L;\Delta_\nu)]_\infty = 0. \tag{10.73}$$

Then $f \in N_L$.

Proof. By Theorem 10.33, (10.73) implies that

$$\liminf_{t\to 0} t^{-m} K_{L,\infty}(t) f = 0,$$

and thus by Theorem 10.30, $f \in N_L$. ■

So far we have been working with the space $\mathcal{PW}(N_L;\Delta)$, as it is the biggest of the L-spline spaces. If we are willing to work with smoother subspaces, it is possible to establish some inverse theorems without the mixing condition on the partitions (cf. the polynomial spline space in Section 6.8). We now give the following analog of Theorem 6.46:

THEOREM 10.35

Let Δ_ν be a sequence of partitions of $[a,b]$ with $\underline{\Delta}_\nu \downarrow 0$ and $\Delta_0 = \{a,b\}$. For each ν let $\mathcal{S}_\nu$ be a linear space of L-splines contained in $\mathcal{PW}(N_L;\Delta_\nu) \cap C^l[a,b]$. Given $f \in C[a,b]$, let $\varepsilon_\nu = d(f,\mathcal{S}_\nu)_\infty$. Then

$$K_{L,\infty}(\underline{\Delta}_\nu) f \leq C_1 \underline{\Delta}_\nu^{l+1} \sum_{r=1}^{\nu} \frac{\varepsilon_r + \varepsilon_{r-1}}{\underline{\Delta}_r^{l+1}}. \tag{10.74}$$

Proof. By Theorem 10.31, for any Δ and any $0<\varepsilon\leqslant\underline{\Delta}$, if $s\in\mathcal{PW}(N_L;\Delta)\cap C^l[a,b]$, then

$$K_{L,\infty}(\varepsilon)s\leqslant C_4\sum_{j=l+1}^{m-1}\varepsilon^j J(D^js), \tag{10.75}$$

where

$$J(D^js)=\max_{1\leqslant i\leqslant k}\text{jump}\left[D^js\right]_{x_i}.$$

This is the analog of (6.86). Now, by exactly the same arguments used to prove (6.87) substituting the Markov inequality (10.5) for the usual one, we can show

$$J\left(D^js_\nu\right)\leqslant C_3\sum_{r=1}^{\nu}\frac{\varepsilon_r+\varepsilon_{r-1}}{\underline{\Delta}_r^j},\qquad j=l+1,\dots,m-1, \tag{10.76}$$

where $s_\nu\in\mathbb{S}$ are such that $\|f-s_\nu\|\leqslant 2\varepsilon_\nu$. Now

$$K_{L,\infty}(\underline{\Delta}_\nu)f\leqslant K_{L,\infty}(\underline{\Delta}_\nu)(f-s_\nu)+K_{L,\infty}(\underline{\Delta}_\nu)s_\nu.$$

Since $K_{L,\infty}(\underline{\Delta}_\nu)(f-s_\nu)\leqslant C_2\|f-s_\nu\|$, combining these results with the same argument used to prove Theorem 6.46, we obtain (10.74). ■

Theorem 10.35 can now be used to characterize the classical Lipschitz and Zygmund spaces in terms of how well they can be approximated by smooth spaces of L-splines without any mixing condition on the sequence of partitions. In particular, it is clear that by using the same kind of arguments as in the proof of Theorem 10.33, we can show that Theorem 6.47 is valid with the space of piecewise polynomials $\mathcal{PP}_m(\Delta)$ replaced by the space of piecewise L-splines $\mathcal{PW}(N_L;\Delta)$.

As in the polynomial spline case, it is not possible to establish a saturation result for arbitrary functions and arbitrary sequences of partitions. We can do something for smooth functions, however.

THEOREM 10.36

Let Δ_ν be a sequence of partitions as in Theorem 10.33. Suppose $f\in C^m[a,b]$ is a function such that

$$d\left[f,\mathcal{PW}(N_L;\Delta_\nu)\right]_\infty\leqslant C_1\bar{\Delta}_\nu^m\psi\left(\bar{\Delta}_\nu\right), \tag{10.77}$$

where ψ is a monotone-increasing function with $\psi(t)\to 0$ as $t\to 0$. Then $f\in N_L$.

Proof. The proof follows along the same lines as that of Theorem 6.48. Here we must replace ω_m by $K_{L,\infty}$. In addition, we need to find a substitute for the Whitney theorem which was used there. By Theorem 10.9 and the estimate (10.50) there exists $g_i \in N_L$ such that

$$\|f-g_i\|_{L_\infty[I_i]} \leqslant C_1 \varepsilon^m \|Lf\|_{L_\infty[I_i]},$$

where $I_i=(x_i-\varepsilon, x_i+\varepsilon)$, $i=1,2,\ldots,k$. Now the remainder of the proof proceeds as before using the Markov inequality for N_L. ■

We devote the remainder of this section to inverse and saturation results for the case where the distance is measured in the p-norm, $1 \leqslant p < \infty$. Our first result is the analog of Theorem 6.49 (cf. Theorem 10.31 for the case of $p=\infty$).

THEOREM 10.37

Let Δ be a partition of the interval $[a,b]$, and suppose $0<\varepsilon \leqslant \underline{\Delta}$. Then for any spline $s \in \mathcal{P}\mathcal{W}(N_L;\Delta)$,

$$K_{L,p}(\varepsilon)s \leqslant C_1 \varepsilon^{1/p} \sum_{j=0}^{m-1} \varepsilon^j J_p(D^j s), \tag{10.78}$$

where

$$J_p(D^j s) = \left(\sum_{i=1}^{k} |\text{jump}\left[D^j s \right]_{x_i}|^p \right)^{1/p}.$$

Proof. The proof is very similar to the proof of Theorem 10.31. Indeed, with g as in (10.69), we have

$$K_{L,p}(\varepsilon)s \leqslant \|s-g\|_{L_p[a,b]} + \varepsilon^m \|Lg\|_{L_p[a,b]}$$

$$\leqslant \left(\sum_{i=1}^{k} \|p_i\|_{L_p[J_i]}^p \right)^{1/p} + \varepsilon^m \left(\sum_{i=1}^{k} \|Lp_i\|_{L_p[J_i]}^p \right)^{1/p}.$$

Now $\|p_i\|_{L_p[J_i]} \leqslant \varepsilon^{1/p} \|p_i\|_{L_\infty[J_i]}$, and the remainder of the proof proceeds as before. ■

We can now give an inverse result for the spaces $\mathcal{P}\mathcal{W}(N_L;\Delta_\nu)$ assuming that Δ_ν satisfies an appropriate p-mixing condition.

THEOREM 10.38

Let Δ_ν be a sequence of quasi-uniform partitions satisfying the p-mixing condition of Definition 6.50. Then for every $f \in L_p[a,b]$,

$$K_{L,p}(\overline{\Delta}_\nu) \leqslant C_1 \sup_{n \geqslant \nu} d[f, \mathcal{P}\mathcal{U}(N_L;\Delta_n)]_p. \tag{10.79}$$

Proof. The proof follows the same lines as the proof of Theorem 6.52 for polynomial splines. Here we replace $\omega_m(f;\overline{\Delta}_\nu)_p$ by $K_{L,p}(\overline{\Delta}_\nu)f$ and use Theorem 10.37 instead of Theorem 6.49. In the estimate of $J_p(D^j s)$ we use the Markov inequality (10.5) for N_L. ■

It is now clear that Theorem 10.38 can be used to characterize the classical Lipschitz and Zygmund spaces in terms of their approximability by L-splines. Indeed, by using the same argument that was used in the proof of Theorem 10.33 to transform results about $K_{L,p}(t)f$ into results about $\omega_m(f;t)_p$, we immediately obtain Theorem 6.54 [minus the assertion (6.110)] with $\mathcal{P}\mathcal{P}_m(\Delta_\nu)$ replaced by $\mathcal{P}\mathcal{U}(N_L;\Delta_\nu)$. The saturation assertion (6.110) of Theorem 6.54 is replaced by the following result:

THEOREM 10.39

Suppose Δ_ν is a sequence of partitions as in Theorem 10.38 and that $f \in L_p[a,b]$ is such that

$$\liminf_{\nu \to \infty} \overline{\Delta}_\nu^{-m} d[f, \mathcal{P}\mathcal{U}(N_L;\Delta_\nu)]_p = 0. \tag{10.80}$$

Then there exists $g \in N_L$ such that $f = g$ almost everywhere.

Proof. Theorem 10.38 implies that

$$\liminf_{t \to 0} t^{-m} K_{L,p}(t)f = 0,$$

and the result follows from Theorem 10.30. ■

We close this section with two results for the nonmixed case. First we have the following analog of Theorems 6.55 and 10.35:

THEOREM 10.40

Let Δ_ν be a sequence of partitions of $[a,b]$ with $\underline{\Delta}_\nu \downarrow 0$ and $\Delta_0 = \{a,b\}$. For each ν let $\mathcal{S}_\nu$ be a linear space of spline functions contained in

$\mathcal{PW}(N_L;\Delta_\nu)\cap C^l[a,b]$. Given $f\in L_p[a,b]$, let $\varepsilon_\nu = d(f,\mathcal{S}_\nu)_p$. Then

$$K_{L,p}(\underline{\Delta}_\nu)f \leq C_1 \underline{\Delta}_\nu^{l+1+1/p} \sum_{r=1}^{\nu} \frac{\varepsilon_r+\varepsilon_{r-1}}{\underline{\Delta}_r^{l+1+1/p}}. \tag{10.81}$$

Proof. The proof parallels that of Theorem 6.55, with the only changes being that it is now based on Theorem 10.37 and that the Markov inequality for N_L must be substituted for the usual one. ■

The inverse result of Theorem 10.40 can now be translated directly into a characterization for certain classical smooth spaces in terms of how well they can be approximated by L-splines with some smoothness. The result is essentially identical to Theorem 6.56. Our last result is a saturation theorem.

THEOREM 10.41

Let Δ_ν be a sequence of partitions with $\bar{\Delta}_\nu \downarrow 0$ and $\Delta_0=\{a,b\}$. Suppose $f\in L_p^m[a,b]$ is such that

$$d[f,\mathcal{PW}(N_L;\Delta_\nu)]_p \leq C_1 \bar{\Delta}_\nu^m \psi(\bar{\Delta}_\nu), \tag{10.82}$$

where ψ is monotone increasing and $\psi(t)\to 0$ as $t\to 0$. Then $f\in N_L$.

Proof. We may follow the proof of Theorem 6.57. Here we use Theorem 10.9 and the estimate (10.50) to construct $g_i \in N_L$ such that

$$\|f-g_i\|_{L_p[I_i]} \leq C_2 \varepsilon^m \|Lf\|_{L_p[I_i]},$$

where $I_i=[x_i^\nu-\varepsilon, x_i^\nu+\varepsilon]$, $i=1,2,\ldots,k$. Arguing as before leads to the assertion that $K_{L,p}(\varepsilon)f=o(\varepsilon^m)$, and the result follows from Theorem 10.30. ■

§ 10.8. TRIGONOMETRIC SPLINES

In this section we consider a special kind of L-spline of considerable importance in applications. Given $m=2r$ let

$$\{v_1,\ldots,v_m\}=\{\cos(x/2),\sin(x/2),\ldots,\cos[(r-1/2)x],\sin[(r-1/2)x]\}, \tag{10.83}$$

Similarly, if $m=2r+1$, let

$$\{u_1,\ldots,u_m\}=\{1,\cos(x),\sin(x),\ldots,\cos(rx),\sin(rx)\} \tag{10.84}$$

and define the m-dimensional space $\mathcal{T}_m$ by

$$\mathcal{T}_m = \begin{cases} \operatorname{span}\{v_1,\ldots,v_m\}, & m=2r \\ \operatorname{span}\{u_1,\ldots,u_m\}, & m=2r+1. \end{cases} \tag{10.85}$$

We now introduce the space of splines of interest in this section. Suppose $\Delta=\{a=x_0<x_1<\cdots<x_{k+1}=b\}$ is a partition of the interval $[a,b]$, and suppose $\mathfrak{M}=(m_1,\ldots,m_k)$ is a vector of integers with $1\leqslant m_i\leqslant m$, $i=1,2,\ldots,k$. Then we define the space of *trigonometric splines of order m with knots $x_1,\ldots,x_k$ of multiplicities $m_1,\ldots,m_k$* by

$$\mathcal{S}(\mathcal{T}_m;\mathfrak{M};\Delta)=\left\{\begin{array}{l} s\text{: there exist } s_0,\ldots,s_k\in\mathcal{T}_m \text{ with } s(x) \\ =s_i(x) \quad \text{for} \quad x\in I_i=[x_i,x_{i+1}), \quad i= \\ 0,1,\ldots,k, \quad \text{and} \quad D^{j-1}s_{i-1}(x_i)= \\ D^{j-1}s_i(x_i), \quad j=1,2,\ldots,m-m_i, \quad i= \\ 1,2,\ldots,k \end{array}\right\}. \tag{10.86}$$

Since the space $\mathcal{T}_m$ is clearly the null space of the linear differential operator

$$L_m=\begin{cases} \left(D^2+(r-1/2)^2\right)\cdots\left(D^2+(1/2)^2\right), & m=2r \\ (D^2+r^2)\cdots(D^2+1)D, & m=2r+1, \end{cases} \tag{10.87}$$

we see that the space of trigonometric splines is in fact a space of L-splines. We can apply the theory of the preceding sections to derive some properties of trigonometric splines. We begin by introducing the Green's function associated with L_m.

THEOREM 10.42

Let

$$G_L(x;y)=G_m(x;y)=(x-y)_+^0\frac{2^{m-1}}{(m-1)!}\left[\sin\left(\frac{x-y}{2}\right)\right]^{m-1}. \tag{10.88}$$

Then

$$D_x^jG_m(x;y)|_{y=x}=\delta_{j,m-1}, \qquad j=0,1,\ldots,m-1. \tag{10.89}$$

Moreover, for $x \geqslant y$,

$$G_m(x;y)=\frac{1}{(m-1)!}\begin{cases} 2\sum_{\nu=1}^{r}(-1)^{\nu-1}\binom{m-1}{r+\nu-1}\sin\left[\left(\frac{2\nu-1}{2}\right)(x-y)\right], & m=2r \\ \binom{m-1}{r}+2\sum_{\nu=1}^{r}(-1)^{\nu}\binom{m-1}{r+\nu}\cos[\nu(x-y)], & m=2r+1. \end{cases} \tag{10.90}$$

and G_m is the Green's function associated with $L=L_m$.

Proof. To prove (10.89), we note that

$$D_xG_m(x;y)=\frac{2^{m-2}}{(m-2)!}\left[\sin\left(\frac{x-y}{2}\right)\right]^{m-2}\cos\left(\frac{x-y}{2}\right). \tag{10.91}$$

The assertion follows. To prove (10.90), we apply known trigonometric identities to $\sin^{m-1}$. Since

$$\sin[j(x-y)]=\sin(jx)\sin(jy)-\cos(jx)\sin(jy)$$

$$\cos[j(x-y)]=\cos(jx)\cos(jy)-\sin(jx)\sin(jy),$$

it follows that G_m belongs to $\mathcal{T}_m$ for $x\geqslant y$, and by Definition 10.6, G_m is the Green's function associated with $L=L_m$. ■

Before proceeding with further properties of trigonometric splines, we observe that (10.90) can be used to identify the adjunct functions associated with $\{u_i\}_1^m$ and $\{v_i\}_1^m$. We have $u_1^*(y)=(m-1 \atop r)/(m-1)!$ and

$$\left.\begin{aligned} v_{2\nu-1}^*(y)&=\frac{2(-1)^{\nu}}{(m-1)!}\binom{m-1}{r+\nu-1}\sin\left(\frac{2\nu-1}{2}\right)y \\ v_{2\nu}^*(y)&=\frac{2(-1)^{\nu-1}}{(m-1)!}\binom{m-1}{r+\nu-1}\cos\left(\frac{2\nu-1}{2}\right)y \end{aligned}\right\} \quad m=2r$$

$$\left.\begin{aligned} u_{2\nu}^*(y)&=\frac{2(-1)^{\nu}}{(m-1)!}\binom{m-1}{r+\nu}\cos(\nu y) \\ u_{2\nu+1}^*(y)&=\frac{2(-1)^{\nu}}{(m-1)!}\binom{m-1}{r+\nu}\sin(\nu y), \end{aligned}\right\} \quad m=2r+1$$

for $\nu=1,2,\ldots,r$. This shows that the space $\mathfrak{U}^*$ spanned by the adjunct functions [which is the null space of $L_m=(-1)^mL_m$] is in fact $\mathcal{T}_m$.

The results of Section 10.3 show that $\mathbb{S}(\mathfrak{T}_m; \mathfrak{M}; \Delta)$ is a linear space of dimension $m+K$ with $K=\sum_{i=1}^{k} m_i$. Using the Green's function G_m of (10.88), we can now apply Theorem 10.15 to construct a one-sided basis for $\mathbb{S}$. The methods of Section 10.3 or 10.4 can now be used to derive a basis of local support splines for $\mathbb{S}$. Here there is some advantage to following Section 10.3, as in this special case it leads to B-splines for which a recurrence relation can be established. In order to apply the results of Section 10.3, we must show that the space $\mathfrak{U}^* = \text{span}\{u_1^*,\ldots,u_m^*\} = \mathfrak{T}_m$ is an ET-space on sufficiently small intervals.

LEMMA 10.43

$\mathfrak{T}_m$ is an ET-space on any interval I of length $|I| < 2\pi$.

Proof. This fact follows from Theorem 2.33 if we can show that

$$Z_I(u) \leqslant m-1, \qquad \text{all nontrivial } u \in \mathfrak{T}_m, \tag{10.92}$$

where Z counts multiple zeros. But this assertion about the zeros of a trigonometric polynomial is a result from classical analysis. ■

Lemma 10.43 can also be established directly by looking at the determinants formed from the functions $\{u_i\}_1^m$ or $\{v_i\}_1^m$; see Remark 10.2. It is easy to see that $|I| < 2\pi$ is sufficient to guarantee that $\mathfrak{T}_m$ is a Tchebycheff system on the interval I. We do not need the ECT-property in the following construction of trigonometric B-splines.

Let $t_1 < t_2 < \cdots < t_{m+1}$ with $t_{m+1} - t_1 < 2\pi$. Then we define the mth *order trigonometric divided difference* of a function f over $t_1,\ldots,t_{m+1}$ by

$$[t_1,\ldots,t_{m+1}]_T f = \begin{cases} \dfrac{4^{-r} D\begin{pmatrix} t_1,\ldots,t_{m+1} \\ u_1,\ldots,u_m, f \end{pmatrix}}{D\begin{pmatrix} t_1,\ldots,t_{m+1} \\ v_1,\ldots,v_{m+1} \end{pmatrix}}, & m = 2r+1 \\[3ex] \dfrac{4^{-r} D\begin{pmatrix} t_1,\ldots,t_{m+1} \\ v_1,\ldots,v_m, f \end{pmatrix}}{D\begin{pmatrix} t_1,\ldots,t_{m+1} \\ u_1,\ldots,u_{m+1} \end{pmatrix}}, & m = 2r. \end{cases} \tag{10.93}$$

The following theorem collects some elementary properties of these divided differences:

THEOREM 10.44

For any $t_1 < t_2 < \cdots < t_{m+1}$ with $t_{m+1} - t_1 < 2\pi$,

$$[t_1, \ldots, t_{m+1}]_T f = \sum_{j=1}^{m+1} \frac{f(t_j)}{\prod_{\substack{i=1 \\ i \neq j}}^{m+1} \sin\left(\frac{t_j - t_i}{2}\right)}. \tag{10.94}$$

More generally, if

$$t_1 \leqslant t_2 \leqslant \cdots \leqslant t_{m+1} = \overbrace{\tau_1, \ldots, \tau_1}^{l_1}, \ldots, \overbrace{\tau_d, \ldots, \tau_d}^{l_d} \tag{10.95}$$

with $\tau_1 < \tau_2 < \cdots < \tau_d$, then

$$[t_1, \ldots, t_{m+1}]_T f = \sum_{i=1}^{d} \sum_{j=1}^{l_i} \alpha_{ij} D^{j-1} f(\tau_i) \tag{10.96}$$

with $\alpha_{il_i} \neq 0$, $i = 1, 2, \ldots, d$. Finally, if u is any element of $\mathfrak{T}_m$, then

$$[t_1, \ldots, t_{m+1}]_T u = 0. \tag{10.97}$$

Proof. The identity (10.94) can be obtained by expanding the determinants in the numerator and denominator of (10.93) using Laplace's expansion and the explicit formulae given in Remark 10.2 for determinants formed from the u's and v's. The formula (10.96) also follows from Laplace's expansion, while $\alpha_{il_i} \neq 0$ is due to the fact that it is a quotient of two determinants formed from the u's and v's which by Lemma 10.43 form an ET-system. Property (10.97) is immediate from the definition (10.93) since when $u \in \mathfrak{T}_m$, then the last column in the numerator is a linear combination of the previous columns. ■

We are now ready to introduce the trigonometric B-splines. Suppose

$$\cdots \leqslant y_{-1} \leqslant y_0 \leqslant y_1 \leqslant y_2 \leqslant \cdots$$

is a sequence of real numbers. Then for any i with $y_i < y_{i+m}$, we define the *trigonometric B-spline* by

$$Q_i^m(x) = [y_i, y_{i+1}, \ldots, y_{i+m}]_T \left[\sin\left(\frac{x-y}{2}\right)\right]_+^{m-1}. \tag{10.98}$$

If $y_i = y_{i+m}$, we define Q_i^m to be identically zero. For later use we observe that

$$Q_i^1(x) = \begin{cases} \dfrac{1}{\sin((y_{i+1}-y_i)/2)}, & y_i \leqslant x < y_{i+1} \\ 0, & \text{otherwise.} \end{cases} \tag{10.99}$$

The following theorem shows that the trigonometric B-splines have properties similar to those of the ordinary polynomial B-splines:

THEOREM 10.45

Suppose

$$y_i \leqslant y_{i+1} \leqslant \cdots \leqslant y_{i+m} = \overbrace{\tau_1,\ldots,\tau_1}^{l_1},\ldots,\overbrace{\tau_d,\ldots,\tau_d}^{l_d}. \tag{10.100}$$

Then Q_i^m as defined in (10.98) is a trigonometric spline with knots at the points $\tau_1,\ldots,\tau_d$ of multiplicities $l_1,\ldots,l_d$. Moreover,

$$Q_i^m(x) > 0 \qquad \text{for } y_i < x < y_{i+m}, \tag{10.101}$$

and

$$Q_i^m(x) = 0 \qquad \text{for } x < y_i \text{ and } x > y_{i+m}. \tag{10.102}$$

Proof. The fact that $Q_i(x)$ is a trigonometric spline follows from the expansion (10.96) combined with the properties of the Green's function given in Theorem 10.42. In particular, these properties combine to show that $Q_i(x)$ has the correct piecewise structure, and that it ties together with the desired smoothness at each knot. The fact that $Q_i(x)$ vanishes for $x < y_i$ is clear from the one-sided nature of $G_m(x;\, y)$. For $x > y_{i+m}$, $Q_i(x)$ is the trigonometric divided difference of a function in $\mathfrak{T}_m$, hence it vanishes also. The positivity assertion (10.101) follows inductively from a recursion relation to be given below. ■

Our next theorem gives an important recurrence relation for computing trigonometric B-splines of order m from trigonometric B-splines of order $m-1$:

THEOREM 10.46

Let $y_i \leqslant y_{i+1} \leqslant \cdots \leqslant y_{i+m}$ with $y_i < y_{i+m} < y_i + 2\pi$. Then

$$Q_i^m(x) = \frac{\sin\left(\dfrac{x-y_i}{2}\right)Q_i^{m-1}(x) + \sin\left(\dfrac{y_{i+m}-x}{2}\right)Q_{i+1}^{m-1}(x)}{\sin\left(\dfrac{y_{i+m}-y_i}{2}\right)}. \tag{10.103}$$

Proof. The proof works through exponential divided differences. See Lyche and Winther [1977]. ■

It is also of interest to have a formula for the derivative of a trigonometric B-spline.

THEOREM 10.47

Let $y_i,\ldots,y_{i+m}$ be as in Theorem 10.46, and suppose $m \geqslant 2$. Then

$$DQ_i^m(x)=\left(\frac{m-1}{2}\right)\frac{\cos\left(\frac{x-y_i}{2}\right)Q_i^{m-1}(x)-\cos\left(\frac{y_{i+m}-x}{2}\right)Q_{i+1}^{m-1}(x)}{\sin\left(\frac{y_{i+m}-y_i}{2}\right)}. \tag{10.104}$$

Proof. See Lyche and Winther [1977]. ■

Our next theorem is a Peano representation for the trigonometric divided difference.

THEOREM 10.48

For all f such that $L_m f \in L_1[y_i,y_{i+m}]$,

$$[y_i,\ldots,y_{i+m}]_T f=\frac{2^{m-1}}{(m-1)!}\int_{y_i}^{y_{i+m}} Q_i^m(y)L_m f(y)\,dy. \tag{10.105}$$

Proof. We simply apply the trigonometric divided difference to the Taylor expansion formula

$$f(y)=u_f(y)+\int G_m(x;y)L_m f(x)\,dx, \tag{10.106}$$

where u_f is the unique element of $\mathfrak{T}_m$ such that

$$D^j u_f(y_i)=D^j f(y_i), \qquad j=0,1,\ldots,m-1.$$ ■

It is also possible to give a trigonometric version of Marsden's identity (cf. Theorem 4.21). First we need to normalize our B-splines. Let

$$N_i^m(x)=\sin\left(\frac{y_{i+m}-y_i}{2}\right)Q_i^m(x). \tag{10.107}$$

THEOREM 10.49

Let $l \leqslant r$ and $y_l < y_{r+1}$. Then for any $y \in \mathbf{R}$

$$\sin\left(\frac{y-x}{2}\right)^{m-1} = \sum_{i=l+1-m}^{r} \varphi_{i,m}(y) N_i^m(x), \qquad \text{all } y_l \leqslant x < y_{r+1}, \quad (10.108)$$

where

$$\varphi_{i,m}(y) = \prod_{\nu=1}^{m-1} \sin\left(\frac{y-y_{i+\nu}}{2}\right).$$

Proof. See Lyche and Winther [1977]. ■

The result (10.108) can now be used to obtain the expansions of the functions in $\mathfrak{T}_m$ in terms of trigonometric B-splines. It is also possible to give the dual basis for the normalized B-splines explicitly.

We close this section by noting that it is also possible to define spaces of periodic trigonometric splines as well as spaces of natural trigonometric splines; compare the discussion for L-splines.

§ 10.9. HISTORICAL NOTES

Section 10.1

The theory of ordinary differential equations is treated in a wealth of classical books, but some of the results quoted here are often not included.

Section 10.2

The Green's function plays a major role in the theory of ordinary differential equations. The notion of adjunct functions can be found in the text by Ince [1944]. Theorem 10.10 should be part of the classical theory but we couldn't find it anywhere. It is proved in Jerome and Schumaker [1976].

Section 10.3

We discuss the historical development of generalized splines in the notes for Section 11.1. Early contributors to the theory of L-splines included Greville [1964b], Ahlberg, Nilson, and Walsh [1964], and Schultz and Varga [1967]. These papers concentrated on the natural L-splines which arise as solutions of appropriate best interpolation problems. Our development of a one-sided basis is based on ideas developed in the articles by Jerome and Schumaker [1971, 1976] for g-splines (cf. Section 8.3).

Section 10.4

Local support L-splines were first constructed by Jerome [1973a]. The fact that one can use the Tchebycheffian B-splines when $\overline{\Delta}$ is sufficiently small was observed by Scherer and Schumaker [1980]. The construction of the dual basis given in Theorem 10.19 is also taken from this paper, as are Theorems 10.20 to 10.23.

Section 10.5

The approximation power of generalized splines was investigated by several authors. The first results were obtained via interpolating splines, see, for example, Ahlberg, Nilson, and Walsh [1967b] and Schultz and Varga [1967]; they were then refined in a long series of later papers. As these results were derived for natural L-splines, they do not apply to the general class of L-splines considered here. The first error bounds for approximation by general classes of L-splines are credited to Jerome [1973a]; see also Johnen and Scherer [1976]. The approach taken here is attributed to Scherer and Schumaker [1980].

Section 10.6

Lower bounds for approximation by generalized splines were given in the paper by Jerome and Schumaker [1974] using the theory of n-widths. Direct proofs of the type used here were later given by Schumaker [1978]. The generalization of the Hobby-Rice theorem given in Lemma 10.27 is also taken from this paper. The proof is based on an idea of Pinkus [1976].

Section 10.7

The results of this section are based on the work of DeVore and Richards [1973a, b], Johnen and Scherer [1976], and Scherer [1976, 1977].

Section 10.8

Trigonometric splines were first considered by Schoenberg [1964b]. He considered the case where m is odd, as well as natural trigonometric splines. The introduction of the trigonometric splines with m even is credited to Lyche and Winther [1977]. Most of the results of this section come from this paper.

§ 10.10 REMARKS

Remark 10.1

The following simple fact about Hermite interpolating polynomials is proved in the article by Swartz and Varga [1972]:

LEMMA 10.50

Let $p \in \mathcal{P}_{2m}$ be a polynomial such that

$$D^j p(0) = a_j, \qquad D^j p(h) = b_j, \qquad j = 0, 1, \ldots, m-1.$$

Then

$$\|D^j p\|_{L_p[0,h]} \leqslant C h^{-j+1/p} \sum_{\nu=0}^{m-1} h^{\nu} \max(|a_\nu|, |b_\nu|).$$

Remark 10.2

If $m = 2r+1$ and $t_1 < t_2 < \cdots < t_m$, then the functions $\{u_i\}_1^m$ defined in (10.84) satisfy

$$D\begin{pmatrix} t_1, \ldots, t_m \\ u_1, \ldots, u_m \end{pmatrix} = 4^{r^2} \prod_{1 \leqslant j < i \leqslant m} \sin\left(\frac{t_i - t_j}{2}\right). \tag{10.109}$$

Similarly, if $m = 2r$, then the $\{v_i\}_1^m$ in (10.84) satisfy

$$D\begin{pmatrix} t_1, \ldots, t_m \\ v_1, \ldots, v_m \end{pmatrix} = 4^{r^2 - r} \prod_{1 \leqslant j < i \leqslant m} \sin\left(\frac{t_i - t_j}{2}\right). \tag{10.110}$$

These formulae can be established by relating them to certain determinants formed exponentials.

11
GENERALIZED SPLINES

In the previous two chapters we have studied several spaces of generalized splines where a large part of the theory of polynomial splines can be carried over. In this chapter we shall look at some more general spaces of splines where we can still say quite a bit about dimension, one-sided bases, local bases, zeros and sign changes. In addition, we introduce a nonlinear space of splines and look briefly at rational and complex splines.

§ 11.1. A GENERAL SPACE OF SPLINES

Let Ω be a partially ordered set, and suppose

$$\Delta=\{x_1<x_2<\cdots<x_k\}$$

is a set of distinct elements in Ω. The set Δ partitions Ω into $k+1$ "intervals"

$$\begin{aligned} I_0&=\{x\in\Omega:\ x<x_1\};\\ I_i&=\{x\in\Omega:\ x_i\leqslant x<x_{i+1}\}, \qquad i=1,2,\ldots,k-1;\\ I_k&=\{x\in\Omega:\ x_k\leqslant x\}. \end{aligned} \tag{11.1}$$

Suppose

$$\mathfrak{U}_i=\operatorname{span}\{u_j^i\}_{j=1}^{n_i} \tag{11.2}$$

are finite dimensional linear spaces of functions defined on I_i, $i=0,1,\ldots,k$. We begin by defining the analog of the space of piecewise polynomials:

$$\mathcal{P}\mathfrak{W}(\mathfrak{U}_0,\ldots,\mathfrak{U}_k;\Delta)=\left\{\begin{array}{l} s\text{: there exist } s_i\in\mathfrak{U}_i \text{ with}\\ s=s_i \text{ on } I_i,\ i=0,1,\ldots,k. \end{array}\right\} \tag{11.3}$$

In order to define a generalized space of splines, we now introduce some linear functionals to be used in forcing the pieces of $\mathcal{P}\mathcal{W}$ to tie together appropriately. Suppose

$$\Gamma = \{\Gamma_{ij}: 0 \leq i < j \leq k\}, \tag{11.4}$$

where

$$\Gamma_{ij} = \{(\underline{\gamma}_\nu^{ij}, \bar{\gamma}_\nu^{ij})\}_{\nu=1}^{r_{ij}}, \qquad 0 \leq i < j \leq k, \tag{11.5}$$

where $\{\underline{\gamma}_\nu^{ij}\}_{\nu=1}^{r_{ij}}$ and $\{\bar{\gamma}_\nu^{ij}\}_{\nu=1}^{r_{ij}}$ are sets of *linear* functionals defined on the spaces $\mathcal{U}_i$ and $\mathcal{U}_j$, respectively.

DEFINITION 11.1

Let

$$\mathcal{S}(\mathcal{U}_0, \ldots, \mathcal{U}_k; \Gamma; \Delta) = \left\{ \begin{array}{l} s \in \mathcal{P}\mathcal{W}(\mathcal{U}_0, \ldots, \mathcal{U}_k; \Delta) \text{ such that} \\ \underline{\gamma}_\nu^{ij} s_i = \bar{\gamma}_\nu^{ij} s_j,\ \nu = 1, 2, \ldots, r_{ij},\ 0 \leq i < j \leq k. \end{array} \right\}$$

We call $\mathcal{S}$ the space of *generalized splines relative to* $\mathcal{U}_0, \ldots, \mathcal{U}_k$, Γ, *and* Δ.

It is clear that $\mathcal{S}$ is a linear space. It consists of all real-valued functions defined on Ω such that in each subinterval I_i, s is a member of $\mathcal{U}_i$ with some additional conditions on how the pieces are tied together. In particular, we require that the pieces in the ith and jth intervals tie together in such a way that a certain linear functional operating on the ith piece has the same value as another (possibly different) linear functional operating on the jth piece.

Definition 11.1 carries the idea of a linear space of splines about as far as it can go. The interval has been replaced by a general partially ordered set; the polynomials have been replaced by (possibly different) general linear spaces in each subinterval; and continuity of derivatives at the knots has been replaced by the matching of general linear functionals.

Before giving several examples to illustrate how the flexibility of Definition 11.1 can be used to describe some rather unusual kinds of spline spaces, we introduce one more piece of notation. In the case where the ties between the pieces are described by continuity of successive derivatives, it will be convenient to use a special notation. In particular, if $\mathcal{R} = (r_1, \ldots, r_k)$ is a vector of positive integers, then we write

$$\mathcal{S}(\mathcal{U}_0, \ldots, \mathcal{U}_k; \mathcal{R}; \Delta) = \left\{ \begin{array}{l} s \in \mathcal{P}\mathcal{W}(\mathcal{U}_0, \ldots, \mathcal{U}_k; \Delta) \text{ such that} \\ D^{j-1} s_{i-1}(x_i) = D^{j-1} s_i(x_i), j = 1, \ldots, r_i,\ i = 1, 2, \ldots, k. \end{array} \right\} \tag{11.6}$$

Definition 11.1 clearly includes all the spaces of splines considered in the previous chapters. It should be noted, however, that the role of $\mathcal{R}$ in this case is in a certain sense the dual of the role of $\mathfrak{M}$ in previous spline spaces. The discrete splines provide an example where the set Ω is not an ordinary interval. In Section 11.7 we introduce some spaces of splines where Ω is a curve in the plane. The Tchebycheffian and L-splines considered in Chapters 9 and 10 provide examples of splines where the pieces are drawn from nonpolynomial spaces. The following example shows how the structure of these pieces can vary from interval to interval:

EXAMPLE 11.2

Let $\Omega=[-1,2]$ and $\Delta=\{0,1\}$. Let $\mathfrak{U}_0=\mathcal{P}_2$, $\mathfrak{U}_1=\text{span}\{e^x\}$, and $\mathfrak{U}_2=\text{span}\{\cos(x),\sin(x)\}$. Suppose $\mathcal{R}=(1,1)$.

Discussion. In this case the space of splines $\mathbb{S}$ defined in (11.6) consists of functions that are linear polynomials in the interval $[-1,0)$, exponential in $[0,1)$, and a linear combination of $\cos(x)$ and $\sin(x)$ in $[1,2]$. The tie conditions require that any spline s in $\mathbb{S}$ must be a continuous function on $[-1,2]$. Some typical elements in $\mathbb{S}$ are shown in Figure 35. ■

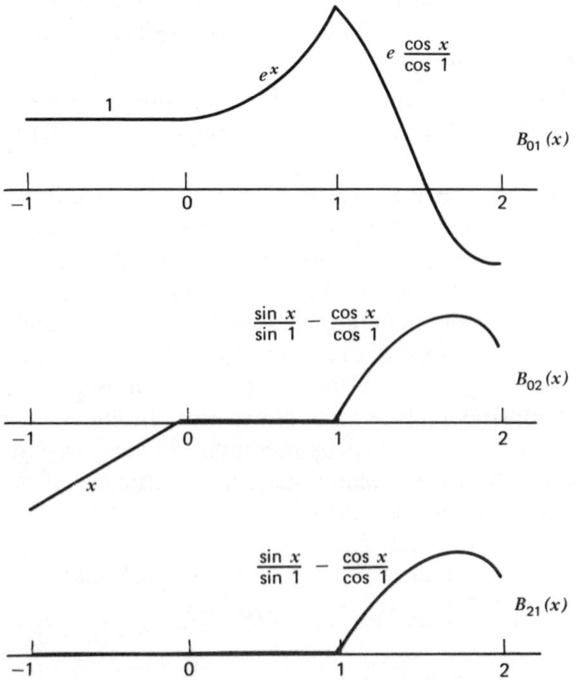

Figure 35. A one-sided basis for the splines in Example 11.2.

The g-splines discussed in Section 8.3 provide an example of where the ties between the pieces of the splines are defined by general linear functionals. We now give a rather different kind of example.

EXAMPLE 11.3

Let $\Omega=[a,b]$, and suppose $\mathcal{U}_i=\mathcal{P}_4$, $i=0,1,\ldots,k$. Let $\Gamma=\{\Gamma_{i-1,i}\}_{i=1}^k$ where

$$\Gamma_{i-1,i}=\{(e_{x_{i-1}-},e_{x_{i-1}+}),(e_{x_i-},e_{x_i+}),(e_{x_{i+1}-},e_{x_{i+1}+})\}$$
$$i=1,2,\ldots,k.$$

Discussion. In this case the space $\mathcal{S}$ of splines consists of piecewise cubic polynomials, where the ties between successive pieces take place at several points rather than at one. In particular, for each $i=0,1,\ldots,k-1$, the pieces s_i and s_{i+1} have to agree at the points x_i, x_{i+1} and x_{i+2}. ∎

We begin our study of the linear space $\mathcal{S}$ by identifying its dimension under some natural conditions on the linear functionals describing Γ which assure that they are compatible with each other.

THEOREM 11.4

Suppose that for each $j=1,2,\ldots,k$, the matrix

$$A_j^+=\begin{bmatrix} A_{0j}^+ \\ A_{1j}^+ \\ \vdots \\ A_{j-1,j}^+ \end{bmatrix} \text{ with } A_{ij}^+=\begin{bmatrix} \bar{\gamma}_1^{ij}u_1^j & \cdots & \bar{\gamma}_1^{ij}u_{n_j}^j \\ \bar{\gamma}_2^{ij}u_1^j & \cdots & \bar{\gamma}_2^{ij}u_{n_j}^j \\ \cdots & & \\ \bar{\gamma}_{r_{ij}}^{ij}u_1^j & \cdots & \bar{\gamma}_{r_{ij}}^{ij}u_{n_j}^j \end{bmatrix} \tag{11.7}$$

is of full rank $r_j=r_{0j}+\cdots+r_{j-1,j}$. Then

$$\dim\mathcal{S}=n_0+\sum_{j=1}^k(n_j-r_j).$$

Proof. The condition on the matrices $A_0^+,\ldots,A_k^+$ can be fulfilled only when $r_j\leqslant n_j$, $j=1,2,\ldots,k$. It is clear that $\mathcal{S}$ is a linear space, and that each $s\in\mathcal{S}$ can be written in the form

$$s(x)=\left\{ s_j(x)=\sum_{p=1}^{n_j}c_{jp}u_p^j(x) \text{ for } x\in I_j,\quad j=0,1,\ldots,k.\right.$$

The conditions tying the pieces together can be written as a linear system of equations on the coefficients of s. Indeed, if we write $\mathbf{c}_j=(c_{j1},\ldots,c_{jn_j})^T$

for the coefficient vector of the jth piece, and define matrices A_{ij}^- as in (11.7) (but using $\underline{\gamma}_\nu^{ij}$ instead of $\bar{\gamma}_\nu^{ij}$), then the required system of equations has the form

$$\begin{bmatrix} A_{01}^- & -A_{01}^+ & 0 & \cdots & & 0 \\ A_{02}^- & 0 & A_{02}^+ & \cdots & & 0 \\ \cdots & & & & & \\ A_{0k}^- & 0 & 0 & \cdots & & -A_{0k}^+ \\ \cdots & & & & & \\ 0 & 0 & 0 & \cdots & A_{k-1,k}^- & -A_{k-1,k}^+ \end{bmatrix} \begin{bmatrix} \mathbf{c}_0 \\ \mathbf{c}_1 \\ \\ \\ \\ \mathbf{c}_k \end{bmatrix} = 0.$$

The matrix of this system has $\Sigma_{j=0}^k n_j$ columns and $\Sigma_{j=1}^k r_j$ rows. As it is easily seen that the hypotheses guarantee that it is of full rank (equal to the number of rows), it follows that the dimensionality of its null space is $\Sigma_0^k n_j - \Sigma_1^k r_j$. This is, of course, also the dimensionality of $\mathcal{S}$. ■

§ 11.2. A ONE-SIDED BASIS

In this section we construct a one-sided basis for the space $\mathcal{S}$, assuming that the hypotheses of Theorem 11.4 are satisfied. Our method of approach will be to mimic the construction of the one-sided basis for the polynomial splines (cf. Theorem 4.5). In particular, we shall first construct $m_0 = n_0$ elements in $\mathcal{S}$ by extending the functions $u_1^0, \ldots, u_{n_0}^0$ to all of Ω. Then associated with each knot x_i, we shall construct $m_i = n_i - r_i$ splines which vanish identically for $x < x_i$, $i = 1, 2, \ldots, k$. The following lemma shows how to construct the first m_0 basis splines:

LEMMA 11.5

There exist splines $B_{01}, \ldots, B_{0m_0}$ in $\mathcal{S}$ so that

$$B_{0j}(x) = u_j^0(x) \qquad \text{for } x \in I_0, \qquad j = 1, 2, \ldots, m_0 = n_0.$$

Proof. Fix $1 \leqslant j \leqslant m_0$. We define B_{0j} to be equal to u_j^0 on I_0. Now to extend this function to $I_1, I_2, \ldots$, and so on, we need to choose coefficients α_{pq}^{0j} so that

$$B_{0j}(x) = \sum_{q=1}^{n_p} \alpha_{pq}^{0j} u_q^p(x) \qquad \text{for } x \in I_p, \qquad p = 0, 1, \ldots, k.$$

To assure that the pieces of B_{0j} in I_0 and I_1 are properly tied together, we should choose the coefficients $\alpha_{11}^{0j},\ldots,\alpha_{1n_1}^{0j}$ so that

$$A_{01}^{+}\begin{bmatrix}\alpha_{11}^{0j}\\ \vdots\\ \alpha_{1n_1}^{0j}\end{bmatrix}=\begin{bmatrix}\underline{\gamma}_1^{01}u_j^0\\ \vdots\\ \underline{\gamma}_{r_{01}}^{01}u_j^0\end{bmatrix}.$$

Since this is a system of $r_1 \leqslant n_1$ equations (of full rank), it can always be solved. Clearly, this process can be continued to extend B_{0j} into the intervals $I_2,\ldots,I_k$. ■

We now construct the m_i splines associated with each knot x_i for our one-sided basis.

LEMMA 11.6

For each $i=1,2,\ldots,k$ there exist splines $B_{i1},\ldots,B_{im_i}$ which vanish identically for $x<x_i$, and which are linearly independent on I_i.

Proof. Fix $1\leqslant i\leqslant k$ and $1\leqslant j\leqslant m_i$. We show how to construct B_{ij}. For $x<x_i$, we define $B_{ij}(x)=0$. To define B_{ij} on the interval I_i, we need to find coefficients $\{\alpha_{iq}^{ij}\}_{q=1}^{n_i}$ so that

$$B_{ij}(x)=\sum_{q=1}^{n_i}\alpha_{iq}^{ij}u_q^i(x) \tag{11.8}$$

ties properly onto the function 0 at x_i. To this end, let $\tilde{A}$ be any n_i by n_i matrix obtained from A_i^+ [cf. (11.7)] by adding m_i rows to it. Since A_i^+ is of full rank, we may assume $\tilde{A}$ has been constructed to be nonsingular. Now we choose the coefficients $\{\alpha_{iq}^{ij}\}$ as the solution of the system

$$\tilde{A}\begin{bmatrix}\alpha_{i1}^{ij}\\ \vdots\\ \alpha_{in_i}^{ij}\end{bmatrix}=\delta(n_i,r_i+j),$$

where $\delta(n_i,r_i+j)$ is a vector of length n_i with all zero components except for the r_i+jth component, which we take to be 1. Now B_{ij} is defined on I_i by (11.8). Using the technique of Lemma 11.5, we can now extend B_{ij} to the intervals $I_{i+1},\ldots,I_k$ to obtain a spline lying in the space $\mathcal{S}$.

It remains to check that the $B_{i1},\ldots,B_{im_i}$ are linearly independent on I_i. To establish this, we shall construct a dual basis for these functions.

Suppose that the νth additional row of $\tilde{A}$ is given by $(\tilde{A}_{r_i+\nu,1},\ldots,\tilde{A}_{r_i+\nu,n_i})$, $\nu=1,2,\ldots,m_i$. Then we define linear functionals $\lambda_1^i,\ldots,\lambda_{m_i}^i$ on $\mathfrak{U}_i$ by

$$\lambda_\nu^i u=\lambda_\nu^i\left(\sum_{p=1}^{n_i}\alpha_p u_p^i\right)=\sum_{p=1}^{n_i}\alpha_p\tilde{A}_{r_i+\nu,p}.$$

By the construction of B_{ij} on I_i, it is clear that $\{\lambda_\nu^i\}_1^{m_i}$ form a dual basis for $\{B_{ij}\}_{j=1}^{m_i}$; that is,

$$\lambda_\nu^i B_{ij}=\delta_{\nu j},\qquad \nu,j=1,2,\ldots,m_i.$$

Now we can easily prove the linear independence of the $B_{i1},\ldots,B_{im_i}$ on I_i. Indeed, if $\sum_{j=1}^{m_i}c_jB_{ij}=0$, then

$$c_\nu=\lambda_\nu^i\left(\sum_{j=1}^{m_i}c_jB_{ij}\right)=0,\qquad \nu=1,2,\ldots,m_i. \qquad \blacksquare$$

It is clear that there is some arbitrariness in the construction of the basis elements in Lemmas 11.5 and 11.6; that is, this construction of one-sided basis elements is not uniquely determined. The following theorem shows that in any case we have constructed a basis for $\mathcal{S}$:

THEOREM 11.7

The set of splines $\{B_{ij}\}_{j=1,i=0}^{m_i\ \ k}$ defined in Lemmas 11.5 and 11.6 form a basis for $\mathcal{S}$.

Proof. Our proof follows the idea of the proof of Theorem 4.5 in the polynomial spline case. First, it is clear by construction that each of the splines B_{ij} belongs to the space $\mathcal{S}$. By Theorem 11.4, we know that the dimension of this space is $m_0+m_1\cdots+m_k$. Thus to complete the proof, we need only check that the functions are linearly independent on Ω. Suppose

$$\sum_{i=0}^{k}\sum_{j=1}^{m_i}c_{ij}B_{ij}(x)=0\qquad\text{for all } x\in\Omega.$$

We need to show that all these coefficients must be zero. First, restricting attention to the interval I_0, we see that $c_{01}B_{01}+\cdots+c_{0m_0}B_{0m_0}=0$ there. But since these functions are assumed to be linearly independent, it follows that the corresponding c's are zero. Now consider x in the interval I_1. We now have $c_{11}B_{11}+\cdots+c_{1m_1}B_{1m_1}=0$ on this interval. By the linear independence of $B_{11},\ldots,B_{1m_1}$ on this interval, we find that these coefficients are also zero. This process can be continued, moving one interval to the right at a time, and the theorem is proved. ■

To illustrate the construction of the one-sided basis in Theorem 11.7, we give several examples.

EXAMPLE 11.8

Let $\mathcal{S}$ be the class of splines discussed in Example 11.2. Construct a one-sided basis.

Discussion. We compute the dimension of this space of splines to be 3. A basis of one-sided splines for $\mathcal{S}$ can be constructed by the process of Lemmas 11.5 and 11.6. For example, we may take the three functions shown in Figure 35. ■

The construction of the one-sided basis is somewhat simplified when all the spaces $\mathcal{U}_i$ are the same.

EXAMPLE 11.9

Construct a one-sided basis for the space of splines defined in Example 11.3.

Discussion. By Theorem 11.4, we find that this space of splines is of dimension $k+4$. As a basis for $\mathcal{S}$, we may take $B_{01}, B_{02}, B_{03}, B_{04}$ to be the powers $1, x, x^2, x^3$, and the remaining basis elements to be the one-sided splines

$$B_{i1}(x) = \begin{cases} 0, & a \leqslant x < x_i, \\ (x - x_{i-1})(x - x_i)(x - x_{i+1}), & x_i \leqslant x \leqslant b. \end{cases}$$

$i = 1, 2, \ldots, k$. A typical one-sided spline of this type is shown in Figure 36. ■

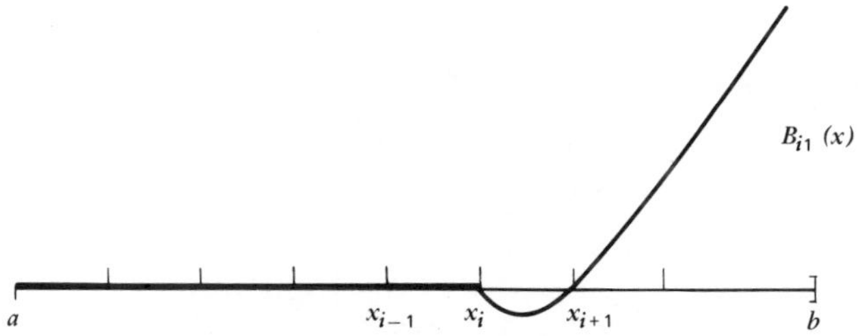

Figure 36. A one-sided spline from Example 11.9.

We have already seen examples of one-sided bases for g-splines in Section 8.3 and for discrete splines in Section 8.5. A one-sided basis for the trigonometric splines was constructed in Section 10.8.

§ 11.3. CONSTRUCTING A LOCAL BASIS

As observed in Chapter 4, the one-sided bases are usually not satisfactory in applications, and it is desirable to construct spline bases with small supports, if possible. In this section we give some general results that are useful for this purpose.

Throughout this section we suppose that the hypotheses of Theorem 11.4 are satisfied. As in Section 4.2, we attempt to construct local support splines as linear combinations of one-sided splines. While it is possible to give results in the general case, the notation is considerably simplified if we restrict our attention to the case where all of the spaces $\mathcal{U}_i$ are the same, and where

$$\Gamma=\{\Gamma_{01},\Gamma_{12},\ldots,\Gamma_{k-1,k}\} \tag{11.9}$$

(so that only adjoining pieces are forced to satisfy some tying condition). In the interest of saving subscripts, we write $m=m_0=n_0=n_1=\cdots=n_k$, and assume

$$\mathcal{U}_i=\operatorname{span}\{u_j\}_1^m,\qquad \text{all } i=0,1,\ldots,k, \tag{11.10}$$

and

$$\Gamma_{i-1,i}=\{(\underline{\gamma}_\nu^i,\bar{\gamma}_\nu^i)\}_{\nu=1}^{r_i},\qquad i=1,2,\ldots,k. \tag{11.11}$$

Suppose now that $\{B_{ij}\}_{j=1,i=0}^{m_i\;\;k}$ is the one-sided basis for $\mathcal{S}$ constructed in Theorem 11.7. It is convenient to renumber these splines with a single subscript. We number them in lexicographical order as

$$\rho_1,\ldots,\rho_{m+K}=B_{01},\ldots,B_{0m_0},B_{11},\ldots,B_{1m_1},\ldots,B_{k1},\ldots,B_{km_k}, \tag{11.12}$$

where $K=\sum_{i=1}^k m_i$. If we introduce the notation

$$y_{m+1}\leqslant y_{m+2}\leqslant\cdots\leqslant y_{m+K}=\overbrace{x_1,\ldots,x_1}^{m_1},\ldots,\overbrace{x_k,\ldots,x_k}^{m_k}, \tag{11.13}$$

then it follows that each of the splines $\rho_i(x)$ vanishes identically for all $x<y_i$, $i=m+1,\ldots,m+K$. In general, we can write

$$\rho_i(x)=\sum_{j=1}^m C_{ij}u_j(x),\qquad i=1,2,\ldots,m, \tag{11.14}$$

and

$$\rho_i(x)=\begin{cases} 0, & x<y_i \\ \sum_{j=1}^{m} C_{ij}u_j(x), & x\geqslant y_i \end{cases}, \tag{11.15}$$

$i=m+1,\ldots,m+K$ for some appropriate coefficients $\{C_{ij}\}$. The one-sided basis in this case is completely described by the matrix $C=(C_{ji})$.

The following lemma is a rather general result on when local support splines can be constructed as linear combinations of one-sided splines. It is the analog of Lemma 8.32.

LEMMA 11.10

Suppose $1\leqslant i_1<i_2<\cdots<i_r\leqslant m+K$, and suppose $\boldsymbol{\delta}=(\delta_1,\ldots,\delta_r)^T$ is a solution of the homogeneous system

$$C\langle i_1,\ldots,i_r\rangle\boldsymbol{\delta}=0, \tag{11.16}$$

where C is the matrix describing the one-sided basis $\{\rho_j\}_1^{m+K}$, and where $C\langle i_1,\ldots,i_r\rangle$ denotes the submatrix of C obtained by taking only columns $i_1,\ldots,i_r$. Then

$$B(x)=\sum_{j=1}^{r}\delta_j\rho_{i_j}(x) \tag{11.17}$$

is a spline that vanishes for $x<y_{i_1}$ (if $i_1>m$), and for $x>y_{i_r}$.

Proof. When $i_1>m$, each of the ρ's appearing in the sum (11.17) vanishes for $x<y_{i_1}$, and it is clear that B has the same property. On the other hand,

$$\rho_{ij}(x)=[u_1(x),\ldots,u_m(x)]C\langle i_j\rangle \qquad \text{for } x>y_{i_r},$$

hence for $x>y_{i_r}$,

$$B(x)=[u_1(x),\ldots,u_m(x)]C\langle i_1,\ldots,i_r\rangle\boldsymbol{\delta}=0. \qquad ■$$

Lemma 11.10 can be used to construct local support splines as linear combinations of one-sided splines. It remains to see whether enough linearly independent local support splines can be constructed in this way to form a basis for $\mathcal{S}$. We have already seen in Example 8.31 that this cannot always be done, even in the case of polynomial g-splines. In this connection we may apply the general algebraic result of Lemma 8.33. It asserts that if $\{\beta_\nu=(\beta_{\nu 1},\ldots,\beta_{\nu,m+K})\}_{\nu=1}^q$ is a set of q linearly independent vectors, then the splines

$$B_\nu=\sum_{j=1}^{m+K}\beta_{\nu j}\rho_j,\qquad \nu=1,2,\ldots,q$$

are linearly independent. Thus the problem reduces to finding $q=m+K$ linearly independent β_ν's to serve as the coefficients of our local support splines.

As in the polynomial g-spline case, the possibility of choosing a set of $m+K$ linearly independent β_ν's which at the same time produce splines with small supports depends heavily on the structure of the matrix C describing the one-sided splines. Theorem 8.34 provides one solution to this problem. It gives a set of conditions on the matrix C that guarantee the existence of a basis for $\mathbb{S}$ which consists of splines with fairly small supports.

§ 11.4. SIGN CHANGES AND WEAK TCHEBYCHEFF SYSTEMS

In this section we obtain bounds on the number of sign changes or zeros a generalized spline can have. As a by-product we will show that a wide class of generalized splines forms a Weak-Tchebycheff space. Conversely, we shall show that most Weak-Tchebycheff spaces are actually spaces of generalized splines.

Throughout this section we assume that $\Omega=[a,b]$ and that $\Delta=\{a=x_0<x_1<\cdots<x_{k+1}=b\}$ is a partition of it into subintervals $I_0,\ldots,I_k$. It is clear that in order to make any assertions about the sign changes or zeros of a generalized spline defined on Ω, we will have to make some kind of assumptions about the nature of the linear spaces $\mathfrak{U}_0,\ldots,\mathfrak{U}_k$ from which the spline pieces are drawn. Our first result deals with the space of generalized piecewise polynomials defined in (11.3).

THEOREM 11.11

Suppose that for each $i=0,1,\ldots,k$, $\mathfrak{U}_i=\{u_{ij}\}_{j=1}^{n_i}$ is a T-system on the subinterval I_i. Then

$$d:=\dim\mathfrak{P}\mathfrak{W}(\mathfrak{U}_0,\ldots,\mathfrak{U}_k;\Delta)=\sum_{i=0}^{k}n_i.$$

Moreover, with $J=\cup_{i=0}^{k}(x_i,x_{i+1})$,

$$Z_J^{\text{sep}}(s)\leqslant d-1-k,\qquad \text{all nontrivial } s\in\mathfrak{P}\mathfrak{W},\tag{11.18}$$

where Z_J^{sep} counts the separated zeros of s on J [cf. (2.53)]. Finally,

$$S^-(s) \leqslant d-1, \qquad \text{all nontrivial } s \in \mathcal{P}\mathcal{W}, \tag{11.19}$$

and thus $\mathcal{P}\mathcal{W}$ is a WT-space.

Proof. The dimension of $\mathcal{P}\mathcal{W}$ follows directly from Theorem 11.4. Since $\mathcal{U}_i$ is a T-system, Theorem 2.21 assures that $Z_{(x_i,x_{i+1})}(s) \leqslant n_i - 1$, each $i = 0, 1, \ldots, k$. Summing these inequalities yields (11.18). The reason we have to use Z^{sep} in the statement (11.18) is that s may vanish identically on some of the intervals $I_0, \ldots, I_k$. Now the number of strong sign changes of s is bounded by the number of zeros of s in J, plus a possible sign change at each of the k knots. Thus (11.19) follows from (11.18). The assertion that $\mathcal{P}\mathcal{W}$ is a WT-space now follows from Theorem 2.39. ∎

Theorem 11.11 shows that the space of generalized piecewise polynomials (where the pieces are drawn from T-systems) is a WT-space. The following theorem shows that the same is true of the space of generalized splines obtained from $\mathcal{P}\mathcal{W}$ by enforcing continuity across the knots:

THEOREM 11.12

Let $\mathcal{U}_0, \ldots, \mathcal{U}_k$ be T-systems. Then

$$\mathbb{S}^* = \mathcal{P}\mathcal{W}(\mathcal{U}_0, \ldots, \mathcal{U}_k; \Delta) \cap C[a,b] \tag{11.20}$$

has dimension $d^* = \sum_{i=0}^{k} n_i - k$. It is a WT-space, and

$$S^-_{[a,b]}(s) \leqslant Z^{\text{sep}}_{[a,b]}(s) \leqslant d^* - 1, \qquad \text{all nontrivial } s \in \mathbb{S}^*. \tag{11.21}$$

Proof. For each $i = 0, 1, \ldots, k$ we have $Z_{I_i}^{\text{sep}}(s) \leqslant n_i - 1$ since $\mathcal{U}_i$ is a T-system on I_i. Adding these inequalities together and taking account of the continuity of s, we obtain (11.21). ∎

Theorem 11.12 provides a large class of examples of WT-spaces. Our next theorem is a kind of converse. It shows that if we rule out certain somewhat pathological cases, essentially all WT-spaces in $C[a,b]$ are subspaces of the spline space $\mathbb{S}^*$ defined in (11.20).

THEOREM 11.13

Let $\mathcal{W}$ be a d dimensional WT-space in $C[a,b]$. Suppose all points in $[a,b]$ are *essential with respect to* $\mathcal{W}$; that is

$$\text{for every } x \in [a,b] \text{ there exists some } s \in \mathcal{W} \text{ with } s(x) \neq 0. \tag{11.22}$$

Suppose, in addition, that there exists a $\delta > 0$ such that

$$s(x) = 0 \text{ on } [c,d] \subseteq [a,b] \text{ implies } d-c > \delta. \tag{11.23}$$

Then there exist $a = x_0 < x_1 < \cdots < x_k < x_{k+1} = b$ and T-spaces $\mathfrak{U}_0, \ldots, \mathfrak{U}_k$ such that

$$\mathfrak{W} \subseteq \mathcal{S}^* = \mathcal{P}\mathfrak{W}(\mathfrak{U}_0, \ldots, \mathfrak{U}_k; \Delta) \cap C[a,b]. \tag{11.24}$$

Proof. In view of the hypotheses, we can choose $a = y_0 < y_1 < \cdots y_l < y_{l+1} = b$ so that $s(x) = 0$ on any subinterval of (y_i, y_{i+1}) implies $s(x) \equiv 0$ on all of $I_i = (y_i, y_{i+1})$, $i = 0, 1, \ldots, l$. By Theorem 2.40 each of the finite dimensional linear spaces $\mathcal{V}_i = \mathfrak{W}|_{\bar{I}_i}$ is a WT-space of dimension n_i on $\bar{I}_i = [y_i, y_{i+1}]$. If each of these is actually a T-space, then we have shown that each element of $\mathfrak{W}$ belongs to $\mathcal{S}^*$ with $\Delta = \{y_i\}_1^k$ and $\mathfrak{U}_i = \mathcal{V}_i$, $i = 0, 1, \ldots, k$.

Suppose now that $\mathcal{V}_i$ is not a T-space on $\bar{I}_i$. In this case we have to divide $\bar{I}_i$ into two parts. First, since no nontrivial element $u \in \mathcal{V}_i$ can vanish on a subinterval of $\bar{I}_i$, Theorem 2.47 asserts that

$$Z_{\bar{I}_i}(u) < \infty \qquad \text{for all nontrivial } u \in \mathcal{V}_i.$$

But then Theorem 2.48 implies that $\mathcal{V}_i$ is a T-system on both of the intervals $[y_i, y_{i+1})$ or $(y_i, y_{i+1}]$. It follows that $\mathcal{V}_i$ is a T-system on both of the intervals $[y_i, \bar{y}_i]$ and $[\bar{y}_i, y_{i+1}]$, where $\bar{y}_i = (y_i + y_{i+1})/2$. This process of dividing the intervals $\bar{I}_i$ into two parts, if necessary, leads to a partition $\Delta = \{x_i\}_1^k$ and a set of T-systems $\mathfrak{U}_0, \ldots, \mathfrak{U}_k$ such that $\mathfrak{W} \subseteq \mathcal{S}^*$. ■

To illustrate what can happen, we give several examples.

EXAMPLE 11.14

Let $\mathfrak{W} = \text{span}\{x, x^2\}$ on [0, 1].

Discussion. It is easily checked that $\mathfrak{W}$ is a WT-space. The point 0 is not an essential point with respect to $\mathfrak{W}$, however. Thus no matter how we attempt to divide the interval, we cannot get a T-system on $[0, x_1]$ as $D\binom{0, x_1}{u_1, u_2}$ will always be 0. ■

EXAMPLE 11.15

Let $\mathfrak{W} = \text{span}\{1, u_2\}$, where

$$u_2(x) = \begin{cases} x+1, & -2 \leqslant x \leqslant -1 \\ 0, & -1 \leqslant x \leqslant 1 \\ x-1, & 1 \leqslant x \leqslant 2. \end{cases} \tag{11.25}$$

Discussion. $\mathcal{W}$ is a WT-space on $[-2,2]$. Choosing $x_1=-1$ and $x_2=1$, we see that $\mathcal{W}|_{I_0}=\mathcal{P}_2=\mathcal{W}|_{I_2}$, while $\mathcal{W}|_{I_1}=\mathcal{P}_1$. Thus the space $\mathcal{W}$ is a two-dimensional subspace of the generalized spline space $\mathcal{S}(\mathcal{P}_2,\mathcal{P}_1,\mathcal{P}_2;\Delta)$, which is of dimension three in this case. ■

The following example illustrates the process of dividing the initial set of intervals $\bar{I}_0,\ldots,\bar{I}_l$ into parts, as was done in the proof of Theorem 11.13.

EXAMPLE 11.16

Let $\mathcal{W}=\text{span}\{u_1,u_2\}$, with u_2 as in (11.25) and

$$u_1(x)=\begin{cases} x^2-1, & -1\leqslant x\leqslant 1 \\ 0, & \text{otherwise}. \end{cases}$$

Discussion. Since any $s\in\mathcal{W}$ can have at most one strong sign change, $\mathcal{W}$ is a WT-system on $[-2,2]$. Following the proof of Theorem 11.13, we would take $y_0=-2$, $y_1=-1$, $y_2=1$, and $y_3=2$. Then s cannot vanish on any part of a subinterval without vanishing identically throughout the subinterval. But $\mathcal{W}|_{\bar{I}_1}$ with $\bar{I}_1=[-1,1]$ is not a T-system (as x^2-1 vanishes at both ends). To achieve a partition so that $\mathcal{W}$ restricted to each subinterval is a T-system, we have to add one more knot; for example, the point 0. Thus $\Delta=\{-2,-1,0,1,2\}$ works. ■

We devote the remainder of this section to a linear space of generalized splines where the pieces are tied together by the continuity of consecutive derivatives at the knots.

THEOREM 11.17

Let $[a,b]$ be partitioned into subintervals $I_0,\ldots,I_k$ by $\Delta=\{a=x_0<x_1<\cdots<x_{k+1}=b\}$. Suppose

$$\mathcal{U}_i=\text{span}\{u_{i,j}\}_{j=1}^{n_i} \qquad \text{is an ECT-space on } \bar{I}_i=[x_i,x_{i+1}] \tag{11.26}$$

for each $i=0,1,\ldots,k$. Finally, let $\mathcal{R}=(r_1,\ldots,r_k)$ be a vector of positive integers with

$$r_i\leqslant \min(n_{i-1},n_i), \qquad i=1,2,\ldots,k. \tag{11.27}$$

Then the space $\mathcal{S}(\mathcal{U}_0,\ldots,\mathcal{U}_k;\mathcal{R};\Delta)$ defined in (11.6) is a linear space of dimension

$$d=\sum_{i=0}^{k} n_i-\sum_{i=1}^{k} r_i.$$

Proof. The assumption that the $\mathcal{U}_i$ are ECT-spaces assures us that the required derivatives exist. The assumption (11.27) assures that the hypotheses of Theorem 11.4 hold, and the result follows. ■

$\mathcal{S}$ can be regarded as a space of generalized T-splines (where the nature of the pieces changes from interval to interval). Our main result concerning the space $\mathcal{S}$ is Theorem 11.18 below. It is a kind of Budan-Fourier theorem giving a rather good bound on the number of zeros a spline in $\mathcal{S}$ can have. At the same time we prove that $\mathcal{S}$ is a WT-space. Before stating this result we need some additional notation. We suppose now that for each $i=0,1,\ldots,k$,

$$L_{i,0},\ldots,L_{i,n_i} \tag{11.28}$$

are the differential operators [cf. (9.5)] associated with the canonical basis for the ECT-system $\mathcal{U}_i$ on I_i. Thus $\mathcal{U}_i$ is the null space of L_{i,n_i}.

Suppose now that s is an element of $\mathcal{S}$. Then for each $i=0,1,\ldots,k$ we define

$$\alpha_i(s)=\min\left\{j\colon L_{i,j}s(x)=0, \qquad \text{all } x\in\bar{I}_i\right\}. \tag{11.29}$$

We can think of $\alpha_i(s)$ as giving the exact order of s on I_i, that is, s can be written as a linear combination of the first α_i functions in the canonical ECT-system basis for $\mathcal{U}_i$. The derivative $L_{i,\alpha_i-1}s$ is nonzero throughout I_i. We also need the notation

$$A(s,x_i)=S^+\left[s(x_i+),-L_{i,1}s(x_i+),\ldots,(-1)^{\alpha_i-1}L_{i,\alpha_i-1}s(x_i+)\right];$$

$$B(s,x_i)=S^+\left[s(x_i-),L_{i,1}s(x_i-),\ldots,L_{i,\alpha_{i-1}-1}s(x_i-)\right). \tag{11.30}$$

THEOREM 11.18

Let $\mathcal{S}=\mathcal{S}(\mathcal{U}_0,\ldots,\mathcal{U}_k;\mathcal{R};\Delta)$ be the space of generalized splines of dimension $d=\sum_{i=0}^k n_i-\sum_{i=1}^k r_i$ discussed in Theorem 11.17. Then

$$Z_J^{\text{sep}}(s)\leqslant d-1-A(s,a)-B(s,b), \tag{11.31}$$

where A and B are defined in (11.30), and where $J=\bigcup_{i=0}^k(x_i,x_{i+1})$. The same bound holds for $S_{[a,b]}^-(s)$, and thus $\mathcal{S}$ is a WT-space on $[a,b]$.

Proof. We proceed by induction on k. For $k=0$ we may apply the Budan-Fourier Theorem 9.12 for ECT-spaces to obtain

$$S_{[a,b]}^-(s)\leqslant Z_{(a,b)}^{\text{sep}}(s)\leqslant d-1-A(a)-B(b). \tag{11.32}$$

Suppose now that the assertion has been proved for splines with $k-1$ knots. We now prove (11.31) for splines with k knots. By the induction hypothesis (applied to the spline s restricted to $[a,x_k]$),

$$Z_{\tilde{J}}^{\text{sep}}(s) \leqslant \sum_{i=0}^{k-1} n_i - \sum_{i=1}^{k-1} r_i - 1 - A(a) - B(x_k),$$

where $\tilde{J} = \bigcup_{i=0}^{k-1}(x_i, x_{i+1})$. The Budan-Fourier Theorem 9.12 for ECT-spaces applied to s on (x_k, x_{k+1}) yields

$$Z_{(x_k,x_{k+1})}^{\text{sep}}(s) \leqslant \alpha_k - 1 - A(x_k) - B(b).$$

Adding these inequalities together, we get

$$Z_{J}^{\text{sep}}(s) \leqslant \sum_{i=0}^{k} n_i - \sum_{i=1}^{k-1} r_i - 1 - A(a) - B(b) - \left[1 + A(x_k) + B(x_k) + n_k - \alpha_k\right]. \tag{11.33}$$

The result will follow if we can show that

$$r_k \leqslant 1 + A(x_k) + B(x_k) + n_k - \alpha_k. \tag{11.34}$$

We consider two cases.

CASE 1. Both α_{k-1} and α_k are smaller than r_k. In this case using (2.48) we see that $A(x_k) + B(x_k) \geqslant \alpha_k - 1$, and (11.34) follows since $r_k \leqslant n_k$.

CASE 2. Either α_{k-1} or α_k is at least r_k. In this case $A(x_k) + B(x_k) \geqslant r_k - 1$, and (11.34) follows since $n_k \geqslant \alpha_k$.

We have proved (11.31).

The proof that the same bound holds for $S_{[a,b]}^{-}(s)$ is similar. We have already established it for $k=0$ in (11.32). Combining the bounds on $[a,x_k)$ with the bound on $(x_k,b]$, we now obtain

$$S_{[a,b]}^{-}(s) \leqslant \sum_{i=1}^{k} n_i - \sum_{i=1}^{k-1} r_i - 1 - A(a) - B(b) \tag{11.35}$$

$$- \left[1 + A(x_k) + B(x_k) + n_k - \alpha_k\right]$$

plus a possible sign change at x_k. If there is no sign change at x_k, the result follows from (11.34). On the other hand, if s changes sign at x_k, then it can only do so by jumping through zero, which means that r_k must have been zero. But then since trivially

$$A(x_k) + B(x_k) + n_k - \alpha_k \geqslant 0, \tag{11.36}$$

we can again advance the induction, and the theorem is proved. ■

§ 11.5. A NONLINEAR SPACE OF GENERALIZED SPLINES

In this section we define generalized splines where the pieces are members of various nonlinear spaces of functions, and where the ties are defined by nonlinear functionals. Let Δ be a partition of the partially ordered set Ω into "intervals" $I_0, I_1, \ldots, I_k$ as in (11.1). Suppose $\mathfrak{U}_0, \mathfrak{U}_1, \ldots, \mathfrak{U}_k$ are spaces of "*finitely parametrizable functions*"; that is for each $i = 0, 1, \ldots, k$ suppose

$$\mathfrak{U}_i = \left\{ \begin{array}{l} u(x) = u(\alpha_1, \ldots, \alpha_{n_i} : x)\text{: for each choice of} \\ \alpha = (\alpha_1, \ldots, \alpha_{n_i}) \text{ in some parameter set } H_i \\ \subseteq \mathbf{R}^{n_i},\ u(x) \text{ is an extended real-valued} \\ \text{function on } I_i \end{array} \right\}. \tag{11.37}$$

For each $0 \leqslant i < j \leqslant k$, let

$$\Gamma_{ij} = \left\{ \left(\underline{\gamma}_\nu^{ij}, \bar{\gamma}_\nu^{ij} \right) \right\}_{\nu=1}^{r_{ij}}, \tag{11.38}$$

where $\{\underline{\gamma}_\nu^{ij}\}_{\nu=1}^{r_{ij}}$ and $\{\bar{\gamma}_\nu^{ij}\}_{\nu=1}^{r_{ij}}$ are sets of (possibly nonlinear) functionals defined on $\mathfrak{U}_i$ and $\mathfrak{U}_j$, respectively. Write

$$\Gamma = \{ \Gamma_{ij} \colon 0 \leqslant i < j \leqslant k \}. \tag{11.39}$$

DEFINITION 11.19

We call

$$\mathbb{S}(\mathfrak{U}_0, \ldots, \mathfrak{U}_k; \Gamma; \Delta) = \left\{ \begin{array}{l} s\text{: there exists } s_i \in \mathfrak{U}_i \text{ with } s = s_i \text{ on } I_i, \\ i = 0, 1, \ldots, k \text{ and moreover, } \underline{\gamma}_\nu^{ij} s_i = \bar{\gamma}_\nu^{ij} s_j,\ \nu \\ = 1, 2, \ldots, r_{ij},\ \text{all } 0 \leqslant i < j \leqslant k \end{array} \right\} \tag{11.40}$$

the *space of generalized splines* (*relative to* $\mathfrak{U}_0, \ldots, \mathfrak{U}_k$, Γ, *and* Δ.)

The space of splines $\mathbb{S}$ consists of all (extended) real-valued functions defined on Ω such that in each subinterval I_i, s belongs to $\mathfrak{U}_i$, and such that the pieces in the ith and jth subintervals are tied together in the sense that certain functionals operating on the ith piece must have the same value as certain functionals operating on the jth piece, all $0 \leqslant i < j \leqslant k$.

It is easy to give examples of nonlinear classes of generalized splines. In the following section we discuss one example that has proved useful for applications. Because of the generality of Definition 11.19, it is difficult to say much about the space $\mathbb{S}$ in general. We have worked with finitely parametrizable functions as in (11.37) in order to have functions that are computer compatible (see the discussion in Section 1.5).

§ 11.6. RATIONAL SPLINES

In this section we give a typical example of a class of nonlinear splines. Let $\Delta = \{a = x_0 < x_1 < \cdots < x_k < x_{k+1} = b\}$ be a partition of the interval $[a,b]$. For each $i = 0, 1, \ldots, k$ let

$$\mathcal{U}_i = \left\{ u(x) = a_i + b_i x + \frac{1}{(x - \xi_i)^2} \right\}, \qquad (11.41)$$

where $a_i, b_i \in \mathbf{R}$ and $x_i < \xi_i < x_{i+1}$. Let $\Gamma = \{\Gamma_i\}_{i=1}^{k}$, where

$$\Gamma_i = \{(e_{x_i-}, e_{x_i+})\}, \qquad i = 1, 2, \ldots, k. \qquad (11.42)$$

DEFINITION 11.20

Given $\mathcal{U}_0, \ldots, \mathcal{U}_k$ and Γ as in (11.41) and (11.42), we call

$$\mathcal{S}(\mathcal{U}_0, \ldots, \mathcal{U}_k; \Gamma; \Delta)$$

a space of *rational splines*.

With this definition, rational splines consist of piecewise rational functions such that in each subinterval there is one simple pole. The pieces are tied together in such a way that the spline is continuous at each of the knots. The splines are extended real-valued functions since they take on the value $+\infty$ at each of the poles ξ_i, $i = 0, 1, \ldots, k$. We graph a typical member of this space in Figure 37.

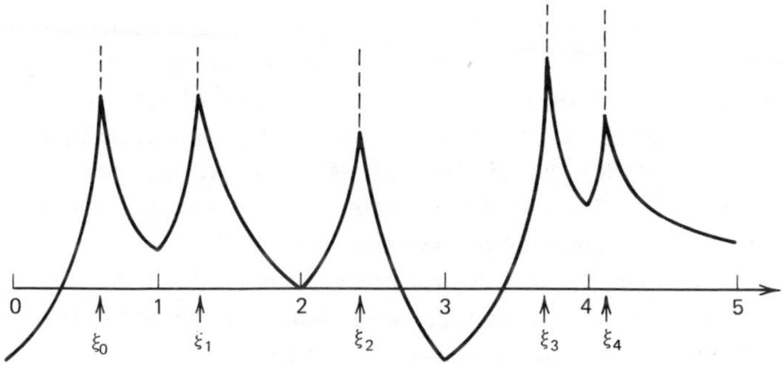

Figure 37. A rational spline.

§ 11.7. COMPLEX AND ANALYTIC SPLINES

So far in this book we have been dealing only with real-valued (or extended real-valued) functions. In this section we show that it is also possible to define a reasonable class of complex-valued splines. Such classes are of interest in the approximation of complex-valued functions on subsets of the complex plane. We begin by defining a complex-valued spline on an interval. To this end we need the space of polynomials of order m (in the real-variable t) with complex coefficients defined by

$$\mathcal{P}_m^{\mathbf{C}} = \left\{ p(t) = \sum_{i=1}^{m} c_i t^{i-1}, \qquad c_1, \ldots, c_m \in \mathbf{C} \right\}, \tag{11.43}$$

where $\mathbf{C}$ is the set of complex numbers.

DEFINITION 11.21

Let $\Delta = \{0 = t_0 < t_1 < \cdots < t_k < t_{k+1} = L\}$ be a partition of the interval $[0, L]$. Let $\mathfrak{M} = (m_1, \ldots, m_k)$ be a vector of integers with $1 \leqslant m_i \leqslant m$, $i = 1, 2, \ldots, k$. Define

$$\mathcal{S}(\mathcal{P}_m^{\mathbf{C}}; \mathfrak{M}; \Delta) = \left\{ \begin{array}{l} s(t)\text{: there exist } s_0, \ldots, s_k \text{ in } \mathcal{P}_m^{\mathbf{C}} \text{ with} \\ s_i = s|_{I_i},\ i = 0, 1, \ldots, k, \text{ and } D^{j-1} s_{i-1}(t_i) = \\ D^{j-1} s_i(t_i),\ j = 1, \ldots, m - m_i,\ i = 1, 2, \ldots, k \end{array} \right\}. \tag{11.44}$$

We call $\mathcal{S}$ the space of *complex splines on* $I = [0, L]$ with *knots at* $t_1, \ldots, t_k$ *of multiplicities* $m_1, \ldots, m_k$.

We emphasize that the complex splines defined above are *complex-valued* functions of a real variable. While this space is defined on an interval $I \subseteq \mathbf{R}$, it is also possible to regard it as a space of functions defined on a curve in $\mathbf{C}$. Indeed, if Ω is a rectifiable Jordan curve in $\mathbf{C}$ of length L, then we can find a function $\alpha(t)$ defined on $[0, L]$ such that $\alpha(t)$ runs over Ω as t runs over $[0, L]$. Then the splines of $\mathcal{S}$ can be regarded as being defined on Ω, and the knots can be thought of as being at the points $P_i = \alpha(t_i)$, $i = 1, 2, \ldots, k$. Then the smoothness requirements involve the directional derivatives at the knots, taken along the curve.

Although complex-valued splines are not covered directly by Theorem 11.4, it is clear that the arguments used there can be applied to establish that the space $\mathcal{S}$ defined in (11.44) is a linear space (over the field of scalars $\mathbf{R}$) of dimension $2(m + K)$, where $K = \sum_{i=1}^{k} m_i$.

THEOREM 11.22

The functions

$$\{t^{j-1}, it^{j-1}\}_{j=1}^{m} \cup \{(t-t_\nu)_+^{m-j}, i(t-t_\nu)_+^{m-j}\}_{j=1,\nu=1}^{m_i \quad k}$$

form a basis for $\mathcal{S}(\mathcal{P}_m^{\mathbf{C}};\mathfrak{M};\Delta)$. (Here i is the complex unit defined by $i=\sqrt{-1}$.)

We now define a space of periodic complex splines on an interval $[0,L]$.

DEFINITION 11.23

We call

$$\mathring{\mathcal{S}}(\mathcal{P}_m^{\mathbf{C}};\mathfrak{M};\Delta)=\{s\in\mathcal{S}(\mathcal{P}_m^{\mathbf{C}};\mathfrak{M};\Delta): s_0=s_k\}, \tag{11.45}$$

where s_0 and s_k are the pieces in the first and last intervals, respectively, the space of *periodic complex splines*.

This space is of dimension $2K$ over the field $\mathbf{R}$. If Ω is a *closed* Jordan rectifiable curve in $\mathbf{C}$ of length L, then we may also regard $\mathring{\mathcal{S}}$ as a space of complex-valued splines defined on Ω. In this case it is of interest to extend the complex-valued splines into the complex plane.

DEFINITION 11.24

Let

$$\mathcal{A}\mathcal{S}_m(\mathfrak{M};\Delta)=\left\{\begin{array}{l} s(z):\ s(z)=\dfrac{1}{2\pi i}\displaystyle\int_0^L \frac{s(\alpha(t))\,d\alpha(t)}{\alpha(t)-z} \quad \text{where} \\ s(t)\in\mathring{\mathcal{S}}(\mathcal{P}_m^{\mathbf{C}};\mathfrak{M};\Delta),\ \text{and } \alpha(t) \text{ is the arc-} \\ \text{length parametrization of } \Omega \end{array}\right\}. \tag{11.46}$$

We call $\mathcal{A}\mathcal{S}_m$ the space of *analytic splines* with knots at $\Delta=\{P_1,\ldots,P_k\}$ of multiplicities $m_1,\ldots,m_k$, where $P_i=\alpha(t_i)$, $i=1,2,\ldots,k$.

The term "analytic spline" is appropriate in this case because our definition of $s(z)$ assures us that it is an analytic function throughout $\mathbf{C}$, with the exception of the points $P_1,\ldots,P_k$ as the following theorem shows.

THEOREM 11.25

Let $s\in\mathcal{A}\mathcal{S}_m(\mathfrak{M};\Delta)$. Then

$$s(z)=p_0(z)+\frac{1}{2\pi i}\sum_{j=1}^{k} p_j(z)\ln\left(\frac{P_j-z}{P_{j-1}-z}\right), \tag{11.47}$$

for all $z \in \mathbf{C} \backslash \Delta$, where $p_0,\ldots,p_k$ are complex polynomials of order m with $p_0 = p_k$. Thus s is analytic in $\mathbf{C}\backslash\Delta$, and has logarithmic branch points at $P_1,\ldots,P_k$.

Proof. Let $p_1(\zeta),\ldots,p_k(\zeta)$ be the polynomials representing s for $\zeta \in \Omega$. In particular, suppose

$$p_j(\Pi_j) = \sum_{\nu=1}^{m} a_{j\nu}(\zeta - P)^{j-1}, \qquad j = 1,2,\ldots,k.$$

Then by the definition of $s(z)$,

$$s(z) = \frac{1}{2\pi i} \sum_{j=1}^{k} \int_{I_j} \frac{p_j(\zeta)\,d\zeta}{\zeta - z} = \sum_{j=1}^{k} \frac{1}{2\pi i} \int_{I_j} \frac{a_{j1}\,d\zeta}{\zeta - z}$$

$$+ \sum_{j=1}^{k} \sum_{\nu=2}^{m} \frac{1}{2\pi i} \int_{I_j} a_{j\nu}(\zeta - z)^{\nu-2}\,d\zeta$$

$$= p_0(z) + \sum_{j=1}^{k} \frac{p_j(z)}{2\pi i} \ln\left(\frac{P_j - z}{P_{j-1} - z}\right). \qquad \blacksquare$$

§ 11.8. HISTORICAL NOTES

Section 11.1

It is as difficult to trace the history of spaces of piecewise nonpolynomial functions as it is to trace the history of piecewise polynomials. Nevertheless, we can mention some early papers where nonpolynomial splines appear. Golomb and Weinberger [1959] found certain nonpolynomial splines as solutions of best interpolation problems. Similar best interpolation problems were discussed (independently) by Ahlberg, Nilson, and Walsh [1964], Schoenberg [1964b], and others. Greville [1964b] was the first to give a constructive treatment of a class of generalized splines. Splines with continuity conditions at several points (i.e., ties between the polynomials in nonadjacent intervals) arose in the piecewise polynomial interpolation methods devised by actuaries in the early 1900s; see Greville [1944] for a survey of some of these methods.

Section 11.2

Greville [1964b] gave a one-sided basis for his class of generalized splines. His approach was adapted by Jerome and Schumaker [1971] to construct one-sided splines for g-splines (cf. Section 8.3), and later to deal with more general spline spaces (Jerome and Schumaker [1976]). The treatment here follows Schumaker [1976c].

Section 11.3

Local support basis elements for nonpolynomial spline spaces were first constructed for nonpolynomial splines by Schoenberg [1964b], where a space of trigonometric splines was treated (see also Section 10.8). Karlin [1968, Chapter 10] gave local support basis splines for a class of T-splines (see also Chapter 10 here). Local support basis splines for the case of g-splines were constructed by Jerome and Schumaker [1971] (see also Section 8.3). These ideas were later extended in the paper by Jerome and Schumaker [1976], which we have followed here.

Section 11.4

That spaces of polynomial splines form WT-systems was already mentioned in the book by Karlin and Studden [1966]. That the same assertion holds for various classes of T-splines was observed by Schumaker [1966], Karlin and Schumaker [1967], and Karlin [1968]. Using a different method of proof, Bartelt [1975] showed that piecing together functions taken from T-spaces leads to a WT-system (our Theorem 11.12). The case where no continuities are enforced does not seem to have been handled in the literature. The converse Theorem 11.13 also originates from Bartelt [1975], although his method of proof required the assumption that $\mathbb{S}$ contain the constant functions. This assumption was removed by Sommer [1980], who gave a proof based on the concept of Tchebycheff rank of a set. The proof given here essentially follows Sommer, using the results of Stockenberg [1977a] on the separated zeros of functions in a WT-space (see Section 2.6). Theorem 11.18 shows that various smoother spaces formed by patching together ECT-systems are also WT-systems. This result is new.

Section 11.5

The prime examples of nonlinear spline spaces are the rational functions and generalizations of them. This development was carried out by Schaback [1973] and by Werner [1974, 1976]. The very general space of nonlinear splines defined here comes from Schumaker [1976c].

Section 11.7

Complex splines were introduced by Ahlberg, Nilson, and Walsh [1967a]. Later papers dealing with complex and analytic splines include Ahlberg [1969] and Ahlberg, Nilson, and Walsh [1969, 1971].

12

TENSOR-PRODUCT SPLINES

In this chapter we construct a space of multidimensional splines by taking the tensor-product of one-dimensional spaces of polynomial splines. Because of the tensor nature of the resulting space, many of the simple algebraic properties of ordinary polynomial splines in one dimension can be carried over. We also develop both direct and inverse approximation theorems.

§ 12.1. TENSOR-PRODUCT POLYNOMIAL SPLINES

We begin by defining the space of tensor-product splines of interest throughout this chapter. First we need to introduce some one-dimensional spline spaces. Let i be an integer with $1 \leqslant i \leqslant d$. Then given an interval $[a_i, b_i]$ and a positive integer m_i, we suppose that

$$\Delta_i = \{ a_i = x_{i,0} < x_{i,1} < \cdots < x_{i,k_i+1} = b_i \} \tag{12.1}$$

is a partition of $[a_i, b_i]$, and that

$$\mathfrak{M}_i = (m_{i,1}, \ldots, m_{i,k_i}), \qquad 1 \leqslant m_{i,j} \leqslant m_i, \qquad j = 1, 2, \ldots, k_i. \tag{12.2}$$

Then by the results of Chapter 4 we know that the spline space $\mathcal{S}(\mathcal{P}_{m_i}; \mathfrak{M}_i; \Delta_i)$ is an $m_i + K_i$ dimensional linear space, where $K_i = \sum_{j=1}^{k_i} m_{ij}$. Suppose

$$\mathcal{S}(\mathcal{P}_{m_i}; \mathfrak{M}_i; \Delta_i) = \operatorname{span}\{ B_{i,j}(x_i) \}_{j=1}^{m_i + K_i}. \tag{12.3}$$

When there is no chance of confusion, we shall write $B_j(x_i)$ instead of $B_{i,j}(x_i)$. We are ready to define tensor-product polynomial splines.

DEFINITION 12.1

We define the space of *tensor-product polynomial splines* by

$$\mathcal{S} = \bigotimes_{i=1}^{d} \mathcal{S}(\mathcal{P}_{m_i}; \mathfrak{M}_i; \Delta_i) = \operatorname{span}\{B_{i_1}(x_1)\cdots B_{i_d}(x_d)\}_{i_1=1,\ldots,i_d=1}^{m_1+K_1,\ldots,m_d+K_d}. \tag{12.4}$$

It is clear from the definition that $\mathcal{S}$ is a linear space of dimension $\Pi_{i=1}^{k}(m_i+K_i)$. Each spline s in $\mathcal{S}$ is a function defined on the set

$$H = \bigotimes_{i=1}^{d} [a_i, b_i] = \{\mathbf{x}=(x_1,\ldots,x_d): a_i \leqslant x_i \leqslant b_i, i=1,\ldots,d\}. \tag{12.5}$$

If $d=2$, H is a rectangle in the plane. We shall call H a rectangle even when $d>2$.
The partition $\boldsymbol{\Delta}=\Delta_1\otimes\cdots\otimes\Delta_d$ subdivides H into smaller rectangles:

$$H_{i_1\cdots i_d} = \{\mathbf{x}: x_{j,i_j} \leqslant x_j < x_{j,i_j+1}, \quad j=1,2,\ldots,d\}. \tag{12.6}$$

The following theorem shows that the tensor-product splines are smooth piecewise polynomials:

THEOREM 12.2

If $s\in\mathcal{S}$, then for each $i=1,2,\ldots,d$ and any fixed $a_j \leqslant x_j \leqslant b_j$, $j=1,2,\ldots, i-1,i+1,\ldots,d$,

$$s(x_1,\ldots,x_{i-1},\cdot,x_{i+1},\ldots,x_d) \in \mathcal{S}(\mathcal{P}_{m_i}; \mathfrak{M}_i; \Delta_i). \tag{12.7}$$

Moreover, for all $0 \leqslant i_j \leqslant k_j$, $\quad j=1,2,\ldots,d$

$$s|_{H_{i_1\cdots i_d}} \in \mathcal{P}_{\mathbf{m}}, \tag{12.8}$$

where

$$\mathcal{P}_{\mathbf{m}} = \bigotimes_{i=1}^{d} \mathcal{P}_{m_i} = \operatorname{span}\{x_1^{\nu_1-1}\cdots x_d^{\nu_d-1}\}_{\nu_1=1,\ldots,\nu_d=1}^{m_1,\ldots,m_d} \tag{12.9}$$

is the space of *tensor-product polynomials of order* $\mathbf{m}=(m_1,\ldots,m_d)$.

Proof. Since each $s\in\mathcal{S}$ can be written in the form

$$s(x_1,\ldots,x_d) = \sum_{i_1=1}^{m_1+K_1}\cdots\sum_{i_d=1}^{m_d+K_d} c_{i_1\cdots i_d} B_{i1}(x_1)\cdots B_{i_d}(x_d),$$

property (12.7) follows immediately. The assertion (12.8) stems from the fact that each of the $B_{i_j}(x_j)$ is a piecewise polynomial of order m_j, $j=1,2,\ldots,d$. ■

The exact smoothness of splines $s\in\mathcal{S}$ can be deduced from (12.7). In particular, since

$$D_{x_1}^{\alpha_1}\cdots D_{x_d}^{\alpha_d}s(x)=\sum_{i_1=1}^{m_1+K_1}\cdots\sum_{i_d=1}^{m_d+K_d}c_{i_1\cdots i_d}D_{x_1}^{\alpha_1}B_{i_1}(x_1)\cdots D_{x_d}^{\alpha_d}B_{i_d}(x_d),$$

all of these partial derivatives exist, and they are continuous inside the subrectangles while their smoothness across the faces between two such subrectangles is controlled by the multiplicity vectors $\mathfrak{M}_1,\ldots,\mathfrak{M}_d$.

§ 12.2. TENSOR-PRODUCT B-SPLINES

In order to work with the space of tensor-product polynomial splines on a digital computer, we need a convenient basis for it. Fortunately, because of the tensor-product nature of the space, it is possible to work with the usual one-dimensional B-splines.

For each $i=1,2,\ldots,d$, suppose $\Delta_{i,e}$ is an extended partition (cf. Definition 4.8) associated with Δ_i; that is

$$\Delta_{i,e}=\{y_{i,j}\}_{j=1}^{K_i+2m_i} \tag{12.10}$$

with

$$y_{i,1}\leqslant\cdots\leqslant y_{i,m_i}\leqslant a_i,\qquad b_i\leqslant y_{i,m_i+K_i+1}\leqslant\cdots\leqslant y_{i,K_i+2m_i} \tag{12.11}$$

and

$$y_{i,m_i+1}\leqslant\cdots\leqslant y_{i,m_i+K_i}=\overbrace{x_{i,1},\ldots,x_{i,1}}^{m_{i,1}}\cdots\overbrace{x_{i,k_i},\ldots,x_{i,k_i}}^{m_{i,k_i}} \tag{12.12}$$

Let $\{N_{i,j}(x_i)\}_{j=1}^{m_i+K_i}$ be the normalized B-splines associated with the extended partition $\Delta_{i,e}$.

DEFINITION 12.3

For each $i_1,\ldots,i_d$ with $1\leqslant i_j\leqslant m_j+K_j$, $j=1,2,\ldots,d$, let

$$N_{i_1\cdots i_d}(x_1,\ldots,x_d)=N_{i_1}^{m_1}(x_1)\cdots N_{i_d}^{m_d}(x_d). \tag{12.13}$$

We call these the *tensor-product B-splines*.

General properties of the tensor-product B-splines can be derived from the corresponding properties of the one-dimensional B-splines. We collect some of the more important of these in the following theorem. Let

$$Y_{i_1\cdots i_d}=\bigotimes_{j=1}^{d}(y_{j,i_j},y_{j,i_j+1}), \tag{12.14}$$

and

$$\overline{Y}_{i_1\cdots i_d}=\bigotimes_{j=1}^{d}\left[y_{j,i_j},y_{j,i_j+1}\right]. \tag{12.15}$$

THEOREM 12.4

For any $1\leqslant i_j\leqslant m_j+K_j$, $j=1,2,\ldots,d$,

$$N_{i_1\cdots i_d}(\mathbf{x})>0 \qquad \text{for } \mathbf{x}=(x_1,\ldots,x_d)\in Y_{i_1\cdots i_d}; \tag{12.16}$$

$$N_{i_1\cdots i_d}(\mathbf{x})=0 \qquad \text{for } \mathbf{x}\notin \overline{Y}_{i_1\cdots i_d}; \tag{12.17}$$

$$\sum_{\nu_1=i_1+1-m_1}^{i_1}\cdots\sum_{\nu_d=i_d+1-m_d}^{i_d}N_{\nu_1\cdots\nu_d}(\mathbf{x})\equiv 1 \qquad \text{for } \mathbf{x}\in Y_{i_1\cdots i_d}. \tag{12.18}$$

Proof. These results follow directly from (12.13) and Theorems 4.17 and 4.20. ■

The shape of tensor-product B-splines is most easily visualized in the case of $d=2$. When $m_1=m_2=2$ they are pyramid-shaped functions with support over four subrectangles. When $m_1=m_2=3$ they are shaped like a hill, and have support over nine subrectangles. See Figure 38 for some typical examples.

Using the tensor-product B-splines, it is a relatively straightforward task to deal with tensor-product polynomial splines on a digital computer. Since

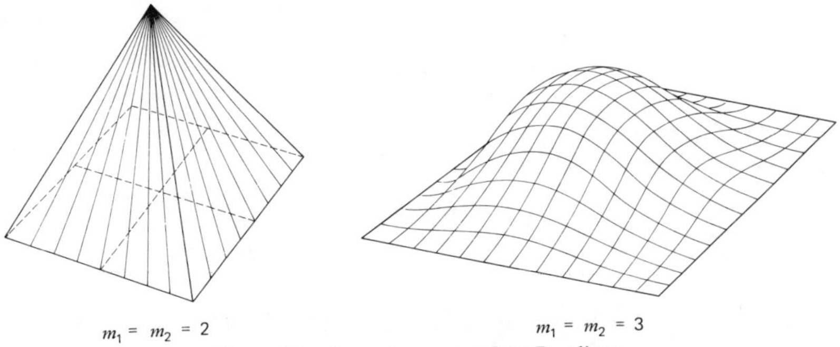

Figure 38. Some tensor-product B-splines.

each $s \in \mathcal{S}$ can be identified with a B-spline series of the form

$$s(x_1,\ldots,x_d)=\sum_{i_1=1}^{m_1+K_1}\cdots\sum_{i_d=1}^{m_d+K_d} c_{i_1\cdots i_d}N_{i_1\cdots i_d}(x_1,\ldots,x_d), \tag{12.19}$$

to *store* a tensor-product spline on the computer, we need only store the coefficients $\{c_{i_1\cdots i_d}\}$.

Given a B-spline expansion (12.19), to *evaluate* it at a point $\mathbf{x}\in H$, we must first find the vector $\mathbf{l}=(l_1,\ldots,l_d)$ such that

$$\mathbf{x}=(x_1,\ldots,x_d)\in \bigotimes_{j=1}^{d}\left[y_{j,l_j},y_{j,l_j+1}\right).$$

Then

$$s(\mathbf{x})=\sum_{i_1=l_1+1-m_1}^{l_1}\cdots\sum_{i_d=l_d+1-m_d}^{l_d} c_{i_1\cdots i_d}N_{i_1}^{m_1}(x_1)\cdots N_{i_d}^{m_d}(x_d). \tag{12.20}$$

This is a sum over just $m_1m_2\cdots m_d$ terms. Each of the required B-spline values can be computed by using Algorithm 5.5 to generate the necessary one-dimensional B-spline values.

Concerning *derivatives* of the B-spline expansion (12.19), we note that by Theorem 5.9,

$$D_{x_1}^{\alpha_1}\cdots D_{x_d}^{\alpha_d}s(x_1,\ldots,x_k)=\sum\cdots\sum c_{i_1\cdots i_d}^{\alpha_1\cdots\alpha_d}N_{i_1}^{m_1-\alpha_1}(x_1)\cdots N_{i_d}^{m_d-\alpha_d}(x_d), \tag{12.21}$$

where the coefficients $\{c_{i_1,\ldots,i_d}^{\alpha_1,\ldots,\alpha_d}\}$ can be computed by the recursions of (5.15) applied to one variable at a time. This can be accomplished numerically using Algorithm 5.10. A similar situation persists for the *antiderivatives* (cf. Theorem 5.17 and Algorithm 5.19).

It is also possible to represent tensor-product polynomial splines as *piecewise polynomials*. In particular, for any $i_1,\ldots,i_d$ with $0\leqslant i_j\leqslant k_j$, $j=1,2,\ldots,d$, we can write

$$s(\mathbf{x})=p_{i_1\cdots i_d}(\mathbf{x}) \qquad \text{for } \mathbf{x}=(x_1,\ldots,x_d)\in Y_{i_1,\ldots,i_d}, \tag{12.22}$$

where

$$p_{i_1\cdots i_d}(x_1,\ldots,x_d)$$
$$=\sum_{\nu_1=0}^{m_1-1}\cdots\sum_{\nu_d=0}^{m_d-1}\frac{D_{x_1}^{\nu_1}\cdots D_{x_d}^{\nu_d}s(x_{i_1},\ldots,x_{i_d})(x_1-x_{i_1})^{\nu_1}\cdots(x_d-x_{i_d})^{\nu_d}}{\nu_1!\cdots\nu_d!}. \tag{12.23}$$

We close this section by giving a dual basis for the set of B-splines.

THEOREM 12.5

For each $i=1,2,\ldots,d$, let $\{\lambda_{i,j}\}_{j=1}^{m_i+K_i}$ be the linear functionals constructed in Theorem 4.41 as a dual basis for $\{N_{i,j}^{m_i}\}_{j=1}^{m_i+K_i}$. Then

$$\{\lambda_{i_1\cdots i_d}=\lambda_{1,i_1}\cdots\lambda_{d,i_d}\}_{i_1=1,\ldots,i_d=1}^{m_1+K_1,\ldots,m_d+K_d} \tag{12.24}$$

form a dual basis for the set of tensor-product B-splines $\{N_{i_1\cdots i_d}\}$. Moreover,

$$|\lambda_{i_1\cdots i_d}f|\leqslant\prod_{j=1}^{d}(2m_j+1)9^{(m_j-1)}h_{i_1\cdots i_d}^{-1/p}\|f\|_{L_p(\tilde{Y}_{i_1\ldots i_d})} \tag{12.25}$$

for every $f\in L_p[\tilde{Y}_{i_1\cdots i_d}]$ where

$$\tilde{y}_{i_1\cdots i_d}=\bigotimes_{j=1}^{d}\left[y_{j,i_j},y_{j,i_j+m_j}\right), \tag{12.26}$$

and

$$h_{i_1\cdots i_d}=\prod_{j=1}^{d}(y_{j,i_j}-y_{j,i_j+m_j}). \tag{12.27}$$

Proof. The fact that (12.24) form a dual basis follows immediately from the fact that

$$\lambda_{i_1\cdots i_d}N_{\nu_1\cdots\nu_d}=(\lambda_{1,i_1}N_{1,\nu_1}^{m_1})\cdots(\lambda_{d,i_d}N_{d,\nu_d}^{m_d})$$

and the duality of the $\{\lambda_{i,j}\}$ with $\{N_{i,j}^{m_i}\}$. The bound (12.25) is an immediate consequence of the bound (4.86) on the individual one-dimensional functionals. ■

§ 12.3. APPROXIMATION POWER OF TENSOR-PRODUCT SPLINES

In this section we give some direct theorems relating the smoothness of a function to how well it can be approximated by tensor-product splines. For convenience we restrict our attention to functions defined on a rectangle. For references to results on more general domains, see Section 12.6.

Throughout we suppose that H is a rectangle as in (12.5), and that $\mathbb{S}$ is the space of tensor-product polynomial splines defined in (12.4). Our aim is to give estimates for

$$d(f,\mathbb{S})_p=\inf_{s\in\mathbb{S}}\|f-s\|_{L_p(H)}.$$

To achieve this, we construct a quasi-interpolant Q mapping $L_p(H)$ into $\mathbb{S}$. Our construction will be based on the B-splines defined in (12.13) and the dual linear functionals given in (12.24). We suppose that the extended partitions $\Delta_{i,e}$ have been chosen so that

$$y_{i,1} = \cdots = y_{i,m_i} = a_i, \qquad b_i = y_{i,m_i+K_i+1} = \cdots = y_{i,K_i+2m_i}, \tag{12.28}$$

$i = 1, 2, \ldots, d$.

THEOREM 12.6

Given any $f \in L_1(H)$, let

$$Qf(\mathbf{x}) = \sum_{i_1=1}^{m_1+K_1} \cdots \sum_{i_d=1}^{m_d+K_d} (\lambda_{i_1 \cdots i_d} f) N_{i_1 \cdots i_d}(x). \tag{12.29}$$

We call Qf the *quasi-interpolant* of f. It is a bounded linear operator mapping $L_1(H)$ onto $\mathbb{S}$ with

$$Qs = s \qquad \text{for all } s \in \mathbb{S}. \tag{12.30}$$

Proof. Property (12.30) is an immediate consequence of the fact that the λ's form a dual basis for the B-splines. To prove the boundedness of Q, suppose that $\mathbf{x} \in Y_{i_1 \cdots i_d}$. Then by Theorem 12.5,

$$\begin{aligned} |Qf(\mathbf{x})| &\leqslant \sum \cdots \sum |\lambda_{i_1 \cdots i_d} f| \, |N_{i_1 \cdots i_d}(\mathbf{x})| \\ &\leqslant C h_{i_1 \cdots i_d}^{-1/p} \|f\|_{L_p(\tilde{Y}_{i_1 \cdots i_d})}. \end{aligned}$$

Integrating over $\mathbf{x} \in Y_{i_1 \ldots i_d}$, we obtain

$$\|Qf\|_{L_p(Y_{i1 \cdots i_d})} \leqslant C \|f\|_{L_p(\tilde{Y}_{i1 \cdots i_d})}. \tag{12.31}$$

If we sum the pth power of this inequality over all $m_j \leqslant i_j \leqslant m_j + K_j$, $j = 1, 2, \ldots, d$, and then take the pth root, we obtain

$$\|Qf\|_{L_p(H)} \leqslant C \|f\|_{L_p(H)}, \tag{12.32}$$

which is the assertion that Q is bounded. ∎

Theorem 12.6 asserts that Q is a bounded linear projector of $L_1(H)$ onto $\mathbb{S}$. We now want to give bounds on how well Qf approximates f. Our next theorem deals with functions f in the tensor Sobolev space (see §13.2)

$$L_p^{\mathbf{r}}(H) = \left\{ f \colon D_{x_1}^{\alpha_1} \cdots D_{x_d}^{\alpha_d} f \in L_p(H), \text{ all } 0 \leqslant \alpha_j \leqslant r_j, j = 1, 2, \ldots, d \right\}$$

where $\mathbf{r}=(r_1,\ldots,r_d)$ is a vector with

$$r_j \leqslant m_j, \qquad j=1,2,\ldots,d. \tag{12.33}$$

To state the theorem, we need the mesh widths

$$\overline{\Delta}_i = \max_{0 \leqslant j \leqslant k_i} (x_{i,j+1} - x_{i,j}), \qquad i=1,2,\ldots,d, \tag{12.34}$$

and

$$\underline{\Delta}_i = \min_{0 \leqslant j \leqslant k_i} (x_{i,j+1} - x_{i,j}), \qquad i=1,2,\ldots,d. \tag{12.35}$$

THEOREM 12.7

Let $\mathbf{r}$ be a vector satisfying (12.33). Then there exists a constant C depending only on $\mathbf{m}, \mathbf{r}, d$, and

$$\gamma = \max_{1 \leqslant i \leqslant d} \frac{\overline{\Delta}_i}{\underline{\Delta}_i} \tag{12.36}$$

such that for all $f \in L_p^{\mathbf{r}}(H)$,

$$\|f - Qf\|_{L_p(H)} \leqslant C \sum_{i=1}^{d} \overline{\Delta}_i^{r_i} \|D_{x_i}^{r_i} f\|_{L_p(H)}, \tag{12.37}$$

$1 \leqslant p \leqslant \infty$.

Proof. Fix $m_j \leqslant i_j \leqslant m_j + K_j$, $j=1,2,\ldots,d$. We obtain first an error bound for $f - Qf$ on the rectangle $Y_{i_1 \cdots i_d}$. By results on multidimensional Taylor expansions (cf. Theorem 13.18 below), there exists a tensor-product polynomial $p_{i_1 \cdots i_d} \in \mathcal{P}_{\mathbf{r}}$ such that

$$\|f - p_{i_1 \cdots i_d}\|_{L_p(\tilde{Y}_{i_1 \cdots i_d})} \leqslant C \sum_{i=1}^{d} \overline{\Delta}_i^{r_i} \|D_{x_i}^{r_i} f\|_{L_p(\tilde{Y}_{i_1 \cdots i_d})}, \tag{12.38}$$

where $\tilde{Y}_{i_1 \cdots i_d}$ is defined in (12.26). The constant C depends only on $\mathbf{m}$, $\mathbf{r}$, d, and γ. Now since Q reproduces polynomials, by (12.31)

$$\begin{aligned}\|f - Qf\|_{L_p(Y_{i_1 \cdots i_d})} &\leqslant \|f - p_{i_1 \cdots i_d}\|_{L_p(Y_{i_1 \cdots i_d})} \\ + \|Q(f - p_{i_1,\ldots,i_d})\|_{L_p(Y_{i_1 \cdots i_d})} &\leqslant C \|f - p_{i_1 \cdots i_d}\|_{L_p(\tilde{Y}_{i_1 \cdots i_d})} \\ &\leqslant C \sum_{i=1}^{d} \overline{\Delta}_i^{r_i} \|D_{x_i}^{r_i} f\|_{L_p(\tilde{Y}_{i_1 \cdots i_d})}.\end{aligned}$$

Taking the pth power of this inequality and summing over all $m_j \leq i_j \leq m_j + K_j$, $j = 1, 2, \ldots, d$, we obtain (12.37). ∎

Theorem 12.7 gives an estimate for $d(f, \mathbb{S})_p$ for functions f in $L_p^{\mathbf{r}}(H)$, $\mathbf{r} \leq \mathbf{m}$. We now convert this to a result involving the tensor modulus of smoothness defined in Example 13.27 below.

THEOREM 12.8

Fix $1 \leq p \leq \infty$. Then there exists a constant C (depending only on $\mathbf{m}$, p, and d) such that for all $f \in L_p(H)$,

$$d(f, \mathbb{S})_p \leq C\omega_{\mathbf{m}}(f; \boldsymbol{\delta})_p, \tag{12.39}$$

where $\boldsymbol{\delta} = (\overline{\Delta}_1, \ldots, \overline{\Delta}_d)$.

Proof. Given any partition of H, we may thin out each of the Δ_i's as in Lemma 6.17 to get new partitions $\Delta_i^* \subseteq \Delta_i$ with $\overline{\Delta}_i^* / \underline{\Delta}_i^* \leq 3$. Then the associated quasi-interpolant Q^* maps into $\mathbb{S}$, and it follows from Theorem 12.7 that

$$d(f, \mathbb{S})_p \leq \|f - Q^* f\|_p \leq C \sum_{i=1}^{d} \overline{\Delta}_i^{*m_i} \|D_{x_i}^{m_i} f\|_{L_p(H)}$$

for all $f \in L_p^{\mathbf{m}}(H)$. But then applying Theorem 13.30, we obtain (12.39).

§ 12.4. INVERSE THEORY FOR PIECEWISE POLYNOMIALS

In this section we establish some inverse theorems which assert that if a function can be approximated to a certain order by piecewise tensor-product polynomials, then the function must possess a certain amount of smoothness. These theorems can then be used to show that there is a saturation phenomenon for approximation by piecewise polynomials. Similar results are given in the following section for splines.

We begin by introducing the spaces of piecewise polynomials of interest. Let H be a rectangle as in (12.5), and let $\Delta_1, \ldots, \Delta_d$ be partitions as in (12.1). Then given a vector $\mathbf{m} = (m_1, \ldots, m_d)$ in Z_+^d, we define the space of *piecewise tensor-product polynomials of order m* by

$$\mathcal{P}\mathcal{W}(\mathcal{P}_{\mathbf{m}}; \boldsymbol{\Delta}) = \left\{ s \colon s|_{H_{i_1 \cdots i_d}} \in \mathcal{P}_{\mathbf{m}},\ 0 \leq i_j \leq k_j,\ j = 1, \ldots, d \right\}, \tag{12.40}$$

where $\mathcal{P}_{\mathbf{m}}$ is defined in (12.9) and $H_{i_1 \cdots i_d}$ are the subrectangles (12.6) of H defined by the partition $\boldsymbol{\Delta} = \Delta_1 \otimes \cdots \otimes \Delta_d$.

As in the one-dimensional case, in order to establish inverse theorems, we have to work with a sequence of partitions $\mathbf{\Delta}^{(\nu)}$. Moreover, in order to get meaningful results, we have to impose some mixing conditions on $\mathbf{\Delta}^{(\nu)}$. We discuss the case of $p=\infty$ first.

DEFINITION 12.9. Mixing Condition

Let

$$\mathbf{\Delta}^{(\nu)}=\Delta_1^{(\nu)}\otimes\cdots\otimes\Delta_d^{(\nu)} \tag{12.41}$$

be a sequence of partitions of the rectangle H. We say that $\mathbf{\Delta}^{(\nu)}$ satisfies the *mixing condition* provided there exists a constant ρ such that for all ν and any $\mathbf{x}\in H$, there is some partition $\mathbf{\Delta}^{(n)}$ with $n\geqslant\nu$ such that $\mathbf{x}$ is contained in the interior of a subrectangle $H_{n,\mathbf{x}}$ defined by $\mathbf{\Delta}^{(n)}$, and

$$d(x_i,\partial H_{n,\mathbf{x}})\geqslant\rho\underline{\Delta}_i^{(n)},\qquad i=1,2,\dots,d, \tag{12.42}$$

where $\partial H_{n,x}$ denotes the boundary of $H_{n,\mathbf{x}}$, and

$$\underline{\Delta}_i^{(\nu)}=\min_{1<j<k_i^{(\nu)}}\left(x_{i,j+1}^{(\nu)}-x_{i,j}^{(\nu)}\right). \tag{12.43}$$

It can be shown by arguments similar to those used in proving Theorem 6.38 that if $\mathbf{\Delta}^{(\nu)}$ is a sequence of partitions as in (12.41), with each $\Delta_i^{(\nu)}$ a uniform partition, then $\mathbf{\Delta}^{(\nu)}$ satisfies the mixing condition. We are ready for our first inverse theorem.

THEOREM 12.10

Let $\mathbf{\Delta}^{(\nu)}$ be a sequence of partitions satisfying the mixing condition. Then there is a constant C such that for all $f\in C(H)$,

$$\omega_{\mathbf{m}}(f;\underline{\mathbf{\Delta}}^{(\nu)})_\infty\leqslant C\sup_{n>\nu}d\left[f,\mathcal{PW}(\mathcal{P}_{\mathbf{m}};\mathbf{\Delta}^{(n)})\right]_\infty, \tag{12.44}$$

where $\omega_{\mathbf{m}}$ is the tensor-product modulus of smoothness discussed in Example 13.27 below, and $\underline{\mathbf{\Delta}}^{(\nu)}=(\underline{\Delta}_1^{(\nu)},\dots,\underline{\Delta}_d^{(\nu)})$.

Proof. Fix $1\leqslant i\leqslant d$ and $\mathbf{x}\in H$. Suppose $H_{n,\mathbf{x}}$ is the subrectangle associated with n and $\mathbf{x}$, as in the definition of the mixing condition. Let

$$h_i\leqslant\rho\frac{\underline{\Delta}_i^{(\nu)}}{m_i}, \tag{12.45}$$

where ρ is the constant in the mixing condition. For convenience set $\varepsilon_\nu = d[f, \mathcal{P}\mathcal{S}(\mathcal{P}_{\mathbf{m}}; \mathbf{\Delta}^{(\nu)})]_\infty$, and suppose $s_\nu \in \mathcal{P}\mathcal{S}(\mathcal{P}_{\mathbf{m}}; \mathbf{\Delta}^{(\nu)})$ is such that

$$\|f - s_\nu\|_{L_\infty(H)} \leqslant 2\varepsilon_\nu, \qquad \text{all } \nu.$$

Then since s_ν is a polynomial in $\mathcal{P}_{\mathbf{m}}$ on $H_{n,\mathbf{x}}$ and h_i is small enough,

$$|\Delta_{h_i}^{m_i} f(\mathbf{x})| = |\Delta_{h_i}^{m_i}(f - s_\nu)(\mathbf{x})| \leqslant C\varepsilon_\nu,$$

where the divided difference is taken on the ith variable. Taking the supremum over all h_i satisfying (12.45) and then over all $\mathbf{x} \in H$, we obtain

$$\omega_{m_i \mathbf{e}_i}(f; \underline{\Delta}^{(\nu)}) \leqslant C \sup_{n \geqslant \nu} \varepsilon_n, \tag{12.46}$$

where $\mathbf{e}_i$ is the unit vector in the ith variable. Summing these inequalities over $i = 1, 2, \ldots, d$, we arrive at (12.44). ■

Theorem 12.10 is the analog of Theorem 6.39 in the one-dimensional case. We can now use it to give a complete inverse theorem in which $\omega_{\mathbf{m}}(f; \mathbf{t})$ is estimated for all vectors $\mathbf{t} = (t_1, \ldots, t_n)$ with $t_i > 0$, $i = 1, \ldots, d$.

THEOREM 12.11

Let $\mathbf{\Delta}^{(\nu)}$, as in (12.41), be a sequence of partitions satisfying the mixing condition and such that $\Delta_i^{(\nu)}$ go steadily to zero in the sense of Definition 6.5. Suppose $f \in C(H)$ and that

$$d\big[f, \mathcal{P}\mathcal{S}(\mathcal{P}_{\mathbf{m}}; \mathbf{\Delta}^{(\nu)})\big]_\infty \leqslant \phi(\overline{\Delta}_1^{(\nu)}, \ldots, \overline{\Delta}_d^{(\nu)}), \tag{12.47}$$

where

$$\overline{\Delta}_i^{(\nu)} = \max_{1 \leqslant j \leqslant k_i^{(\nu)}} \left(x_{i,j+1}^{(\nu)} - x_{i,j}^{(\nu)}\right), \qquad i = 1, 2, \ldots, d, \tag{12.48}$$

and where $\phi(t_1, \ldots, t_d)$ is a function that is monotone in each variable. Then

$$\omega_{m_i \mathbf{e}_i}(f; \mathbf{t}) \leqslant C_1 \phi(\mathbf{t}), \qquad i = 1, 2, \ldots, d, \tag{12.49}$$

all $\mathbf{t} = (t_1, \ldots, t_d) > 0$.

Proof. The assertion follows directly from (12.46) and simple properties of $\omega_{\mathbf{m}}$ (cf. Theorems 13.23–13.24). See Theorem 6.41 for the one-dimensional case. ■

Theorem 12.7 above shows that for sufficiently smooth functions [$f\in L_\infty^{\mathbf{m}}(H)$ suffices],

$$d\left[f, \mathcal{P}\mathcal{W}(\mathcal{P}_{\mathbf{m}};\boldsymbol{\Delta}^{(\nu)})\right]_\infty = \mathcal{O}\left[\left(\bar{\Delta}_1^{(\nu)}\right)^{m_1} + \cdots + \left(\bar{\Delta}_d^{(\nu)}\right)^{m_d}\right].$$

The following saturation theorem asserts that unless f is a tensor-product polynomial, this is the maximal order of convergence obtainable.

THEOREM 12.12

Let $\boldsymbol{\Delta}^{(\nu)}$ be a sequence of partitions as in Theorem 12.11. Suppose $f\in C(H)$ is such that

$$d\left[f, \mathcal{P}\mathcal{W}(\mathcal{P}_{\mathbf{m}};\boldsymbol{\Delta}^{(\nu)})\right]_\infty \leqslant \left(\min_{1\leqslant i\leqslant d}\left(\bar{\Delta}_i^{(\nu)}\right)^{m_i}\right)\psi(\bar{\boldsymbol{\Delta}}^{(\nu)}), \tag{12.50}$$

where $\psi(\mathbf{t})$ is a function which is monotone in each variable and with $\psi(\mathbf{t})\to 0$ as $\|\mathbf{t}\|\to 0$. Then $f\in\mathcal{P}_{\mathbf{m}}$.

Proof. The hypothesis (12.50) coupled with Theorem 12.11 implies that

$$\omega_{m_i\cdot\mathbf{e}_i}(f;\mathbf{t}) \leqslant C(t_i^{m_i})\psi(\mathbf{t}), \qquad i=1,2,\ldots,d.$$

Now let $\mathbf{t}_n = ((1/n)^{m_1},\ldots,(1/n)^{m_d})$, $n=1,2,\ldots$. We conclude that

$$\sum_{i=1}^{d}\frac{\omega_{m_i\cdot\mathbf{e}_i}(f;\mathbf{t}_n)}{(1/n)^{m_i}}\to 0 \qquad \text{as } n\to\infty.$$

But then by Theorem 13.26 if follows that $f\in\mathcal{P}_{\mathbf{m}}$. ■

So far we have been dealing with approximation in the uniform norm. In order to give analogous results for approximation in p-norms, we first have to introduce an appropriate p-mixing condition.

DEFINITION 12.13. *p*-Mixing Condition

Let $\boldsymbol{\Delta}^{(\nu)}$ be a sequence of partitions of the rectangle H. We say that $\boldsymbol{\Delta}^{(\nu)}$ satisfies the p-mixing condition provided there exists a constant ρ such that for every ν there is a corresponding set of numbers $\{a_{i,j}^{(\nu)}\}_{i=1,\ j=\nu}^{d\ \ \infty}$ with $\sum_{j=\nu}^{\infty}a_{i,j}^{(\nu)}=1$ such that for every $\mathbf{x}\in H$,

$$\sum_{j=\nu}^{\infty} a_{i,j}^{(\nu)} d\left(x_i, \partial\boldsymbol{\Delta}^{(j)}\right)^{m_ip+1} \geqslant \rho\left(\underline{\Delta}_i^{(\nu)}\right)^{m_ip+1}, \tag{12.51}$$

$i=1,2,\ldots,d$, where $\partial\boldsymbol{\Delta}^{(j)}$ denotes the boundary of $\boldsymbol{\Delta}^{(j)}$.

The p-mixing condition is satisfied by equidistant partitions with $a_{i,j}^{(\nu)} = 1/\nu$, $i=1,2,\ldots,d$, $j=\nu,\ldots,2\nu-1$. We can now give an inverse theorem.

THEOREM 12.14

Let $1 \leqslant p \leqslant \infty$ and suppose that the sequence of partitions $\boldsymbol{\Delta}^{(\nu)}$ satisfies the p-mixing condition. Then there exists a constant C such that for every $f \in L_p(H)$,

$$\omega_{\mathbf{m}}(f;\underline{\boldsymbol{\Delta}}^{(\nu)})_p \leqslant C \sup_{n \geqslant \nu} d\left[f, \mathcal{P}\mathcal{W}(\mathcal{P}_{\mathbf{m}};\boldsymbol{\Delta}^{(n)})\right]_p. \tag{12.52}$$

Discussion. The proof of this theorem proceeds along exactly the same lines as the proof of Theorem 6.52 in the one-dimensional case. ■

By the same kind of arguments used above, we can translate Theorem 12.14 into an inverse theorem for $\omega_{\mathbf{m}}(f;\mathbf{t})_p$ and a saturation result. Since the details are not substantially different, we only state the results.

THEOREM 12.15

Fix $1 \leqslant p \leqslant \infty$ and let $\boldsymbol{\Delta}^{(\nu)}$ be a sequence of partitions as in Theorem 12.14. Suppose $\Delta_i^{(\nu)}$ are partitions that go to zero steadily in the sense of Definition 6.5. Finally, suppose $f \in L_p(H)$ is a function such that

$$d\left[f, \mathcal{P}\mathcal{W}(\mathcal{P}_{\mathbf{m}};\boldsymbol{\Delta}^{(\nu)})\right]_p \leqslant \phi(\overline{\Delta}_1^{(\nu)},\ldots,\overline{\Delta}_d^{(\nu)}), \tag{12.53}$$

where $\overline{\Delta}_i^{(\nu)}$ are defined in (12.48) and $\phi(t_1,\ldots,t_d)$ is a function that is monotone in each variable. Then

$$\omega_{m_i\mathbf{e}_i}(f;\mathbf{t})_p \leqslant C_1\phi(\mathbf{t}), \qquad i=1,2,\ldots,d \tag{12.54}$$

for all $\mathbf{t} > 0$.

Finally, we have the following saturation result:

THEOREM 12.16

Let $\boldsymbol{\Delta}^{(\nu)}$ be as in Theorem 12.15, and suppose that $f \in L_p(H)$ is such that

$$d\left[f, \mathcal{P}\mathcal{W}(\mathcal{P}_{\mathbf{m}};\boldsymbol{\Delta}^{(\nu)})\right]_p \leqslant \left(\min_{1 \leqslant i \leqslant d} \left(\overline{\Delta}_i^{(\nu)}\right)^{m_i}\right)\psi(\overline{\boldsymbol{\Delta}}^{(\nu)}), \tag{12.55}$$

where ψ is as in Theorem 12.12. Then $f \in \mathcal{P}_{\mathbf{m}}$.

§ 12.5. INVERSE THEORY FOR SPLINES

In this section we establish inverse theorems similar to those given in §12.4, but where the space of piecewise polynomials $\mathcal{P}\mathcal{W}(\mathcal{P}_{\mathbf{m}};\mathbf{\Delta}^{(\nu)})$ is replaced by a space of splines belonging to $C^{\mathbf{l}}(H)$ for some vector $\mathbf{l}$. In this case (as in the one dimensional case), we can dispense with mixing conditions and obtain results for general partitions. In order to simplify the notation, we restrict our attention to the case $d=2$ of two variables. To further simplify matters, we assume that H is the unit square $[0,1]\times[0,1]$.

We begin with an inverse theorem for the case of uniform approximation.

THEOREM 12.17

Let $\mathbf{\Delta}^{(\nu)}=\Delta_1^{(\nu)}\otimes\Delta_2^{(\nu)}$ be a sequence of partitions of H with $\Delta_1^{(0)}=\Delta_2^{(0)}=[0,1]$ and with $\underline{\Delta}_1^{(\nu)}\downarrow 0$ and $\underline{\Delta}_2^{(\nu)}\downarrow 0$. Suppose that for each ν, s_ν is a spline in the space

$$\mathcal{P}\mathcal{W}(\mathcal{P}_{\mathbf{m}};\mathbf{\Delta}^{(\nu)})\cap C^{\mathbf{l}}(H), \tag{12.56}$$

where $\mathbf{l}=(l_1,l_2)$ is a given vector with $l_i<m_i-1$, $i=1,2$. Then there exists a constant C such that for every function $f\in C(H)$,

$$\omega_{m_1\mathbf{e}_1}(f;\underline{\mathbf{\Delta}}^{(\nu)})\leqslant C(\underline{\Delta}_1^{(\nu)})^{l_1+1}\sum_{r=1}^{\nu}\frac{\varepsilon_r+\varepsilon_{r+1}}{(\underline{\Delta}_1^{(r)})^{l_1+1}}, \tag{12.57}$$

where $\varepsilon_r=\|f-s_r\|_\infty$. A similar inequality holds for the modulus of smoothness $\omega_{m_2\mathbf{e}_2}(f;\underline{\mathbf{\Delta}}^{(\nu)})$.

Proof. By Theorem 6.44, for any $0\leqslant y\leqslant 1$ and any $0<\varepsilon\leqslant\underline{\Delta}_1^{(\nu)}$,

$$\omega_{m_1}\big[f(\cdot,y);\varepsilon\big]\leqslant C\sum_{j=l_1+1}^{m_1-1}\varepsilon^j J\big[D_{x_1}^j s_\nu(\cdot,y)\big],$$

where

$$J\big[D_{x_1}^j s(\cdot,y)\big]=\max_{1\leqslant i\leqslant k_1}\big[D_{x_1}^j s_\nu(x_i+,y)-D_{x_1}^j s_\nu(x_i-,y)\big].$$

Now by applying Theorem 6.45 for each fixed y and afterward taking the supremum over all $0\leqslant y\leqslant 1$, the same arguments as were used in the proof of Theorem 6.46 yield (12.57). ■

Theorem 12.17 can now be converted to a result giving estimates for $\omega_{m_i\mathbf{e}_i}(f;\mathbf{t})$ for all vectors $\mathbf{t}>0$.

THEOREM 12.18

Let $\mathbf{\Delta}^{(\nu)} = \Delta_1^{(\nu)} \otimes \Delta_2^{(\nu)}$ be a sequence of partitions of H such that $\Delta_1^{(\nu)}$ and $\Delta_2^{(\nu)}$ each satisfy the hypotheses of Theorem 6.47. Suppose $\mathbf{l}$ is a vector with $l_i < m_i - 1$, $i = 1, 2$, and that $\mathbb{S}_\nu$ is a sequence of spline spaces belonging to $\mathcal{P}\mathcal{W}(\mathcal{P}_m; \mathbf{\Delta}^{(\nu)}) \cap C^{\mathbf{l}}(H)$. Then for $i = 1, 2$,

$$d(f, \mathbb{S}_\nu)_\infty = \mathcal{O}\big(\overline{\Delta}_i^{(\nu)}\big)^{\alpha_i}, \qquad \alpha_i < l_i + 1 \tag{12.58}$$

implies

$$\omega_{m_i \mathbf{e}_i}(f; \mathbf{t}) = \mathcal{O}(t_i^{\alpha_i}). \tag{12.59}$$

If (12.58) holds for both $i = 1$ and $i = 2$, then

$$\omega_{\mathbf{m}}(f; \mathbf{t}) = \mathcal{O}(t_1^{\alpha_1} + t_2^{\alpha_2}). \tag{12.60}$$

Proof. The first assertion follows from Theorem 12.17 and the same arguments used to prove Theorem 6.47. Adding the statements (12.59) for $i = 1, 2$ leads to (12.60). ■

We conclude this section with some analogous theorems for the case of $1 \leqslant p < \infty$.

THEOREM 12.19

Let $\mathbf{\Delta}^{(\nu)}$ be a sequence of partitions of H and let s_ν be a sequence of splines as in Theorem 12.17. Then there exists a constant C such that for every function $f \in L_p(H)$,

$$\omega_{m_1 \mathbf{e}_1}(f; \underline{\mathbf{\Delta}}^{(\nu)})_p \leqslant C(\underline{\Delta}_1^{(\nu)})^{l_1 + 1 + 1/p} \sum_{r=1}^{\nu} \frac{\varepsilon_r + \varepsilon_{r+1}}{(\underline{\Delta}_1^{(r)})^{l_1 + 1/p}}. \tag{12.61}$$

A similar inequality holds for $\omega_{m_2 \mathbf{e}_2}(f; \underline{\mathbf{\Delta}}^{(\nu)})$.

Proof. By the arguments of Theorem 6.55, we can now show that

$$\omega_{m_1 \mathbf{e}_1}(f; \varepsilon)_p \leqslant C \varepsilon_1^{1/p} \sum_{j = l_1 + 1}^{m_1 - 1} \varepsilon_1^j \tilde{J}_p\big(D_{x_1}^j s\big),$$

where

$$\tilde{J}_p(\phi) = \Big(\int_0^1 |J_p(\phi(\cdot, y)|^p \, dy \Big)^{1/p},$$

and

$$J_p[\phi(\cdot,y)] = \left[\sum_{i=1}^{k_1} |\phi(x_i+,y) - \phi(x_i-,y)|^p \right]^{1/p}.$$

The $\tilde{J}_p$'s can be estimated by methods similar to those used in Theorem 6.55 in the one-dimensional case, and the result follows. ■

Theorem 12.19 and arguments similar to those used in the proof of Theorem 6.47 (cf. also Theorem 6.56) lead to the following inverse theorem:

THEOREM 12.20

Let $\Delta^{(\nu)}$ be a sequence of partitions as in Theorem 12.18, and let $\mathbb{S}_\nu$ be a sequence of spaces of splines in $\mathcal{P}\mathcal{W}(\mathcal{P}_{\mathbf{m}};\Delta^{(\nu)}) \cap C^{\mathbf{l}}(H)$. Then for $i=1,2$,

$$d(f,\mathbb{S}_\nu)_p = \mathcal{O}\left(\overline{\Delta}_i^{(\nu)}\right)^{\alpha_i}, \qquad \alpha_i < l_i + 1 \tag{12.62}$$

implies

$$\omega_{m_i \mathbf{e}_i}(f;\mathbf{t})_p = \mathcal{O}(t_i^{\alpha_i}). \tag{12.63}$$

If (12.62) holds for both $i=1$ and $i=2$, then

$$\omega_{\mathbf{m}}(f;\mathbf{t})_p = \mathcal{O}(t_1^{\alpha_1} + t_2^{\alpha_2}). \tag{12.64}$$

§ 12.6. HISTORICAL NOTES

Section 12.1

The idea of constructing finite dimensional linear spaces of functions of several variables by taking tensor products of spaces of functions of one variable is an old and well-established technique in analysis. Tensor-product splines appeared first explicitly as the solution of a certain best interpolation problem in the article by deBoor [1962]. Although they are firmly entrenched as part of the spline lore, tensor-product splines are treated explicitly in only a relatively small number of papers.

Section 12.3

The first results on the approximation power of tensor-product splines were estimates on the interpolation error for best interpolating splines—cf.

deBoor [1962] and Ahlberg, Nilson, and Walsh [1965b]. Quasi-interpolants based on tensor-product B-splines have been studied by several authors—see, for example, Aubin [1967, 1972], Fix and Strang [1969], DiGuglielmo [1969], Babuska [1970], deBoor and Fix [1973], Strang and Fix [1973], Munteanu and Schumaker [1974], and Lyche and Schumaker [1975]. All of these papers have given error bounds for their quasi-interpolants in terms of the total modulus of continuity. Dahmen, DeVore, and Scherer [1980] noticed that the estimates need only depend on the tensor modulus of smoothness. Their paper also contains direct theorems for more general domains Ω (where Ω satisfies a rather complicated and quite restrictive form of the cone condition).

Section 12.4

The inverse theorems given here are based on the work of Dahmen, DeVore, and Scherer [1980]. Their paper contains similar results for the more general piecewise polynomial spaces $\mathcal{P}\mathcal{W}(\mathcal{P}_\Lambda;\mathbf{\Delta})$, where Λ is a general set of multi-indices. They also have been able to deal with nontensor partitions.

Section 12.5

The results of this section are also based on the work of Dahmen, DeVore, and Scherer [1980]. These authors were the first to see that the correct inverse theorems should involve coordinate moduli of smoothness. Earlier, Munteanu and Schumaker [1974] had given some inverse theorems involving total modulus of smoothness.

13 SOME MULTIDIMENSIONAL TOOLS

In this chapter we collect a number of results that are useful in dealing with functions of several variables. The topics treated include Sobolev spaces, multidimensional polynomials, Taylor theorems, moduli of smoothness, and K-functionals.

§ 13.1. NOTATION

The main problem in dealing with functions of more than one variable is to find a suitable notation. The standard solution to this problem is to use vector notation. Given a positive integer d, we write

$$\mathbf{R}^d = \{\mathbf{x}: \mathbf{x} = (x_1, \ldots, x_d), \quad x_i \in \mathbf{R}, i = 1, 2, \ldots, d\} \tag{13.1}$$

for the usual *Euclidean vector space*. We shall consistently use boldface letters for vectors. Given two vectors $\mathbf{a}$ and $\mathbf{b}$ in $\mathbf{R}^d$, we shall use the standard notations

$$\|\mathbf{a}\| = \left(\sum_{i=1}^{d} a_i^2 \right)^{1/2}; \tag{13.2}$$

$$\mathbf{a} \leqslant \mathbf{b} \quad \text{if and only if } a_i \leqslant b_i,\ i = 1, 2, \ldots, d; \tag{13.3}$$

$$\mathbf{a} < \mathbf{b} \quad \text{if and only if } \mathbf{a} \leqslant \mathbf{b} \quad \text{and } a_i < b_i, \text{ some } 1 \leqslant i \leqslant d; \tag{13.4}$$

$$\mathbf{a} + \mathbf{b} = (a_1 + b_1, \ldots, a_d + b_d). \tag{13.5}$$

$$\mathbf{0} = (0, \ldots, 0), \qquad \mathbf{1} = (1, \ldots, 1), \tag{13.6}$$

$$\mathbf{e}_i = \text{unit vector in the } i\text{th direction} = (0, \ldots, 1, \ldots, 0). \tag{13.7}$$

In order to be able to subscript quantities depending on d variables, it is convenient to introduce the sets $Z_+ = \{\text{nonegative integers}\}$ and

$$Z_+^d = \{\boldsymbol{\alpha}\colon \boldsymbol{\alpha} = (\alpha_1, \ldots, \alpha_d),\ \alpha_i \in Z_+, i = 1, 2, \ldots, d\}. \tag{13.8}$$

The set Z_+^d is often referred to as the set of *multi-indices*. If $\boldsymbol{\alpha} \in Z_+^d$, we write

$$|\boldsymbol{\alpha}| = \alpha_1 + \cdots + \alpha_d. \tag{13.9}$$

In dealing with vectors and with multi-indices, we shall observe the following standard conventions:

$$\mathbf{x}^\theta = x_1^\theta \cdots x_d^\theta, \qquad \mathbf{x} \in \mathbf{R}^d,\ \theta \in \mathbf{R}, \tag{13.10}$$

$$\mathbf{x}^{\boldsymbol{\alpha}} = x_1^{\alpha_1} \cdots x_d^{\alpha_d}, \qquad \mathbf{x} \in \mathbf{R}^d,\ \boldsymbol{\alpha} \in Z_+^d; \tag{13.11}$$

$$D^{\boldsymbol{\alpha}} = D_{x_1}^{\alpha_1} \cdots D_{x_d}^{\alpha_d}, \qquad \boldsymbol{\alpha} \in Z_+^d, \tag{13.12}$$

where D_{x_i} stands for the derivative in the ith variable.

Suppose now that Ω is a *bounded open set* in $\mathbf{R}^d$. Given any such Ω, we define

$$C(\Omega) = \{f\colon f \text{ is continuous on } \Omega\}, \tag{13.13}$$

$$C^m(\Omega) = \{f\colon D^{\boldsymbol{\alpha}} f \in C(\Omega) \text{ for all } 0 \leqslant |\boldsymbol{\alpha}| \leqslant m\}. \tag{13.14}$$

$$C^{\mathbf{m}}(\Omega) = \{f\colon D^{\boldsymbol{\alpha}} f \in C(\Omega), \text{ all } \mathbf{0} \leqslant \boldsymbol{\alpha} \leqslant \mathbf{m}\}, \tag{13.15}$$

$\mathbf{m} \in Z_+^d$.

Finally, we also need the classical Lebesgue spaces. Given any $1 \leqslant p \leqslant \infty$, we define

$$L_p(\Omega) = \left\{ \begin{array}{l} f\colon f \text{ is a measurable real-valued} \\ \text{function on } \Omega \text{ with } \|f\|_{L_p(\Omega)} < \infty \end{array} \right\}, \tag{13.16}$$

where

$$\|f\|_{L_p(\Omega)} = \begin{cases} \int_\Omega |f(\mathbf{x})|^p \, d\mathbf{x}, & 1 \leqslant p < \infty \\ \operatorname*{ess\,sup}_{\mathbf{x} \in \Omega} |f(\mathbf{x})|, & p = \infty \end{cases}. \tag{13.17}$$

§ 13.2. SOBOLEV SPACES

The aim in this section is to introduce the analogs of the Sobolev spaces $L_p^m[a,b]$ which were so useful in the one-dimensional theory. Suppose Ω is a bounded open set in $\mathbf{R}^d$, and let

$$C_0^\infty(\Omega)=\left\{\begin{array}{l} f\in C^\infty(\Omega)\text{: } f \text{ has support on a} \\ \text{compact subset } \Omega_f\subseteq\Omega \end{array}\right\}. \tag{13.18}$$

We call $C_0^\infty(\Omega)$ the *set of test functions*. It is possible to put a topology on this set in such a way that convergence of a sequence $\phi_n\in C_0^\infty(\Omega)$ to a function $\phi\in C_0^\infty(\Omega)$ means

1. There exists a compact set Ω_ϕ such that support $(\phi_n-\phi)\subseteq\Omega_\phi$ for all n;
2. $D^{\boldsymbol{\alpha}}\phi_n(x)\to D^{\boldsymbol{\alpha}}\phi(x)$ uniformly on Ω_ϕ for all $\boldsymbol{\alpha}\in Z_+^d$.

We denote this topological vector space by $D(\Omega)$.

DEFINITION 13.1. Distributions

We call

$$D'(\Omega)=\{\lambda\text{: } \lambda \text{ is a bounded linear functional on } D(\Omega)\} \tag{13.19}$$

the *set of distributions on* Ω.

Clearly $D'(\Omega)$ is itself a topological vector space. Distributions are not themselves functions defined on Ω. On the other hand, every function $f\in L_1(\Omega)$ can be associated with a distribution $\tilde{f}$ on Ω in a natural way by defining

$$\tilde{f}(\phi)=\int_\Omega \phi(x)f(x)\,dx, \qquad \text{all } \phi\in D(\Omega).$$

The converse of this statement is not true; there are distributions that are not associated with functions in $L_1(\Omega)$. If $\tilde{f}$ is a distribution that is associated with a function $f\in L_p(\Omega)$, then it is common practice to abuse the notation somewhat and to write $\tilde{f}\in L_p(\Omega)$.

Although distributions are not functions in the ordinary sense, it is possible to define a meaningful derivative of a distribution. Given a distribution f and a multi-index $\boldsymbol{\alpha}$, we define the $\boldsymbol{\alpha}$th derivative of f to be the distribution $D^{\boldsymbol{\alpha}}f$ whose effect on test functions is described by the formula

$$(D^{\boldsymbol{\alpha}}f)(\phi)=(-1)^{|\boldsymbol{\alpha}|}f(D^{\boldsymbol{\alpha}}\phi), \qquad \text{all } \phi\in D(\Omega).$$

Before introducing the Sobolev spaces, we need one more definition.

DEFINITION 13.2

Let A be a set of multi-indices. Then we say that A is *regular* provided for some nonnegative integers $r_1,\ldots,r_d$,

$$r_i \cdot \mathbf{e}_i \in A, \qquad i=1,2,\ldots,d, \tag{13.20}$$

and

$$\text{if } \boldsymbol{\alpha}\in A, \text{ then there is no } \boldsymbol{\beta}\in A \text{ with } \boldsymbol{\alpha}<\boldsymbol{\beta}. \tag{13.21}$$

DEFINITION 13.3. Sobolev Space

Let A be a regular set of multi-indices, and define

$$|f|_{L_p^A(\Omega)} = \sum_{\alpha\in A} \|D^{\alpha} f\|_{L_p(\Omega)}, \tag{13.22}$$

and

$$\|f\|_{L_p^A(\Omega)} = \|f\|_{L_p(\Omega)} + |f|_{L_p^A(\Omega)}. \tag{13.23}$$

We call

$$L_p^A(\Omega) = \{ f\in L_p(\Omega) \colon \|f\|_{L_p^A(\Omega)} < \infty \} \tag{13.24}$$

the *Sobolev space associated with A and p*.

With some work it can be shown that $L_p^A(\Omega)$ is a Banach space with respect to the norm (13.23). We now give two especially important examples.

EXAMPLE 13.4. Classical Sobolev Space

Let $A=\{\boldsymbol{\alpha}\colon |\boldsymbol{\alpha}|=r\}$, where r is any positive integer. We call

$$W_p^r(\Omega) = L_p^A(\Omega) \tag{13.25}$$

the classical *Sobolev space*.

Discussion. In this case

$$\|f\|_{W_p^r(\Omega)} = \|f\|_{L_p(\Omega)} + |f|_{W_p^r(\Omega)},$$

where

$$|f|_{W_p^r(\Omega)} = \sum_{|\alpha|=r} \|D^\alpha f\|_{L_p(\Omega)}.$$

For more on the classical Sobolev spaces, see, for example, Adams [1975]. ■

The next example introduces the space that we used in discussing the approximation power of tensor-product polynomial splines.

EXAMPLE 13.5 Tensor Sobolev Spaces

Let $A = \{r_i \cdot \mathbf{e}_i\}_{i=1}^d$, where $r_1, \ldots, r_d$ are positive integers, and where $\mathbf{e}_i$ denotes the unit vector in the *ith* direction. We call

$$L_p^{\mathbf{r}}(\Omega) = L_p^A(\omega), \quad \mathbf{r} = (r_1, \ldots, r_d) \tag{13.26}$$

a *tensor Sobolev space.*

Discussion. Here

$$\|f\|_{L_p^{\mathbf{r}}(\Omega)} = \|f\|_{L_p(\Omega)} + |f|_{L_p^{\mathbf{r}}(\Omega)}$$

with

$$|f|_{L_p^{\mathbf{r}}(\Omega)} = \sum_{i=1}^d \|D_{x_i}^{r_i} f\|_{L_p(\Omega)}.$$ ■

We conclude this section with an imbedding theorem which states that the space $L_p^A(\Omega)$ is a subset of classical Sobolev spaces. To prove the result, it is necessary to make some kind of assumption on Ω to prevent difficulties near its boundary.

DEFINITION 13.6

Let Ω be a bounded open subset of $\mathbf{R}^d$. Then we say that Ω is *star shaped* provided there exists an open ball

$$B = B_{\mathbf{x}_0, \rho} = \{\mathbf{x} : \|\mathbf{x} - \mathbf{x}_0\| < \rho\}$$

such that for every $\mathbf{x} \in B$ and every $\mathbf{y} \in \Omega$, the line from $\mathbf{x}$ to $\mathbf{y}$ lies in Ω.

Clearly every bounded open *convex* set Ω is star shaped. Figure 39 shows some common nonconvex star-shaped domains as well as some domains that are not star shaped.

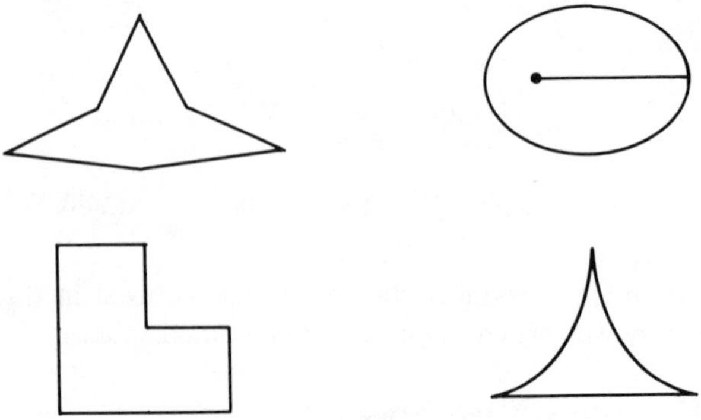

Figure 39. (a) Star-shaped domains. (b) Non-star-shaped domains.

THEOREM 13.7

Let Ω be a star-shaped domain, and suppose A is a regular set of multi-indices (cf. Definition 13.2). Let $r_1,\ldots,r_d$ be the integers appearing in (13.20). Then $L_p^A(\Omega)\subseteq W_p^k(\Omega)$ for all $k<\min_i(r_i)$. More precisely, there exists a constant C such that

$$\|f\|_{W_p^k(\Omega)}\leqslant C\|f\|_{L_p^A(\Omega)},\qquad \text{all } f\in L_p^A(\Omega). \tag{13.27}$$

Discussion. See Smith [1970], page 66. ∎

§ 13.3. POLYNOMIALS

DEFINITION 13.8

Given a set Λ of multi-indices in Z_+^d, we define the associated set of *polynomials* in $\mathbf{x}\in\mathbf{R}^d$ by

$$\mathcal{P}_\Lambda=\operatorname{span}\{\mathbf{x}^{\boldsymbol{\alpha}}: \boldsymbol{\alpha}\in\Lambda\}. \tag{13.28}$$

It is clear that $\mathcal{P}_\Lambda$ is a linear space whose dimension is equal to the number of indices in the set Λ. While $\mathcal{P}_\Lambda$ makes sense for arbitrary sets Λ, there are two special cases of particular interest. We highlight them in the following two examples:

EXAMPLE 13.9. The Space of Tensor-Product Polynomials

Let $m_1,\ldots,m_d$ be positive integers, $\Lambda=\{\boldsymbol{\alpha}:\ \alpha_i<m_i,\ i=1,2,\ldots,d\}$, and

consider

$$\mathcal{P}_\Lambda = \mathcal{P}_{\mathbf{m}} = \otimes_{i=1}^d \mathcal{P}_{m_i} = \operatorname{span}\left\{ x_1^{i_1-1} \cdots x_d^{i_d-1} \right\}_{i_1=1,\ldots,i_d=1}^{m_1 \qquad m_d}.$$

Discussion. This space is of dimension $m_1 \cdots m_d$. For quick reference we list the basis functions for several spaces of tensor-product polynomials in two variables in Table 10. ■

Table 10. Some Tensor-Product Polynomial Spaces

m_1	m_2	Dimension	Basis Functions	Name
2	2	4	$1, x, y, xy$	Bilinear
3	3	9	$1, x, y, xy, x^2, y^2$ xy^2, x^2y, x^2y^2	Biquadratic
4	4	16	$1, x, y, xy, x^2, y^2$ $xy^2, xy^3, x^2y, x^2y^2,$ $x^2y^3, x^3, y^3, x^3y,$ x^3y^2, x^3y^3	Bicubic

EXAMPLE 13.10. Polynomials of Total Order m

Let m be a positive integer, and let $\Lambda = \{\boldsymbol{\alpha}\colon |\boldsymbol{\alpha}| < m\}$. Then

$$\mathcal{P}_\Lambda = \mathcal{P}_m^{(d)} = \operatorname{span}\{\mathbf{x}^{\boldsymbol{\alpha}}\colon |\boldsymbol{\alpha}| < m\}.$$

Discussion. The dimension of this space depends on both m and d. For $d = 2$ it is given by $m(m+1)/2$. For some specific examples, see Table 11. ■

Table 11. Some Spaces of Polynomials of Total Order

m	Dimension	Basis Functions	Name
1	1	1	Constants
2	3	$1, x, y$	Linear
3	6	$1, x, y, xy, x^2, y^2$	Quadratic
4	10	$1, x, y, xy, x^2, y^2, xy^2, x^2y, x^3, y^3$	Cubic

§ 13.4. TAYLOR THEOREMS AND THE APPROXIMATION POWER OF POLYNOMIALS

In this section we develop some rather general Taylor expansions for functions of several variables, and we use them to derive results on the approximation power of multidimensional polynomials.

DEFINITION 13.11. Taylor Expansion

Given any set Λ of multi-indices and any function f such that the required derivatives exist at $\mathbf{y}$, let

$$T_{\mathbf{y}}^{\Lambda}f(\mathbf{x}) = \sum_{\alpha \in \Lambda} \frac{(\mathbf{x}-\mathbf{y})^{\alpha} D^{\alpha}f(\mathbf{y})}{\alpha!}, \tag{13.29}$$

where

$$\alpha! = \alpha_1! \cdots \alpha_d!,$$

and

$$(\mathbf{x}-\mathbf{y})^{\alpha} = (x_1-y_1)^{\alpha_1} \cdots (x_d-y_d)^{\alpha_d}.$$

We call $T_{\mathbf{y}}^{\Lambda}f$ the *Taylor expansion of f about the point* $\mathbf{y}$. It is a polynomial in the space $\mathcal{P}_{\Lambda}$.

The most important example of a Taylor expansion of the form (13.29) is the classical Taylor expansion:

EXAMPLE 13.12. Classical Taylor Expansion

Let $\Lambda = \{|\alpha| < m\}$. Then

$$T_{\mathbf{y}}^{m}f(\mathbf{x}) = T_{\mathbf{y}}^{\Lambda}f(\mathbf{x}) = \sum_{|\alpha|<m} \frac{D^{\alpha}f(\mathbf{y})(\mathbf{x}-\mathbf{y})^{\alpha}}{\alpha!}. \tag{13.30}$$

Discussion. The classical Taylor expansion produces a polynomial of total order m (cf. Example 13.10). If Ω is a bounded open set such that the line from $\mathbf{x}$ to $\mathbf{y}$ lies in Ω, then it is known (cf., e.g., Edwards [1965]) that

$$f(\mathbf{x}) - T_{\mathbf{y}}^{m}f(\mathbf{x}) = \sum_{|\alpha|=m} \frac{(\mathbf{x}-\mathbf{y})^{\alpha}}{\alpha!} \int_0^1 s^{m-1} D^{\alpha}f[\mathbf{x}+s(\mathbf{y}-\mathbf{x})]\, ds. \tag{13.31}$$ ∎

The Taylor expansions defined in (13.29) apply only to functions whose derivatives are defined at the point y. For our applications we need a Taylor expansion that applies to a wider class of functions.

DEFINITION 13.13

Suppose Ω is a star-shaped domain with respect to a ball B as in Definition 13.6. Let ψ be a function in $C_0^{\infty}(B)$ with $\int_B \psi = 1$. Then given any $f \in$

$L_1(B)$, we define its *Taylor expansion with respect to* ψ *and* Λ by

$$T_\psi^\Lambda f(\mathbf{x}) = \sum_{\alpha \in \Lambda} \int_B (-1)^{|\alpha|} D^\alpha \left[\psi(\mathbf{y}) \frac{(\mathbf{x}-\mathbf{y})^\alpha}{\alpha!} \right] f(\mathbf{y})\, d\mathbf{y}. \tag{13.32}$$

For sufficiently smooth functions f, $T_\psi^\Lambda f$ is closely related to $T_\mathbf{y}^\Lambda f$. In fact, repeated application of Green's formula shows that

$$T_\psi^\Lambda f(\mathbf{x}) = \int_B \psi(\mathbf{y})\, T_\mathbf{y}^\Lambda f(\mathbf{x})\, d\mathbf{y}.$$

It is useful to have a special notation for the two most important Taylor expansions. If m is a positive integer, we write

$$T_\psi^m f(\mathbf{x}) = \sum_{|\alpha| < m} \int_B (-1)^{|\alpha|} D^\alpha \left[\psi(\mathbf{y}) \frac{(\mathbf{x}-\mathbf{y})^\alpha}{\alpha!} \right] f(\mathbf{y})\, d\mathbf{y}. \tag{13.33}$$

We call this the *total Taylor expansion* since it produces a polynomial of total order m. If $\mathbf{m}$ is a vector in Z_+^d, then we write

$$T_\psi^\mathbf{m} f(\mathbf{x}) = \sum_{\alpha < \mathbf{m}} \int_B (-1)^{|\alpha|} D^\alpha \left[\psi(\mathbf{y}) \frac{(\mathbf{x}-\mathbf{y})^\alpha}{\alpha!} \right] f(\mathbf{y})\, d\mathbf{y}. \tag{13.34}$$

We call this the *tensor Taylor expansion* since it produces a tensor-product polynomial.

The following continuity assertion for T will be useful later:

THEOREM 13.14

For all $f, g \in L_1(B)$,

$$\| T_\psi^\Lambda f - T_\psi^\Lambda g \|_{W_\infty^{m-1}(\Omega)} \leq C \| f - g \|_{L_1(B)}, \tag{13.35}$$

where C is a constant depending only on m, d, ψ, and $|\Omega|$, and where $|\Omega|$ is the diameter of Ω defined by $|\Omega| = \sup\{\|\mathbf{y}-\mathbf{x}\|: \mathbf{x}, \mathbf{y} \in \Omega\}$.

Proof. This estimate is clear from the fact that

$$T_\psi^\Lambda (f-g)(\mathbf{x}) = \sum_{\alpha \in \Lambda} \int_B (-1)^{|\alpha|} (f-g)(\mathbf{y}) D^\alpha \left[\psi(\mathbf{y}) \frac{(\mathbf{x}-\mathbf{y})^\alpha}{\alpha!} \right] d\mathbf{y}. \qquad \blacksquare$$

Our aim now, is to study the remainder term $R_\psi^\Lambda f = f - T_\psi^\Lambda f$, but first we need a definition.

DEFINITION 13.15

We say that the set Λ of multi-indices is *complete* provided that

$$\text{if } \boldsymbol{\beta}\in\Lambda, \text{ then all } \boldsymbol{\alpha} \text{ with } \boldsymbol{\alpha}<\boldsymbol{\beta} \text{ are also in } \Lambda. \tag{13.36}$$

Given a complete set Λ, we define the *boundary* of Λ by

$$\partial\Lambda=\{\boldsymbol{\beta}: \boldsymbol{\beta}\notin\Lambda, \text{ but all } \boldsymbol{\alpha} \text{ with } \boldsymbol{\alpha}<\boldsymbol{\beta} \text{ are in } \Lambda\}. \tag{13.37}$$

We note that the sets Λ corresponding to total Taylor expansions and to tensor Taylor expansions [cf. (13.33) and (13.34)] are both complete. In particular, we have

$$\Lambda=\{\boldsymbol{\alpha}: |\boldsymbol{\alpha}|<m\} \text{ implies } \partial\Lambda=\{\boldsymbol{\alpha}: |\boldsymbol{\alpha}|=m\}, \tag{13.38}$$

$$\Lambda=\{\boldsymbol{\alpha}: \boldsymbol{\alpha}<\mathbf{m}=(m_1,\ldots,m_d)\} \text{ implies } \partial\Lambda=\{m_i\mathbf{e}_i\}_1^d. \tag{13.39}$$

THEOREM 13.16. Generalized Taylor Remainder Theorem

Let Λ be a complete set of multi-indices. Then for any function $f\in C^\infty(\Omega)$,

$$f(\mathbf{x})=T_\psi^\Lambda f(\mathbf{x})+R_\psi^\Lambda f(\mathbf{x}), \qquad \text{all } \mathbf{x}\in\Omega, \tag{13.40}$$

where

$$R_\psi^\Lambda f(\mathbf{x})=\sum_{\boldsymbol{\alpha}\in\partial\Lambda}\int_\Omega K_{\boldsymbol{\alpha}}(\mathbf{x},\mathbf{y})\,D^{\boldsymbol{\alpha}}f(\mathbf{y})\,d\mathbf{y}, \tag{13.41}$$

and $K_{\boldsymbol{\alpha}}(\mathbf{x},\mathbf{y})$ are certain kernel functions with

$$|D_{\mathbf{x}}^{\boldsymbol{\beta}}D_{\mathbf{y}}^{\boldsymbol{\gamma}}K_{\boldsymbol{\alpha}}(\mathbf{x},\mathbf{y})|\leqslant C\,\|\mathbf{x}-\mathbf{y}\|^{|\boldsymbol{\alpha}|-|\boldsymbol{\beta}|-|\boldsymbol{\gamma}|-d}. \tag{13.42}$$

Here C is a constant that depends only on $\boldsymbol{\alpha}$, $\boldsymbol{\beta}$, $\boldsymbol{\gamma}$, d, and Ω.

Proof. We prove the result first for the case of total polynomials; that is, $\Lambda=\{|\boldsymbol{\alpha}|<m\}$ and $\partial\Lambda=\{|\boldsymbol{\alpha}|=m\}$. If we multiply both sides of the classical Taylor expansion (13.31) by $\psi(\mathbf{y})$ and integrate over B, we obtain

$$f(\mathbf{x})-T_\psi^\Lambda f(\mathbf{x})=m\sum_{|\boldsymbol{\alpha}|=m}\int_B\frac{\psi(\mathbf{y})(\mathbf{x}-\mathbf{y})^{\boldsymbol{\alpha}}}{\boldsymbol{\alpha}!}\int_0^1 s^{m-1}D^{\boldsymbol{\alpha}}f[\mathbf{x}+s(\mathbf{y}-\mathbf{x})]\,ds\,d\mathbf{y}.$$

Now, by using Fubini's theorem and the change of variables $\mathbf{z}=\mathbf{x}+s(\mathbf{y}-\mathbf{x})$, we obtain

$$\begin{aligned}
&\int_B \psi(\mathbf{y})(\mathbf{x}-\mathbf{y})^{\alpha}\int_0^1 s^{m-1}D^{\alpha}f[\mathbf{x}+s(\mathbf{y}-\mathbf{x})]\,ds\,d\mathbf{y}\\
&\qquad=\int_0^1\int_B \psi(\mathbf{y})(\mathbf{x}-\mathbf{y})^{\alpha}s^{m-1}D^{\alpha}f[\mathbf{x}+s(\mathbf{y}-\mathbf{x})]\,d\mathbf{y}\,ds\\
&\qquad=\int_0^1\int_B \psi[(\mathbf{x}+s^{-1}(\mathbf{z}-\mathbf{x})](\mathbf{x}-\mathbf{z})^{\alpha}s^{-1}D^{\alpha}f(\mathbf{z})s^{-d}\,d\mathbf{z}\,ds\\
&\qquad=\int_B (\mathbf{x}-\mathbf{z})^{\alpha}D^{\alpha}f(\mathbf{z})\left(\int_0^1 \psi[\mathbf{x}+s^{-1}(\mathbf{z}-\mathbf{x})]s^{-d-1}\,ds\right)d\mathbf{z}\\
&\qquad=\frac{\alpha!}{m}\int_B k_{\alpha}(\mathbf{x},\mathbf{z})D^{\alpha}f(\mathbf{z})\,d\mathbf{z},
\end{aligned}$$

where

$$k_{\alpha}(\mathbf{x},\mathbf{y})=\frac{m(\mathbf{x}-\mathbf{y})^{\alpha}k(\mathbf{x},\mathbf{y})}{\alpha!}$$

$$k(\mathbf{x},\mathbf{y})=\int_0^1 s^{-d-1}\psi[\mathbf{x}+s^{-1}(\mathbf{y}-\mathbf{x})]\,ds.$$

The use of Fubini's theorem is justified since

$$\begin{aligned}
|k(\mathbf{x},\mathbf{z})|&=\left|\int_0^1 \psi[\mathbf{x}+s^{-1}(\mathbf{z}-\mathbf{x})]s^{-d-1}\,ds\right|\\
&=\left|\int_{|\mathbf{z}-\mathbf{x}|/|\Omega|}^1 \psi[\mathbf{x}+s^{-1}(\mathbf{z}-\mathbf{x})]s^{-d-1}\,ds\right|\leqslant C|\mathbf{x}-\mathbf{z}|^{-d}, \qquad (13.43)
\end{aligned}$$

where $|\Omega|$ is the diameter of Ω (cf. Theorem 13.14), and where $C=|\Omega|^d\|\psi\|_{L_\infty(B)}/d$. A similar argument shows that the derivatives of k_{α} satisfy (13.42), and the theorem is proved in the case of total polynomials.

Suppose now that Λ is a complete set of multi-indices. Let $m=1+\max_{\alpha\in\Lambda}|\alpha|$. We claim that

$$\text{if } |\boldsymbol{\beta}|\leqslant m \text{ and } \boldsymbol{\beta}\notin\Lambda, \text{ then there exists } \tilde{\boldsymbol{\beta}}\in\partial\Lambda \text{ with } \tilde{\boldsymbol{\beta}}\leqslant\boldsymbol{\beta}. \qquad (13.44)$$

Indeed, suppose $|\boldsymbol{\beta}|\leqslant m$ and $\boldsymbol{\beta}\notin\Lambda$. Then if $\boldsymbol{\beta}\notin\partial\Lambda$, there must be some $\boldsymbol{\beta}_2<\boldsymbol{\beta}$ with $\boldsymbol{\beta}_2\notin\Lambda$. If $\boldsymbol{\beta}_2\notin\partial\Lambda$, we can find $\boldsymbol{\beta}_3\notin\Lambda$ with $\boldsymbol{\beta}_3<\boldsymbol{\beta}_2$, and so on. At some step we must find $\tilde{\boldsymbol{\beta}}=\boldsymbol{\beta}_n$ such that all $\boldsymbol{\alpha}<\boldsymbol{\beta}_n$ belong to Λ (and so

$\tilde{\boldsymbol{\beta}}\in\partial\Lambda$), for otherwise we arrive at $\mathbf{0}\notin\Lambda$ which contradicts property (13.36).

We are ready to prove (13.40). By the result for total polynomials we know that

$$f(\mathbf{x})=T_\psi^\Lambda f(\mathbf{x})+\sum_{|\boldsymbol{\beta}|<m}\int\frac{(\mathbf{x}-\mathbf{y})^{\boldsymbol{\beta}}}{\boldsymbol{\beta}!}\psi(\mathbf{y})D^{\boldsymbol{\beta}}f(\mathbf{y})\,d\mathbf{y}+\sum_{|\boldsymbol{\beta}|=m}\int k_{\boldsymbol{\beta}}(\mathbf{x},\mathbf{y})D^{\boldsymbol{\beta}}f(\mathbf{y})\,d\mathbf{y}.$$

Now if $|\boldsymbol{\beta}|=m$, then by (13.44) we can write $D^{\boldsymbol{\beta}}f(\mathbf{y})=D^{\boldsymbol{\beta}-\tilde{\boldsymbol{\beta}}}D^{\tilde{\boldsymbol{\beta}}}f$ and integrate $|\boldsymbol{\beta}-\tilde{\boldsymbol{\beta}}|$ times by parts to obtain

$$\int k_{\boldsymbol{\beta}}(\mathbf{x},\mathbf{y})D^{\boldsymbol{\beta}}f(\mathbf{y})\,d\mathbf{y}=(-1)^{|\boldsymbol{\beta}-\tilde{\boldsymbol{\beta}}|}\int\left[D_{\mathbf{y}}^{\boldsymbol{\beta}-\tilde{\boldsymbol{\beta}}}k_{\boldsymbol{\beta}}(\mathbf{x},\mathbf{y})\right]D^{\tilde{\boldsymbol{\beta}}}f(\mathbf{y})\,d\mathbf{y}$$

with $\tilde{\boldsymbol{\beta}}\in\partial\Lambda$. Similarly, if $|\boldsymbol{\beta}|<m$ and $\boldsymbol{\beta}\notin\Lambda$, then

$$\int\psi(\mathbf{y})\frac{(\mathbf{x}-\mathbf{y})^{\boldsymbol{\beta}}}{\boldsymbol{\beta}!}D^{\boldsymbol{\beta}}f(\mathbf{y})\,d\mathbf{y}=(-1)^{|\boldsymbol{\beta}-\tilde{\boldsymbol{\beta}}|}\int D_{\mathbf{y}}^{\boldsymbol{\beta}-\tilde{\boldsymbol{\beta}}}\left[\frac{\psi(\mathbf{y})(\mathbf{x}-\mathbf{y})^{\boldsymbol{\beta}}}{\boldsymbol{\beta}!}\right]D^{\tilde{\boldsymbol{\beta}}}f(\mathbf{y})\,d\mathbf{y}$$

with $\tilde{\boldsymbol{\beta}}\in\partial\Lambda$. It follows that (13.40) holds with appropriate $K_{\boldsymbol{\alpha}}(\mathbf{x},\mathbf{y})$ satisfying (13.42). ∎

We can now give an explicit error bound for $f-T_\psi^\Lambda f$ when f is a sufficiently smooth function.

THEOREM 13.17

Let Λ be a complete set of multi-indices. Then for any $0<\varepsilon<1$ there exists a constant C (depending only on d, ψ, Λ, ε, and Ω) such that

$$\|f-T_\psi^\Lambda f\|_{L_q(\Omega)}\leqslant C\sum_{\boldsymbol{\alpha}\in\partial\Lambda}\|D^{\boldsymbol{\alpha}}f\|_{L_p(\Omega)} \tag{13.45}$$

for all $f\in L_p^{\partial\Lambda}(\Omega)$ and for all $1\leqslant p,q\leqslant\infty$ such that

$$\varepsilon\leqslant\max\left[\left\lfloor\frac{|\boldsymbol{\alpha}|}{d}\right\rfloor,\frac{1}{q}-\frac{1}{p}+\frac{|\boldsymbol{\alpha}|}{d},\ \min\left(1-\frac{1}{p},\frac{1}{q}\right)\right],\qquad \boldsymbol{\alpha}\in\partial\Lambda. \tag{13.46}$$

Proof. In view of Theorem 13.16, we need to show that

$$\left\|\int K_{\boldsymbol{\alpha}}(\mathbf{x},\mathbf{y})D^{\boldsymbol{\alpha}}f(\mathbf{y})\,d\mathbf{y}\right\|_q\leqslant C\|D^{\boldsymbol{\alpha}}f\|_p \tag{13.47}$$

for all $\boldsymbol{\alpha}\in\partial\Lambda$. Fix $\boldsymbol{\alpha}\in\partial\Lambda$. There are three cases.

CASE 1. ($|\alpha| \geqslant d$). In this case $\lfloor |\alpha|/d| \rfloor \geqslant 1$, and we have to prove the result for all p and q. Now if $f \in C^{\infty}(\Omega) \cap L_1^{\partial\Lambda}(\Omega)$, then applying the Hölder inequality to the integral in (13.47) and using the fact that the $K_{\alpha}(\mathbf{x},\mathbf{y})$ are uniformly bounded [cf. (13.42)], we see that (13.47) holds for $q=\infty, p=1$. Since $C^{\infty}(\Omega) \cap L_1^{\partial\Lambda}(\Omega)$ is dense in $L_1^{\partial\Lambda}(\Omega)$, we have proved (13.47) for $q=\infty, p=1$ and for all elements of $L_1^{\partial\Lambda}(\Omega)$. The result now follows for all p and q by Hölder's inequality.

CASE 2. ($|\alpha|<d$ and $1/q-1/p+|\alpha|/d \geqslant \varepsilon$). It suffices to prove the result for $q=p<\infty$ since all other cases follow by Hölder's inequality. It will be enough to work with $f \in C^{\infty}(\Omega) \cap L_p^{\partial\Lambda}(\Omega)$ since this space is dense in $L_p^{\partial\Lambda}(\Omega)$. Let

$$g_{\alpha}(\mathbf{x}) = \begin{cases} \|\mathbf{x}\|^{|\alpha|-d}, & \|\mathbf{x}\| \leqslant |\Omega| \\ 0, & \text{otherwise} \end{cases},$$

and suppose that we define $D^{\alpha}f(\mathbf{x})$ to be zero for $\mathbf{x}$ outside of Ω. Then it is clear from (13.42) that for all $\mathbf{x} \in \Omega$,

$$\left|\int K_{\alpha}(\mathbf{x},\mathbf{y}) D^{\alpha}f(\mathbf{y})\,dy\right| \leqslant C\int g_{\alpha}(\mathbf{x}-\mathbf{y}) D^{\alpha}f(\mathbf{y})\,dy.$$

Now using Young's inequality,

$$\left\|\int K_{\alpha}(\mathbf{x},\mathbf{y}) D^{\alpha}f(\mathbf{y})\,dy\right\|_q \leqslant C\| g_{\alpha}\|_{L_s(\Omega)} \|D^{\alpha}f\|_{L_p(\Omega)},$$

where $1/s = 1-(1/p)+(1/q) \geqslant 1-(|\alpha|/d)+\varepsilon$. Thus (13.47) follows since

$$\| g_{\alpha}\|_{L_s(R^d)} \leqslant C(d)\left[\frac{1}{(|\alpha|-d)s+d}\right]^{1/s}.$$

CASE 3. ($|\alpha|<d$, $1/q=1/p-|\alpha|/d$, and $\min(1-1/p), 1/q] \geqslant \varepsilon$). In this case p must be less than ∞ (since otherwise $1/q=-|\alpha|/d$, which is impossible). Then for any $f \in C^{\infty}(\Omega) \cap L_p^{\partial\Lambda}(\Omega)$ we have

$$\left|\int K_{\alpha}(\mathbf{x},\mathbf{y}) D^{\alpha}f(\mathbf{y})\,dy\right| \leqslant C\Phi_{|\alpha|}(|D^{\alpha}f|),$$

where

$$\Phi_{|\alpha|}(g)(\mathbf{x}) = \int g(\mathbf{y}) \|\mathbf{x}-\mathbf{y}\|^{|\alpha|-d}\,dy$$

is the so-called Riesz potential, and where we have assumed that $D^{\alpha}f$ has been extended to $\mathbf{R}^d$ as in Case 2. The estimate (13.47) now follows from standard estimates on the Riesz potential. ■

The condition (13.46) is satisfied for all but a small set of p and q. Indeed, if $|\boldsymbol{\alpha}| \geqslant d$, then there is no restriction on p and q. If $|\boldsymbol{\alpha}| < d$, we may describe the set of p and q satisfying (13.46) geometrically; it consists of all p and q except for those such that $(1/p, 1/q)$ lies in the two cross-hatched parallelopipeds in Figure 40. Each of these two parallelopipeds has width depending on ε, and, in fact reduce to the points $(1, 1-|\boldsymbol{\alpha}|/d)$ and $(|\boldsymbol{\alpha}|/d, 0)$ as $\varepsilon \to 0$.

It is convenient to have a version of Theorem 13.17 where the dependence on the size of the domain Ω is given explicitly. To measure the size of Ω, we introduce

$$\boldsymbol{\delta} = (\delta_1, \ldots, \delta_d), \qquad \text{where } \delta_i = \sup\{|y_i - x_i| : \mathbf{x}, \mathbf{y} \in \Omega\}. \tag{13.48}$$

The vector $\boldsymbol{\delta}$ contains the dimensions of the smallest rectangle enclosing Ω.

THEOREM 13.18

Let Λ be a complete set of multi-indices. Then for any $0 < \varepsilon < 1$ there exists a constant C (depending only on d, ψ, Λ, and ε) such that

$$\|f - T_\psi^\Lambda f\|_{L_q(\Omega)} \leqslant C \boldsymbol{\delta}^{1/p - 1/q} \sum_{\boldsymbol{\alpha} \in \partial\Lambda} \boldsymbol{\delta}^{\boldsymbol{\alpha}} \|D^{\boldsymbol{\alpha}} f\|_{L_p(\Omega)} \tag{13.49}$$

for all $f \in L_p^{\partial\Lambda}(\Omega)$ and for all $1 \leqslant q \leqslant p \leqslant \infty$ satisfying (13.46).

Proof. First suppose $p = q$. Then if Ω is the unit cube with $\boldsymbol{\delta} = (1, 1, \ldots, 1)$, (13.49) follows directly from Theorem 13.17. The assertion for a general Ω

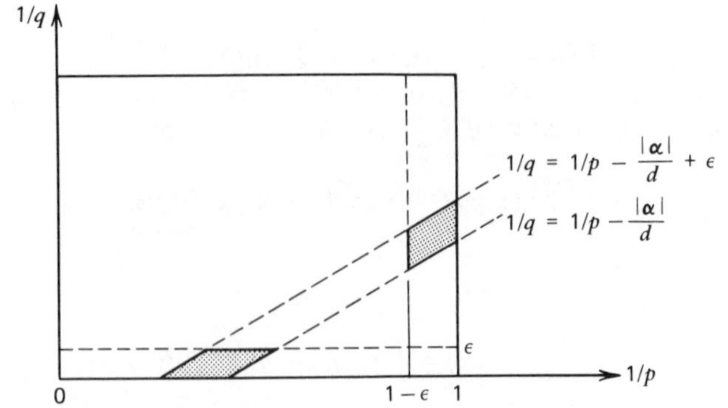

Figure 40. The set of p and q satisfying (13.46).

now follows by a simple change of variables. Finally, to get the result for $p \neq q$, we simply observe that by Hölder's inequality

$$\|\phi\|_q \leqslant \delta^{1/q-1/p}\|\phi\|_p \tag{13.50}$$

for any ϕ. ■

Theorem 13.18 shows that the Taylor expansion provides a polynomial which approximates a given smooth function quite well. We now show that, in fact, the derivatives of the Taylor polynomial are also good approximations to the derivatives of f. First we need the following commutativity result:

THEOREM 13.19

Given any multi-index $\boldsymbol{\beta}$,

$$D^{\boldsymbol{\beta}} T_{\psi}^{\Lambda} f = T_{\psi}^{\Lambda-\boldsymbol{\beta}} D^{\boldsymbol{\beta}} f, \qquad \text{all } f \in L_1(B). \tag{13.51}$$

Proof. First consider $f \in C^{\infty}(B)$. Then since $D^{\boldsymbol{\beta}}\mathbf{x}^{\boldsymbol{\alpha}} = 0$ for all $\boldsymbol{\alpha} < \boldsymbol{\beta}$, we have

$$\begin{aligned} D_{\mathbf{x}}^{\boldsymbol{\beta}} T_{\psi}^{\Lambda} f(x) &= \sum_{\substack{\boldsymbol{\alpha} \in \Lambda \\ \boldsymbol{\alpha} > \boldsymbol{\beta}}} \int \psi(\mathbf{y}) D_{\mathbf{y}}^{\boldsymbol{\alpha}} f(\mathbf{y}) \frac{(\mathbf{x}-\mathbf{y})^{\boldsymbol{\alpha}-\boldsymbol{\beta}}}{(\boldsymbol{\alpha}-\boldsymbol{\beta})!} d\mathbf{y} \\ &= \sum_{\boldsymbol{\delta} \in \Lambda - \boldsymbol{\beta}} \int \psi(\mathbf{y}) D_{\mathbf{y}}^{\boldsymbol{\beta}+\boldsymbol{\delta}} f(\mathbf{y}) \frac{(\mathbf{x}-\mathbf{y})^{\boldsymbol{\delta}}}{\boldsymbol{\delta}!} d\mathbf{y} \\ &= T_{\psi}^{\Lambda-\boldsymbol{\beta}}(D^{\boldsymbol{\beta}} f)(\mathbf{x}). \end{aligned}$$

The assertion then follows for general distributions since $D_{\mathbf{x}}^{\boldsymbol{\beta}}$ and T_{ψ}^{Λ} are bounded operators on $L_1(B)$ while $C^{\infty}(B)$ is dense in $L_1(B)$. ■

THEOREM 13.20

Let Λ be a complete set of multi-indices, and let $0 < \varepsilon < 1$. Then there exists a constant C (depending only on d, ε, ψ, and Λ) such that for all $f \in L_p(\Omega)$,

$$\|D^{\boldsymbol{\beta}}(f - T_{\psi}^{\Lambda} f)\|_q \leqslant C\delta^{1/q-1/p} \sum_{\boldsymbol{\alpha} \in \partial(\Lambda-\boldsymbol{\beta})} \delta^{\boldsymbol{\alpha}} \|D^{\boldsymbol{\alpha}} D^{\boldsymbol{\beta}} f\|_p \tag{13.52}$$

for any $\boldsymbol{\beta}$ and for all $1 \leqslant q \leqslant p \leqslant \infty$ satisfying (13.46) with $\partial\Lambda$ replaced by $\partial(\Lambda - \boldsymbol{\beta})$.

Proof. In view of Theorem 13.19,

$$\|D^{\beta}(f-T_{\psi}^{\Lambda}f)\|_p=\|R_{\psi}^{\Lambda-\beta}D^{\beta}f\|_p.$$

Now it is easy to see that if Λ is complete, then so is $\Lambda-\boldsymbol{\beta}$. The assertion (13.52) then follows from Theorem 13.18 coupled with (13.51). Here we note that $\boldsymbol{\beta}+\partial(\Lambda-\boldsymbol{\beta})\subseteq\partial\Lambda$ so that $f\in L_p^{\partial\Lambda}(\Omega)$ implies $D^{\beta}f\in L_p^{\partial(\Lambda-\beta)}(\Omega)$. ■

We close this section by noting that it is possible to give bounds on how well polynomials approximate smooth functions in terms of certain moduli of smoothness. We state the following result here. (The definition of the modulus of smoothness can be found in Section 13.5 below).

THEOREM 13.21

Let Λ be a complete set of multi-indices. Then for any $0<\varepsilon<1$ there exists a constant C (depending only on d, ε, ψ, and Λ) such that

$$d(f,\mathcal{P}_{\Lambda})_q\leqslant C\boldsymbol{\delta}^{1/q-1/p}\omega_{\partial\Lambda}(f;\boldsymbol{\delta})_p \tag{13.53}$$

for all $f\in L_p(\Omega)$ and all $1\leqslant q\leqslant p\leqslant\infty$ satisfying (13.46).

Proof. For $p=q$ the result follows from Theorem 13.18 coupled with Theorem 13.30. The result for general q then follows from (13.50). ■

§ 13.5. MODULI OF SMOOTHNESS

In this section we study moduli of smoothness similar to those discussed in Section 2.8 for functions of one variable. Given a multi-index $\boldsymbol{\alpha}$ and any vector $\mathbf{h}>\mathbf{0}$, let

$$\Delta_{\mathbf{h}}^{\boldsymbol{\alpha}}=\Delta_{h_1}^{\alpha_1}\cdots\Delta_{h_d}^{\alpha_d}. \tag{13.54}$$

In applying $\Delta_{\mathbf{h}}^{\boldsymbol{\alpha}}$ to a function of d variables, we assume that $\Delta_{h_i}^{\alpha_i}$ applies to the ith variable.

DEFINITION 13.22

Given $1\leqslant p\leqslant\infty$ and a multi-index $\boldsymbol{\alpha}$, we define the ***α**-modulus of smoothness in the p-norm* of a function $f\in L_p(\Omega)$ by

$$\omega_{\boldsymbol{\alpha}}(f;\mathbf{t})_p=\sup_{\mathbf{h}<\mathbf{t}}\|\Delta_{\mathbf{h}}^{\boldsymbol{\alpha}}f\|_{L_p(\Omega_{\boldsymbol{\alpha},\mathbf{h}})},\qquad \mathbf{t}>\mathbf{0}, \tag{13.55}$$

where

$$\Omega_{\boldsymbol{\alpha},\mathbf{h}} = \{\mathbf{x} \in \Omega \colon \mathbf{x} + \boldsymbol{\alpha} \otimes \mathbf{h} \in \Omega\}$$

$$\boldsymbol{\alpha} \otimes \mathbf{h} = (\alpha_1 h_1, \ldots, \alpha_d h_d).$$

The following theorem collects some elementary properties of $\omega_{\boldsymbol{\alpha}}(f;\boldsymbol{\tau})_p$:

THEOREM 13.23

Let $1 \leqslant p \leqslant \infty$. Then

$$\omega_{\boldsymbol{\alpha}}(f;\mathbf{t})_p \leqslant \omega_{\boldsymbol{\alpha}}(f;\boldsymbol{\tau}), \qquad \text{all } \mathbf{t} \leqslant \boldsymbol{\tau}; \tag{13.56}$$

$$\omega_{\boldsymbol{\alpha}}(f+g;\mathbf{t})_p \leqslant \omega_{\boldsymbol{\alpha}}(f;\mathbf{t})_p + \omega_{\boldsymbol{\alpha}}(g;\mathbf{t})_p; \tag{13.57}$$

$$\omega_{\boldsymbol{\alpha}}(f;\boldsymbol{\lambda}\otimes\mathbf{t})_p \leqslant \lceil \boldsymbol{\lambda} \rceil^{\boldsymbol{\alpha}} \omega_{\boldsymbol{\alpha}}(f;\mathbf{t})_p, \qquad \lceil \boldsymbol{\lambda} \rceil^{\boldsymbol{\alpha}} = \lceil \lambda_1 \rceil^{\alpha_1} \cdots \lceil \lambda_d \rceil^{\alpha_d}; \tag{13.58}$$

$$\omega_{\boldsymbol{\alpha}}(f;\mathbf{t})_p \leqslant 2^{\boldsymbol{\alpha}} \|f\|_{L_p(\Omega)}; \tag{13.59}$$

$$\omega_{\boldsymbol{\alpha}+\boldsymbol{\beta}}(f;\mathbf{t})_p \leqslant 2^{\boldsymbol{\beta}} \omega_{\boldsymbol{\alpha}}(f;\mathbf{t})_p; \tag{13.60}$$

$$\omega_{\boldsymbol{\alpha}}(f;\mathbf{t})_p \leqslant \mathbf{t}^{\boldsymbol{\alpha}} \|D^{\boldsymbol{\alpha}} f\|_{L_p(\Omega)}; \tag{13.61}$$

$$\omega_{\boldsymbol{\alpha}+\boldsymbol{\beta}}(f;\mathbf{t})_p \leqslant C \mathbf{t}^{\boldsymbol{\beta}} \omega_{\boldsymbol{\alpha}}(D^{\boldsymbol{\beta}} f;\mathbf{t})_p. \tag{13.62}$$

Proof. These results follow by arguments similar to those used in the one-dimensional case. For example, using the formula (4.56) for the divided difference of a function of one variable, we can write

$$\Delta_{\mathbf{h}}^{\boldsymbol{\alpha}} f(\mathbf{x}) = \mathbf{h}^{\boldsymbol{\alpha}-1} \int_{\mathbf{R}^d} N^m\left(\frac{u_1}{h_1}\right) \cdots N^m\left(\frac{u_d}{h_d}\right) D^{\boldsymbol{\alpha}} f(\mathbf{x}+\mathbf{u})\, d\mathbf{u}, \tag{13.63}$$

where N^m is the normalized B-spline of order m defined in (4.47). Then applying the Minkowski inequality (cf. Remark 2.2) leads to (13.61). ■

While moduli of smoothness of the form $\omega_{\boldsymbol{\alpha}}(f;\mathbf{t})_p$ will be useful as building blocks, by themselves they are not adequate to fully describe the smoothness of a function f. Indeed, it is clear that if some $\alpha_i > 0$, then $\omega_{\boldsymbol{\alpha}}(f;\mathbf{t})_p = 0$ for all $\mathbf{t} > \mathbf{0}$ whenever f is a function that is constant in the x_i variable. But such a function can behave arbitrarily badly in the other variables.

In order to get a useful modulus of smoothness for measuring the smoothness of a function of several variables, we must take a linear combination of the moduli $\omega_{\boldsymbol{\alpha}}$ defined above for various values $\boldsymbol{\alpha}$. In particular, in view of the above discussion, we need to include enough different moduli to control the smoothness in each variable. Given any set A of multi-indices, we define

$$\omega_A(f;\mathbf{t})_p = \sum_{\boldsymbol{\alpha}\in A} \omega_{\boldsymbol{\alpha}}(f;\mathbf{t})_p. \tag{13.64}$$

The following theorem gives several properties of ω_A:

THEOREM 13.24

Let $1 \leqslant p \leqslant \infty$. Then

$$\omega_A(f;\mathbf{t})_p \leqslant \omega_A(f;\boldsymbol{\tau})_p, \qquad \text{all } \mathbf{t} \leqslant \boldsymbol{\tau}; \tag{13.65}$$

$$\omega_A(f+g;\mathbf{t})_p \leqslant \omega_A(f;\mathbf{t})_p + \omega_A(g;\mathbf{t})_p; \tag{13.66}$$

$$\omega_A(f;\boldsymbol{\lambda}\otimes\mathbf{t})_p \leqslant \lceil\boldsymbol{\lambda}\rceil^{\boldsymbol{\beta}}\omega_A(f;\mathbf{t})_p, \qquad \boldsymbol{\beta} = \min\{\boldsymbol{\beta}\colon \boldsymbol{\alpha} \leqslant \boldsymbol{\beta}, \text{ all } \boldsymbol{\alpha}\in A\}; \tag{13.67}$$

$$\omega_A(f;\mathbf{t})_p \leqslant \Big(\sum_{\boldsymbol{\alpha}\in A} 2^{\boldsymbol{\alpha}}\Big)\|f\|_{L_p(\Omega)}; \tag{13.68}$$

$$\omega_A(f;\mathbf{t})_p \leqslant \sum_{\boldsymbol{\alpha}\in A} \mathbf{t}^{\boldsymbol{\alpha}}\|D^{\boldsymbol{\alpha}}f\|_{L_p(\Omega)} \qquad \text{if } f \in L_p^A(\Omega). \tag{13.69}$$

Proof. These results follow by summing over $\boldsymbol{\alpha}\in A$ in the inequalities of Theorem 13.23. ∎

We turn now to the question of how to choose A. The following example shows that if $\omega_A(f;\mathbf{t})_p$ is to control the smoothness of f in each direction, then it will have to contain a positive multiple of each of the unit vectors $\mathbf{e}_1,\ldots,\mathbf{e}_d$.

EXAMPLE 13.25

Suppose that for some $1 \leqslant i \leqslant d$, $r\cdot\mathbf{e}_i \notin A$ for all $r>0$.

Discussion. Let f be a function of the variable x_i only. Then for any $\boldsymbol{\alpha}\in A$ there must be $j\neq i$ with $\alpha_j>0$. But then $\omega_{\boldsymbol{\alpha}}(f;\mathbf{t})_p \equiv 0$ and so $\omega_A(f;\mathbf{t})_p \equiv 0$ for all $\mathbf{t}$. But f can be an arbitrarily nasty function of x_i. ∎

In view of property (13.60) of moduli of smoothness, if a set A of multi-indices contains the vector $\boldsymbol{\alpha}$, then there is no need for A to include any indices $\boldsymbol{\beta}$ with $\boldsymbol{\beta} > \boldsymbol{\alpha}$. This observation together with Example 13.25 suggests that the sets A of most interest are those that are regular in the sense of Definition 13.2.

To state our next theorem we need to introduce a certain space of polynomials associated with a set A. Given A, let

$$\mathring{A} = \{\boldsymbol{\beta}: \boldsymbol{\beta} < \boldsymbol{\alpha} \text{ for all } \boldsymbol{\alpha} \in A\}. \tag{13.70}$$

It is clear that $\mathring{A}$ is always complete in the sense of Definition 13.15. Conversely, if Λ is a complete set, then $\partial\Lambda$ is regular and $\partial\mathring{\Lambda} = \Lambda$.

The following theorem shows that there is a special relationship between $\omega_A(f; \mathbf{t})_p$ and the class of polynomials $\mathcal{P}_{\mathring{A}}$:

THEOREM 13.26

Suppose A is a regular set of multi-indices. Then

$$f \in \mathcal{P}_{\mathring{A}} \text{ implies } \omega_A(f; \mathbf{t})_p = 0, \text{ all } \mathbf{t} > 0. \tag{13.71}$$

Conversely, if $1 \leqslant p < \infty$ and $f \in L_p(\Omega)$ is such that

$$\sum_{\boldsymbol{\alpha} \in A} \frac{\omega_{\boldsymbol{\alpha}}(f; \mathbf{t}_\nu)_p}{\mathbf{t}_\nu^{\alpha}} \to 0 \text{ for some sequence } \mathbf{t}_\nu \text{ with } \|\mathbf{t}_\nu\| \to 0, \tag{13.72}$$

then $f \in \mathcal{P}_{\mathring{A}}$. The same result holds with $p = \infty$ if $f \in C(\Omega)$.

Proof. The first assertion is obvious since for every $\boldsymbol{\alpha} \in A$, $\Delta_{\mathbf{h}}^{\boldsymbol{\alpha}} x^{\boldsymbol{\beta}} = 0$ for all $\boldsymbol{\beta} \in \mathring{A}$. To prove the second assertion, we note that since A is regular it must contain the vectors $r_i\mathbf{e}_i$, $i = 1, 2, \ldots, d$ for some positive integers $r_1, \ldots, r_d$. This implies (cf. Timan [1963]) that f is at most a polynomial of order r_i in the ith variable, $i = 1, 2, \ldots, d$. Suppose

$$f(\mathbf{x}) = \sum_{\boldsymbol{\beta}} a_{\boldsymbol{\beta}} \mathbf{x}^{\boldsymbol{\beta}}.$$

Then $a_{\boldsymbol{\beta}} = 0$ for all $\boldsymbol{\beta} \geqslant \mathbf{r} = (r_1, \ldots, r_d)$. We now show that if $\boldsymbol{\alpha} \in A$, then $a_{\boldsymbol{\beta}} = 0$ for all $\boldsymbol{\alpha} \leqslant \boldsymbol{\beta} \leqslant \mathbf{r}$. We accomplish this by using (13.60) to compare $\omega_{\boldsymbol{\beta}}$ and $\omega_{\boldsymbol{\alpha}}$. We have

$$\|a_{\boldsymbol{\beta}}\|_p = \|\Delta_{\mathbf{t}_\nu}^{\boldsymbol{\beta}} f\|_p \leqslant \omega_{\boldsymbol{\beta}}(f; \mathbf{t}_\nu)_p \leqslant C\omega_{\boldsymbol{\alpha}}(f; \mathbf{t}_\nu)_p \to 0 \qquad \text{as } \nu \to \infty.$$

We conclude that $a_{\boldsymbol{\beta}} = 0$, and the inductive proof is complete. We have shown that the only nonzero coefficients are $a_{\boldsymbol{\beta}}$ with $\boldsymbol{\beta} < \boldsymbol{\alpha}$ for all $\boldsymbol{\alpha} \in A$. But these are just the $a_{\boldsymbol{\beta}}$ with $\boldsymbol{\beta} \in \mathring{A}$. ■

We illustrate the above discussion by introducing the modulus of smoothness that we used for our results on the approximation power of tensor-product splines.

EXAMPLE 13.27. Coordinate Modulus of Smoothness

Let $\mathbf{r}=(r_1,\ldots,r_d)\in Z_+^d$, and define

$$\omega_{\mathbf{r}}(f;\mathbf{t})_p=\sum_{i=1}^{d}\omega_{r_i\mathbf{e}_i}(f;\mathbf{t})_p. \tag{13.73}$$

Discussion. Here $A=\{r_i\mathbf{e}_i\}_1^d$. It is easy to see that

$$\mathcal{P}_{\dot{A}}=\bigotimes_{i=1}^{d}\mathcal{P}_{r_i}=\operatorname*{span}_{\alpha<\mathbf{r}}\{\mathbf{x}^\alpha\}.$$ ■

§ 13.6. THE K-FUNCTIONAL

In this section we introduce an alternate way of measuring the smoothness of a function in $L_p(\Omega)$. We shall see that for nice sets Ω it is equivalent with the modulus of smoothness of the preceding section.

DEFINITION 13.28. The K-Functional

Let $1\leqslant p\leqslant\infty$, and suppose A is a regular set of multi-indices. Then for every $\mathbf{t}>\mathbf{0}$ we define the *K-functional* $K_{A,p}(\mathbf{t})$ by

$$K_{A,p}(\mathbf{t})f=\inf_{g\in L_p^A(\Omega)}\left(\|f-g\|_{L_p(\Omega)}+\sum_{\alpha\in A}\mathbf{t}^\alpha\|D^\alpha g\|_{L_p(\Omega)}\right). \tag{13.74}$$

For each choice of the vector $\mathbf{t}$ with $\mathbf{t}>\mathbf{0}$, $K_{A,p}(\mathbf{t})$ is a nonlinear functional defined on $L_p(\Omega)$. It measures the smoothness of f in terms of how well f can be approximated by functions $g\in L_p^A(\Omega)$ while keeping a control on the size of the derivatives of g. $K_{A,p}(\mathbf{t})$ has many properties analogous to those of the one-dimensional K-functional treated in Section 2.9.

THEOREM 13.29

Let Ω be a rectangle in $\mathbf{R}^d$. Then there exist constants C_1 and C_2 such that

$$\omega_A(f;\mathbf{t})_p\leqslant C_2K_{A,p}(\mathbf{t})f, \tag{13.75}$$

and

$$K_{A,p}(f,\mathbf{t}) \leqslant C_1 \omega_A(f;\mathbf{t})_p \tag{13.76}$$

for all $f \in L_p(\Omega)$ and all $\mathbf{t} > \mathbf{0}$.

Proof. See Dahmen, DeVore, and Scherer [1980]. ■

We close this section with an analog of Theorem 2.68 which allows us to establish estimates in terms of $\omega_A(f;\mathbf{t})_p$ by looking only at functions in $L_p^A(\Omega)$.

THEOREM 13.30

Let Ω be a rectangle, and suppose A is a regular set of multi-indices. Let $\mathbb{S}$ be a set of functions in $L_p(\Omega)$ such that for each $g \in L_p^A(\Omega)$ there is an element $s_g \in \mathbb{S}$ with

$$\|g - s_g\|_{L_p(\Omega)} \leqslant C_0 + C_1 \sum_{\alpha \in A} \mathbf{t}^{\alpha} \|D^{\alpha} g\|_{L_p(\Omega)}, \tag{13.77}$$

where C_0, C_1 are constants depending only on A, p, and Ω. Then there exists a constant C_2 depending on the same quantities such that for each $f \in L_p(\Omega)$ there exits $s_f \in \mathbb{S}$ with

$$\|f - s_f\|_{L_p(\Omega)} \leqslant C_0 + C_2 \omega_A(f;\mathbf{t})_p. \tag{13.78}$$

Proof. Let $f \in L_p(\Omega)$. Then for any $g \in L_p^A(\Omega)$,

$$\|f - s_g\|_p \leqslant \|f - g\|_p + \|g - s_g\|_p \leqslant C_0 + \max(C_1, 1)\Big(\|g - f\|_p + \sum_{\alpha \in A} \mathbf{t}^{\alpha} \|D^{\alpha} g\|_p \Big).$$

Since the K-functional is defined as an infimum, if we vary g in $L_p^A(\Omega)$, we can find some $g^* \in L_p^A(\Omega)$ so that

$$\|f - s_{g^*}\|_p \leqslant C_0 + 2\max(C_1, 1) K_{A,p}(\mathbf{t}) f.$$

Using (13.76) we obtain (13.78). ■

§ 13.7. HISTORICAL NOTES

Section 13.1

The multivariate notation used here is quite standard—see for example, Adams [1975].

Section 13.2

Sobolev spaces play an important role in many areas of analysis. For an excellent modern treatment, see Adams [1975]. The book of Sobolev [1963] is also useful.

Section 13.3

The use of multidimensional polynomials in numerical analysis and approximation theory has a long and rich history. Although the tensor-product polynomials and the total-order polynomials appear more frequently, there has also been considerable interest in other polynomial spaces where certain powers are missing, particularly in the construction of finite elements. See, for example, Strang and Fix [1973].

Section 13.4

Multidimensional Taylor expansions for functions of several variables are a part of classical analysis, and they can be found in various forms in practically any book on the multivariate calculus. The development of similar theorems for distributions is part of modern functional analysis. The Taylor expansion in Theorem 13.16 in the case where Λ is the set of polynomials of total order m is usually referred to as the Sobolev integral indentity (see the book by Sobolev [1963] for the original derivation); it was an important tool in establishing certain imbedding theorems, and for the general study of Sobolev spaces. The development of such expansions for general classes of polynomials is more recent—motivated no doubt by the desire to analyze the approximation power of certain finite element approximation methods. Our proof of Theorem 13.16 follows Dupont and Scott [1978, 1980].

The problem of estimating how well a smooth function (or more generally a distribution) can be approximated by a polynomial has been considered by many authors. Most of the early results were based on the Sobolev integral identity, and dealt only with polynomials of total order (cf. e.g., Sobolev [1963]). Sobolev's ideas were reformulated and applied to analyze finite element approximation methods by Bramble and Hilbert [1970, 1971]. Brudnyi [1970] obtained estimates involving an appropriate modulus of smoothness; he treated tensor-product polynomials as well as total order ones. Results for more general spaces of polynomials have been obtained only recently; see Dupont and Scott [1978, 1980] and Dahmen, DeVore, and Scherer [1980]. We have followed the approach of the former paper because it is more constructive. Our terminology and notation is a mix of the two papers, however.

Section 13.5

For a summary of results on moduli of smoothness for functions of several variables, see Timan [1963]. Only the total modulus and the tensor modulus (cf. Examples 13.27) are discussed. The idea of working with moduli corresponding to a general set A as in (13.64) is credited to Dahmen, DeVore, and Scherer [1980].

Section 13.6

The results of this section on K-functionals for general Sobolev spaces are attributed to Dahmen, DeVore, and Scherer [1980].

§ 13.8. REMARKS

Remark 13.1

As suggested in Section 13.2, some of the finer properties of Sobolev spaces require an assumption on the domain and/or its boundary. We have elected to work with the star-shaped domains because the condition is easy to state and understand, yet still general enough to include most domains of interest. There is a heirarchy of cone conditions which is often invoked in dealing with Sobolev spaces, and many of the results quoted here hold under such conditions—see Adams [1975].

Remark 13.2

We have stated Theorem 13.29 on the equivalence of the modulus of smoothness and the K-functional only for rectangles. Dahmen, DeVore, and Scherer [1980] have established the same equivalence for other kinds of domains. To do so, they were obliged to introduce an appropriate "cone condition" on the domain (this cone condition is more restrictive than the usual ones, and it is also more restrictive than the star-shaped property).

REFERENCES

Achieser, N. I.

[1956] *Theory of Approximation*, Ungar, New York.

Adams, R. A.

[1975] *Sobolev Spaces*, Academic Press, New York.

Ahlberg, J. H.

[1969] "Splines in the complex plane," in Schoenberg [1969b], pp. 1–27.

Ahlberg, J. H. and E. N. Nilson

[1963] "Convergence properties of the spline fit," *SIAM J.* **11**, 95–104.

[1966] "The approximation of linear functionals," *SIAM J. Numer. Anal.*, **3**, 173–182.

Ahlberg, J. H., E. N. Nilson, and J. L. Walsh

[1964] "Fundamental properties of generalized splines," *Proc. Nat. Acad. Sci. U.S.A.*, **52**, 1412–1419.

[1965a] "Best approximation and convergence properties of higher-order spline approximations," *J. Math. Mech.*, **14**, 231–244.

[1965b] "Extremal, orthogonality, and convergence properties of multidimensional splines," *J. Math. Anal. Appl.*, **12**, 27–48.

[1965c] "Convergence properties of generalized splines," *Proc. Nat. Acad. Sci. U.S.A.*, **54**, 344–350.

[1967a] "Complex cubic splines," *Trans. Am. Math. Soc.*, **129**, 391–413.

[1967b] *The Theory of Splines and Their Applications*, Academic Press, New York.

[1969] "Properties of analytic splines (1). Complex polynomial splines," *J. Math. Anal. Appl.*, **27**, 262–278.

[1971] "Complex polynomial splines on the unit circle," *J. Math. Anal. Appl.*, **33**, 234–257.

Arthur, D. W.

[1974] "Multivariate spline functions. I. Construction, properties, and computation," *J. Approximation Theory*, **12**, 396–411.

[1975] "Multivariate spline functions. II. Best error bounds," *J. Approximation Theory*, **15**, 1–10.

Asker, B.

[1962] "The spline curve, a smooth interpolating function used in numerical design of ship lines," *BIT*, **2**, 76–82.

Astor, P. H., and C. S. Duris

[1974] "Discrete L-splines," *Numer. Math.*, **22**, 393–402.

Atkinson, K.

[1968] "On the order of convergence of natural cubic spline interpolation problems," *SIAM J. Numer. Anal.*, **5**, 89–101.

Atteia, M.

[1965a] "Spline-fonctions généralisées," *C. R. Acad. Sci. Paris Ser. A*, **261**, 2149–2152.

[1965b] "Généralisation de la définition et des propriétés des spline-fonctions," *C. R. Acad. Sci. Paris Ser. A*, **260**, 3550–3553.

[1966a] "Étude de certains noyaux et théorie des fonctions spline en Analysis Numérique," dissertation, Grenoble.

[1966b] "Existence et détermination des fonctions splines á plusieurs variables," *C. R. Acad. Sci. Paris Ser. A*,**262**, 575–578.

Aubin, J. -P.

[1967] "Behavior of the error of the approximate solution of boundary value problems for linear elliptic operators by Galerkins and finite difference methods," *Ann. Scuola Norm. Pisa*, **21**, 599–637.

[1972] *Approximation of Elliptic Boundary-Value Problems*, Wiley, New York.

Aumann, G.

[1963] "Approximation by step functions," *Proc. Am. Math. Soc.*, **14**, 477–482.

Babadshanov, S. B., and V. M. Tihomirov

[1967] "Diameter of a function class in L_p space, $p > 1$," (Russian), *A. N. Usb. SSR. Ser. Fiz.-Mat.*, **2**, 24–30.

Babuska, I.

[1970] "Approximation by hill functions," *Comment. Math. Univ. Carolinae*, **11**, 787–811.

Bak, J. and D. J. Newman

[1972] "Müntz-Jackson Theorems in $L^p[0,1]$ and $C[0,1]$," *Am. J. Math.*, **154**, 437–457.

Bartelt, M. W.

[1975] "Weak Chebychev sets and splines," *J. Approximation Theory*, **14**, 30–37.

Bastien, R. and J. Dubuc

[1976] "Systèmes Faibles de Tchebycheff et Polynomes de Bernstein," *Canad. J. Math.*, **28**, 653–658.

Baum, A. M.

[1976a] "An algebraic approach to simply hyperbolic splines on the real line," *J. Approximation Theory*, **17**, 189–199.

[1976b] "Double hyperbolic splines on the real line," *J. Approximation Theory*, **18**, 174–188.

Beckenbach, E. F., and R. Bellman

[1961] *Inequalities*, Cambridge University Press.

Bellman, R., and R. Roth

[1969] "Curve fitting by segmented straight lines," *J. Am. Statist. Assoc.*, **64**, 1079–1084.

Berger, S. A., and W. C. Webster

[1963] "An application of linear programming to the fairing of ships lines," in *Recent Advances in Mathematical Programming*, R. L. Graves and P. Wolfe, Eds., McGraw-Hill, New York, pp. 241–253.

Berger, S. A., W. C. Webster, R. A. Tapia, and D. A. Atkins

[1966] "Mathematical ship lofting," *J. Ship Research*, **10**, 203–222.

Bergh, J., and J. Peetre

[1974] "On the spaces V_p," *Bollettino U.M.I.*, **10**, 632–648.

Bernstein, S. N.

[1926] *Lecons sur les Propriétés Extremales et la Meillure Approximation des Fonctions Analytiques d'une Variable Realle*, Gauthier-Villars, Paris.

Birkhoff, G. D.

[1906] "General mean value and remainder theorems with applications to mechanical differentiation and quadrature," *Trans. Am. Math. Soc.*, **7**, 107–136.

Birkhoff, G.

[1967] "Local spline approximation by moments," *J. Math. Mech.*, **16**, 987–990.

[1969] "Piecewise bicubic interpolation and approximation in polygons," in Schoenberg [1969b], pp. 185–221.

Birkhoff, G., and C. deBoor

[1964] "Error bounds for spline interpolation," *J. Math. Mech.* **13**, 827–836.

[1965] "Piecewise polynomial interpolation and approximation," in Garabedian [1965], pp. 164–190.

Birkhoff, G. and H. Garabedian

[1960] "Smooth surface interpolation," *J. Math. and Phys.*, **39**, 258–268.

Birkhoff, G., M. H. Schultz, and R. S. Varga

[1968] "Piecewise Hermite interpolation in one and two variables with applications to partial differential equations," *Numer. Math.*, **11**, 232–256.

Birman, M. S. and M. Z. Solomjak

[1966] "Approximation of the functions of the classes W_p^α by piecewise polynomial functions" (Russian), *Soviet Math. Dokl.*, **7**, 1573–1577; see also *Dokl. Akad. Nauk SSSR*, **171**, 1015–1018.

[1967] "Piecewise polynomial approximations of functions of the classes W_p^α" (Russian), *Math. USSR Sb.*, **2**, 295–371; see also *Mat. Sb.*, **73**, 331–355.

Böhmer, K.

[1974] *Spline-Funktionen*, Teubner-Studienbücher, Stuttgart.

Böhmer, K., G. Meinardus, and W. Schempp

[1974] *Spline Funktionen*, Eds., Bibliographisches Institut, Zurich.

[1976] *Spline Functions, Karlsruhe 1975*, Eds., Springer-Verlag Lecture Notes 501, Heidelberg.

deBoor, C.

[1962] "Bicubic spline interpolation," *J. Math. Phys.*, **41**, 212–218.

[1963] "Best approximation properties of spline functions of odd degree," *J. Math. Mech.*, **12**, 747–749.

[1966] "The method of projections, etc.," PhD. Dissertation, University of Michigan.

[1968a] "On the convergence of odd-degree spline interpolation," *J. Approximation Theory*, **1**, 452–463.

[1968b] "On local spline approximation by moments," *J. Math. Mech.*, **17**, 729–736.

[1968c] "On uniform approximation by splines," *J. Approximation Theory*, **1**, 219–235.

[1972] "On calculating with B-splines," *J. Approximation Theory*, **6**, 50–62.

[1973a] "The quasi-interpolant as a tool in elementary polynomial spline theory," in Lorentz [1973], pp. 269–276.

[1973b] "Good approximation by splines with variable knots," in Meir and Sharma [1973], pp. 57–72.

[1974a] "A smooth and local interpolation with "small" kth derivative," in *Numerical Solutions for Ordinary Differential Equations*, A. Aziz, ed., Academic Press, New York, pp. 177–197.

[1974b] "A remark concerning perfect splines," *Bull. Am. Math. Soc.*, **80**, 723–727.

[1975] "How small can one make the derivatives of an interpolating function," *J. Approximation Theory*, **13**, 105–116.

[1976a] "Total positivity of the spline collocation matrix," *Indiana Univ. J. Math.*, **25**, 541–551.

[1976b] "Splines as linear combinations of B-splines. A survey," in Lorentz, Chui, and Schumaker [1976], pp. 1–47.

[1976c] "On local linear functionals which vanish at all B-splines but one," in Law and Sahney [1976], pp. 120–145.

[1977] "Subroutine package for calculating with B-splines," *SIAM J. Numer. Anal.*, **14**, 441–472.

[1978] *A Practical Guide to Splines*, Springer-Verlag, New York.

deBoor, C. and G. J. Fix

[1973] "Spline approximation by quasi-interpolants," *J. Approximation Theory*, **8**, 19–45.

deBoor, C., T. Lyche, and L. L. Schumaker

[1976] "On calculating with B-splines II. Integration." in *Numerische Methoden der Approximations Theorie*, Band 3, ISNM Vol. 30, Birkhauser, pp. 123–146.

deBoor, C., and R. E. Lynch

[1966] "On splines and their minimum properties," *J. Math. Mech.*, **15** ,953–969.

deBoor, C., and A. Pinkus

[1977] "Backward error analysis for totally positive linear systems," *Numer. Math.*, **27**, 485–490.

deBoor, C., and I. J. Schoenberg

[1976] "Cardinal interpolation and spline functions VIII. The Budan-Fourier theorem for splines and applications," in Böhmer, Meinardus, and Schempp [1976], pp. 1–79.

Braess, D.

[1971] "Chebyshev approximation by spline functions with free knots," *Numer. Math.*, **17**, 357–366.

[1974] "Rationale Interpolation, Normalität und Monosplines," *Numer. Math.*, **22**, 219–232.

[1975] "On the degree of approximation by spline functions with free knots," *Aequationes Math.*, **12**, 80–81.

Bramble, J. H., and S. R. Hilbert

[1970] "Estimation of linear functionals on Sobolev spaces with application to Fourier transforms and spline interpolation," *SIAM J. Numer. Anal.*, **7**, 112–124.

[1971] "Bounds for a class of linear functionals with applications to Hermite interpolation," *Numer. Math.*, **16**, 362–369.

Brudnyi, Ju. A.

[1964] "On a theory of local best approximation" (Russian), *Kazan. Gos. Usen. Zap.*, **124**, 43–49.

[1970] "A multidimensional analog of a theorem of Whitney," (Russian) *Math. USSR. Sb.*, **11**, 157–170.

[1971] "Piecewise polynomial approximation and local approximations," *Soviet Math. Dokl.*, **12**, 1591–1594.

[1974a] "Local approximation and differential properties of functions of several variables" (Russian), *Uspehi Mat. Nauk*, **29**, 163–164.

[1974b] "Spline approximation and functions of bounded variation," *Soviet Math. Dokl.*, **15**, 518–521.

[1976] "Piecewise polynomial approximation, embedding theorem and rational approximation," in Schaback and Scherer [1976], pp. 73–98.

Brudnyi, Ju. A., and I. E. Gopengauz

[1963] "Approximation by piecewise polynomial functions" (Russian), *Izv. Akad. Nauk SSSR Ser. Mat.*, **27**, 723–746.

Burchard, H. G.

[1968] "Interpolation and approximation by generalized convex functions," Dissertation, Purdue.

[1973] "Extremal positive splines with applications," in Lorentz [1973], pp. 291–294.

[1974] "Splines (with optimal knots) are better," *J. Applicable Analysis*, **3**, 309–319.

[1977] "On the degree of convergence of piecewise polynomial approximation on optimal meshes II," *Trans. Am. Math. Soc.*, **234**, 531–559.

Burchard, H. G., and D. F. Hale

[1975] "Piecewise polynomial approximation on optimal meshes," *J. Approximation Theory*, **14**, 128–147.

Burenkov, V. I.

[1974] "Sobolev's integral representation and Taylor's Formula," *Trudy Mat. Inst. Steklow*, **131**, 33–38.

Burkhill, H.

[1961] "On Riesz and Riemann summability," *Proc. Cambridge Philos. Soc.*, **57**, 55–60.

Butler, G. J., and F. B. Richards

[1972] "An L_p saturation theorem for splines," *Canad. J. Math.*, **24**, 957–966.

Butterfield, K. R.

[1976] "The computation of all the derivatives of a B-spline basis," *J. Inst. Math. Appl.*, **17**, 15–25.

Butzer, P. L., and H. Berens

[1967] *Semi-groups of Operators and Approximation*, Springer-Verlag, Berlin.

Butzer, P. L., and K. Scherer

[1969] "On the fundamental approximation theorems of D. Jackson, S. N. Bernstein, and theorems of M. Zamansky and S. B. Steckin," *Aequationes Math.*, **3**, 170–185.

[1970] "On fundamental theorems of approximation theory and their dual versions," *J. Approximation Theory*, **3**, 87–100.

Cavaretta, A. S.

[1975] "Oscillatory and zero properties for perfect splines and monosplines," *J. Analyse Math.*, **28**, 41–59.

Cheney, E. W.

[1966] *Introduction to Approximation Theory*, McGraw Hill, New York.

Cheney, E. W., and F. Schurer

[1968] "A note on the operators arising in spline approximation," *J. Approximation Theory*, **1**, 94–102.

[1970] "Convergence of cubic spline interpolants," *J. Approximation Theory*, **3**, 114–116.

Chernik, N. E.

[1975] "Approximation by splines with fixed knots," (Russian) *Proc. Steklov. Inst. Math.*, **138**, 174–199.

Ciesielski, Z.

[1975] "Spline bases in function spaces," in Ciesielski and Musielak [1975], pp. 49–54.

Ciesielski, Z., and J. Musielak

[1975] *Approximation Theory*, Eds., Reidel Publishing Co., Holland (also PWN—Polish Scientific Publishers, Warsaw).

Conte, S., and C. deBoor

[1972] *Elementary Numerical Analysis*, 2nd edition, McGraw Hill, New York.

Cox, M. G.

[1971] "Curve fitting with piecewise polynomials," *J. Inst. Math. Appl.*, **8**, 36–52.

[1972] "The numerical evaluation of B-splines," *J. Inst. Math. App.*, **10**, 134–149.

Curry, H. B.

[1947] Review, *Math. Tables & Other Aids to Computation*, **2**, 167–169, 211–213.

Curry, H. B., and I. J. Schoenberg

[1947] "On spline distributions and their limits: the Pólya distribution functions," Abstract 380t, *Bull. Am. Math. Soc.*, **53**, 1114.

[1966] "On Pólya frequency functions IV: The fundamental spline functions and their limits," *J. Analyse Math.*, **17**, 71–107.

Dahmen, G., R. DeVore, and K. Scherer

[1980] "Multidimensional spline approximation," *SIAM J. Numer. Anal.*, in press.

Davis, P. J.

[1963] *Interpolation and Approximation*, Blaisdell, New York.

delaVallée Poussin, C. J.

[1919] *Lecons sur l'Approximation des Fonctions d'une Variable Realle*, Gauthier-Villars, Paris.

Demko, S.

[1977] "Local approximation properties of spline projections," *J. Approximation Theory*, **19**, 176–185.

Demko, S., and R. S. Varga

[1974] "Extended L_p error bounds for spline and L-spline interpolation," *J. Approximation Theory*, **12**, 242–264.

DeVore, R.

[1968] "On Jackson's Theorem," *J. Approximation Theory*, **1**, 314–318.

[1972] *Approximation of Continuous Functions by Positive Linear Operators*, Springer Lecture Notes in Mathematics, No. 293, Springer, Berlin.

[1976] "Degree of approximation," in Lorentz, Chui, and Schumaker [1976], pp. 117–161.

[1977a] "Monotone approximation by polynomials," *SIAM J. Math. Anal.*, **8**, 906–921.

[1977b] "Monotone approximation by splines," *SIAM J. Math. Anal.*, **8**, 891–905.

[1977c] "Pointwise approximation by polynomials and splines," in *Constructive Function Theory, Akad. Nauk SSSR*, Moscow, pp. 132–141.

DeVore, R. A., and F. Richards

[1973a] "The degree of approximation by Chebyshevian splines," *Trans. Am. Math. Soc.*, **181**, 401–418.

[1973b] "Saturation and inverse theorems for spline approximation," in Meir and Sharma [1973], pp. 73–82.

DeVore, R. A., and K. Scherer

[1976] "A constructive theory for approximation by splines with an arbitrary sequence of knot sets," in Schaback and Scherer [1976], pp. 167–183.

Di Guglielmo, F.

[1969] "Construction d'approximation des espaces de Sobolev sur des reseaux en simplexes," *Calcolo*, **6**, 279–331.

Dodson, D. S.

[1972] "Optimal order approximation by polynomial spline functions," Ph.D. thesis, Purdue University, West Lafayette.

Domsta, J.

[1972] "A theorem on B-splines," *Studia Math.*, **41**, 291–314.

[1976] "A theorem on B-splines II. The periodic case," *Bull. Acad. Polon. Sci. Ser. Sci. Math. Astron. Phys.*, **24**, 1077–1084.

Dupont, T., and R. Scott

[1978] "Constructive polynomial approximation," in *Recent Advances in Numerical Analysis*, C. deBoor and G. Golub, Eds., Academic Press, New York.

[1980] "Polynomial approximation of functions in Sobolov spaces," *Math. of Comp.* **34**, 441–463.

Dunford, N., and J. T. Schwartz

[1957] *Linear Operators, Part I*, Wiley Interscience, New York.

Eagle, A.

[1928] "On the relation between the Fourier constants of a periodic function and the coefficients determined by harmonic analysis," *Phil. Mag.*, **5**, 113–132.

Edwards, R. E.

[1965] *Functional Analysis, Theory and Applications*, Holt, Rinehart and Winston, New York.

El Kolli, A.

[1974] "Niéme épaisseur dans les espaces de Sobolev," *J. Approximation Theory*, **10**, 268–294.

Faber, G.

[1914] "Uber die interpolatorische Darstellung Stetiger Funktionen" *Deutsch. Math. Verein*, **23**, 192–210.

Favard, J.

[1940] "Sur l'interpolation," *J. Math. Pures Appl.* (9), **19**, 281–306.

Ferguson, D.

[1973] "Sufficient conditions for the Peano kernel to be one sign," *SIAM J. Numer. Anal.*, **10**, 1047–1054.

[1974] "Sign changes and minimal support properties of Hermite-Birkhoff splines with compact support," *SIAM J. Numer. Anal.*, **11**, 769–779.

Fisher, S. D., and J. W. Jerome

[1975] "Minimum Extremals in Function Spaces," Springer Verlag Lecture Notes 479, Heidelberg.

Fitzgerald, C. H., and L. L. Schumaker

[1969] "A differential equation approach to interpolation at extremal points," *J. Analyse Math.*, **22**, 117–134.

Fix, G., and G. Strang

[1969] "Fourier analysis of the finite element method in Ritz-Galerkin theory," *Studies Appl. Math.*, **48**, 265–273.

Fowler, A. H., and C. W. Wilson

[1963] "Cubic spline, a curve fitting routine," Report Y-1400, Oak Ridge.

Freud, G.

[1970] "Some questions that are related to approximation by spline functions and polynomials" (Russian), *Studia Sci. Math. Hungar.*, **5**, 161–171.

Freud, G., and V. Popov

[1969] "On approximation by spline functions," *Proc. Conf. on Constructive Theory of Functions*, G. Alexits and S. B. Stechkin, Eds., Akademia Kiado, Budapest, pp. 163–172.

Friedman, A.

[1969] *Partial Differential Equations*, Holt, Rinehart and Winston, New York.

Gaffney, P. W.

[1974] "The calculation of indefinite integrals of B-splines," CSS 16, AERE Harwell.

Gaier, D.

[1970] "Saturation bei Spline-Approximation und Quadratur," *Numer. Math.*, **16**, 129–140.

Garabedian, H. L.

[1965] *Approximation of Functions*, Elsevier, Amsterdam.

Glaeser, G.

[1967] "Prolongement extremal de fonctions differentiables," Publ. Sect. Math. Faculté des Sciences Rennes, France.

[1973] "Prolongement extremal de fonctions differentiables d'une variable," *J. Approximation Theory*, **8**, 249–261.

Gluskin, E.

[1974] "On a problem concerning diameters," *Sov. Math. Dokl.* **15**, 1592–1596.

Golomb, M.

[1968] "Approximation by periodic spline interpolants on uniform meshes," *J. Approximation Theory*, **1**, 26–65.

Golumb, M., and H. Weinberger

[1959] "Optimal approximation and error bounds," in *On Numerical Approximation*, R. E. Langer, Ed., University of Wisconsin Press, pp. 117–190.

Greville, T. N. E.

[1944] "The general theory of osculatory interpolation," *Trans. Actuar. Soc. Am.*, **45**, 202–265.

[1964a] "Numerical procedures for interpolation by spline functions," *SIAM J. Numer. Anal.*, **1**, 53–68.

[1964b] "Interpolation by generalized spline functions," Math. Research Center Tech. Summary Report #476, U.S. Army, University of Wisconsin.

[1969a] *Theory and Applications of Spline Functions*, Ed., Academic Press, New York.

[1969b] "Introduction to spline functions," in Greville [1969a], pp. 1–35.

Hall, C. A.

[1968] "On error bounds for spline interpolation," *J. Approximation Theory*, **1**, 209–218.

Hardy, G. H., J. E. Littlewood, and G. Pólya

[1959] *Inequalities*, Cambridge University Press

Hedstrom, G. W., and R. S. Varga

[1971] "Application of Besov spaces to spline approximation," *J. Approximation Theory*, **4**, 295–327.

Hildebrandt, F. B.

[1956] *Introduction to Numerical Analysis*, McGraw-Hill, New York.

Holladay, J. C.

[1957] "A smoothest curve approximation," *Math. Tables Aids Computation*, **11**, 233–243.

Holmes, R. B.

[1972] *A Course on Optimization and Best Approximation*, Lecture Notes in Mathematics, Springer, Heidelberg.

Ince, E. L.

[1944] *Ordinary Differential Equations*, Dover, New York.

Isaacson, E., and H. B. Keller

[1966] *Analysis of Numerical Methods*, Wiley, New York.

Ismagaov, R. S.

[1974] "Diameter of sets in normal linear spaces and the approximation of functions by trigonometric polynomials," (Russian), *Uspeki, Mat. Nauk*, **29**, 161–178 translated in *Russian Math Surveys*, **29**, 179–186.

Jackson, D.

[1930] *The Theory of Approximation*, AMS Vol. XI, Colloq. Publ. Providence, Rhode Island.

Jerome, J. W.

[1973a] "On uniform approximation by certain generalized spline functions," *J. Approximation Theory*, **7**, 143–154.

[1973b] "Topics in multivariate approximation theory," in Lorentz [1973], pp. 151–198.

Jerome, J. W., and L. L. Schumaker

[1969] "Characterizations of functions with higher order derivatives in L_p," *Trans. Am. Math. Soc.*, **143**, 363–371.

[1971] "Local bases and computation of g-splines," *Methoden und Varfahren der Math. Phys.*, **5**, 171–199.

[1973] "Characterizations of absolute continuity and essential boundedness for higher order derviatives," *J. Math. Analy. Appl.*, **42**, 452–465.

[1974] "On the distance to a class of generalized splines," in *Linear Operators and Approximation II*, ISNM 25, Birkhauser Verlag, Basel, pp. 503–517.

[1976] "Local support bases for a class of spline functions," *J. Approximation Theory*, **16**, 16–27.

Jetter, K.

[1976] "Nullstellen von Splines," in Schaback and Scherer [1976], pp. 291–304.

Johnen, H.

[1972] "Inequalities connected with the moduli of smoothness," *Mat. Vestnik*, **9**, 289–303.

Johnen, H., and K. Scherer

[1976] "Direct and inverse theorems for best approximation by Λ-splines," in Böhmer, Meinardus, and Schempp [1976], pp. 116–131.

[1977] "On the equivalence of the *K*-functional and moduli of continuity and some applications," in Schempp and Zeller [1977], pp. 119–140.

Johnson, R. S.

[1960] "On monosplines of least deviation," *Trans. Am. Math. Soc.*, **96**, 458–477.

Jones, R. C., and L. A. Karlovitz

[1970] "Equioscillation under nonuniqueness in the approximation of continuous functions," *J. Approximation Theory*, **3**, 138–145.

Kahane, J. P.

[1961] *Teoria Constructiva de Functiones*, Course Notes, University de Buenos Aires.

Karlin, S.

[1968] *Total positivity, Volume* I, Stanford University Press.

[1969] "The fundamental theorem of algebra for monosplines satisfying certain boundary conditions and applications to optimal quadrature formulas," in Schoenberg [1969b], pp. 467–484.

[1971] "Total positivity, interpolation by splines, and Green's functions of differential operators," *J. Approximation Theory*, **4**, 91–112.

[1976a] "A global improvement theorem for polynomial monosplines," in Karlin, Michelli, Pinkus and Schoenberg, [1976], pp. 67–82.

[1976b] "Oscillatory perfect splines and related extremal problems," in Karlin, Michelli, Pinkus, and Schoenberg [1976], pp. 371–460.

Karlin, S., and J. Karon

[1968] "A variation diminishing generalized spline approximation method," *J. Approximation Theory*, **1**, 255–268.

[1970] "A remark on B-splines," *J. Approximation Theory*, **3**, 455.

Karlin, S., and J. W. Lee

[1970] "Periodic boundary-value problems with cyclic totally positive Green's functions with applications to periodic spline theory," *J. Differential Equations*, **8**, 374–396.

Karlin, S., and C. Micchelli

[1972] "The fundamental theorem of algebra for monosplines satisfying boundary conditions," *Israel J. Math.*, **11**, 405–451.

Karlin, S., A. Micchelli, A. Pinkus, and I. J. Schoenberg

[1976] *Studies in Spline Functions and Approximation Theory*, Academic Press, New York.

Karlin, S., and A. Pinkus

[1976a] "Interpolation by splines with mixed boundary conditions," in Karlin, Michelli, Pinkus, and Schoenberg [1976], pp. 305–325.

[1976b] "Divided differences and other non-linear existence problems at extremal points," in Karlin, Michelli, Pinkus, and Schoenberg [1976], pp. 327–352.

Karlin, S., and L. L. Schumaker

[1967] "The fundamental theorem of algebra for Tchebycheffian monosplines," *J. Analyse Math.*, **20**, 233–270.

Karlin, S., and W. J. Studden

[1966] *Tchebycheff Systems: With Applications in Analysis and Statistics*, Interscience, New York.

Karlin, S., and Z. Ziegler

[1966] "Chebyshevian spline functions," *SIAM J. Numer. Anal.*, **3**, 514–543.

Karon, J. M.

[1968] "The sign-regularity properties of a class of Green's functions for ordinary differential equations and some related results," Dissertation, Stanford University.

[1969] "The sign-regularity properties of a class of Green's functions for ordinary differential equations," *J. Differential Equations*, **6**, 484–502.

Kashin, B. S.

[1977a] "The widths of certain finite dimensional sets and classes of smooth functions," *Izv. Akad. Nauk. SSSR. Ser. Math.*, **41**, 334–335.

[1977b] "Orders of the widths of certain classes of smooth functions," *Uspeki. Math. Nauk.*, **32**, 191–192.

Krein, M., and G. Finkelstein

[1939] "Sur les fonctions de Green completement non-negatives des opérateurs diffé rentiels ordinaires," *Dok. Akad. Nauk SSSR*, **24**, 220–223.

Kolmogorov, A. N.

[1936] "Annäherung von Funktionen einer gegebenen Funktionenklasse," *Ann. of Math.*, **31** 107–111.

Laurent, P. J.

[1972] *Approximation et Optimisation*, Hermann, Paris.

Law, A. G., and B. N. Sahney

[1976] *Theory of Approximation with Applications*, Eds., Academic Press, New York.

Lawson, C. L.

[1964] "Characteristic properties of the segmented rational min-max approximation problem," *Numer. Math.*, **6**, 293–301.

Löfstrom, J.

[1967] "Some theorems on interpolation spaces with applications to approximation in L_p," *Math. Ann.*, **172**, 176–196.

[1970] "Besov spaces in theory of approximation," *Ann. Mat. Pura. Appl.*, **85**, 93–184.

Loginov, A. S.

[1969] "Approximation of continuous functions by broken lines" (Russian), *Mat. Zametki* , **6**, 149–160; see also *Math. Notes*, **6**, 549–555.

[1970] "Estimates for the approximation by polygonal lines of continuous functions of class HW," (Russian) *Vestnik Moskov Univ. Ser. I. Mat. Meh.*, **25**, 47–55.

Lorentz, G. G.

[1953] *Bernstein Polynomials*, University of Toronto Press.

[1960] "Lower bounds for the degree of approximation," *Trans. Am. Math. Soc.*, **97**, 25–34.

[1966] *Approximation of Functions*, Holt, Rinehart and Winston, New York.

[1973] *Approximation Theory*, Ed. in cooperation with H. Berens, E. W. Cheney, and L. L. Schumaker, Academic Press, New York.

[1975] "Zeros of splines and Birkhoff's kernel," *Math. Z.*, **142**, 173–180.

Lorentz, G. G., C. K. Chui, and L. L. Schumaker

[1976] *Approximation Theory II*, Eds., Academic Press, New York.

Louboutin, R.

[1967a] "Sur une bonne partition de l'unité," in Glaeser [1967].

[1967b] *Le Prolongateur de Whitney, Volume II*, Publ. Sect. Math. Faculté des Sciences Rennes, France.

Love, A. E. H.

[1944] *A Treatise on the Mathematical Theory of Elasticity*, Dover, New York.

Lyche, T.

[1975] "Discrete polynomial spline approximation methods," Dissertation, University of Texas.

[1976] "Discrete polynomial spline approximation methods," in Böhmer, Meinardus, and Schempp [1976], pp. 144–176.

[1978] "A note on the condition numbers of the B-spline bases" *J. Approximation Theory*, **22**, 202 – 205.

Lyche, T., and L. L. Schumaker

[1973] "Computation of smoothing and interpolating natural splines via local bases," *SIAM J. Numer. Anal.*, **10**, 1027–1038.

[1975] "Local spline approximation methods," *J. Approximation Theory*, **15**, 294–325.

[1976] "On determinants of spline functions," SFB Report #107, University of Bonn, 1976.

Lyche, T., L. L. Schumaker, and K. Sepehrnoori

[1980] "FORTRAN subroutines for computing smoothing and interpolating splines," *Trans. on Math. Software*, in press.

Lyche, T., and R. Winther

[1979] "A stable recurrence relation for trigonometric B-splines," *J. Approximation Theory*, **25**, 266–279.

Maclaren, D. H.

[1958] "Formulas for fitting a spline curve through a set of points," Boeing Appl. Math. Report #2.

Makovoz, Yu. N.

[1972] "A method of obtaining lower estimates for diameters of sets in Banach spaces," (Russian), *Mat. Sb.*, **87**, 136–146 translated in *Math USSR Sb.*, **16**, 139–146.

Malozemov, V. N.

[1966] "On the deviation of broken lines," *Vestnik Leningrad Univ.*, **21**, 150–153.

[1967] "Polygonal interpolation" (Russian), *Mat. Zametki*, **1**, 537–540, see also *Math. Notes*, **1**, 355–357.

Mangasarian, O. L., and L. L. Schumaker

[1971] "Discrete splines via mathematical programming," *SIAM J. Control*, **9**, 174–183.

[1973] "Best summation formulae and discrete splines," *SIAM J. Numer. Anal.*, **10**, 448–459.

Marsden, M. J.

[1970] "An identity for spline functions with applications to variation-diminishing spline approximation," *J. Approximation Theory*, **3**, 7–49.

Marsden, M. J., and I. J. Schoenberg

[1966] "On variation diminishing spline approximation methods," *Mathematica (Cluj)*, **8**, 61–82.

Mayorov, V. E.

[1975] "Discretization of the problem of diameters" (Russian) *Uspeki Mat. Nauk*, **30**, 179–180.

McClure, D.

[1975] "Nonlinear segmented function approximation and analysis of line patterns," *Quart. Appl. Math.*, **33**, 1–37.

Meinardus, G.

[1967] *Approximation of Functions; Theory and Numerical Methods*, Springer-Verlag, Heidelberg.

[1974] "Bemerkung zur Theorie der B-Splines," in Böhmer, Meinardus, and Schemp [1974], pp. 165–175.

Meir, A., and A. Sharma

[1969] "On uniform approximation by cubic splines," *J. Approximation Theory*, **2**, 270–274.

[1973] *Spline Functions and Approximation theory*, Eds., ISNM Vol. 21, Birkhauser-Verlag, Basel.

Melkman, A. A.

[1974a] "The Budan-Fourier theorem for splines," *Israel J. Math.*, **19**, 256–263.

[1974b] "Interpolation by splines satisfying mixed boundary conditions," *Israel J. Math.*, **19**, 369–381.

[1977] "Splines with maximal zero sets," *J. Math. Anal. Appl.*, **61**, 739–751.

Méray, C

[1884] "Observations sur la légitime de l'interpolation," *Ann. Sci. Ecole Norm Sup.*, **1**, 165–176.

[1886] "Nouvelles examples d'interpolation illusoires," *Bull. Sci. Math.*, **20**, 266–270.

Meyers, L. F., and A. Sard

[1950a] "Best interpolation formulas," *J. Math. and Phys.*, **29**, 198–206.

[1950b] "Best approximate integration formulas," *J. Math. and Phys.*, **29**, 118–123.

Micchelli, C.

[1972] "The fundamental theorem of algebra for monosplines with multiplicities," in *Linear Operators and Approximation*, P. L. Butzer, J. Kahane, and B. S. Nagy, Eds., Basel, pp. 419–430.

Micchelli, C. A., and A. Pinkus

[1977] "Moment theory for weak Tchebychev systems with applications to monosplines, quadrature formulae, and best one-sided L_1 approximation by spline functions with fixed knots," *SIAM Math. Anal.*, **8**, 206–230.

Micchelli, C., and T. J. Rivlin

[1977] *Optimal Estimation in Approximation Theory*, Plenum Press, New York.

Milne-Thompson, L. M.

[1960] *The Calculus of Finite Differences*, Macmillan, London.

Mühlbach, A.

[1973] "A recurrence formula for generalized divided differences and some applications," *J. Approximation Theory*, **9**, 165–172.

Munteanu, M. J., and L. L. Schumaker

[1974] "Some multidimensional spline approximation methods," *J. Approximation Theory*, **10**, 23–40.

Natanson, I. P.

[1964] *Constructive Function Theory*, Ungar, New York.

Nemeth, A. B.

[1966] "Transformations of the Chebyshev systems," *Mathematica (Cluj)*, **8**, 315–333.

[1969] "About the extension of the domain of the Chebycheff systems defined on intervals of the real axis," *Mathematica (Cluj)*, **11**, 307–310.

[1975] "A geometrical approach to conjugate point classification for linear differential equations," *Rev. Anal. Num. Theor. Approx.*, **4**, 137–152.

Nitsche, J.

[1969a] "Sätze vom Jackson-Bernstein-Typ für die Approximation mit Spline-Funktionen," *Math. Z.*, **109**, 97–106.

[1969b] "Umkehrsätze für Spline-Approximationen," *Compositio Math.*, **21**, 400–416.

Oswald, P.

[1978] "Ungleichungen vom Jackson-typ für die algebraische best Approximation in L_p," *J. Approximation Theory*, **23**, 113–136.

Pence, D.

[1976] "Best approximation by constrained splines," Dissertation, Purdue University.

Peetre, J.

[1963] *A Theory of Interpolation of Normed Spaces*, Lecture Notes, Brazilia.

[1964] "Espaces d'interpolation, généralisations, applications," *Rend. Sem. Math. Fis. Milano*, **34**, 133–164.

Peterson, I.

[1962] "On a piecewise polynomial approximation" (Russian), *Eesti Nsv Tead. Akad. Toimetised Fuus. Mat.*, **11**, 24–32.

Phillips, G. M.

[1968] "Estimate of the maximum error in best polynomial approximations," *Comput. J.*, **11**, 110–111.

[1970] "Error estimates for best polynomial approximation," in Talbot [1970], 1–6.

Phillips, J. L., and R. J. Hanson

[1974] "Gauss quadrature rules with B-spline weight functions," *Math. Comp.*, **28**, 666.

Pinkus, A.

[1976] "A simple proof of the Hobby-Rice theorem," *Proc. AMS*, **60**, 82–84.

Popov, V. A.

[1974] "On the connection between rational and spline approximation," *Compt. Rendus Acad. Bulgarae Sci.*, **27**, 623–626.

[1975a] "Direct and converse theorem for spline approximation with free knots," *Bulgaricae Mathematicae Publ.*, **1**, 218–224.

[1975b] "On approximation of absolutely continuous functions by splines," *Mathematica (Cluj)*, **8**, 1299–1301.

Popov, V. A., and G. Freud

[1971] "Lower error bounds in the theory of spline functions," *Studia Sci. Math. Hungar.*, **6**, 387–391.

Popov, V. A., and B. H. Sendov

[1970] "Classes characterized by best-possible approximation by spline functions," (Russian) *Mat. Zametki*, **8**, 137–148; translated in *Math. Notes*, **8**, 550–557.

[1972] "On approximation of functions by spline functions and rational functions," in *Constructive Function Theory*, Bulgarian Academy of Sciences, pp. 89–92.

Popov, V. A., and J. Szabados

[1974] "On a general localization theorem and some applications in the theory of rational approximation," *Acta Math. Acad. Sci. Hungar.*, **25**, 165–170.

Popoviciu, T.

[1959] "Sur le reste dans certaines formules linéaires d'approximation de l'analyse," *Mathematica (Cluj)*, **1**, 95–142.

Potapov, M. K.

[1956] "Some questions on best approximation in the L_p-metric" (Russian), Kand. Dissertation, Moscow.

[1961] "On approximation by polynomials in L_p," (Russian) *Issedovaia po. Sovr. Prob. Konstr. Teori Funksi*, Moscow, pp. 64–69.

Prenter, P. M.

[1975] *Splines and Variational Methods*, Wiley, New York.

Quade, W. and L. Collatz

[1938] "Zur Interpolationstheorie der reellen periodischen Funktionen," *Aka. Wiss. (Math-Phys. Kl)*, **30**, 383–429.

Ralston, A.

[1965] *A First Course in Numerical Analysis*. McGraw-Hill, New York.

Ream, N.

[1961] "Note on approximation of curves by line segments," *Math Comp.*, **15**, 418–419.

Reddien, G. W., and L. L. Schumaker

[1976] "On a collocation method for singular two-point boundary value problems," *Numer. Math.*, **25**, 427–432.

Reinsch, C. H.

[1967] "Smoothing by spline functions," *Numer. Math.*, **10**, 177–183.

Rice, J. R.

[1964] *The Approximation of Functions*, Vol. I, *Linear Theory*, Addison-Wesley, Reading, Massachusetts.

[1969a] "On the degree of convergence of nonlinear spline approximation," in Schoenberg [1969b], pp. 349–365.

[1969b] *The Approximation of Functions*, Vol. II., *Nonlinear and Multivariate Theory*, Addison-Wesley, Reading, Massachusetts.

Richards, F.

[1972] "On the saturation class for spline functions," *Proc. Am. Math. Soc.*, **33** (2), 471–476.

[1973] "Best bounds for the uniform periodic spline interpolation operator," *J. Approximation Theory*, **7**, 302–317.

Richardson, C. H.

[1954] *An Introduction to the Calculus of Finite Differences*, van Nostrand, New York.

Rivlin, T. J.

[1969] *An Introduction to Approximation of Functions*, Blaisdell, Waltham, Massachusetts.

[1974] *The Chebyshev Polynomials*, Wiley, New York.

Runge, C.

[1901] "Uber die Darstellung willkürlicher Funktionen und die Interpolation zwischen aquidistanten Ordinaten," *Z. Angew. Math. Phys.*, **46**, 224–243.

[1904] *Theorie und Praxis der Reihen*, Teubner, Leibniz.

[1931] "Über imperische Funktionen," *Math. Phys.*, **46**, 224–243.

Rutman, M. A.

[1965] "Integral representation of functions forming a Markov series," (Russian) *Dokl. Akad. Nauk SSSR*, **164**, 989–992.

Sacks, J., and D. Ylvisaker

[1966] "Designs for regression with correlated errors," *Ann. Math. Stat.*, **37**, 66–89.

[1968] "Designs for regression with correlated errors: many parameters," *Ann. Math. Stat.*, **39**, 49–69.

[1969] "Statistical designs and integral approximation," Proc. 12th Bienn. Sem. Canad. Math. Congr., pp. 115–136.

[1970] "Designs for regression with correlated errors, III," *Ann. Math. Stat.*, **41**, 2057–2074.

Sard, A.

[1949] "Best approximate integration formulas, best approximation formulas," *Am. J. Math.*, **71**, 80–91.

[1963] *Linear Approximation*, AMS Survey Vol. 9, Providence, Rhode Island.

Sard, A., and S. Weintraub

[1971] *A Book of Splines*, Wiley, New York.

Schaback, R.

[1973] "Spezielle rationale Splinefunktionen," *J. Approximation Theory*, **7**, 281–292.

Schaback, R., and K. Scherer

[1976] *Approximation Theory*, Eds., Springer Lecture Notes 556, Heidelberg.

Schempp, W., and K. Zeller

[1977] *Constructive Theory of Functions of Several Variables*, Eds., Springer-Verlag Lecture Notes 571, Heidelberg.

Scherer, K.

[1970a] "Über die beste Approximation von L^p-Funktionen durch Splines," in *Proc. of Conference on Constructive Function Theory*, Varna, pp. 277–286.

[1970b] "On the best approximation of continuous functions by splines," *SIAM J. Numer. Anal.*, **7**, 418–423.

[1974a] "Characterization of generalized Lipschitz classes by best approximation with splines," *SIAM J. Numer. Anal.*, **11**, 283–304.

[1974b] "Stetigkeitsmoduli und beste Approximation durch polynomiale Splines," in Böhmer, Meinardus, and Schempp [1974], pp. 289–302.

[1976] "Some inverse theorems for best approximation by Λ-splines," in Lorentz, Chui, and Schumaker [1976], pp. 549–555.

[1977] "Optimal degree of approximation by splines," in Micchelli and Rivlin [1977], pp. 139–158.

Scherer, K., and L. L. Schumaker

[1980] "A dual basis for Tchebycheffian B-splines and applications," *J. Approximation Theory*, in press.

Schmidt, E., P. Lancaster, and D. Watkins

[1975] "Bases of splines associated with constant coefficient differential operators," *SIAM J. Numer. Anal.*, **12**, 630–645.

Schoenberg, I. J.

[1946a] "Contributions to the problem of approximation of equidistant data by analytic

functions, Part A: On the problem of smoothing of graduation, a first class of analytic approximation formulae," *Quart. Appl. Math.*, **4**, 45–99.

[1946b] "Constributions to the problem of approximation of equidistant data by analytic functions, Part B: On the problem of osculatory interpolation, a second class of analytic approximation formulae." *Quart. Appl. Math.*, **4**, 112–141.

[1958] "Spline functions, convex curves and mechanical quadrature," *Bull. Am. Math. Soc.*, **64**, 352–357.

[1964a] "On interpolation by spline functions and its minimal properties," *Int. Ser. Numer. Anal.*, **5**, 109–129.

[1964b] "On trigonometric spline interpolation," *J. Math. Mech.* **13**, 795–826.

[1964c] "Spline interpolation and the higher derivatives," *Proc. Nat. Acad. Sci. U.S.A.*, **51**, 24–28.

[1967] "On spline functions," in *Inequalities*, O. Shisha, Ed. Academic Press, New York, pp. 255–291.

[1968] "On the Ahlberg-Nilson extension of spline interpolation: the g-splines and their optimal properties," *J. Math. Anal. Appl.*, **21**, 207–231.

[1969a] "Cardinal interpolation and spline functions," *J. Approximation Theory*, **2**, 167–206.

[1969b] *Approximations with Special Emphasis on Spline functions*, Ed., Academic Press, New York.

[1971] "The perfect B-splines and a time optimal control problem," *Israel J. Math.*, **10**, 261–274.

[1972a] "Cardinal interpolation and spline functions. II., Interpolation of data of power growth," *J. Approximation Theory*, **6**, 404–420.

[1972b] "Cardinal interpolation and spline functions, IV. The exponential Euler splines", in ISNM Volume **20** Birkhauser, pp. 382–404.

[1973] *Cardinal Spline Interpolation*, CBMS **12**, SIAM, Philadelphia.

Schoenberg, I. J., and A. Whitney

[1949] " Sur la positivité des déterminants de translations des fonctions de fréquence de Pólya avec une application a une problème d'interpolation par les fonctions 'spline'," *Compt. Rend.*, **228**, 1996–1998.

[1953] "On Polya frequency functions III. The positivity of translation determinants with application to the interpolation problem by spline curves," *Trans. Am. Math. Soc.*, **74**, 246–259.

Scholz, R.

[1974] "Abschätzungen linearer Durchmesser in Sobolev und Besov Räumen," *Manuscripta Math.*, **11**, 1–14.

Schultz, M. H.

[1969a] "Approximation theory of multivariate spline functions in Sobolev spaces," *SIAM J. Numer. Anal.*, **6**, 570–582.

[1969b] "L_∞-multivariate approximation theory," *SIAM J. Numer. Anal.*, **6**, 161–183.

[1973a] "Error bounds for a bivariate interpolation scheme," *J. Approximation Theory*, **8**, 189–194.

[1973b] *Spline Analysis*, Prentice-Hall, Englewood Cliffs, New Jersey.

Schultz, M. H., and R. S. Varga

[1967] "L-splines," *Numer. Math.*, **10**, 345–369.

Schumaker, L. L.

[1966] "On some approximation problems involving Tchebycheff systems and spline functions," Dissertation, Stanford University.

[1968a] "Uniform approximation by Tchebycheffian spline functions," *J. Math. Mech.*, **18**, 369–377.

[1968b] "Uniform approximation by Chebyshev spline functions II. Free Knots," *SIAM J. Numer. Anal.*, **5**, 647–656.

[1969] "On the smoothness of best spline approximation," *J. Approximation Theory*, **2**, 410–418.

[1973] "Constructive aspects of discrete polynomial spline functions," in Lorentz [1973], pp. 469–476.

[1976a] "On Tchebycheffian spline functions," *J. Approximation Theory*, **18**, 278–303.

[1976b] "Zeros of spline functions and applications," *J. Approximation Theory*, **18**, 152–168.

[1976c] "Toward a constructive theory of generalized spline functions," in Böhmer, Meinardus, and Schempp [1976], pp. 265–331.

[1978] "Lower bounds for spline approximation," Banach Center Publ. **4**, pp. 213–223.

Schwartz, J. T.

[1969] *Nonlinear Functional Analysis*, Gordon and Breach, New York.

Schweikert, D. G.

[1966a] "The spline in tension (hyperbolic spline) and the reduction of extraneous inflection points," Dissertation, Brown University.

[1966b] "An interpolation curve using a spline in tension," *J. Math. and Phys.*, **45**, 312–317.

Schwerdtfeger, H.

[1960] "Notes on numerical analysis, II. Interpolation and curve fitting by sectionally linear functions," *Canad. Math. Bull.*, **3**, 41–57.

[1961] "Notes on numerical analysis III. Further remarks on sectionally linear functions," *Canad. Math. Bull.*, **4**, 53–55.

Segethova, J.

[1972] "Numerical construction of the hill functions," *SIAM J. Numer. Anal.*, **9**, 199–204.

Sendov, B. L., and V. A. Popov

[1970a] "Approximation of curves by piecewise polynomial curves," *Dokl. Bulg. Akad. Nauk*, **23**, 639–643.

[1970b] "The approximation of spline functions," *C. R. Acad. Bulgare Sci.*, **23**, 755–758.

Shah, J. M.

[1970] "Two-dimensional polynomial splines," *Numer. Math.*, **15**, 1–14.

Shapiro, H. S.

[1971] *Topics in Approximation Theory*, Springer-Verlag, Lecture Notes 187, Heidelberg.

Sharma, A., and A. Meir

[1966] "Degree of approximation of spline interpolation," *J. Math. Mech.*, **15**, 759–768.

Shisha, O.

[1973] "A characterization of functions having Zygmund's property," *J. Approximation Theory*, **9**, 395–397.

[1974a] "Characterization of smoothness properties of functions by means of their degree of approximation by splines," *J. Approximation Theory*, **12**, 365–371.

[1974b] "On the degree of approximation by step functions," *J. Approximation Theory*, **12**, 435–436.

[1975] "On saturation with splines," *J. Approximation Theory*, **13**, 491–494.

Smith, K.

[1970] "Formulas to represent functions by their derivatives," *Math. Annal.*, **188**, 53–77.

Smoluk, A.

[1964] "On the approximation with piecewise functions" (Polish), *Zeszyty Nauk. Wyz. Szkol. Ekon. Wroclawiu*, **21**, 123–157.

Sobolev, S. L.

[1963] *Applications of Functional Analysis in Mathematical Physics*, Am. Math. Soc., Providence, Rhode Island.

Solomiak, M. Z., and V. M. Tihomirov

[1967] "Geometric characteristics of the imbedding of the classes W_p^α in C," (Russian) *Invest. Vozov. Mat.*, **10**, 76–82.

Sommer, M.

[1976] "Weak-Chebychev spaces and splines," Dissertation, Erlangen.

[1979] "Continuous selections of the metric projection for 1-Chebychev spaces," *J. Approximation Theory*, **26**, 46–53.

[1980a] "Nonexistence of continuous selections of the metric projection for a class of Weak Chebychev systems," *Trans. Am. Math. Soc.*, **260**, 403–410.

[1980b] "Characterization of continuous selections of the metric projection for generalized spline functions," *SIAM J. Math. Anal.*, **11**, 23–40.

[1980c] "Weak Chebyshev spaces and Best L_1-approximation", in press.

Sommer, M., and H. Strauss

[1977] "Eigenschaften von schwach Tschebyscheffschen Räumen," *J. Approximation Theory*, **21**, 257–268.

Späth, H.

[1969] "Exponential spline interpolation," *Computing*, **4**, 225–233.

[1971] "Two-dimensional exponential splines," *Computing*, **7**, 364–369.

[1973] *Spline-Algorithmen zur Konstruktion glatter Kurven and Flächen*, R. Oldenburg Verlag, München, Germany. English translation by W. D. Hoskins and H. W. Sager, *Spline Algorithms for Curves and surfaces*, Utilitas Mathematical Publ. Inc., Winnipeg, Manitoba, 1974.

Stechkin, S. D., and Yu. N. Subbotin

[1978] *Splines in Numerical Mathematics* (Russian), Izd. Nauka, Moscow.

Stockenberg, B.

[1977a] "On the number of zeros of functions in a weak Tchebycheff-space," *Math. Z.*, **156**, 49–57.

[1977b] "Subspaces of weak and oriented Tchebyshev-spaces," *Manuscripta Math.*, **20**, 401–407.

Stone, H.

[1961] "Approximation of curves by line segments," *Math. Comp.*, **15**, 40–47.

Storchai, V. F.

[1968] "The deviation of polygonal functions in the L_p metric," (Russian) *Mat. Zametki*, **5**, 31–37; see also, *Math. Notes*, **5**, 21–25.

Storzenko, E. A.

[1976] "A Jackson theorem in the space $L^p(R^k)$, $0<p<1$," (Russian) *Dokl. Akad. Nauk USSR*, **229**, 554–557.

[1977] "On the approximation of functions in L_p, $0<p<1$, by algebraic polynomials" (Russian), *Izv. Akad. Nauk SSSR Ser. Mat.*, **41**, 652–662.

Strang, G., and G. J. Fix

[1973] *An Analysis of the Finite Element Method*, Prentice Hall, Englewood Cliffs, New Jersey.

Subbotin, Yu. N.

[1967] "Best approximation of a class of functions by another class" (Russian), *Mat. Zametki*, **2**, 495–504; see also *Math Notes*, **2**, 792–797.

[1970a] "Approximation of functions of class W^nH^p by m-order splines" (Russian), *Dokl. Akad. Nauk SSSR*, **195**; see also *Soviet Math. Dokl.* **11**, 1626–1628.

[1970b] "Diameter of class W^rL in $L(0,2\pi)$ and spline function approximation" (Russian), *Mat. Zametki*, **7**, 43–52; see also *Math Notes*, **7**, 256–260.

[1971] "Approximation by spline functions and estimates of diameters" (Russian), *Proc. Steklov. Inst. Math.*, **109**, 39–67.

Subbotin, Yu. N., and N. I. Chernykh

[1970] "Order of the best spline approximations of some classes of functions," *Math Notes*, **7**, 20–26.

Swartz, B.

[1968] "$0(h^{2n+2-1})$ bounds on some spline interpolation errors," *Bull. Am. Math. Soc.*, **74**, 1072–1078.

[1970] "$0[h^{k-j}\omega(D^kf;h)]$ bounds on some spline interpolation errors," Dissertation, New York University.

Swartz, B., and R. S. Varga

[1972] "Error bounds for spline and L-spline interpolation," *J. Approximation Theory*, **6**, 6–49.

Szegö, G.

[1939] *Orthogonal Polynomials*, AMS Colloq. Publ. Vol. 23, Providence, Rhode Island.

Talbot, A.

[1970] *Approximation Theory*, Ed., Academic Press, London.

Theilheimer, F., and W. Starkweather

[1961] "The fairing of ship lines on a high-speed computer," *Numerical Tables Aids Computation*, **15**, 338–355.

Tihomirov, V. M.

[1965] "Some problems of approximation theory," *Soviet Math. Dokl.*, **6**, 202–205.

[1969] "Best methods of approximation and interpolation of differentiable functions in the space $C[-1,1]$" (Russian), *Mat. Sb.*, **80**, 290–304; see also *Math. USSR-Sb*, **9**, 275–289.

[1970] "Some problems in approximation theory" (Russian), Dissertation, Moscow University, 1970.

[1971] "Some problems in approximation theory" (Russian), *Mat. Zametki*, **9**, 593 – 607; see also *Math Notes*, **9**, 343–350.

Timan, A. F.

[1963] *Theory of Approximation of Functions of a Real Variable*, MacMillan, New York.

Todd, J.

[1963] *Introduction to the Constructive Theory of Functions*, Birkhauser, Basel.

Varga, R. S.

[1969] "Error bounds for spline interpolation," in Schoenberg [1969b], pp. 367–388.

van Rooij and F. Schurer

[1974] "A bibliography on spline functions," in Böhmer, Meinardus and Schempp [1974], pp. 315–415.

Volkov, V. I.

[1958] "Some properties of Chebyshev systems" (Russian), *Ucen. Zap. Kalinin. Gos. Ped. Inst.*, **26**, 41–48.

Walsh, J. L.

[1960] *Interpolation and Approximation by Rational Functions in the Complex Domain*, AMS Colloq. Pub. 20, Providence, Rhode Island.

Walsh, J. L., J. H. Ahlberg, and E. N. Nilson

[1962] "Best approximation properties of the spline fit," *J. Math. Mech.*, **11**, 225–234.

Werner, H.

[1962] "Ein Satz über diskrete Tschebyscheff-Approximation bei gebrochen linearen Funktionen," *Numer. Math.*, **4**, 154–157.

[1974] "Tschebyscheff-Approximation mit einer Klasse rationaler Spline-Funktionen," *J. Approximation Theory*, **10**, 74–92.

[1976] "Approximation by regular splines with free knots," in Lorentz, Chui, and Schumaker [1976], pp. 567–573.

Whitney, H.

[1957] "On functions with n-th bounded differences," *J. Math. Pure Appl*, **36**, 60–95.

Young, J. D.

[1968] "Numerical applications of hyperbolic spline functions," *The Logistics Review*, **4**, 17–22.

[1969] "Generalization of segmented spline fitting of third order," *The Logistics Review*, **5**, 33–40.

Zalik, R. A.

[1975] "Existence of Tchebycheff extensions," *J. Math. Anal. Applic.*, **51**, 68–75.

[1977a] "Integral representation of Tchebycheff systems," *Pacific J. Math.*, **68**, 553–568.

[1977b] "On transforming a Tchebycheff system into a complete Tchebycheff system," *J. Approximation Theory*, **20**, 220–222.

Zielke, R.

[1972] "A remark on periodic Tchebyshev systems," *Manuscripta Math.*, **7**, 325–329.

[1973] "On transforming a Tchebyshev-system into a Markov-system," *J. Approximation Theory*, **9**, 357–363.

[1974] "Alternation properties of Tchebycheff systems and the existence of adjoined functions," *J. Approximation Theory*, **10**, 172–184.

[1975] "Tchebycheff systems that cannot be transferred into a Markov system," *Manuscripta Math.*, **17**, 67–71.

[1979] *Discontinuous Cebysev Systems*, Springer-Verlag Lecture Notes 707, Heidelberg.

Zwart, P. B.

[1973] "Multi-variate splines with non-degenerate partitions," *SIAM J. Numer. Anal.*, **10**, 665–673.

Zygmund, A.

[1959] *Trigonometric Series, Volumes I and II*, Cambridge University Press.

INDEX

14 891 | 10 | 82.